Nitric Oxide in Biology and Medicine

Nitric oxide (NO) is one of the most important molecules in biological systems, and there has been a tremendous growth in research on NO and its properties over the last decade. Although one of the simplest biologically-active molecules in nature, NO has found its way into nearly every area of biology and medicine. Nitric oxide plays a critical role in many aspects of human physiology ranging from memory, diabetes, cancer, and drug addiction. Nitric oxide is also the key neurotransmitter mediating erectile function, a major pathophysiological mediator of inflammation and host defense, and the principal endogenous regulator of blood flow and thrombosis. These major discoveries have stimulated intense and extensive research into a widespread array of fields including chemistry, molecular biology, and gene therapy. The newly discovered role of NO in pathophysiology and disease has resulted in the development of new therapeutic strategies.

This series will contain books that review particular areas of NO research and will provide useful summaries of our current state of knowledge into the role of NO in biology and medicine. Researchers and graduate students in physiology, pharmacology, and cell and molecular biology who are interested in the role of NO in living systems will find the books in this series to be the essential guide to the subject.

Nitric Oxide and Free Radicals in Peripheral Neurotransmission

Stanley Kalsner, Ph.D.
Editor

With 61 Figures

Birkhäuser
Boston • Basel • Berlin

Stanley Kalsner, Ph.D.
Department of Physiology and Pharmacology
The Sophie Davis School of Biomedical Education
City University of New York Medical School
New York, NY 10031
USA

Library of Congress Cataloging-in-Publication Data

Nitric oxide and free radicals in peripheral neurotransmission / edited by Stanley Kalsner.
p. cm.—(Nitric oxide in biology and medicine)
Includes bibliographical references and index.
ISBN 0-8176-4070-3 (hardcover: alk. paper)
1. Neurotransmission. 2. Nitric oxide—Physiological effect. 3. Nerves, Peripheral. 4. Free radicals (Chemistry)—Physiological effect. I. Kalsner, Stanley. II. Series.
QP364.5.N56 2000
573.8'54—dc21 99-052761

Printed on acid-free paper.

Birkhäuser

ISBN 0-8176-4070-3
ISBN 3-7643-4070-3 SPIN 10765034

Typeset by Best-set Typesetter Ltd., Hong Kong.
Printed and bound by Maple-Vail Book Manufacturing Group, York, PA.
Printed in the United States of America.

9 8 7 6 5 4 3 2 1

Contents

Preface

The origins of physiology and pharmacology are historically intertwined with those of peripheral nerve and organ relationships. This is exemplified by the early interest of investigators in the internal milieu, homeostasis, and the extrinsic control of organs by nerves, culminating in the concepts of neurochemical transmission, specific receptors for endogenous agonists, and the mechanisms of neurosecretion, cardiovascular function, and smooth muscle contraction.

Autonomic physiology and pharmacology cover almost all phases of the neurotransmission process in the periphery and also delve into the roles of putative transmitters and neuromodulators. The field encompasses an intricate and vast array of biological systems, from the smooth muscle relaxation of the urinary bladder to the mediation of blood hormone levels. We have moved very far from the reigning dictum that one nerve uses one transmitter, and that we could choose only between acetylcholine and the catecholamines. In a broad redefinition of the transmitter concept we now accept as integral to neurotransmission multiple transmitters and neuromodulators. Most recently, the role and function of nitric oxide and free radicals in neurotransmission in the periphery have moved to center stage and form the basis of this volume.

The chapters presented here range from general considerations of nitrergic transmission, in its broadest sense, to elaboration of our current understanding of the role of nitric oxide in transmission in individual peripheral organs, including its role as a backup, or alternate, as well as chief transmitter. The influence of nitric oxide and related compounds on the more conventional modes of autonomic transmission are also considered, as are the modes of combination of nitrergic compounds with receptors, the kinds and categories of receptors, response termination and tissue desensitization, and the identity of the sequence of steps culminating in the effector response. The emerging role of abnormal nitric oxide function in disease states is also treated where appropriate, as are potential avenues of therapy.

The necessity of an integrative approach to many of the current unresolved issues and the understanding that anatomy, biochemistry, or physiology alone cannot resolve pharmacological problems is made abundantly clear in several of the articles included in this volume. In the selection of authors for each of the topics under review a significant criterion was that they be particularly talented investigators active in the mainstream of the research they describe, and that they possess the skill and foresight to analyze cogently not only recent developments but also the historical thrust of their respective areas. It is unavoidable that, to some extent, the choices reflect my own points of view. Undoubtedly some readers may argue with the emphases made and the conclusions reached on individual topics. It is hoped that other opportunities will provide a forum for alternative opinions.

New York, New York *Stanley Kalsner*

Contributors

Karl-Erik Andersson
Lund University Hospital
Department of Clinical Pharmacology
S-221 85 Lund
Sweden

Joseph S. Beckman
University of Alabama at Birmingham
Department of Anesthesiology
Birmingham, AL 35233-1924
USA

Maria G. Belvisi
Rhône-Poulenc Rorer Research and Development
Pharmacology Department
Dagenham, Essez RM10 7XS
United Kingdom

Norman E. Cameron
University of Aberdeen
Diabetes Complications Research Lab
Institute of Medical Sciences
Aberdeen AB25 2ZD
Scotland

Tom M. Cocks
The University of Melbourne
Department of Pharmacology
Parkville 3052
Australia

Mary A. Cotter
University of Aberdeen
Diabetes Complications Research Lab
Institute of Medical Sciences
Aberdeen AB25 2ZD
Scotland

Alvaro G. Estévez
University of Alabama at Birmingham
Department of Anesthesiology
Birmingham, AL 35233-1924
USA

Axel Fischer
Charité, Humboldt University
Division of Allergy Research
Virchow Campus
Berlin
Germany

Gert Folkerts
Utrecht Institute for Pharmaceutical Sciences
Department of Pharmacology and Pathophysiology
3508 TB Utrecht
The Netherlands

Andrés Kamaid
University of Alabama at Birmingham
Department of Anesthesiology
Birmingham, AL 35233-1924
USA

Alan Gibson
King's College London
Biomedical Sciences
Hodgkin Building
London SE1 9RT
England

Thomas D. Giles
Louisiana State University Medical Center
School of Medicine in New Orleans
Section of Cardiology
1542 Tulane Ave.
New Orleans, LA 70112-2822
USA

Stan S. Greenberg (deceased)
Louisiana State University Medical Center
School of Medicine in New Orleans
Section of Cardiology
1542 Tulane Ave.
New Orleans, LA 70112-2822
USA

Inger Jansen-Olesen
Department of Neurology
University of Copenhagen
Glostrup Hospital
DK-2600 Glostrup
Copenhagen
Denmark

Stanley Kalsner
Professor and Chair
City University of New York Medical School
Department of Physiology and Pharmacology
138th Street and Convent Ave.
New York, NY 10031
USA

Hiromu Kawasaki
Department of Clinical Pharmaceutical Science
Graduate School of Natural Science and Technology
Okayama University
1-1-1 Tsushimanaka,
Okayama 700-8530
Japan

Kathleen D. Keef
Department of Physiology and Cell Biology
University of Nevada School of Medicine
Anderson Medical Sciences Bldg. 352
Reno, NV 89557
USA

Tony J-F. Lee
Department of Pharmacology
Southern Illinois University School of Medicine
P.O. Box 19629
Springfield, IL 62794-9629
USA

R.A. Lefebvre
Chair
Heymans Institute of Pharmacology
University of Gent Medical School
Gent B-9000
Belgium

Raj Mishra
Department of Pharmacology
Southern Illinois University
School of Medicine
P.O. Box 19629
Springfield, IL 62794-9629
USA

Leonid L. Moroz
Department of Neuroscience
University of Florida
The Whitney Laboratory
St. Augustine, FL 32086-8623
USA

Richard B.R. Muijsers
Utrecht Institute for
Pharmaceutical Sciences
Department of Pharmacology and
Pathophysiology
3508 TB Utrecht
The Netherlands

Frans P. Nijkamp
Utrecht Institute for
Pharmaceutical Sciences
Department of Pharmacology and
Pathophysiology
3508 TB Utrecht
The Netherlands

Jes Olesen
Department of Neurology
University of Copenhagen
Glostrup Hospital
DK-2600 Glostrup
Copenhagen
Denmark

Jie Ouyang
Louisiana State University
Medical Center
School of Medicine in New
Orleans
Section of Cardiology
1542 Tulane Ave.
New Orleans, LA 70112-2822
USA

Ilari Paakkari
Institute of Biomedicine
University of Helsinki
Department of Pharmacology and
Toxicology
FIN-00014
Finland

Ilkka Pörsti
Institute of Biomedicine
University of Helsinki
Department of Pharmacology and
Toxicology
FIN-00014
Finland

Asko Riutta
Institute of Biomedicine
University of Helsinki
Department of Pharmacology and
Toxicology
FIN-00014
Finland

Kenton M. Sanders
Department of Physiology and
Cell Biology
University of Nevada School of
Medicine
Anderson Medical Sciences
Bldg. 352
Reno, NV 89557
USA

Stavros Selemidis
The University of Melbourne
Department of Pharmacology
Parkville 3052
Australia

Makoto Takenaga
Department of Clinical
Pharmaceutical Science
Graduate School of Natural
Science and Technology
Okayama University
1-1-1 Tsushimanaka,
Okayama 700-8530
Japan

Lars Lykke Thomsen
Department of Neurology
University of Copenhagen
Glostrup Hospital
DK-2600 Glostrup
Copenhagen
Denmark

Liliana Viera
University of Alabama at Birmingham
Department of Anesthesiology
Birmingham, AL 35233-1924
USA

Xinfang Zhao
Louisiana State University Medical Center
School of Medicine in New Orleans
Section of Cardiology
1542 Tulane Ave.
New Orleans, LA 70112-2822
USA

1
On the Origin and Early Evolution of Neuronal Nitric Oxide Signaling: A Comparative Analysis

LEONID L. MOROZ

> Yet NO is not an obscure chemical and certainly no newcomer to the life sciences, as often stated in hyperbole. Early in evolution NO took its role as a central player in bacterial bioenergetics and the global N cycle vital to all organisms.
>
> —W. Zumft, 1993

Introduction

Gaseous transmission in the nervous system is considered a separate and unique mechanism of interneuronal communication (Figure 1.1). In contrast to "classical" neurotransmitters, the radical messenger nitric oxide (NO) is synthesized and released without special storage and transfer mechanisms. It is one of the smallest and most diffusible signal molecules known, and freely crosses membrane barriers to exert its effects through direct binding and/or reaction with its target proteins (Stamler et al., 1992). Its diffusion from a point source can affect targets a relatively large distance away from its origin (Wood and Garthwaite, 1994). It becomes apparent that the concept of a NO microenvironment with transient NO gradients would be a more accurate way to describe the actual situation. Nitric oxide is also a highly reactive radical that can give rise to toxic secondary radical species (Stamler et al., 1992). Overproduction of NO contributes to both nonimmune defense mechanisms and numerous pathological conditions (Dawson and Dawson, 1996; Moncada and Higgs, 1995).

The number of proven and proposed physiological functions for NO has grown explosively in the last several years (Dawson and Snyder, 1994; Garthwaite and Boulton, 1995; Griffin and Stuehr, 1995; Moncada and Higgs, 1995; Moncada et al., 1991; Snyder et al., 1998). Although the signaling function of NO was initially proposed in 1986–1987, by the end of 1999 there were more than 30,000 reports in the literature concerning the distribution and various roles of NO and its synthetic enzyme. The exponentially growing data in this field clearly indicate that NO is involved in

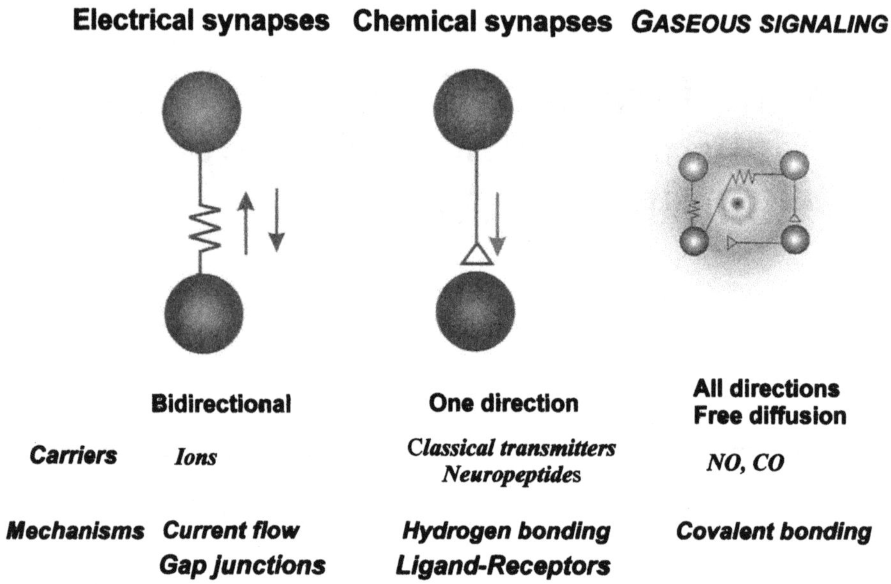

FIGURE 1.1. Three types of signaling in nervous systems: electrical synapses, chemical neurotransmission, and gaseous signaling. The role of CO as a neuronal messenger has been recently proposed but not investigated in great detail (see Snyder et al., 1998 for a review).

literally hundreds of systemic and cellular functions; NO synthases (NOS) can be present in practically all animal tissues; and NO-generating systems are widely distributed in virtually all major animal groups and even among nonanimal kingdoms.

The reasons for the evolutionary choice of this versatile yet dangerous molecule as a modulator of physiological processes in animal tissues are not yet apparent. The implication is that in spite of the substantial progress in the area, a large part of our understanding of nitrergic physiology is yet to come.

In this chapter I will briefly overview comparative aspects of NO signaling, with a focus on the distribution and function of NO in invertebrate animals. Historically, studies of relatively primitive and simple organisms have revealed the fundamental principles in more complex systems. This should be true for NO signaling as well. Nitric oxide-related regulatory mechanisms may be as old as cellular organization itself, so that "ancestral" functions of NO in primitive organisms have probably been well preserved across four billion years of biological evolution. Unfortunately, the comparative physiology and biochemistry of NO-mediated pathways are poorly understood. As a result, speculations about the origin and evolutionary trends in neuronal NO signaling are still hypothetical.

Comparative Aspects of NO Synthesis

The questions of the origin and evolution of NO-mediated signaling in the animal kingdom are directly linked to the basic principles of nitrogen metabolism in living cells and rest firmly on our knowledge of these pathways in lower eukaryotes and prokaryotes. It is not surprising that the major components of NO-generating pathways are very similar between phylogenetically distant animal groups, nonanimal taxa, and even prokaryotes.

At least three mechanisms of enzymatic NO synthesis can be found in living systems. As an alternative pathway, NO can be produced nonenzymatically from nitrites in various cells and tissues under certain chemical conditions.

Enzymatic NO Synthesis: Nitric Oxide Synthase

In all animal tissues the enzymatic synthesis of NO proceeds according to the following reaction:

$$\text{L-Arginine} + O_2 \rightarrow NO + \text{L-citrulline} \qquad [1]$$

$$NO \rightarrow NO_2^- \rightarrow NO_3^- \qquad [2]$$

where nitrites (NO_2^-) and nitrates (NO_3^-) are the sequential products of NO oxidation.

At least three groups of NOS have been described in mammalian tissues. Neuronal NOS (nNOS; type I) and endothelial NOS (eNOS; type III) are Ca^{2+}-dependent and constitutive isoforms (Knowles and Moncada, 1994). Both of them were demonstrated in neuronal (Dinerman et al., 1994) and non-neuronal tissues and various peripheral tissues. The inducible Ca^{2+}-independent NOS (iNOS; type II) was first described in activated macrophages. In general, it cannot be detected in nonactivated tissues, but it can be expressed after appropriate stimulation [e.g., in the presence of lipopolysaccharides or in the response to potentially damaging stimuli (Griffin and Stuehr, 1995)]. Inducible nitric oxide synthase is primarily involved in defense reactions and cytotoxicity (Nathan, 1992) and can be expressed in many cell types, including neurons and glial cells. All of these three isoforms have been characterized biochemically and cloned from several mammalian species (Griffin and Stuehr, 1995; Brenman et al., 1997; Huang and Lo, 1998). Homologous enzymes have also been cloned from insects (Champagne et al., 1995; Luckhart et al., 1998; Luckhart and Rosenberg, 1999; Nighorn et al., 1998; Regulski and Tully, 1995; Yuda et al., 1996) and mollusks (Kornev et al., 1998, 1999); they were mainly characterized as a neuronal-like type of NOS.

In reactions catalyzed by NOS, L-arginine serves as a universal metabolic precursor of NO (Griffin and Stuehr, 1995), but this versatile amino acid is also crucial in the urea cycle and in synthesis of polyamines, glutamate, creatine, and armatine (Figure 1.2). Through L-argininosuccinate as an intermediate, the NO/citrulline cycle can be coupled to the citric acid cycle and, therefore, to cellular bioenergetics (see Wu and Morris, 1998, for a recent review on L-arginine metabolism).

An L-arginine–NO pathway has recently been described in plants (Cueto et al., 1996; Delledonne et al., 1998; Durner et al., 1998; Durner and Klessig, 1999; Bolwell, 1999), fungi (Kanadia et al., 1998; Ninnemann and Maier, 1996; Tao et al., 1996, 1997; Werner-Felmayer et al., 1994), and even in the bacterial species *Nocardia* (Chen and Rosazza, 1994, 1995).

In *Nocardia* a NOS-like enzyme was purified (the molecular mass of the homodimer protein is about 110kDa) and was shown to require the same cosubstrates and cofactors as animal NOS [i.e., O_2, NADPH, Ca^{2+}, tetrahydrobiopterin (BH_4), flavin mononucleotide (FMN), and flavin adenine dinucleotide (FAD)]. N^G-Hydroxy-L-arginine was an intermediate, L-citrulline was a coproduct, and competitive inhibitors of NO synthesis used for mammalian NOS also suppressed L-citrulline formation in *Nocardia*.

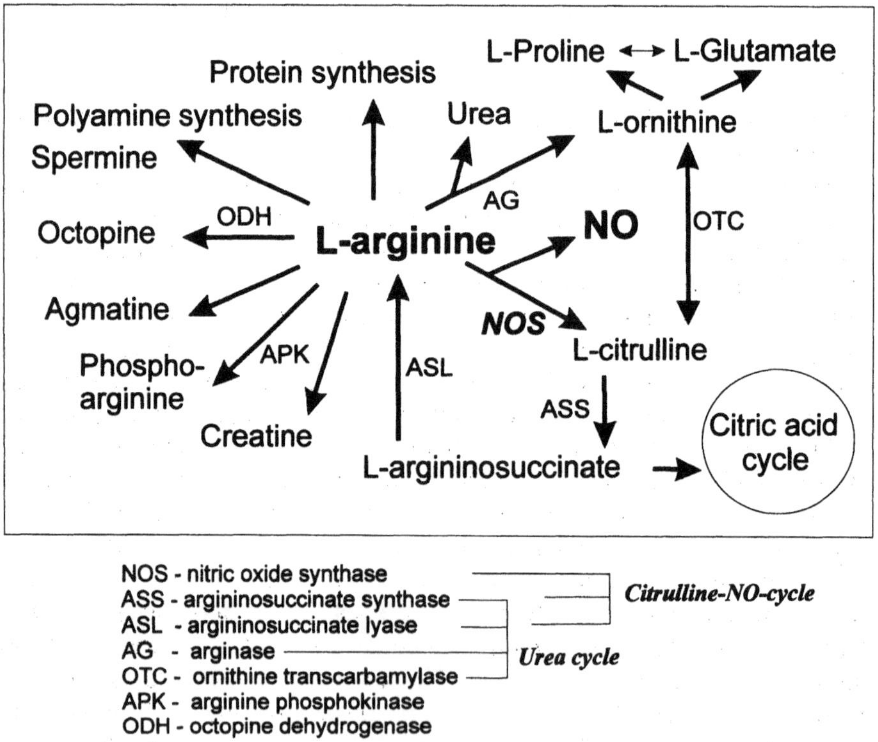

FIGURE 1.2. Schematic diagram of several of the major biochemical pathways involved in the arginine/citrulline cycle showing the interactions of NO-citrulline and the urea cycles.

Nitric oxide synthesis in some plant tissues was found to be Ca^{2+}-dependent (Cueto et al., 1996; Delledonne et al., 1998), whereas in others it was Ca^{2+}-independent (Cueto et al., 1996), resembling the inducible type of NOS of mammals. Since NOS-like enzymes from nonanimal tissues have not been cloned or purified for detailed analysis, it is highly speculative to discuss potential homology and similarity between these proteins and NOS in animals.

Enzymatic NO Synthesis: The Nitrogen Cycle

Nitric oxide is also produced enzymatically in the so-called nitrogen cycle (Figure 1.3). The cycle consists of nitrogen fixation (the reduction of molecular nitrogen to ammonia), nitrification (oxidation of ammonia to nitrates), and denitrification (the reduction of nitrates to other nitrogen oxides, including NO, and back to N_2). The denitrification and associated NO formation initially described in prokaryotes were recently discovered in lower eukaryotes, such as fungi (Kobayashi et al., 1996; Kobayashi and Shoun, 1995; Shoun et al., 1992) and higher plants (Payne et al., 1997).

Components of the denitrification pathway and detection of NO production by bacteria were discovered as early as the 1860s and 1870s; in the 1950s NO formation was directly confirmed in experiments with ^{15}N. Major details of these pathways have been described elsewhere (Ferguson, 1998; Hollocher and Hibbs, 1996; Payne et al., 1997; Zumft, 1993, 1997). Briefly, NO is produced from nitrites by nitrite reductases, a group of metal-containing enzymes. There are two kinds of nitrite reductases: the cytochrome cd_1 type of bacterial reductase, a soluble periplasmic enzyme with both *c*- and *d*-type heme, and the Cu-containing enzymes that can be either soluble proteins (gram-negative bacteria) or membrane proteins (gram-positive bacteria lacking periplasmic space and an outer membrane). Nitric oxide itself can be further reduced to N_2O by NO reductases or by NO dioxygenase. The latter is a flavohemoglobin recently found in *Escherichia coli* (Gardner et al., 1998).

Heme-containing enzymes involved in the denitrification pathways are quite distinct from mammalian NOS. Nevertheless, new discoveries of denitrification in fungi may shed light on potential relationships between denitrification pathways in prokaryotes, cellular bioenergetics in lower eukaryotes, and the origin of NOS. Nitrite reductase from the fungus *Fusarium oxysporum* is similar to the Cu-type of the enzyme in gram-positive bacteria, but it is located in the intermembrane space of the mitochondrion in an area that is considered analogous to the periplasmic space in bacteria. This enzyme seems to be involved in anaerobic ATP production. Interestingly, the next enzyme in the denitrification pathway of *Fusarium*—NO reductase—is a unique cytochrome P_{450} type protein (Park et al., 1997). Similar P_{450} types of NO reductase were discovered in other fungi such as *Cylindrocarpon tonkinese* (Kobayashi et al., 1996; Usuda et al., 1995), *Strep-*

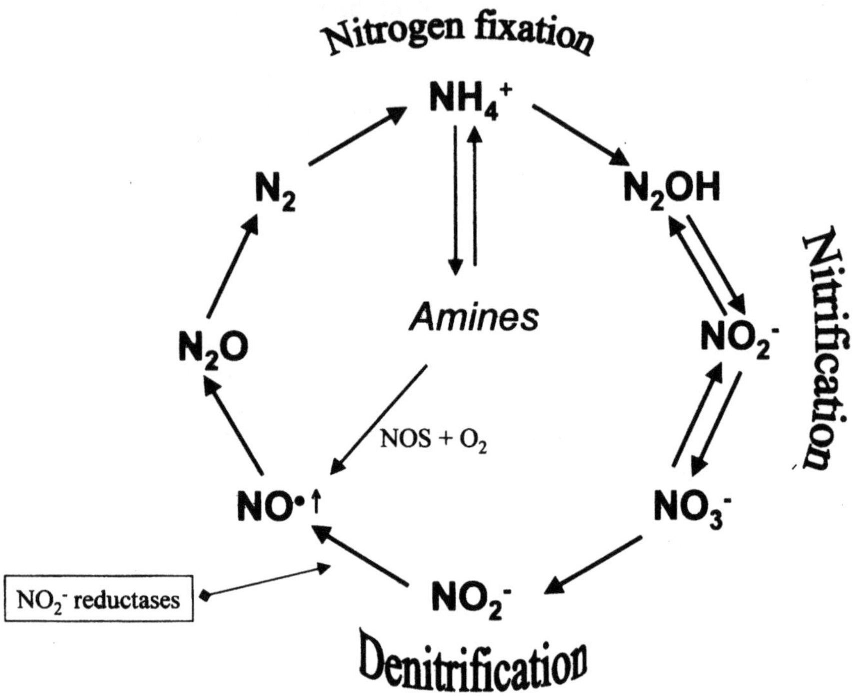

FIGURE 1.3. Simplified sequence of the nitrogen cycle reactions. The reactions of nitrification and denitrification are associated with the electron transport systems and coupled to the cellular energetic and ATP production. NH_3 is widely used in the synthesis of various amine-containing biomolecules and, therefore, the NOS-catalyzed oxidation of L-arginine can be considered as a heterotrophic nitrification route, in addition to the bacterial autotrophic nitrification (Hollocher and Hibbs, 1996). Nitrification and NO synthesis by NOS are aerobic pathways, where molecular oxygen is required for the oxidation of L-arginine and NH_3. In contrast, denitrification usually occurs under either anaerobic or hypoxic conditions. Additionally, nitrates can be reversibly reduced to ammonia in a nitrate/nitrite type of anaerobic respiration (see Hollocher and Hibbs, 1996; Zumft, 1997) for a review of these and associated pathways).

tomyces thioluteus (Shoun et al., 1998), and the yeasts *Trichosporon cutaneum*, *Fellomyces fuzhounensis*, and *Candida* sp. (Tsuruta et al., 1998). As a matter of fact, cytochrome P_{450} reductase is the only known enzyme group that shares substantial homology to all isoforms of the mammalian NOS.

Primitive denitrifying bacteria (similar to the present *Paracoccus denitrificans*) can be considered as a common ancestral symbiotic prototype of the eukaryotic mitochondrion. One can speculate, therefore, that in primitive eukaryotes mitochondria can still preserve their denitrifying capabilities and even associated respiratory functions, whereas more "advanced" groups lose this capacity in evolution (see Kobayashi et al., 1996).

It is of interest that in mammalian tissues under anaerobic conditions nitrites can also be reduced to NO by mitochondrial cytochrome *c* oxidase (Meyer, 1973; Walters and Taylor, 1965), and some other heme-containing proteins (Doyle et al., 1981). This, in addition to NOS and the denitrification pathways, is a potential third mechanism of enzymatic NO synthesis in biological systems.

Nonenzymatic NO Formation: Abiotic Reduction of Nitrites

The fact that a nitrate solution can generate NO without any enzyme is crucial for the comparative analysis of NO synthesis in biological tissues. Abiotic NO formation is well known from the chemistry of NO_x species, but until recently it was not considered to be an endogenous source of NO in animals (Duncan et al., 1997; Weitzberg and Lundberg, 1998; Zwier et al., 1999).

In an acidic environment, nitrites can be easily converted to NO according to the equation:

$$NO_2^- + 2H^+ \rightarrow NO^{\cdot} + H_2O \qquad [3]$$

The actual mechanisms are more complicated (see for details Butler et al., 1995; Feelisch and Stamler, 1996; Kharitonov et al., 1994; Lewis and Deen, 1994; Saran and Bors, 1994; Wink et al., 1996; Wink and Mitchell, 1998; Samouilov et al., 1998), and without a description of secondary pathways the sequence can be presented as follows:

$$2NO_2^- + 2H^+ \rightleftarrows 2HONO \rightarrow H_2O + N_2O_3 \rightarrow NO^{\cdot} + NO_2 + H_2O \qquad [4]$$

Ascorbate (Archer et al., 1975; Mirvish et al., 1972) and some reducing compounds, such as NADPH, L-cysteine, reduced gluthatione, and other thiols, have been reported to stimulate NO formation from nitrites (Feelisch, 1991, 1993; Feelisch and Noack, 1987; Scorza et al., 1997).

Nonenzymatic NO production in vivo was originally demonstrated in the human oral cavity from dietary nitrates (Duncan et al., 1995) and in the gut (Benjamin et al., 1994). In both cases NO concentrations were sufficient to be involved in primary nonimmune defense reactions and, probably, in the control of digestive functions, such as mucosal blood flow, motility, and possibly secretion and absorption (Duncan et al., 1997). Estimated nitrite concentrations were between 0.1 and 100 μM in different parts of the digestive system, approaching a concentration of 1 mM in saliva following a high nitrate/nitrite test meal (Duncan et al., 1995). Similarly, nonenzymatic NO formation has been demonstrated in human skin (Weller et al., 1996), with a suggested physiological role in the inhibition of infection by pathogenic microorganismic fungi as well as in modulation of cutaneous T-cell function, skin blood flow, and keratinocyte differentiation. The last function was independently confirmed by Vallette et al. (1998).

Large quantities of NO (similar to or even higher than those produced by NOS) can be formed in ischemic heart tissues by a mechanism that is not enzyme-dependent (Zweier et al., 1995a,b). The authors conclude that this NO formation is a consequence of acidification, which serves to chemically reduce the large pool of nitrites present within tissue; NO production is not blocked by inhibitors of NOS. The mean nitrite concentration of the ischemic myocardium (12 μM) was shown to be sufficient to generate the detected amount of NO. It was concluded that enzyme-independent NO formation not only contributes to the process of postischemic injury, but also eliminates the protective effect of NOS inhibitors (Zweier et al., 1995b).

Nonenzymatic NO Formation: Nitrite Photolysis

Photochemical generation of NO from nitrites is a notable component of the nitrogen cycle in the earth's biosphere. Nitrite absorbs maximally at 356 nm, and the process can be presented as follows (Zafiriou et al., 1980):

$$NO_2^- + H_2O + h\nu\ (295 \leq \lambda \leq 410\,\text{nm}) \rightarrow NO^{\cdot} + {}^{\cdot}OH + OH^- \qquad [1]$$

The reaction occurs naturally in the surface layers of the world's oceans, predominantly in the central equatorial areas. Nitric oxide formed by this mechanism may play an important role in marine ecosystems and must be considered as a source of NO in the atmosphere (Zafiriou et al., 1980). The measured partial pressure of NO (pNO) in the air was less than 8×10^{-12} atm, compared with pNO calculated for surface sea water of 7×10^{-8} atm (McFarland et al., 1979). The estimated concentrations of NO in the surface film of tropical waters depend strongly on the nitrite distribution; they could be in the subnanomolar to nanomolar range during the day and drop to a practically undetectable level after sunset. It is attractive to speculate that the involvement of NO in the regulation of circadian rhythms observed in higher animals (Ding et al., 1994) might be traced back to the earlier day–night prebiotic conditions in the oceanic waters. For instance, in a unicellular marine alga, the dinoflagellate *Goniaulax polyedra*, nitrites (and potentially NO) are important endogenous regulators of the biological clocks (Roenneberg and Rehman, 1996).

Distribution and Functions of NO Signaling in Nonanimal Taxa

Table 1.1 summarizes the present data related to the identification of NOS in primitive eukaryotes, plants, and fungi.

TABLE 1.1. The distribution and functions of NO in prokaryotes, lower eukaryotes and plants.

Systematic group, species	Location	Properties and type of NO-generating system	Proposed function(s) of NO	References
Prokaryotes				
Denitrifying bacteria and blue-green algae	Periplasmatic space, cellular membrane	NO_2^- respiration, nitrification, and denitrification	An intermediate product of nitrogen oxidation	Zumft, 1993
Rhodobacter sphaeroides	Periplasmatic space	Nitrite reductase system	Regulation of transcription, activation of gene expression	Kwiatkowski and Shapleigh, 1996
Nocardia sp.	Soluble fraction, cytoplasm (?)	NO synthase-like enzyme, Ca^{2+}-NADPH-, H_4B-, FAD-, FMN-dependent	?	Chen and Rosazza, 1994, 1995
Lower Eukaryotes				
Slime molds				
Dictyostelium discoideum	?	Inducible NOS-like, Ca^{2+}-independent NO synthase	Development, cell differentiation, and aggregation	Tao et al., 1996, 1997
Physarum polycephalum	?	Constitutive Ca^{2+}-independent NOS-like	Cell cycle control (?)	Werner-Felmayer et al., 1994
Protozoa				
Tetrachimena termophila	?	?	Cell cycle control, cGMP-dependent	Christensen et al., 1996
Trypanosoma cruzi	?	Ca^{2+}/calmodulin-, NADPH-, FAD-, dependent NOS	Locomotion (cGMP/NMDA-dependent); parasite-host interactions	Paveto et al., 1995; Pereira et al., 1997)

TABLE 1.1. *Continued*

Systematic group, species	Location	Properties and type of NO-generating system	Proposed function(s) of NO	References
Fungi				
Fusarium oxysporum	Mitochondrion	Nitrite reductase system	Denitrification, respiration, ATP synthesis	Kobayashi et al., 1996; Kobayashi and Shoun, 1995
Cylindrocarpon tonkinense	Mitochondrion	Nitrite reductase system	Denitrification, respiration, ATP synthesis	Kobayashi et al., 1996; Usuda et al., 1995
Neurospora crassa	?	Constitutive NOS-like	Differentiation, light-induced conidiation	Ninnemann and Maier, 1996
Saccharomyces cerevisiae (yeasts)	?	Constitutive NOS-like, calmodulin-dependent	?	Kanadia et al., 1998
Plants				
Arabidopsis thaliana	Soybean cytosolic fraction	Ca^{2+}-dependent NOS-like	Antibacterial, cGMP-dependent	Delledonne et al., 1998
Nicotiana tabacum	Leaf cells	?	Antiviral defense, cGMP-dependent	Durner et al., 1998
Lupinus albus	Roots, nodules	Ca-dependent/Ca-independent NOS-like	?	Cueto et al., 1996

Prokaryotes

The early data relating to the production of NO during denitrification considered it solely as a short-lived transient intermediate. It must be realized that denitrifying bacteria can release this gas extracellularly [e.g., to the soil (Yoshinari, 1993)], where accumulations can be large enough to express functional effects in other target cells. The measured steady-state NO concentrations during denitrification are in the range of 1–65 nM and are comparable to those proposed for mammalian tissues. Values as high as 0.5 μM were calculated for intracellular NO (Goretski et al., 1990). These concentrations of NO can have potentially cytostatic and even cytotoxic effects, and can be used as a component of natural defense mechanisms between various bacterial groups.

On the other hand, there is evidence that in some prokaryotes endogenously produced NO is involved in true signaling pathways. For example, in *Rhodobacter* NO up-regulates the expression of certain genes related to NO metabolism (Kwiatkowski and Shapleigh, 1996). No induction of these genes was observed with either cAMP or cGMP analogues, suggesting that NO-mediated effects were cyclic nucleotide-independent. It is important that in a related *Rhodobacter* species that has no nitrite reductase and, therefore, is not able to make NO on its own, exogenously applied NO also regulates transcription of selected genes (Kwiatkowski et al., 1997). This finding implies that NO produced by other bacteria can be used as a genuine intercellular messenger. In this status NO is a good candidate for a signal molecule involved in symbiotic interactions. Additionally, both pathogenic and symbiotic denitrifying bacteria in higher eukaryotes can significantly contribute to a pool of endogenous NO produced by NOS in the host, as well as being an important factor in parasite–host interactions.

Lower Eukaryotes

Similarly to prokaryotes, fungal cells are able to generate NO as an intermediate in the denitrification process. In some species (e.g., *Fusarium* and *Cylindrocarpon*), nitrite and nitrate reduction is strongly coupled to the anaerobic synthesis of ATP with relatively high energy yields (Kobayashi et al., 1996). In others, NO production is associated with NOS-like enzymes involved in the regulation of morphogenesis and development (Table 1.1).

In slime molds (*Dictyostelium* and *Physarum*), nutritional starvation causes a cascade of processes resulting in the arrest of cell growth, then the initiation of cell aggregation and differentiation. Cyclic AMP acts as an intercellular chemotactic signal for the unicellular amoebae and triggers its aggregation. Endogenous NO produced during the initial phase of the developmental cycle can prevent the initiation of the cAMP pulses and, therefore, has a negative effect on the aggregation (Tao et al., 1996, 1997).

The presence of glucose increased NO production, whereas NO scavengers, the NOS inhibitor, *N*-iminoethylornithine (L-NIO), and glucose deprivation facilitate the aggregation. It was suggested that NO generated in the interphase period prevents the adenylyl cyclase from producing cAMP but does not alter cAMP receptor-stimulated activity. Furthermore, in *Dictyostelium* NO can suppress glycolysis by selective modification of GAPDH (glyceraldehyde-3-phosphate dehydrogenase) via an ADP-ribosylation mechanism (Tao et al., 1997). Since NO production is Ca-independent, Tao et al. (1997) speculated that *Dictyostelium* expresses inducible-like NOS activity. Although cGMP-producing enzymes were demonstrated in this species, NO failed to activate soluble guanylyl cyclase or alter the intracellular cGMP level (Tao, 1992).

In *Neurospora* NO suppresses light-induced conidial differentiation (Ninemann and Maier, 1996). Nitric oxide synthase inhibitors stimulate conidiation in the same way that light does, and they are highly effective both in darkness and in the light. The light treatment itself decreased cGMP levels (Ninemann and Maier, 1996). Thus, it was hypothesized that in darkness a higher NOS activity could raise the intracellular cGMP.

Nitric oxide activated the locomotion and guanylyl cyclase activity in the parasitic euglenozoan kinetoplastid *Trypanosoma cruzi*. The effects were mimicked by L-arginine, L-glutamate, and *N*-methyl-D-aspartate (NMDA), which also were shown to up-regulate endogenous NOS-like enzyme and intracellular cGMP. The parasite's motility was also enhanced by a cGMP analogue and blocked by a competitive NOS inhibitor, suggesting the coupling of functional NO- and GMP-dependent pathways controlling a locomotory behavior (Pereira et al., 1997). Nitric oxide synthase-related enzyme was partially purified from soluble extracts of *Trypanosoma*. The enzyme was Ca^{2+}/calmodulin-dependent and required NADPH, tetrahydrobiopterin, and FAD as cofactors (Paveto et al., 1995).

Plants

Practically all possible mechanisms of NO synthesis have been found in plant tissues. Some reports indicate the presence of denitrification pathways in higher plants (reviewed by Payne et al., 1997). In addition, plants can encounter NO produced by symbiotic microorganisms in soil as well as NO produced nonenzymatically from nitrites. Finally, "classical" NO synthesis from L-arginine was recently demonstrated in several species. Regardless of the source of NO, the cytotoxicity and associated nonimmune defense against various infections is one of the most well-confirmed functions of NO in plant tissues. Nitric oxide is able to inhibit the cytochrome oxidase in plant mitochondria (Millar and Day, 1996) and, therefore, contribute to both nonspecific toxicity and regulation of cellular respiration.

Initially, NOS-like activity was found in the legume *Lupinus albus* (Cueto et al., 1996). The enzyme activity in roots was Ca^{2+}-dependent, whereas such dependence was not observed in nodules. The authors speculate that the Ca^{2+}-independent NOS in nodules can be induced by liposaccharides from the symbiotic *Rhizobium*, thus having a role in new plant nodule formation resulting from the bacterial infection.

Since the putative NOS activity was located in phloem cells, a recognized signal transport system, the signaling functions of NO in plants might be even more widespread.

The role of NO in plant immunity was proved in two model plant preparations: tobacco cells and soybean cells (Delledonne et al., 1998; Durner et al., 1998). In tobacco, NO enhanced resistance against viral infection by selective induction of defense-related genes (Durner et al., 1998). Remarkably, these genes were also induced by cGMP and cyclic ADP ribose. The later can release Ca^{2+} from intracellular stores via ryanodyne-like receptors. Furthermore, the pathogen-stimulated NOS is associated with cGMP transients. In *Arabidopsis*, bacterial pathogens triggered rapid and localized cell death, a hypersensitive apoptosis-like response associated with the production of superoxide (O_2^-) and hydrogen peroxide (H_2O_2). The hypersensitive response plays an important role in plant defense mechanisms by limiting the spread of the infection. Nitric oxide both potentiates the induction of hypersensitive cell death and leads to transcriptional activation of defense genes crucial for other immune reactions. In contrast, the inhibitors of NOS promoted the disease and bacterial growth (Delledonne et al., 1998).

Thus, it becomes apparent that two major recognized functions of NO in mammals—nonimmune defense systems and regulation of gene expression and differentiation—could trace their origins back to the prokaryotic world and, later, could be well preserved in various nonmetazoan eukaryotes such as protozoans, plants, and fungi. The same basic functions can be observed in practically all major invertebrate groups and serve an important role in neuronal mechanisms.

Distribution and Functions of NO Signaling in Invertebrate Animals

Table 1.2 summarizes the data available for various invertebrates. Although representatives of many invertebrate phyla are included in the table, detailed information about NO signaling is mainly limited to representatives of two relatively advanced groups: arthropods and mollusks (see also recent reviews by Bicker, 1998; Colasanti and Venturini, 1998; Jacklet, 1997; Martinez, 1995; Moroz and Gillette, 1996a; Moroz et al., 1999; Muller, 1997). It is not surprising that in these groups one can find practically all the major NO functions described for mammals. Non-neuronal functions of NO, such as various defense mechanisms, cytotoxicity, and effects on development, also seem to be well conserved in evolution and probably would be

TABLE 1.2. Distribution and functions of NO in invertebrates.

Systematic group, species	Location	Neuronal (nNOS) vs non-neuronal NOS (nnNOS)	Properties and type of NOS	Proposed function(s), cGMP-dependence of NO-mediated effects	References
Cnidaria (Jellyfishes, polyps)					
Hydra vulgaris	?	?	Ca^{2+}-dependent, calmodulin-independent	Feeding and chemosensory systems; cGMP-dependent	Colasanti et al., 1997
Aiptasia diaphana	Nematocysts	nnNOS	Constitutive, Ca^{2+}-dependent	Defense reactions, activation of nematocyst discharges	Salleo et al., 1996
Aglantha digitale	Neurons	nNOS-like	?	Feeding and locomotion, cGMP-dependent	Moroz et al., 1997
Plathyhelmintes (Flat worms)					
Dugesia tigrina	Pharynx, neuronal tissues	nNOS-like	Ca^{2+}-/calmodulin-independent	Feeding (?)	Eriksson, 1996
Hymenolepis diminuta (tapeworm)	Neuronal tissues	nNOS-like	?	?	Gustafsson et al., 1996
Nemathyhelmintes (Round worms)					
Ascaris suum	Muscles, neurons	nNOS and nnNOS-like	?	Neuromodulation (?)	Bascal et al., 1995, 1996; Bowman et al., 1995
Annelida					
Hirudo medicinalis (medicinal leech)	CNS	nNOS and nnNOS-like	Constitutive, Ca^{2+}-dependent	?	Leake et al., 1996; Leake and Moroz, 1996

H. medicinalis	Immunocytes	nnNOS-like	?	Immune system	Salzet et al., 1997
Echinoderms					
Marthasterias glacialis (starfish)	Gut	nNOS (?)	?	Feeding (?)	Martinez et al., 1994
Asterias rubens	?	?	?	Cardiac stomach (smooth muscle) relaxation (cGMP)	Elphick and Melarange, 1998
Mollusca					
Mytilus edulis (oyster)	Microglia	nnNOS	Morphine/naloxon-sensitive NO release	Nonimmune defense	Liu et al., 1996
Viviparus ater	Hemocytes	nnNOS	Constitutive LPS-activated, Ca^{2+}-independent	Nonimmune defense	Conte and Ottaviani, 1995
Ilyanassa obsoleta	CNS	nNOS	?	Development, metamorphosis	Lin and Leise, 1996
Lymnaea stagnalis	CNS, peripheral tissues	nNOS	Constitutive, Ca^{2+}-dependent	Feeding program, synaptic transmission, cGMP-dependent	Elphick et al., 1995a; Korneev et al., 1998; Moroz et al., 1993, 1994a,b; Park et al., 1998; Korneev et al., 1998, 1999
Helix pomatia	CNS, peripheral tissues	nNOS	Constitutive, Ca^{2+}-dependent	?	Huang et al., 1997; Teyke, 1996
H. aspersa	CNS, peripheral tissues	nNOS	?	?	Cooke et al., 1994; Sanchez-Alvarez et al., 1994; Pisu et al., 1999
Limax maximus	CNS			Olfaction processing	Gelperin, 1994a,b

TABLE 1.2. *Continued*

Systematic group, species	Location	Neuronal (nNOS) vs non-neuronal NOS (nnNOS)	Properties and type of NOS	Proposed function(s), cGMP-dependence of NO-mediated effects	References
Aplysia californica	CNS, peripheral tissues	nNOS	Constitutive, calmodulin-dependent	Feeding, neuronal plasticity synaptic transmission, cGMP-dependent	Cruz et al., 1997; Floyd et al., 1998; Jacklet, 1995; Jacklet and Gruhn, 1994; Meulemans et al., 1995; Moroz et al., 1996; Mothet et al., 1996a,b; Lewin and Walters, 1999; Koh and Jacklet, 1999; Fossiez et al., 1999
Pleurobranchaea californica	CNS, peripheral tissues	nNOS	Constitutive, calmodulin-dependent	feeding	Cruz et al., 1997; Floyd et al., 1998; Moroz et al., 1996; Moroz and Gillette, 1996b
Phestilla sibogae	CNS, peripheral tissues	nNOS	?	Sensory systems, development, metamorphosis	Budko et al., 1996; Meleshkevitch et al., 1997
Octopus vulgaris	CNS, peripheral tissues	nNOS		Learning	Robertson et al., 1994, 1995a,b, 1996
Loligo opalescens	CNS, peripheral tissues	NNOS/nnNOS	?	?	Gilly et al., 1995
Loligo bleeker	CNS, peripheral tissues	nNOS	?	?	Kimura et al., 1997

Sepia officinalis	CNS, peripheral tissues	nNOS	?	?	Palumbo et al., 1997
Arthropoda					
Limulus polyphemus (horseshoe crab)	Hemocytes	nnNOS	Constitutive, Ca^{2+}-independent	Hemocyte aggregation	Radomski et al., 1991
Cambarellus montezumae (crayfish)	CNS	?	?	?	Talavera et al., 1995
Pacifastacus leniusculus (crayfish)	CNS	nNOS	Constitutive, Ca^{2+}-dependent	?	Johansson and Carlberg, 1994
Homarus americanus (lobster)	CNS	nNOS	Constitutive, Ca^{2+}-dependent (?)	Development, olfactory processing (?)	Scholz et al., 1998
Arachnida (Spiders) 10 species	CNS	nNOS			Meyer, 1994
Insects					
Drosophila melanogaster (fruit fly)	CNS, peripheral tissues	nNOS	Constitutive, Ca^{2+}-dependent	Olfaction, vision Development	Muller, 1994; Regulski and Tully, 1995 Gibbs and Truman, 1998
Drosophila melanogaster	CNS, peripheral tissues	nnNOS	Constitutive, Ca^{2+}-dependent	Endothelial functions Development, hypoxic responses	Dow et al., 1994 Kuzin et al., 1996 Winerove and O'Farrell, 1999
Apis mellifera (honeybee)	CNS, peripheral tissues	nNOS	Constitutive, Ca^{2+}-dependent	Learning, memory	Muller, 1994, 1996; Muller and Hildebrandt, 1995

TABLE 1.2. *Continued*

Systematic group, species	Location	Neuronal (nNOS) vs non-neuronal NOS (nnNOS)	Properties and type of NOS	Proposed function(s), cGMP-dependence of NO-mediated effects	References
Locusta migratoria (locust)	CNS peripheral tissues	nNOS	Constitutive, Ca^{2+}-dependent	Olfaction, vision Development	Elphick et al., 1995b Muller, 1996; Muller and Bicker, 1994; Truman et al., 1996
Schistocerca gregaria	CNS, peripheral tissues	nNOS	Constitutive, Ca^{2+}-dependent	Olfaction, vision	Elphick et al., 1993, 1995b
Manduca sexta	CNS	nNOS	Constitutive, Ca^{2+}-dependent	Olfaction, vision	Grueber and Truman, 1999; Nighorn et al., 1998; Stengl and Zintl, 1996
Rhodnius prolixus (blood-sucking bug)	Salivary glands	nnNOS	Constitutive, Ca^{2+}-dependent	Feeding	Yuda et al., 1996
Bombyx mori (silkworm)	Peripheral tissues	nnNOS	Constitutive, Ca^{2+}-dependent and Ca^{2+}-independent	?	Andersen et al., 1997; Champagne et al., 1995; Choi et al., 1995; Nussenzveig et al., 1995; Ribeiro and Walker, 1994

observed in representatives of many other taxa. On the other hand, the involvement of NO in neuronal signaling and the distribution of putative NOS-containing (nitrergic) neurons is difficult to predict based on the existing data. Many taxa crucial for evolutionary analysis either have not been studied (e.g., placozoans, sponges, and lower chordates) or have been poorly investigated (e.g., cnidarians, flatworms, and nematodes).

Can NO be found in the nervous systems of all invertebrate groups, or is neuronal NO signaling a relatively new invention characteristic of higher invertebrate phyla?

Cnidarians (jellyfishes, polyps, sea anemones, and corals) are the metazoan group with one of the most primitive neuronal organizations, which is considered to be closely related to the earliest nervous systems. The initial screening for NOS using the NADPH-diaphorase technique (as a histochemical marker for NOS; see Bredt et al., 1991; Dawson et al., 1991; Hope et al., 1991) in several cnidarians (Elofsson et al., 1993) failed to demonstrate a neuronal location for NOS. In contrast, selected non-neuronal tissues showed a prominent NADPH-diapharase (NADPH-d) reactivity in various coelenterates. Recently we continued the screening, testing more than 20 different cnidarians. In most of the cases, we again failed to detect NOS. However, there were two possible exceptions: the holoplanktonic hydromedusa *Aglantha digitale* (Moroz et al., 1997) and the scyphoid medusa *Aurelia aurita* (unpublished observations). In *Aglantha*, putative NOS-containing/NADPH-d reactive sensorimotor neurons were located in the tentacles. It appears that NO activates a cGMP-dependent type of locomotion characteristic of the feeding behavior in the jellyfish. In the other jellyfish, *Aurelia*, putative nitrergic neurons were located in rhopalia (specialized sensory structures at the edge of the umbrella), but their functions are unknown.

Although NADPH-d histochemistry did not reveal any neuronal staining in the freshwater polyp *Hydra* (Elofsson et al., 1993), it was shown that NO is involved in the initiation of feeding patterns (Colasanti et al., 1995, 1997). Specifically, NO can participate in triggering tentacle movements during the polyp's feeding response. Reduced gluthatione, a standard activator of feeding patterns, increased the production of cGMP; the effect was abolished by preincubation of *Hydra* with a NOS inhibitor, N^G-nitro-L-arginine methylester (L-NAME). However, a cGMP analogue had no effect on the feeding response in this animal. The presence of Ca^{2+}-dependent NOS was also demonstrated in biochemical experiments on both *Hydra* (Colasanti et al., 1997) and an anthozoan *Aiptasia* (Salleo et al., 1996). In *Aiptasia*, Salleo et al. showed that a non-neuronal NOS is involved in defense reactions associated with the induction of Ca^{2+}-dependent nematocyst discharges by stimulated aconia.

Evidence for the presence of neuronal NOS in another group of lower invertebrates (flatworms and nematodes) is also contradictory. The neuronal localization of NOS/NADPH-d activity was demonstrated in two species of flatworms, the freshwater planarian *Dugesia tigrina* (Eriksson, 1996) and the tapeworm *Hymenolepis diminuta* (Gustafsson et al., 1996), and the nematode *Ascaris suum* (Bascal et al., 1995, 1996), but not in the closely related *Caenorhabdites elegans* (Elofsson et al., 1993). Initial screening of the genome of *C. elegans* also failed to detect the presence of NOS genes (Hudson and O'Shea, 1998). In the planarian *Dugesia gonocephala*, we also failed to detect selectively labeled NADPH-d-reactive neurons (Elofsson et al., 1993). However, a non-neuronal location of fixative-resistant NADPH-d activity was observed.

Thus, one might speculate that in terms of the evolution of neuronal NO signaling, the lower invertebrate groups may represent the transient stage of the involvement of NO in neuronal processing, where putative nitrergic neurons can be found in one group but be absent in another.

In more "advanced" invertebrate groups, both neuronal and non-neuronal NOS have been clearly demonstrated, and various functions of NO in different invertebrates have been proposed (Table 1.2). Considered as a whole, the data from different invertebrates suggest that at least two neuronal functions of NO can be highly conserved across the representatives of major phyla and, therefore, can be well preserved in evolution. These functions are the involvement of NO in sensory processing, mainly related to chemosensory and olfactory systems, and the involvement of NO in the control of feeding. For example, NO donors selectively activate feeding motor patterns in representatives of phylogenetically very distant animal groups, such as various mollusks [*Lymnaea* (Moroz et al., 1993), *Pleurobranchaea*, *Tritonia*, and *Clione* (unpublished observations)], and cnidarians [*Aglantha* (Moroz et al., 1997)]. Recent comparative studies on vertebrates, including mammals, provide additional support to the idea that NO is a ubiquitous messenger in feeding systems. A role for NO in chemosensory processing also can be found in representatives of various phyla, such as coelenterates, mollusks, arthropods, and chordates (see Table 1.2).

Hypothetical Trends in the Evolution of NO Signaling

Comparative analysis and the presence of NO signaling in both prokaryotes and eukaryotes strongly support the idea that NO may have played a crucial role even in the early stages of the evolution of life (Feelisch and Martin, 1995; Franchini et al., 1995; Moncada and Martin, 1993). The radical nature and the presence of different redox forms of NO contribute significantly to the omnipresent characteristics of NO-dependent mechanisms. It also appears that all functions of NO are closely related and interact with each other (Figure 1.4). As a result, many physiological and pathophysiological features of NO can be derived from its two major functional activ-

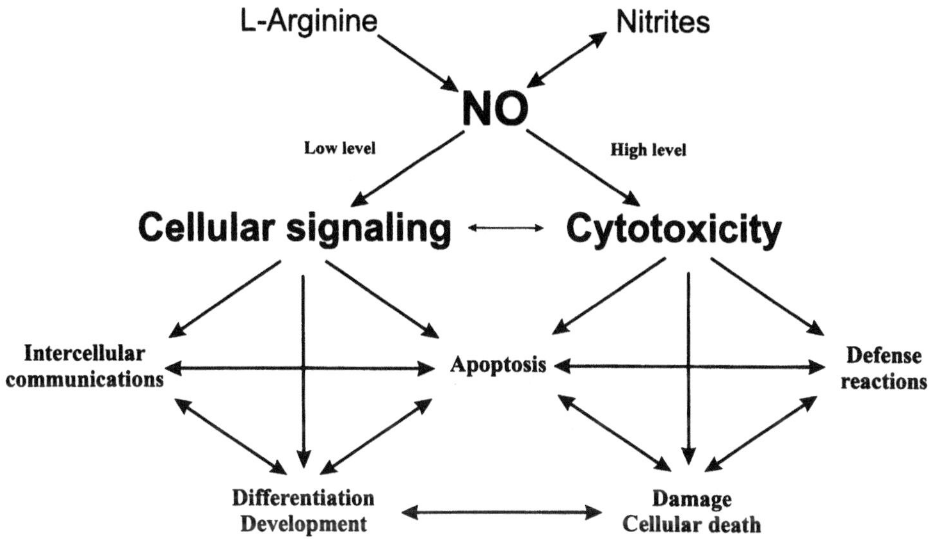

FIGURE 1.4. The major functions of NO signaling in living cells and their potential relationships. In general, a shift between the toxic effects of NO and its genuine signaling functions can be determined by the actual steady-state NO concentrations and the local chemical microenvironment within and around the target cells.

ities: nonimmune defense and intra- and intercellular signaling. The transformations between these functions are primarily determined by local steady-state concentrations of NO. Apparently, these two functions could trace their origin to the very first primordial protocells. Yet in the prokaryotic world, both cytotoxic and signaling characteristics of NO-mediated pathways might contribute to intracellular communications and symbiotic interactions. The role of NO in the development and aggregation processes in higher eukaryotes might also be traced back to the origin of multicellular organization and, probably, to the very dawn of the eukaryotic organization itself (Figure 1.5).

On the other hand, specific neuronal NO signaling seems to be a relatively new evolutionary invention. It might be very likely that neuronal NOS and nitrergic neurons were "introduced" independently in different invertebrate groups. Thus, at first glance, somewhat contradictory data related to the presence of NOS in lower invertebrates can reflect the different stages of the involvement of NO in neuronal NO signaling. It appears that the role of NO in chemosensory processing and feeding can be one of the first neuronal functions for NO in primitive animals. Feeding and chemosensory mechanisms are two complementary components of the same functional system, where the detection of potential food stimuli is coupled to the appropriate feeding response.

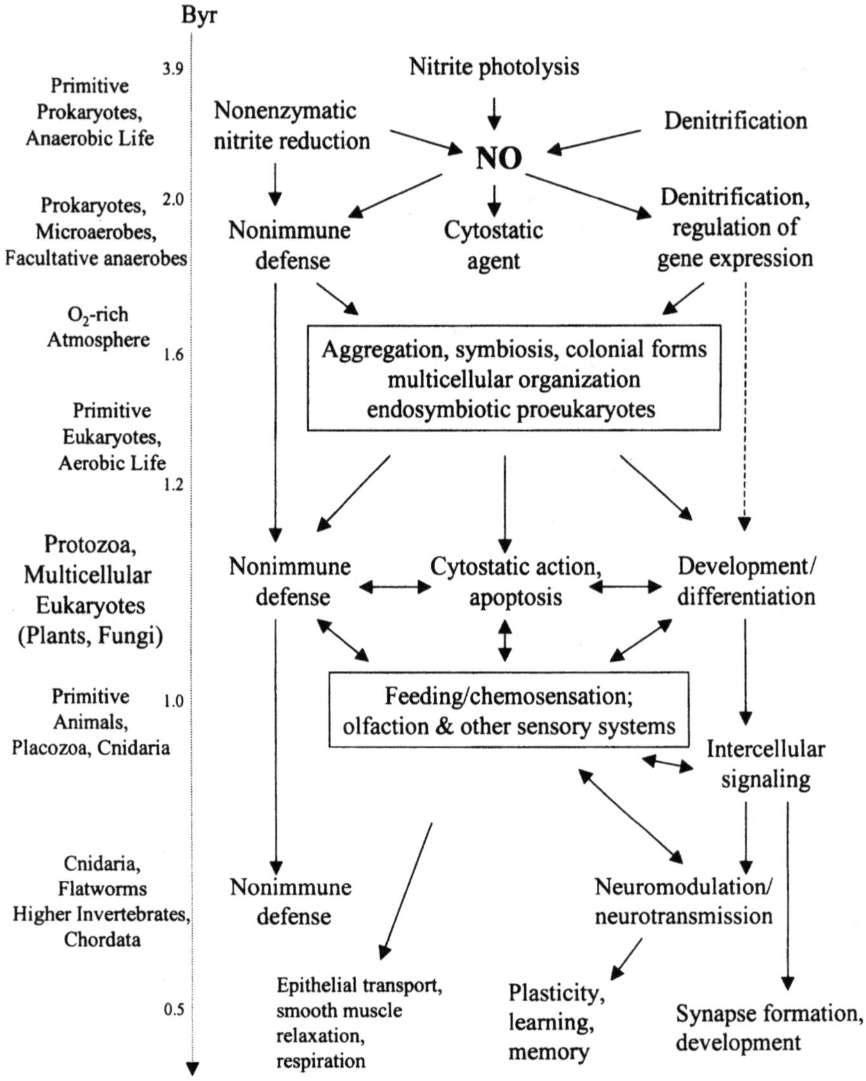

FIGURE 1.5. Hypothetical trends in the evolution of neuronal NO signaling.

Other questions are why would NO be associated with feeding and chemosensory mechanisms, and what would be the potential connections to the non-neuronal functions of NO in primitive eukaryotes? A reasonable explanation may be the development of neuronal signaling from the strategy of using NO in nonimmune defense mechanisms. Indeed, the detection and consumption of food by any animal species is a real exposure both to possibly damaging stimuli and to potential bacterial or viral infections; in other words, feeding and chemosensory structures can be considered as the primary gates for pathogen invasion. At the dawn of metazoan evolution, various microorganisms were the only food source for primitive animals. Acting as a universal bacteriostat and cytotoxic molecule, NO is, indeed, a good candidate to be incorporated into the feeding system. The formation of a protective barrier using this toxic gas would also help to limit a competitive bacterial assimilation of the food as well as to prevent potential infection associated with long-term digestion.

Nonenzymatic NO synthesis is a complementary and crucial component of the digestion process. In a highly acidic microenvironment, endogenous nitrites, plus nitrate/nitrite contamination from the food, serve as important sources of enzyme-independent NO formation. Furthermore, under certain conditions in the digestive tract, the requirement for NO can be fulfilled without NOS by combined chemical NO synthesis from nitrites and nitrate/nitrite reductions by heme-containing proteins or by symbiotic denitrifying prokaryotes.

During evolution, the initial cytotoxic role of NO associated with relatively high steady-state concentrations could be transformed into control of selective feeding responses followed by the regulation of neuronal signaling in sensory and effector systems. For example, in the gastrointestinal tract, NO is involved in the control of smooth muscle motility and peristaltic movements. Association with the gut coelomic sinuses (or with more specialized transport systems) would create the background for the origin of NO signaling in the excretory and circulatory systems of higher invertebrates and vertebrates.

Alternatively, a role for NO in such neuronal functions as synaptogenesis (Ogilvie, Schilling, Billings leg, and Schmidt, 1995), growth cone motility (Hess et al., 1993; Renteria and Constantine-Paton, 1996; Wagenen and Rehder, 1999), and neurodevelopment (Peunova and Enikolopov, 1995; Lin and Leise, 1996; Truman et al., 1996; Cramer et al., 1998; Gibbs and Truman, 1998; Wildemann and Bicker, 1999), as well as the NO-mediated apoptosis and inhibition of neuronal respiration, could be linked to the "ancestral" nonspecific defense mechanisms that have been well conserved during the evolution of nervous systems.

Future Directions

At present, the comparative physiology of NO signaling is in its initial phase, providing the very first and quite superficial overview toward our understanding of its basic design, mechanisms, and evolution. As the obvious next step, cloning and functional characterization of NOS from invertebrate models and, most important, from representatives of nonmetazoan taxa, will lead to the major breakthrough. Significant discoveries can also be expected in the analysis of alternative NOS-independent pathways, the role of NO in ecological adaptations, and the contribution of NO to symbiotic interactions.

Acknowledgments. This research was supported in part by INTAS (93-3504) and International Research Scholars Program (75195-540101; HHMI) and NIH R01-NS39103 and R01-MH60261 grants to L.L.M.

References

Andersen, J.F., Champagne, D.E., Weichsel, A., Ribeiro, J.M., Balfour, C.A., Dress, V., and Montfort, W.R. 1997. Nitric oxide binding and crystallization of recombinant nitrophorin I, a nitric oxide transport protein from the blood-sucking bug *Rhodnius prolixus*. *Biochemistry* 36:4423–4428.

Archer, M.C., Tannenbaum, S.R., Fan, T.Y., and Weisman, M. 1975. Reaction of nitrite with ascorbate and its relation to nitrosamine formation. *J. Natl. Cancer. Inst.* 54:1203–1205.

Bascal, Z.A., Montgomery, A., Holden-Dye, L., Williams, R.G., Thorndyke, M.C., and Walker, R.J. 1996. NADPH diaphorase activity in peptidergic neurones of the parasitic nematode, *Ascaris suum*. *Parasitology* 112:125–134.

Bascal, Z.A., Montgomery, A., Holden-Dye, L., Williams, R.G., and Walker, R.J. 1995. Histochemical mapping of NADPH diaphorase in the nervous system of the parasitic nematode *Ascaris suum*. *Parasitology* 110:625–637.

Benjamin, N., O'Driscoll, F., Dougall, H., Duncan, C., Smith, L., Golden, M., and McKenzie, H. 1994. Stomach NO synthesis. *Nature* 368:502.

Bicker, G. 1998. NO news from insect brain. *Trends Neurosci.* 21:349–355.

Bolwell, G.P. 1999. Role of active oxygen species and NO in plant defence responses. *Curr. Opin. Plant Biol.* 2:287–294.

Bowman, J.W., Winterrowd, C.A., Friedman, A.R., Thompson, D.P., Klein, R.D., Davis, J.P., Maule, A.G., Blair, K.L., and Geary, T.G. 1995. Nitric oxide mediates the inhibitory effects of SDPNFLRFamide, a nematode FMRFamide-related neuropeptide, in *Ascaris suum*. *J. Neurophysiol* 74:1880–1888.

Bredt, D.S., Glatt, C.E., Hwang, P.M., Fotuhi, M., Dawson, T.M., and Snyder, S.H. 1991. Nitric oxide synthase protein and mRNA are discretely localized in neuronal population of the mammalian CNS together with NADPH diaphorase. *Neuron* 7:615–624.

Brenman, J.E., Xia, H., Chao, D.S., Black, S.M., and Bredt, D.S. 1997. Regulation of neuronal nitric oxide synthase through alternative transcripts. *Dev. Neurosci.* 19:224–231.

Budko, D.Y., Moroz, L.L., Hadfield, M., and Gillette, R. 1996. Nitric oxide synthesis in central and peripheral tissues of the tropical sea-slug, *Phestilla sibogae* (Nudibranchia, Aeolidiida). *Soc. Neurosci. Abstr.* 22:P363.

Butler, A.R., Flitney, F.W., and Williams, D.L.H. 1995. NO, nitrosonium ions, nitroxide ions, nitrosothiols and iron-nitrosyls in biology: a chemist's perspective. *Trends Pharmacol. Sci.* 16:18–22.

Champagne, D.E., Nussenzveig, R.H., and Ribeiro, J.M. 1995. Purification, partial characterization, and cloning of nitric oxide-carrying heme proteins (nitrophorins) from salivary glands of the blood-sucking insect *Rhodnius prolixus*. *J. Biol. Chem.* 270:8691–8695.

Chen, Y., and Rosazza, J.P. 1994. A bacterial nitric oxide synthase from a *Nocardia* species. *Biochem. Biophys. Res. Commun.* 203:1251–1258.

Chen, Y., and Rosazza, J.P. 1995. Purification and characterization of nitric oxide synthase (NOSNoc) from a *Nocardia* species. *J. Bacteriol.* 177:5122–5128.

Choi, S.K., Choi, H.K., Kadono-Okuda, K., Taniai, K., Kato, Y., Yamamoto, M., Chowdhury, S., Xu, J., Miyanoshita, A., Debnath, N.C., Asaoka, A., and Yamakawa, M. 1995. Occurrence of novel types of nitric oxide synthase in the silkworm, *Bombyx mori*. *Biochem. Biophys. Res. Commun.* 207:452–459.

Christensen, S.T., Kemp, K., Ouie, H., and Rasmussen, L. 1996. Cell death, survival and proliferatiion in *Tetrachimena termophila*. Effects of insulin, sodium nitroprusside, 8-bromo cyclic GMP, NG-methyl-L-arginine and methylene blue. *Cell. Biol. Internat.* 20:653–666.

Colasanti, M., Lauro, G., and Venturini, G. 1995. NO in *Hydra* feeding response. *Nature* 374:505.

Colasanti, M., and Venturini, G. 1998. Nitric oxide in invertebrates. *Mol. Neurobiol.* 17:157–174.

Colasanti, M., Venturini, G., Merante, A., Musci, G., and Lauro, G.M. 1997. Nitric oxide involvement in *Hydra vulgaris* very primitive olfactory-like system. *J. Neurosci.* 17:493–499.

Conte, A., and Ottaviani, E. 1995. Nitric oxide synthase activity in molluscan hemocytes. *FEBS Lett.* 365:120–124.

Cooke, I.R.C., Edwards, S.L., and Anderson, C.R. 1994. The distribution of NADPH-diaphorase activity and immunoreactivity to nitric oxide synthase in the nervous system of the pulmonate mollusc *Helix aspersa*. *Cell Tissue Res.* 277:565–572.

Cramer, K.S., Leamey, C.A., and Sur, M. 1998. Nitric oxide as a signaling molecule in visual system development, *Prog. Brain Res.* 118:101–114.

Cruz, L., Moroz, L.L., Gillette, R., and Sweedler, J.V. 1997. Nitrite and nitrate levels in individual molluscan neurons: single-cell capillary electrophoresis analysis. *J. Neurochem* 69:110–115.

Cueto, M., Hernandez-Perera, O., Martin, R., Bentura, M.L., Rodrigo, J., Lamas, S., and Golvano, M.P. 1996. Presence of nitric oxide synthase activity in roots and nodules of *Lupinus albus*. *FEBS Lett.* 398:159–164.

Dawson, T.M., Bredt, D.S., Fotuhi, M., Hwang, P.M., and Snyder, S.H. 1991. Nitric oxide synthase and neuronal NADPH diaphorase are identical in brain and peripheral tissues. *Proc. Natl. Acad. Sci. USA* 88:7797–7801.

Dawson, T.M., and Dawson, V.L. 1996. Nitric oxide synthase: role as a transmitter/mediator in the brain and endocrine system. *Annu. Rev. Medicine* 47: 219–227.

Dawson, T.M., and Snyder, S.H. 1994. Gases as biological messengers: nitric oxide and carbon monoxide in the brain. *J. Neurosci.* 14:5147–5159.

Delledonne, M., Xia, Y., Dixon, R.A., and Lamb, C. 1998. Nitric oxide functions as a signal in plant disease resistance. *Nature* 394:585–588.

Dinerman, J.L., Dawson, T.M., Schell, M.J., Snowman, A., and Snyder, S.H. 1994. Endothelial nitric oxide synthase localized to hippocampal pyramidal cells: implications for synaptic plasticity. *Proc. Natl. Acad. Sci. USA* 91:4214–4218.

Ding, J.M., Chen, D., Weber, E.T., Faiman, L.E., Rea, M.A., and Gillette, M.U. 1994. Resetting the biological clock: mediation of nocturnal circadian shifts by glutamate and NO. *Science* 266:1713–1717.

Dow, J.A.T., Maddrell, S.H.P., Davies, S.-A., Skaer, N.J.V., and Kaizer, K. 1994. A novel role for the nitric oxide-cGMP signaling pathway: the control of epithelial function in *Drosophila*. *Am. J. Physiol.* 35:R1716–R1719.

Doyle, M.P., Pickering, R.A., DeWeert, T.M., Hoekstra, J.W., and Pater, D. 1981. Kinetics and mechanism of the oxidation of human deoxyhemoglobin by nitrites. *J. Biol. Chem.* 256:12393–12398.

Duncan, C., Dougall, H., Johnston, P., Green, S., Brogan, R., Leifert, C., Smith, L., Golden, M., and Benjamin, N. 1995. Chemical generation of nitric oxide in the mouth from enterosalivary circulation of dietary nitrate. *Nature Medicine* 1:546–551.

Duncan, C., Li, H., Dykhuizen, R., Frazer, R., Johnston, P., MacKnight, G., Smith, L., Lamza, K., McKenzie, H., Batt, L., Kelly, D., Golden, M., Benjamin, N., and Leifert, C. 1997. Protection againist oral and gastrointestinal diseases: importance of dietary nitrate uptake, oral nitrate reduction and enterosalivary nitrate circulation. *Comp. Biochem. Physiol.* 118A:939–948.

Durner, J., and Klessig, D.F. 1999. Nitric oxide as a signal in plants. *Curr. Opin. Plant Biol.* 2:369–374.

Durner, J., Wendehenne, D., and Klessig, D.F. 1998. Defense gene induction in tobacco by nitric oxide, cyclic GMP, and cyclic ADP-ribose. *Proc. Natl. Acad. Sci. USA* 95:10328–10333.

Elofsson, R., Carlberg, M., Moroz, L., Nezlin, L., and Sakharov, D. 1993. Is nitric oxide (NO) produced by invertebrate neurones? *Neuroreport* 4:279–282.

Elphick, M.R., Green, I.C., and O'Shea, M. 1993. Nitric oxide synthesis and action in an invertebrate brain. *Brain. Res.* 619:344–346.

Elphick, M.R., Kemenes, G., Staras, K., and O'Shea, M. 1995a. Behavioural role for nitric oxide in chemosensory activation of feeding in a mollusc. *J. Neurosci.* 15:7653–7664.

Elphick, M.R., and Melarange, R. 1998. Nitric oxide function in an echinoderm. *Biol. Bull.* 194:260–266.

Elphick, M.R., Rayne, R.C., Riveros-Moreno, V., Moncada, S., and O'Shea, M. 1995b. Nitric oxide synthesis in locust olfactory interneurones. *J. Exp. Biol.* 198:821–829.

Eriksson, K.S. 1996. Nitric oxide synthase in the pharynx of the planarian *Dugesia tigrina*. *Cell Tissue Res.* 286:407–410.

Feelisch, M. 1991. Biotransformation of organic nitrates to nitric oxide by vascular smooth muscle and endothelial cells. *Biophys. Biochem. Res. Commun.* 180:286–293.

Feelisch, M. 1993. Biotransformation to nitric oxide of organic nitrates in comparison to other nitrovasodilators. *Eur. Heart J.* 14 Suppl 1:123–132.

Feelisch, M., and Kelm, M. 1991. Biotransformation of organic nitrates to nitric oxide by vascular smooth muscle and endothelial cells. Biochem. Biophys. *Res. Commun.* 180:286–293.

Feelisch, M., and Martin, J.E. 1995. The early role of nitric oxide in evolution. *Trends Ecol. Evol.* 10:496–499.

Feelisch, M., and Noack, E.A. 1987. Correlation between nitric oxide formation during degeneration of organic nitrates and activation of guanylate cyclase. *Eur. J. Pharmacol.* 139:19–30.

Feelisch, M., and Stamler, J.S. 1996. Donors of nitrogen oxides. In: M. Feelisch and J. S. Stamler (eds.) *Methods in nitric oxide research*, pp. 71–115. John Wiley & Sons, Chichester, New York.

Ferguson, S.J. 1998. Nitrogen cycle enzymology. *Curr. Opin. Chem. Biol.* 2:182–193.

Floyd, P.D., Moroz, L.L., Gillette, R., and Sweedler, J.V. 1998. Capillary electrophoresis analysis of nitric oxide synthase related metabolites in single identified neurons. *Anal. Chem.* 70:2243–2247.

Fossier, P., Blanchard, B., Ducrocq, C., Leprince, C., Tauc, L., and Baux, G. 1999. Nitric oxide transforms serotonin into an inactive form and this affects neuromodulation. *Neuroscience* 93:597–603.

Franchini, A., Conte, A., and Ottaviani, E. 1995. Nitric oxide: an ancestral immunocyte effector molecule. *Adv. Neuroimmunol.* 5:463–478.

Gardner, P.R., Gardner, A.M., Martin, L.A., and Salzman, A.L. 1998. Nitric oxide dioxygenase: an enzymic function for flavohemoglobin. *Proc. Natl. Acad. Sci. USA* 95:10378–10383.

Garthwaite, J., and Boulton, C.L. 1995. Nitric oxide signaling in the central nervous system. *Annu. Rev. Physiol.* 57:683–706.

Gelperin, A. 1994a. Nitric oxide mediates network oscillations of olfactory interneurons in a terrestrial mollusc. *Nature* 369:61–63.

Gelperin, A. 1994b. Nitric oxide, odour processing and plasticity. *Neth. J. Zool.* 44:159–169.

Gibbs, S.M., and Truman, J.W. 1998. Nitric oxide and cyclic GMP regulate retinal patterning in the optic lobe of *Drosophila*. *Neuron* 20:83–93.

Gilly, W.F., Moroz, L.L., and Gillette, R. 1995. Redox and nitric oxide (NO)-dependent modulation of potassium channels of squid stellate ganglion neurons. *Soc. Neurosci. Abstr.* 21:720.11, P1832.

Goretski, J., Zafiriou, O.C., and Hollocher, T.C. 1990. Steady-state nitric oxide concentrations during denitrification. *J. Biol. Chem.* 1990:11535–11538.

Griffin, O.W., and Stuehr, D.J. 1995. Nitric oxide synthases: properties and catalytic mechanism. *Annu. Rev. Physiol.* 57:707–736.

Grueber, W.B., and Truman, J.W. 1999. Development and organization of a nitric-oxide-sensitive peripheral neural plexus in larvae of the moth, *Manduca sexta*. *J. Comp. Neurol.* 404:127–141.

Gustafsson, M.K., Lindholm, A.M., Terenina, N.B., and Reuter, M. 1996. NO nerves in a tapeworm. NADPH-diaphorase histochemistry in adult *Hymenolepis diminuta*. *Parasitology* 113:559–565.

Hess, D.T., Patterson, S.I., Smith, D.S., and Skene, J.H. 1993. Neuronal growth cone collapse and inhibition of protein fatty acylation by nitric oxide. *Nature* 366:562–565.

Hollocher, T.C., and Hibbs, J.B.J. 1996. Enzymes of bacteria, plants and fungi that process free nitrogen oxides. In: M. Feelisch and J.S. Stamler (eds.) *Methods in nitric oxide research*, pp. 119–146. John Wiley & Sons, Chichester, New York.

Hope, B.T., Michael, G.J, Knigge, K.M., and Vincent, S.R. 1991. Neuronal NADPH-diaphorase is a nitric oxide synthase. *Proc. Natl. Acad. Sci. USA* 88:2811–2814.

Huang, S., Kerschbaum, H.H., Engel, E., and Hermann, A. 1997. Biochemical characterization and histochemical localization of nitric oxide synthase in the nervous system of the snail, *Helix pomatia*. *J. Neurochem.* 69:2516–2528.

Huang, P.L., and Lo, E.H. 1998. Genetic analysis of NOS isoforms using nNOS and eNOS knockout animals. *Prog. Brain Res.* 118:13–25.

Hudson, M., and O'Shea, M. 1998. Is nitric oxide a signaling molecule in *Caenorabditis elegans*? *Soc. Neurosci. Abstr.* 24:356.

Jacklet, J.W. 1995. Nitric oxide is used as an orthograde cotransmitter at identified histaminergic synapses. *J. Neurophysiol.* 74:891–895.

Jacklet, J. 1997. Nitric oxide signaling in invertebrates. *Invert. Neurosci.* 3:1–14.

Jacklet, J.W., and Gruhn, M. 1994. Co-localization of NADPH-diaphorase and myomodulin in synaptic glomeruli of *Aplysia*. *NeuroReport* 5:1841–1844.

Johansson, K.U.I., and Carlberg, M. 1994. NADPH-diaphorase histochemistry and nitric oxide synthase activity in deutocerebrum of the crayfish, *Pacifastacus leniusculus* (Crustacea, Decapoda). *Brain Res.* 649:36–42.

Kanadia, R.N., Kuo, W.N., McNabb, M., and Botchway, A. 1998. Constitutive nitric oxide synthase in *Saccharomyces cerevisiae. Biochem. Mol. Biol. Int.* 45:1081–1087.

Kharitonov, V.G., Sundquist, A.R., and Sharma, V.S. 1994. Kinetics of nitric oxide autoxidation in aqueous solution. *J. Biol. Chem.* 269:5881–5883.

Kimura, T., Shouno, O., and Matsumoto, G. 1997. NADPH-diaphorase containing cells and fibers in the central nervous system of squid, *Loligo bleekeri* Keferstein. *Life Sci.* 61:2375–2381.

Knowles, R.G., and Moncada, S. 1994. Nitric oxide synthases in mammals. *Biochem. J.* 298:249–258.

Kobayashi, M., Matsuo, Y., Takimoto, A., Suzuki, S., Maruo, F., and Shoun, H. 1996. Denitrification, a novel type of respiratory metabolism in fungal mitochondrion. *J. Biol. Chem.* 271:16263–16267.

Kobayashi, M., and Shoun, H. 1995. The copper-containing dissimilatory nitrite reductase involved in the denitrifying system of the fungus *Fusarium oxysporum. J. Biol. Chem.* 270:4146–4151.

Koh, H.Y., and Jacklet, J.W. 1999. Nitric oxide stimulates cGMP production and mimics synaptic responses in metacerebral neurons of *Aplysia. J. Neurosci.* 19:3818–3826.

Korneev, S.A., Park, J.H., and O'Shea, M. 1999. Neuronal expression of neural nitric oxide synthase (nNOS) protein is suppressed by an antisense RNA transcribed from an NOS pseudogene. *J. Neurosci.* 19:7711–7720.

Korneev, S.A., Piper, M.R., Picot, J., Phillips, R., Korneeva, E.I., and O'Shea, M. 1998. Molecular characterization of NOS in a mollusc: expression in a giant modulatory neuron. *J. Neurobiol.* 35:65–76.

Kuzin, B., Roberts, I., Peunova, N., and Enikolopov, G. 1996. Nitric oxide regulates cell proliferation during *Drosophila* development. *Cell* 87:639–649.

Kwiatkowski, A., Laratta, W.P., Toffanin, A., and Shapleigh, J.P. 1997. Analysis of the role of the nnrR gene product in the response of *Rhodobacter sphaeroides* 2.4.1 to exogenous nitric oxide. *J. Bacteriol.* 179:5618–5620.

Kwiatkowski, A.V., and Shapleigh, J.P. 1996. Requirement of nitric oxide for induction of genes whose products are involved in nitric oxide metabolism in *Rhodobacter sphaeroides* 2.4.3. *J. Biol. Chem.* 271:24382–24388.

Leake, L.D., Davis, M.P., Chen, D., and Moroz, L.L. 1996. A unique model for the analysis of neuronal nitric oxide signaling: the leech CNS. In: J. Salanki, K. S-Rozsa, and K. Elekes (eds.) *Neurobiology of invertebrates: simple and complex regulatory systems*, pp. 135–143. Akademia Kiado, Budapest.

Leake, L.D., and Moroz, L.L. 1996. Putative nitric oxide synthase (NOS)-containing cells in the central nervous system of the leech, *Hirudo medicinals*: NADPH-diaphorase histochemistry. *Brain Res.* 723:115–124.

Lewin, M.R., and Walters, E.T. 1999. Cyclic GMP pathway is critical for inducing long-term sensitization of nociceptive sensory nneurons. *Nature Neurosci.* 2:18–23.

Lewis, R.S., and Deen, W.M. 1994. Kinetics of the reaction of nitric oxide with oxygen in aqueous solutions. *Chem. Res. Toxicol.* 7:568–574.

Lin, M.F., and Leise, E.M. 1996. NADPH-diaphorase activity changes during gangliogenesis and metamorphosis in the gastropod mollusc *Ilyanassa obsoleta*. *J. Comp. Neurol.* 374:194–203.

Liu, Y., Shenouda, D., Bilfinger, T.V., Stefano, M.L., Magazine, H.I., and Stefano, G.B. 1996. Morphine stimulates nitric oxide release from invertebrate microglia. *Brain Res.* 722:125–131.

Luckhart, S., and Rosenberg, R. 1999. Gene structure and polymorphism of an invertebrate nitric oxide synthase gene. *Gene* 232:25–34.

Luckhart, S., and Vodovotz, Y., Cui, L., and Rosenberg, R. 1998. The mosquito *Anopheles stephensi* limits malaria parasite development with inducible synthesis of nitric oxide. *Proc. Natl. Acad. Sci. USA* 95:5700–5705.

Martinez, A. 1995. Nitric oxide synthase in invertebrates. *Histochem. J.* 27:770–776.

Martinez, A., Riveros-Moreno, V., Polak, J.M., Moncada, S., and Sesma, D. 1994. Nitric oxide (NO) synthase immunoreactivity in the starfish *Marthasterias glacialis*. *Cell Tissue Res.* 275:599–603.

McFarland, M., Kley, D., Drummond, J.W., Schmeltekopf, A.L., and Winkler, R.M. 1979. Nitric oxide measurements in the equatorial Pacific region. *Geophys. Res. Lett.* 6:605–609.

Meleshkevitch, E.A., Budko, D.Y., Norby, S.W., Moroz, L.L., and Hadfield, M.G. 1997. Nitric oxide-dependent modulation of the metamorphosis in mollusc *Phestilla sibogae* (Gastropoda, Nudibranchaea). *Soc. Neurosci. Abstr.* 23: 487.P1233.

Meulemans, A., Mothet, J.P., Schirar, A., Fossier, P., Tauc, L., and Baux, G. 1995. A nitric oxide synthase activity is involved in the modulation of acetylcholine release in *Aplysia* ganglion neurons: a histological, voltammetric and electrophysiological study. *Neuroscience* 69:985–995.

Meyer, D.J. 1973. Interaction of cytochrome oxidases aa3 and d with nitrite. *Nat. New Biol.* 245:276–277.

Meyer, W. 1994. NADPH diaphorase (nitric oxide synthase) in the central nervous system of spiders (Arachnida: Araneida). *Neurosci. Lett.* 165:105–108.

Millar, A.H., and Day, D.A. 1996. Nitric oxide inhibits the cytochrome oxidase but not the alternative oxidase of plant mitochondria. *FEBS Lett.* 398:155–158.

Mirvish, S.S., Wallcave, L., Eagen, M., and Shubik, P. 1972. Ascorbate-nitrite reaction: possible means of blocking the formation of carcinogenic N-nitroso compounds. *Science* 177:65–68.

Moncada, S., and Higgs, E.A. 1995. Molecular mechanisms and therapeutic strategies related to nitric oxide. *FASEB J.* 9:1319–1330.

Moncada, S., and Martin, J.F. 1993. Evolution of nitric oxide. *Lancet* 341:1511.

Moncada, S., Palmer, M.J., and Higgs, E.A. 1991. Nitric oxide: physiology, pathophysiology and pharmacology. *Pharmacol. Rev.* 43:109–144.

Moroz, L.L., Bulloch, A.G.M., Lukowiak, K., and Syed, N. 1994a. Putative NO-synthesizing neurons of *Lymnaea* in vivo and in vitro. *Neth. J. Zool.* 44:535–549.

Moroz, L.L., Chen, D., Gillette, M.U., and Gillette, R. 1996. Nitric oxide synthase activity in the molluscan CNS. *J. Neurochem.* 66:873–876.

Moroz, L.L., and Gillette, R. 1996a. From Polyplacophora to Cephalopoda: comparative analysis of nitric oxide signalling in Mollusca. In: J. Salanki, K. S-Rozsa, and K. Elekes (eds.) *Neurobiology of invertebrates: simple and complex regulatory systems*, pp. 169–182. Akademia Kiado, Budapest.

Moroz, L.L., and Gillette, R. 1996b. NADPH-diaphorase localization in the CNS and peripheral tissues of the predatory sea-slug *Pleurobranchaea californica. J. Comp. Neurol.* 367:607–622.

Moroz, L.L., Gillette, R., and Sweedler, J.V. 1999. Single-cell analyses of nitrergic neurons in simpler nervous systems. *J. Exp. Biol.* 202:333–341.

Moroz, L.L., Meech, R.W., and Mackie, G.O. 1997. The distribution and action of NO in the jellyfish *Aglantha digitale. J. Physiol.* 504:13P.

Moroz, L.L., Park, J.H., and Winlow, W. 1993. Nitric oxide activates buccal motor patterns in *Lymnaea stagnalis. Neuroreport* 4:643–646.

Moroz, L.L., Winlow, W., Turner, R.W., Bulloch, A.G., Lukowiak, K., and Syed, N.I. 1994b. Nitric oxide synthase-immunoreactive cells in the CNS and periphery of *Lymnaea. Neuroreport* 5:1277–1280.

Mothet, J.P., Fossier, P., Tauc, L., and Baux, G. 1996a. NO decreases evoked quantal ACh release at a synapse of *Aplysia* by a mechanism independent of Ca^{2+} influx and protein kinase G. *J. Physiol. (Lond.)* 493:769–784.

Mothet, J.P., Fossier, P., Tauc, L., and Baux, G. 1996b. Opposite actions of nitric oxide on cholinergic synapses: which pathways? *Proc. Natl. Acad. Sci. USA* 93:8721–8726.

Muller, U. 1994. Ca^{2+}/calmodulin-dependent nitric oxide synthase in *Apis mellifera* and *Drosophila melanogaster. Eur. J. Neurosci.* 6:1362–1370.

Muller, U. 1996. Inhibition of nitric oxide synthase impairs a distinct form of long-term memory in the honeybee, *Apis mellifera. Neuron* 16:541–549.

Muller, U. 1997. The nitric oxide system in insects. *Prog. Neurobiol.* 51:363–381.

Muller, U., and Bicker, G. 1994. Calcium activated release of nitric oxide and cellular distribution of nitric oxide synthesizing neurons in the nervous system of the locust. *J. Neurosci.* 14:7521–7528.

Muller, U., and Hildebrandt, H. 1995. The nitric oxide/cGMP system in the antennal lobe of *Apis mellifera* is implicated in integrative processing of chemosensory stimuli. *Eur. J. Neurosci.* 7:2240–2248.

Nathan, C. 1992. Nitric oxide as a secretory product of mammalian cells. *FASEB J.* 6:3051–3064.

Nighorn, A., Gibson, N.J., Rivers, D.M., Hildebrand, J.G., and Morton, D.B. 1998. The nitric oxide-cGMP pathway may mediate communication between sensory afferents and projection neurons in the antennal lobe of *Manduca sexta. J. Neurosci.* 18:7244–7255.

Ninnemann, H., and Maier, J. 1996. Indications for the occurrence of nitric oxide synthases in fungi and plants and the involvement in photoconidiation of *Neurospora crassa. Photochem. Photobiol.* 64:393–398.

Nussenzveig, R.H., Bentley, D.L., and Ribeiro, J.M.C. 1995. Nitric oxide loading of the salivary nitric oxide-carrying hemoproteins (nitrophorins) in the blood-sucking bug *Rhodnius prolixus. J. Exp. Biol.* 198:1093–1098.

Ogilvie, P., Schilling, K., Billingsley, M.L., and Schmidt, H.H. 1995. Induction and variants of neuronal nitric oxide synthase type I during synaptogenesis. *FASEB J.* 9:799–806.

Palumbo, A., Di Cosmo, A., Gesualdo, I., and d'Ischia, M. 1997. A calcium-dependent nitric oxide synthase and NMDA R1 glutamate receptor in the ink gland of *Sepia officinalis*: a hint to a regulatory role of nitric oxide in melanogenesis? *Biochem. Biophys. Res. Commun.* 235:429–432.

Park, S.Y., Shimizu, H., Adachi, S., Nakagawa, A., Tanaka, I., Nakahara, K., Shoun, H., Obayashi, E., Nakamura, H., Iizuka, T., and Shiro, Y. 1997. Crystal structure of nitric oxide reductase from denitrifying fungus *Fusarium oxysporum*. *Nat. Struct. Biol.* 4:827–832.

Park, J.H., Straub, V.A., and O'Shea, M. 1998. Anterograde signaling by nitric oxide: characterization and in vitro reconstitution of an identified nitrergic synapse. *J. Neurosci.* 18:5463–5476.

Paveto, C., Pereira, C., Espinosa, J., Montagna, A.E., Farber, M., Esteva, M., Flawia, M.M., and Torres, H.N. 1995. The nitric oxide transduction pathway in *Trypanosoma cruzi*. *J. Biol. Chem.* 270:16576–16579.

Payne, W.J., Liu, M.Y., Bursakov, S.A., and Le Gall, J. 1997. Microbial and plant metabolism of NO. *Biofactors* 6:47–52.

Pereira, C., Paveto, C., Espinosa, J., Alonso, G., Flawia, M.M., and Torres, H.N. 1997. Control of *Trypanosoma cruzi* epimastigote motility through the nitric oxide pathway. *J. Eukaryot. Microbiol.* 44:155–156.

Peunova, N., and Enikolopov, G. 1995. Nitric oxide triggers a switch to growth arrest during differentiation of neuronal cell. *Nature* 375:68–73.

Pisu, M.B., Conforti, E., Fenoglio, C., Necchi, D., Scherini, E., and Bernocchi, G. 1999. Nitric oxide-containing neurons in the nervous ganglia of *Helix aspersa* during rest and activity: Immunocytochemical and enzyme histochemical detection. *J. Comp. Neurol.* 409:274–284.

Radomski, M.W., Martin, J.F., and Moncada, S. 1991. Synthesis of nitric oxide by the haemocytes of the American horseshoe crab (*Limulus polyphemus*). *Phil. Trans. R. Soc. Lond.* 334B:129–133.

Regulski, M., and Tully, T. 1995. Molecular and biochemical characterization of dNOS: a *Drosophila* Ca^{2+}/calmodulin-dependent nitric oxide synthase. *Proc. Natl. Acad. Sci. USA* 92:9072–9076.

Renteria, R.C., and Constantine-Paton, M. 1996. Exogenous nitric oxide causes collapse of retinal ganglion cell axonal growth cones in vitro. *J. Neurobiol.* 29:415–428.

Ribeiro, J.M.C., and Walker, F.A. 1994. High affinity histamine-binding and antihistaminic activity of the salivary nitric oxide-carrying heme protein (nitrophorin) of *Rhodnius prolixus*. *J. Exp. Med.* 180:2251–2257.

Robertson, J.D., Bonaventura, J., and Kohm, A.P. 1994. Nitric oxide is required for tactile learning in *Octopus vulgaris*. *Proc. R. Soc. Lond. B.* 256:269–273.

Robertson, J.D., Bonaventura, J., and Kohm, A. 1995a. Nitric oxide synthase inhibition blocks Octopus touch learning without producing sensory or motor dysfunction. *Proc. R. Soc. Lond.* 261B:167–172.

Robertson, J.D., Bonaventura, J., Kohm, A., and Hiscat, M. 1996. Nitric oxide is necessary for visual learning in *Octopus vulgaris*. *Proc. R. Soc. Lond. B. Biol. Sci.* 263:1739–1743.

Robertson, J.D., Moroz, L.L., and Gillette, R. 1995b. Central and peripheral distribution of putative nitric oxide synthase (NOS)-containing cells in *Octopus vulgaris*. *Soc. Neurosci. Abstr.* 21:P630.

Roenneberg, T., and Rehman, J. 1996. Nitrate, a nonphotic signal for the circadian system. *FASEB J.* 10:1443–1447.

Salleo, A., Giovanni, M., Barra, P., and Calabrese, L. 1996. The discharge mechanism of aconitial nematocytes involves the release of nitric oxide. *J. Exp. Biol.* 199:1261–1267.

Salzet, M., Salzet-Raveillon, B., Cocquerelle, C., Verger-Bocquet, M., Pryor, S.C., Rialas, C.M., Laurent, V., and Stefano, G.B. 1997. Leech immunocytes contain proopiomelanocortin: nitric oxide mediates hemolymph proopiomelanocortin processing. *J. Immunol.* 159:5400–5411.

Samouilov, A., Kuppusamy, P., and Zweier, J.L. 1998. Evaluation of the magnitude and rate of nitric oxide production from nitrite in biological systems. *Arch. Biochem. Biophys.* 357:1–7.

Sanchez-Alvarez, M., Leon-Olea, M., Talavera, E., Pelicer, F., Sanchez-Islas, E., and Martinez-Lorenzana, G. 1994. Distribution of NADPH-diaphorase in the perioesophageal ganglia of the snail, *Helix aspersa. Neurosci. Lett.* 169:51–55.

Saran, M., and Bors, W. 1994. Pulse radiolysis for investigation of nitric oxide related reactions. *Meth. Enzymol.* 233:20–34.

Scholz, N.L., Chang, E.S., Graubard, K., and Truman, J.W. 1998. The NO/cGMP pathway and the development of neural networks in postembryonic lobsters. *J. Neurobiol.* 34:208–226.

Scorza, G., Pietraforte, D., and Minetti, M. 1997. Role of ascorbate and protein thiols in the release of nitric oxide from S-nitroso-albumin and S-nitroso-glutathione in human plasma. *Free Radic. Biol. Med.* 22:633–642.

Shoun, H., Kano, M., Baba, I., Takaya, N., and Matsuo, M. 1998. Denitrification by actinomycetes and purification of dissimilatory nitrite reductase and azurin from *Streptomyces thioluteus. J. Bacteriol.* 180:4413–4415.

Shoun, H., Kim, D.H., Uchiyama, H., and Sugiyama, J. 1992. Denitrification by fungi. *FEMS Microbiol. Lett.* 94:277–281.

Snyder, S.H., Jaffrey, S.R., and Zakhary, R. 1998. Nitric oxide and carbon monoxide: parallel roles as neural messengers. *Brain Res. Rev.* 26:167–175.

Stamler, J.S., Singel, D.J., and Loscalzo, J. 1992. Biochemistry of nitric oxide and its redox activated forms. *Science* 258:1898–1902.

Stengl, M., and Zintl, R. 1996. NADPH-diaphorase activity in the antennae of the hawkmoth *Manduca sexta. J. Exp. Biol.* 199:1063–1072.

Talavera, E., Martinez-Lorenzana, G., Leon-Olea, M., Sanchez-Alvarez, M., Sanchez-Islas, E., and Pellicer, F. 1995. Histochemical distribution of NADPH-diaphorase in the cerebral ganglion of the crayfish *Cambarellus montezumae. Neurosci. Lett.* 187:177–180.

Tao, Y., Howlett, A., and Klein, C. 1992. Nitric oxide-releasing compounds inhibit *Dictyostelium Discoideum* Aggregation without altering cGMP production. *FEBS Lett.* 314:49–52.

Tao, Y., Howlett, A., and Klein, C. 1996. Nitric oxide inhibits the initiation of cAMP pulsing in *D. discoideum* without altering receptor-activated adenylate cyclase. *Cell Signal.* 8:26–34.

Tao, Y.P., Misko, T.P., Howlett, A.C., and Klein, C. 1997. Nitric oxide, an endogenous regulator of *Dictyostelium discoideum* differentiation. *Development* 124:3587–3595.

Teyke, T. 1996. Nitric oxide, but not serotonin, is involved in acquisition of food-attraction conditioning in the snail *Helix pomatia. Neurosci. Lett.* 206:29–32.

Truman, J.W., De Vente, J., and Ball, E.E. 1996. Nitric oxide-sensitive guanylate cyclase activity is associated with the maturational phase of neuronal development in insects. *Development* 122:3949–3958.

Tsuruta, S., Takaya, N., Zhang, L., Shoun, H., Kimura, K., Hamamoto, M., and Nakase, T. 1998. Denitrification by yeasts and occurrence of cytochrome P450nor in *Trichosporon cutaneum*. *FEMS Microbiol Lett.* 168:105–110.

Usuda, K., Toritsuka, N., Matsuo, Y., Kim, D.H., and Shoun, H. 1995. Denitrification by the fungus *Cylindrocarpon tonkinense*: anaerobic cell growth and two isozyme forms of cytochrome P-450nor. *Appl. Environ. Microbiol.* 61:883–889.

Vallette, G., Tenaud, I., Branka, J.-E., Jarry, A., Sainte-Marie, I., Dreno, B., and Laboisse, C.L. 1998. Control of growth and differentiation of normal human epithelial cells through the manipulation of reactive nitrogen species. *Biochem. J.* 331:713–717.

Wagenen, S., and Rehder Van, V. 1999. Regulation of neuronal growth cone filopodia by nitric oxide. *J. Neurobiol.* 39:168–185.

Walters, C.L., and Taylor, A. 1965. The reduction of nitrite by skeletal muscle mitochondria. *Biochem. Biophys. Acta* 96:522–524.

Weitzberg, E., and Lundberg, J.O. 1998. Nonenzymatic nitric oxide production in humans. *Nitric Oxide* 2:1–7.

Weller, R., Pattullo, S., Smith, L., Golden, M., Ormerod, A., and Benjamin, N. 1996. Nitric oxide is generated on the skin surface by reduction of sweat nitrate. *J. Invest. Dermatol.* 107:327–331.

Werner-Felmayer, G., Golderer, G., Werner, E.R., Grobner, P., and Wachter, H. 1994. Pteridine biosynthesis and nitric oxide synthase in *Physarum polycephalum*. *Biochem. J.* 304:105–111.

Wildemann, B., and Bicker, G. 1999. Developmental expression of nitric oxide/cyclic GMP synthesizing cells in the nervous system of *Drosophila melanogaster. J. Neurobiol.* 38:1–15.

Wingrove, J.A., and O'Farrell, P.H. 1999. Nitric oxide contributes to behavioral, cellular, and developmental responses to low oxygen in *Drosophila*. *Cell* 98:105–114.

Wink, D.A., Grisham, M.B., Mitchell, J.B., and Ford, P.C. 1996. Direct and indirect effects of nitric oxide in chemical reactions relevant to biology. *Meth. Enzymol.* 268:12–31.

Wink, D.A., and Mitchell, J.B. 1998. Chemical biology of nitric oxide: insights into regulatory, cytotoxic, and cytoprotective mechanisms of nitric oxide. *Free Radic. Biol. Med.* 25:434–456.

Wood, J., and Garthwaite, J. 1994. Models of the diffusional spread of nitric oxide: implications for neural nitric oxide signaling and its pharmacological properties. *Neuropharmacology* 33:1235–1244.

Wu, G., and Morris, S.M. Jr, 1998. Arginine metabolism: nitric oxide and beyond. *Biochem. J.* 336:1–17.

Yoshinari, T. 1993. Nitrogen oxides flux in tropical soils. *Trends Ecol. Evol.* 8:155–156.

Yuda, M., Hirai, M., Miura, K., Matsumura, H., Ando, K., and Chinzei, Y. 1996. cDna cloning, expression and characterization of nitric-oxide synthase from the salivary glands of the blood-sucking insect *Rhodnius prolixus*. *Eur. J. Biochem.* 242:807–812.

Zafiriou, O.C., McFarland, M., and Bromund, R.H. 1980. Nitric oxide in sea water. *Science* 207:637–639.

Zumft, W.G. 1993. The biological role of nitric oxide in bacteria. *Arch. Microbiol.* 160:253–264.

Zumft, W.G. 1997. Cell biology and molecular basis of denitrification. *Microbiol. Mol. Biol. Rev.* 61:533–616.
Zweier, J.L., Samouilov, A., and Kuppusamy, P. 1999. Non-enzymatic nitric oxide synthesis in biological systems. *Biochim. Biophys. Acta.* 1411:250–262.
Zweier, J.L., Wang, P., Samouilov, A., and Kuppusamy, P. 1995a. Enzymatic/non-enzymatic formation of nitric oxide. *Nature Medicine* 1:1103–1104.
Zweier, J.L., Wang, P., Samouilov, A., and Kuppusamy, P. 1995b. Enzyme-independent formation of nitric oxide in biological tissues. *Nature Medicine* 1:804–809.

2
The Nitrergic Transmitter of the Anococcygeus: Lessons and Insights

ALAN GIBSON

Introduction

The anococcygeus muscle was a relatively late addition to the arsenal of smooth muscle preparations utilized in biomedical research, with the first description of the pharmacological properties of the rat anococcygeus being published in 1972, from the laboratories of Professor John Gillespie at the University of Glasgow, Scotland (Gillespie, 1972). Despite its recent utilization, however, the anococcygeus muscle has provided some of the most crucial experimental evidence contributing to the establishment of the concept that nitric oxide (NO) functions as a neurotransmitter in the peripheral nervous system, and to the development of our understanding of the mechanisms underlying such so-called nitrergic neurotransmission (Rand, 1992; Gibson et al., 1995; Rand and Li, 1995a,b). This chapter traces the experimental path that led to the identification of NO as a neurotransmitter in the anococcygeus, and highlights some of the areas of controversy that have developed along the way.

The Inhibitory Nonadrenergic, Noncholinergic Response

Following on from the first description of the rat anococcygeus (Gillespie, 1972), the pharmacological profiles of the corresponding tissue in the cat (Gillespie and McGrath, 1974), rabbit (Creed et al., 1977), dog (Dehpour et al., 1980), and mouse (Gibson and Wedmore, 1981) have been reported. Although some notable species variations in drug sensitivity were apparent (Gillespie, 1980), a common feature of the muscles from all species was that, depending on the experimental conditions, field stimulation could result in either contraction or relaxation. Early experiments revealed that the contractions were due to the activation of sympathetic nerves, with the released noradrenaline acting on postsynaptic α-adrenoceptors, but the nature of the transmitter responsible for the relaxations remained obscure. Using the rat anococcygeus, it was shown that the field stimulation-induced relaxations survived destruction of the sympathetic nerve terminals with 6-hydroxydopamine (Gibson and Gillespie, 1973) and that the spinal origin of the inhibitory nerves was different from that of the sympathetics (Gillespie and McGrath, 1973). These observations, together with ultrastructural examination of the nerve terminal profiles (Gillespie and Lullman-Rauch, 1974; Gibbins and Haller, 1979), ruled out the possibility that the muscle received only a single nerve supply that released both noradrenaline and the inhibitory substance as cotransmitters; clearly the muscle was dually innervated. Thus, the relaxant response to field stimulation was considered to be a further example of nonadrenergic, noncholinergic (NANC) neurotransmission in the peripheral nervous system (Burnstock, 1986).

Figure 2.1 reproduces the first published illustration of these NANC relaxations in the rat anococcygeus and indeed highlights the type of experiment that first uncovered the NANC innervation in this tissue. The adrenergic neuron-blocking drug guanethidine was being used to characterize the motor sympathetic response; low concentrations produced the expected inhibition of contraction, but higher concentrations raised the tone of the tissue and reversed the previous motor effect of field stimulation to relaxation. Guanethidine possessed the two important properties necessary to observe consistent NANC relaxations: inhibition of the sympathetic nerves and increased muscle tone, the latter effect resulting from the indirect sympathomimetic activity of guanethidine at these concentrations. It is an important lesson of scientific investigation that the identification of the muscle itself (Gillespie, 1997) and of its NANC innervation (Gillespie, 1972)

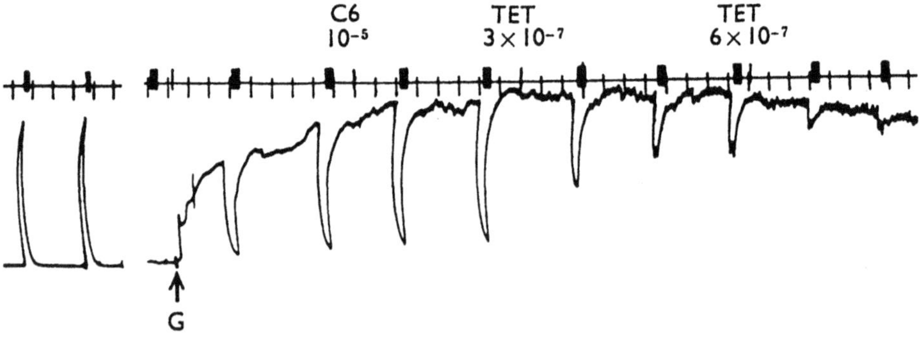

FIGURE 2.1. One of the original published tracings showing NANC relaxations of the rat anococcygeus muscle. Guanethidine (G) at 30 μM raises the tone of the muscle and converts the previous motor responses during field stimulation to relaxations. These relaxations are resistant to the ganglion-blocking drug hexamethonium (C6) but are almost abolished by tetrodotoxin (TTX). From Gillespie, J.S. 1972.

resulted from the pursuance of unexpected results that arose from projects designed to investigate other topics (the sympathetic innervation of the terminal colon and the sympathetic supply to the anococcygeus muscle, respectively).

Although the anococcygeus has been used extensively as an experimental model in several areas of pharmacological research, including denervation/drug-induced supersensitivity (Gibson and Pollock, 1973; Doggrell and Vincent, 1981; Doggrell and Waldron, 1982a,b), α-adrenoceptors (Docherty and Starke, 1981; Coates and Weetman, 1983; Gibson and Yu, 1983; McGrath, 1984), muscarinic receptors (Doggrell, 1981; Oriowo, 1983; Gross et al., 1997; Weiser et al., 1997), and the biophysics of smooth muscle contraction (Gillis et al., 1988; Walker et al., 1988; Boland et al., 1992), by far the greatest experimental effort has been directed toward identification of the NANC transmitter.

Early NANC Transmitter Candidates

ATP

Evidence in support of the proposal that ATP might be the NANC transmitter of the anococcygeus included the demonstration of quinacrine fluorescence-positive nerve fibers running to the muscle (Burnstock et al., 1978; Olson and Alund, 1979), ultrastructural studies (Iijima, 1983), and the release of ATP from the rat anococcygeus during nerve stimulation (Burnstock et al., 1978). Against the proposal was the variability of the functional response to ATP across species, with relaxations being readily observed in the rabbit but contractions predominating in the rat and mouse; these contractions were shown to be due to prostaglandin production, since they were reversed to relaxations in the presence of indomethacin (Burnstock et al., 1978; Gibson and Wedmore, 1981). In the rabbit anococcygeus, the photoaffinity-label ATP antagonist $ANAPP_3$ had no effect on NANC relaxations but produced a small inhibition of those to exogenous ATP (Sneddon et al., 1982). Similarly, the P_1 purinoceptor antagonist 8-(*p*-sulphophenyl)theophylline reduced relaxations to both adenosine and ATP without affecting NANC relaxations (Graham and Sneddon, 1993). Overall, therefore, the case that ATP was the NANC transmitter in the anococcygeus was relatively weak.

Neuropeptides

Several neuropeptides have been shown to relax the anococcygeus muscle, including bradykinin (Gillespie and McKnight, 1978; Mudumbi and Leighton, 1994a), urotensin II (Gibson et al., 1984; Larsen et al., 1985), vasoactive intestinal peptide (VIP) (Gibson and Tucker, 1982), and peptide histidine isoleucine (Blank et al., 1986). Of these, most attention was paid to VIP, since it relaxed the anococcygeus of all species studied and there was clear histological evidence for VIP-positive nerve fibers running to, and coursing within, the muscle (Gibson and Tucker, 1982; Larsen et al., 1985; Dail et al., 1990). However, whereas relaxations to exogenous VIP were reduced by VIP antiserum or α-chymotrypsin, NANC relaxations were unaffected; further, NANC responses were normal in tissues desensitized to exogenous VIP (Carvajal et al., 1986). Thus, although the anococcygeus receives a substantial VIP innervation [approximately 50% of the parasympathetic fibers are VIP-positive (Dail et al., 1990)], the peptide does not appear to be involved in the NANC relaxations. One possibility suggested by the anatomical distribution of the VIP-positive nerve terminals is that the peptide regulates blood flow through the muscle (Dail et al., 1990).

Inhibitory Factor

Although the exact physiological function of the anococcygeus has not been fully determined, its close association with the retractor penis has suggested that it is part of the erectile machinery (for discussion of the anatomy of the muscle see Gillespie, 1972, 1997; Larsen et al., 1985; Dail et al., 1990). Based on this, Gillespie and his colleagues assumed that the relaxant NANC transmitters of the anococcygeus and the retractor penis would be the same, and therefore set out to extract possible transmitter candidates from both tissues. They confirmed previous work by Ambache et al. (1975) that a thermolabile relaxant substance could be extracted from the retractor penis (Gillespie and Martin, 1980); moreover, a similar material was obtained from the rat anococcygeus. Further studies, mainly using the inhibitory factor (IF) from the retractor penis, revealed that the IF existed in two forms—a stable, inactive form that could be converted to the labile, active form by exposure to acid (Gillespie and Martin, 1980; Gillespie et al., 1981). It soon became clear that IF had much in common with the endothelium-derived relaxing factor (EDRF) that had been described by Furchgott and Zawadski in 1980. Both substances were extremely labile, inhibited by drugs that reduced the guanylate cyclase/cyclic GMP pathway, and potentiated by the cyclic GMP phosphodiesterase inhibitor zaprinast (Bowman and Gillespie, 1982; Bowman et al., 1982; Bowman and Drummond, 1984; Martin et al., 1985, 1986; Griffith et al., 1985). This similarity, and the identification of EDRF as NO, eventually led to the proposal that the inactive form of IF from the bovine retractor penis was extracted nitrite that was converted to the active NO by acid exposure (Furchgott, 1987; Martin et al., 1988). The experiments with IF, together with the observation that NANC relaxations of the mouse anococcygeus were reduced by the guanylate cyclase inhibitor *N*-methylhydroxylamine and were potentiated by zaprinast (Gibson and Mirzazadeh, 1989), strongly suggested that the NANC transmitter of the retractor penis and anococcygeus resembled EDRF and was likely to be another example of a physiological pathway employing an endogenous nitrate as chemical mediator. Definitive proof was obtained when selective inhibitors of nitric oxide synthase (NOS) became available.

Nitric Oxide as the NANC Transmitter

Presynaptic Mechanisms

The experiments that clearly established the involvement of the L-arginine-NO pathway in NANC neurotransmission in the rat and mouse anococcygeus were reported independently by four groups in 1989–90 (Gillespie et al., 1989; Li and Rand, 1989a; Ramagopal and Leighton, 1989; Gibson et al., 1990). In each case it was shown that inhibitors of NOS, either N^G-nitro-L-arginine (L-NOARG) or N^G-monomethyl-L-arginine (L-NMMA), greatly reduced the NANC relaxations and that this inhibition was stereoselective, with neither D-NOARG nor D-NMMA having any effect. Further, the inhibition of NANC transmission could be reversed by addition of excess L-, but not D-, arginine. An example of the type of experimental evidence presented in these papers, in this case the mouse anococcygeus, is given in Figure 2.2. Similar observations were subsequently made in other NANC-innervated smooth muscles, including the rabbit anococcygeus (Graham and Sneddon, 1993; Kasakov et al., 1995), and the neurotransmission process was termed "nitrergic" (Rand, 1992). Although these experiments indicated that NOS was activated during nitrergic field stimulation, they did not pinpoint the anatomical location of the enzyme. It was still possible that the NOS was located postsynaptically in the smooth muscle cell and was activated by release of an as yet uncharacterized transmitter. Resolution of this question was provided by immunofluorescence studies using NOS antibodies and by NADPH diaphorase histochemistry (Schmidt et al., 1992; Brave et al., 1993c; Dail et al., 1993; Song et al., 1993; Kasakov et al., 1994), which showed that the NOS was located in nerve fibers within the tissue; these nerves were distinct from the motor sympathetic supply, since they were resistant to 6-hydroxydopamine (Brave et al., 1993c). Enzyme activity was also detected biochemically in homogenates of the rat anococcygeus by measuring the conversion of ^{3}H-L-arginine to ^{3}H-L-citrulline (Mitchell et al., 1991). The enzyme was located in both soluble and particulate forms and was dependent for its activity on tetrahydrobiopterin, NADPH, and calcium. In the original experiments with the NOS inhibitors mentioned above, addition of L-arginine by itself to the muscles had no obvious effect, but it produced a rapid recovery of nitrergic relaxations when added during NOS inhibition. This suggested that the enzyme is not active in quiescent nitrergic nerves but becomes active when the nerve is stimulated, probably due to the calcium entering the cell during depolarization. The channels through which the calcium gains entry can be

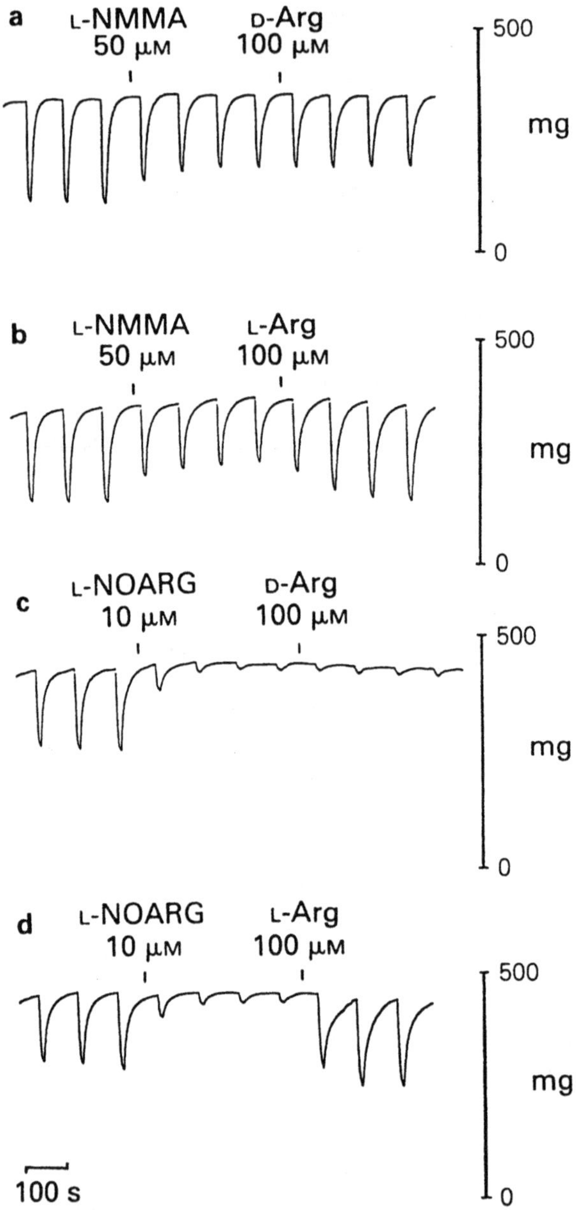

FIGURE 2.2. One of the original published tracings showing that NANC relaxations of the mouse anococcygeus are blocked by NOS inhibitors. The relaxations to field stimulation are blocked by either N^{G}-monomethyl-L-arginine (L-NMMA) (a,b) or N^{G}-nitro-L-arginine (L-NOARG) (c,d), and the inhibition is reversed by addition of L-arginine (L-Arg) (b,d) but not D-arginine (D-Arg) (a,c). From Gibson, A., et al. 1990.

inhibited by ω-conotoxin and are therefore likely to be N-type voltage-operated calcium channels (De Luca et al., 1990; Mudumbi and Leighton, 1994b), although it is of note that ω-conotoxin blocks nitrergic transmission in the rat and mouse anococcygeus when tone is raised by guanethidine or clonidine but not when carbachol is used as the contractile agent; why this should be the case has not yet been explained. As shown in Figure 2.2, NOS inhibitors are effective within two or three stimulation trains after addition of the drug to the bath, which suggests that there is no preformed store of transmitter but that it is synthesized on demand. It is generally assumed that the NO then passively diffuses out through the nerve terminal membrane. However, it is again noteworthy that there is good ultrastructural evidence that the nitrergic nerves contain synaptic vesicles (Gillespie and Lullman-Rauch, 1974; Gibbins and Haller, 1979); whether these synaptic vesicles are stimulated to release their contents during nerve activation and, if so, what the released substances might be (NO-adduct, cotransmitter, or antioxidant) remain to be determined.

Few presynaptic receptor types modulating nitrergic transmission have been identified in the anococcygeus, although in the rat there is evidence for the existence of inhibitory muscarinic receptors (possibly M_1) and excitatory nicotinic receptors (Li and Rand, 1989b; Rand and Li, 1991). As mentioned earlier, the anococcygeus muscles of all species studied receive a dual innervation, with motor sympathetic and inhibitory nitrergic nerve supply. To date, there is no real evidence for noradrenaline, released from the sympathetic neurons, modulating the activity of the nitrergic fibers. On the other hand, sympathetic contractions to field stimulation are enhanced in the presence of NOS inhibitors, suggesting that the nitrergic nerves can inhibit sympathetic activity (Li and Rand, 1989a; Gibson et al., 1990; Kasakov et al., 1994); this inhibition appears to be mediated postsynaptically by simple physiological antagonism, since NOS inhibitors do not increase noradrenaline release (Brave et al., 1993a; Kasakov et al., 1994).

One interesting recent observation is that capsaicin produced concentration-dependent relaxations of the rat anococcygeus muscle (Davies et al., 1998). These relaxations were greater in low (0.5 mM) than in high (2.5 mM) calcium solution, desensitized after one exposure to capsaicin, and were blocked by tetrodotoxin or ruthenium red; they were also blocked by NOS inhibitor and hemoglobin. The authors suggest that a fraction of the nitrergic nerves in the anococcygeus may be sensory, and that such fibers may contribute to the relaxations observed with field stimulation. Although there is little doubt that a large proportion of the NOS-containing fibers in the anococcygeus are motor and have their origin in the pelvic ganglia (Dail et al., 1993; Song et al., 1993), the role of sensory nitrergic nerves in the function of the muscle requires further investigation.

A summary of the presynaptic nitrergic mechanisms in the anococcygeus is given in Figure 2.3.

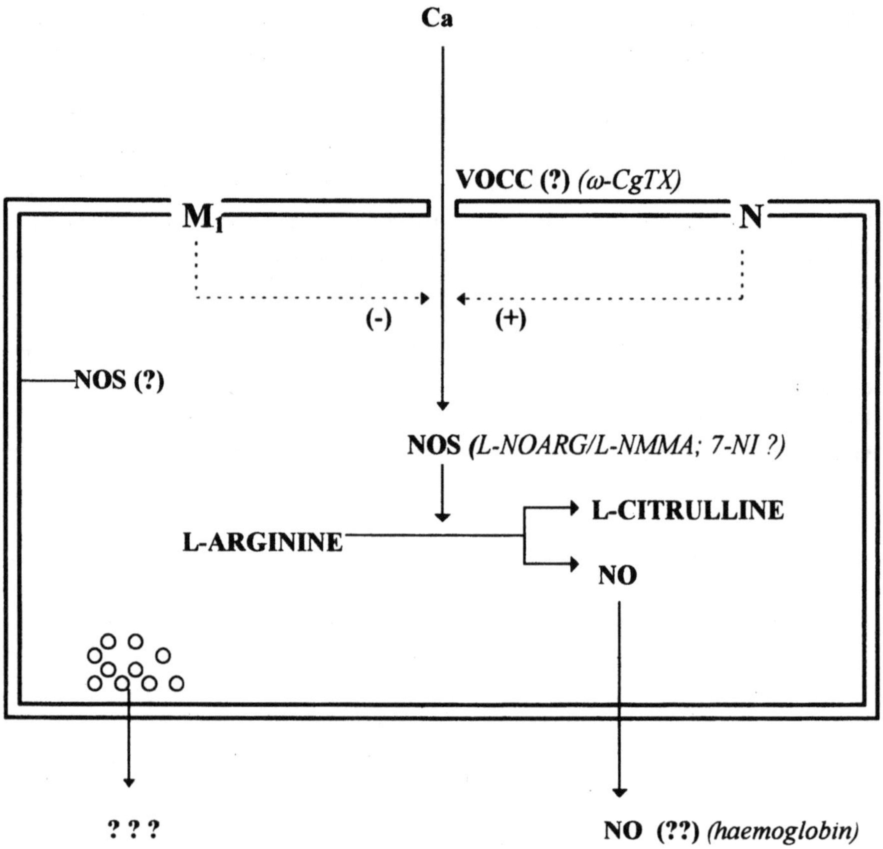

FIGURE 2.3. Diagram showing some of the presynaptic mechanisms at a nitrergic synapse. When an action potential reaches the terminal region, Ca^{2+} enters through voltage-operated calcium channels (VOCCs) to activate neuronal nitric oxide synthase (NOS), which converts L-arginine to L-Citrulline and nitric oxide (NO). Ca^{2+} entry through the VOCCs may be reduced by muscarinic receptor (possibly M_1) activation or by ω-conotoxin (ω-CgTX), and may be enhanced by nicotinic receptor (N) activation. NOS activity can be reduced by competitive inhibitors such as N^G-nitro-L-arginine (L-NOARG) and N^G-monomethyl-L-arginine (L-NMMA). The NO diffuses into the junctional gap where it may be sequestered by hemoglobin. Some remaining questions still to be answered (?) include the subtype of muscarinic receptor involved; the type of VOCC involved; the relative roles for particulate and soluble NOS; and the role of the synaptic vesicles in the neurotransmission process. In addition, the NOS inhibitor 7-nitroindazole (7-NI) does not appear to influence nitrergic transmission in the muscle. The debate on the chemical identity of the substance actually released by the nerve (??) is discussed in the text.

Is Nitric Oxide Released as a Free Radical or in Some Other Form?

Although the functional studies with NOS inhibitors and the immunocytochemistry with NOS antibodies established unequivocally that the enzyme was located in the nitrergic nerves and was responsible for synthesis of the neurotransmitter, several early observations were difficult to reconcile with the proposal that free radical NO was the substance actually released into the junctional gap, and that EDRF and the nitrergic transmitter were identical. Gillespie and Sheng (1989) found that hemoglobin, which binds and inactivates NO, reduced nitrergic relaxations of the rat anococcygeus and EDRF-mediated relaxations of rabbit aorta. However, although hemoglobin-equivalent suspensions of erythrocytes inhibited the relaxant effects of EDRF on blood vessels, they had no effect on nitrergic transmission, suggesting that the nitrergic transmitter was not as diffusible as EDRF through biological membranes. In addition, a number of drugs were found to inhibit relaxations of the anococcygeus muscle to authentic NO but to have little or no effect on relaxations elicited by nitregic field stimulation; such drugs included hydroquinone (Hobbs et al., 1991), carboxy-PTIO[2-(4-carboxyphenyl)-4,4,5,5,-tetramethylimidazoline-1-oxyl-3-oxide] (Rand and Li, 1995c), hydroxocobalamin (Rajayanagam et al., 1993), and superoxide anion generators (Lilley and Gibson, 1995). Again, most of these agents were effective inhibitors of EDRF-induced relaxations of blood vessels. Subsequent experiments revealed that the differential effects obtained with hydroxocobalamin were related to the light conditions under which the experiments were conducted (La et al., 1997), but the observations with the radical scavengers (hydroquinone and carboxy-PTIO) and the superoxide anion generators caused considerable debate about the exact nature of the nitrergic transmitter (Gibson et al., 1995); several explanations were proffered.

One possibility was that the substance actually released from the nitrergic nerves was not free radical NO but rather a NO-adduct, and that incorporation of the NO into this adduct molecule provided protection from scavengers and superoxide anions. The most favored candidates were nitrosothiols, mainly nitrosoglutathione and nitrosocysteine; indeed, it was demonstrated that the original IF from the bovine retractor penis was likely to be a nitrosothiol generated by the acidification of nitrite and thiols in the tissue extract (Kerr et al., 1993). However, although nitrosothiols produced quite potent relaxations of the anococcygeus and other nitrergically innervated tissues, none of them completely paralleled the properties of the nitrergic transmitter (Gibson et al., 1992; Rand and Li, 1993), and the formation of nitrosothiols under physiological conditions was questioned (Liu et al., 1994).

Another explanation for the differential effects of superoxide anions and radical scavengers on endogenous and exogenous NO was provided by

Wood and Garthwaite (1994), who described mathematical models for the diffusional spread of NO released from nerve terminals. They demonstrated that the rapid rate of diffusion of NO over short distances (up to 100 μm) meant that it was relatively immune to attack; scavengers would have to lower the half-life of NO to submillisecond levels to exert any influence on its biological activity over such distances. It has been argued that the rate of interaction of NO with certain scavengers, such as carboxy-PTIO (Rand and Li, 1995c) and superoxide anions (Ellis et al., 1998), is fast enough to influence the biological actions of endogenous NO. It seems therefore that the kinetic explanation for the immunity of endogenous NO from attack may hold for some, but not all, scavenger molecules.

A third explanation, which has now received considerable experimental support, was that free radical NO was indeed the substance actually released from the nitrergic nerves, but that it was protected from scavenger attack by the presence of "guardian" molecules within the tissue (Brave et al., 1993b). In the bovine retractor penis (Martin et al., 1994), superoxide anion generators had little effect on nitrergic relaxations under normal conditions, but following inhibition of Cu/Zn superoxide dismutase (Cu/Zn SOD), activity within the muscle nitrergic responses became highly susceptible to block by superoxide anions. Similar observations followed in the mouse and rat anococcygeus (Lilley and Gibson, 1995; Liu et al., 1997) and the rat gastric fundus (Lefebvre, 1996; De Man et al., 1996). Further, Cu/Zn SOD was colocalized with NOS in nitrergic nerves in the rat anococcygeus (Liu et al., 1997). This seemed to explain the differential effects observed with superoxide anion generators; Cu/Zn SOD activity provided protection for neurotransmitter NO released within the tissue, but exogenous NO was still vulnerable to attack while in the bathing medium. Clearly, however, tissue Cu/Zn SOD could not explain the differential effects of drugs such as hydroquinone and carboxy-PTIO, which did not act via superoxide anion generation, although it was possible that other physiological antioxidant mechanisms might act to protect NO. Indeed, it was shown that a number of antioxidants were able to protect relaxations of the mouse anococcygeus in response to exogenous NO: ascorbic acid (vitamin C) protected against superoxide anions, hydroquinone, and carboxy-PTIO; α-tocopherol (vitamin E) and uric acid protected against carboxy-PTIO; and glutathione protected against hydroquinone (Lilley and Gibson, 1996, 1997). Since ascorbic acid produced the most wide-ranging protection, and since there is strong evidence that it has an important role as an antioxidant in the central nervous system (Grunewald, 1993), its presence in, and possible release by, the anococcygeus was investigated. Indeed, the rat anococcygeus was found to release significant amounts of ascorbic acid under resting conditions, and this release was enhanced by depolarization (Lilley and Gibson, 1997). The depolarization-induced increase in release was inhibited in calcium-free conditions and by tetrodotoxin but was unaffected by 6-hydroxydopamine. Taken together, these results suggested that neurons distinct from the sympathetic terminals provide at least one source of ascorbic acid in the rat anococcygeus. Obvious questions that require answers are: What neurons are involved in the release of ascorbic acid? How much ascorbic acid is released? Does it reach sufficient concentrations in the junctional gap to provide protection for neurotransmitter NO? Interestingly, uric acid was also released by the rat anococcygeus, although in this case the release was not enhanced by depolarization (Lilley and Gibson, 1997). Uric acid has been considered as another important extracellular antioxidant (Ames et al., 1981), and its release by the anococcygeus may indicate the presence of xanthine oxidase activity within the muscle, which could contribute to endogenous superoxide anion generation (Ellis et al., 1998).

Overall, therefore, it now seems likely that free radical NO is the substance actually released by the nitrergic nerves in the anococcygeus; the lack of effect of radical scavengers and superoxide anions on nitrergic transmission probably results from a combination of the rapid diffusion kinetics of NO over short distances and the presence within the tissue of antioxidant mechanisms that provide the correct redox environment to allow NO to act as a neurotransmitter. If so, certain disease states that produce oxidative stress may lead to malfunction of nitrergic transmission by exhausting these protective mechanisms; it has already been demonstrated that nitrergic mechanisms are depressed in anococcygeus muscles from diabetic rats (Way and Reid, 1994). Since the anococcygeus muscle forms part of the erectile machinery and is found in both male and female animals (Gibson and Gillespie, 1973), it may prove a useful model with which to probe the mechanisms underlying erectile dysfunction in both sexes.

Nerve-Derived Hyperpolarizing Factor?

While the debate on the nature of the nitrergic transmitter of the anococcygeus seems to be reaching a settlement, a further complication has arisen recently with the proposal that, under certain conditions, a non-NO hyperpolarizing factor might contribute to NANC relaxations of the rat anococcygeus (Selemidis and Cocks, 1997). This substance, termed nerve-derived hyperpolarizing factor (NDHF), becomes the dominant NANC relaxant transmitter in the rat anococcygeus when low concentrations of contractile agent (less than EC_{40}) are used to induce tone in the preparation. Under these conditions, NANC relaxations are resistant to NOS inhibitors but sensitive to inhibition by L-type calcium channel blocking drugs such as nifedipine. However, no evidence for such a factor was obtained in the mouse anococcygeus (Fonseca et al., 1998) or rat gastric fundus (Lefebvre, 1998) under experimental conditions that should have favored its detection. One interesting observation, which has been known for some years, is that relatively small increases in extracellular potassium concentrations can induce relaxations of the anococcygeus, due partly to the release of the

NANC transmitter (Gibson and James, 1977) and partly to activation of the sodium pump in the smooth muscle cells, resulting in hyperpolarization (Ishii and Shimo, 1980). Further work is required to establish the significance of the putative hyperpolarizing factor to NANC relaxations in the anococcygeus and other smooth muscle tissues.

Postsynaptic Mechanisms

There is clear and convincing evidence that the target molecule for neurotransmitter NO in the anococcygeus smooth muscle cell is soluble guanylate cyclase, and therefore that cyclic GMP acts as a second messenger in the relaxation pathway activated by nitrergic stimulation. Thus, nitrergic relaxations are blocked by inhibitors of soluble guanylate cyclase (Gibson and Mirzazadeh, 1989; Cellek et al., 1996; Fonseca et al., 1998), associated with increased cyclic GMP levels within the tissue (Mirzazadeh et al., 1981; Cellek et al., 1996), potentiated by inhibitors of type V phosphodiesterase (Gibson and Mirzazadeh, 1989; Cellek et al., 1996), and mimicked by cell-permeable analogues of cyclic GMP (Mirzazadeh et al., 1981; Cellek et al., 1996). Although it is generally assumed that the cyclic GMP so formed then activates G-kinase to phosphorylate target proteins, there is very little hard experimental evidence to decide this one way or the other; equally, the molecular mechanisms beyond G-kinase activation that bring about the smooth muscle relaxation are still uncertain.

The majority of studies into the mechanisms underlying nitrergic relaxation in muscle have been carried out on the mouse anococcygeus, but it is not certain that the experimental results obtained can be transposed to

FIGURE 2.4. A model for the mechanisms of contraction and relaxation in the mouse anococcygeus. Receptor agonists such as carbachol generate production of inositol 1,4,5-trisphosphate (IP_3), which releases Ca from the sarcoplasmic reticulum (SR). This Ca stimulates calcium-activated chloride channels to open giving rise to a chloride current [I(ClCa)], resulting in membrane depolarization and opening of voltage-operated calcium channels (VOCCs); subsequent depletion of the SR signals the opening of store-operated channels (SOCs) through which Ca enters to support sustained contraction. The effects of Ca on the contractile proteins are enhanced by sensitization mechanisms that may involve protein kinase C (PKC) or rho-associated kinase (Rho K). NO released from the nitrergic nerves diffuses into the smooth muscle cell to activate guanylate cyclase. The GMP so produced then causes relaxation, possibly by way of G-kinase. The mechanisms by which GMP effects relaxation may be inhibition of store-operated calcium entry; activation of calcium sequestration by the SR (which may result in inhibition of the SOC); or inhibition of the calcium sensitization/contractile protein pathways. The precise mechanisms have yet to be determined. Cyclic GMP is metabolized by GMP specific phosphodiesterase (PDE V).

other species. Early experiments revealed that relaxations in some species (e.g., the rabbit) are associated with strong hyperpolarizations (Creed and Gillespie, 1977), whereas little change in membrane potential is observed in others [e.g., the rat (Creed et al., 1975)]. Nonetheless, since our understanding of the molecular mechanisms of contraction/relaxation is greatest in the mouse anococcygeus, this will now be described in more detail.

Contractions of the mouse anococcygeus in response to agonists acting on phospholipase C-linked receptors are highly dependent on extracellular

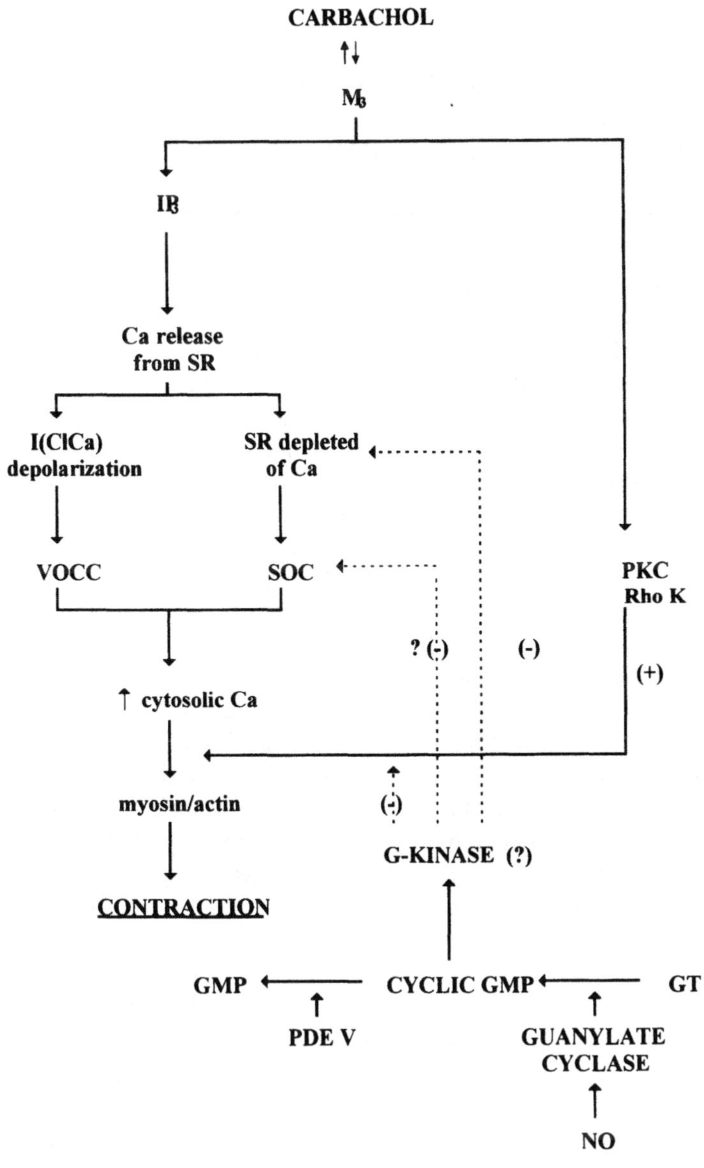

calcium but are only partially reduced by inhibitors of L-type voltage-operated calcium channels (VOCCs) such as nifedipine (Gibson et al., 1994b); since the nifedipine-resistant contraction is abolished by the general calcium entry blocker SKF96365, it seems that the calcium necessary for sustained contractions gains access to the smooth muscle cell through both VOCCs and some other calcium entry mechanism. Patch-clamp studies show that depolarization of the cell is due to initial calcium release from the sarcoplasmic reticulum (SR) opening calcium-activated chloride channels (Byrne and Large, 1987a,b; Wayman et al., 1996a); this depolarization results in the opening of VOCCs through which calcium enters the cell to provide the nifedipine-sensitive component of the contractile response. A clue to the nature of the non-VOCC component was provided by the observation that inhibitors of the SR Ca^{2+}-ATPase, such as cyclopiazonic acid (CPA) and thapsigargin (Tg), cause strong and sustained contractions of the mouse anococcygeus (Gibson et al., 1994b). By preventing reuptake of the calcium that continually leaks out of the SR, these drugs lead to store depletion, which has been shown in many cell types to activate store-operated calcium channels in the plasma membrane—so-called capacitative calcium entry (Putney, 1990; Berridge, 1995). Indeed, several agents that deplete the SR by a variety of mechanisms (carbachol, CPA, caffeine, ryanodine) activate a small, nonselective cation current in mouse anococcygeus smooth muscle cells, and the pharmacology of this current is similar to that of nifedipine-resistant contractions in the whole muscle (Wayman et al., 1996a, 1998). Based on these findings, a model for excitation–contraction coupling in the mouse anococcygeus has been proposed in which receptor agonists stimulate the production of inositol 1,4,5-trisphosphate (IP_3) and diacylglycerol (DAG). The IP_3 initiates calcium release from the SR, which leads to the opening of calcium-activated chloride channels; the resultant depolarization opens VOCCs, allowing calcium to enter the cell. Subsequent depletion of the SR is the signal for the opening of store-operated channels (SOCs) through which calcium can enter the cell to sustain contraction and refill the store on removal of the agonist. In parallel with the increase in calcium entry through VOCCs and SOCs, receptor activation promotes calcium sensitization, perhaps through diacylglycerol-induced activation of protein kinase C or via the rho-associated kinase pathway (Gibson et al., 1998; Wayman et al., 1998). A summary of this model is given in Figure 2.4; important questions still to be answered include the nature of the signal between the SR and the plasma membrane that opens the SOCs, and the molecular structure of the SOCs themselves.

So how can the NO/cyclic GMP system interact with this excitation–contraction coupling process to effect relaxation? Inhibition of IP_3 generation or of its receptors on the SR seems unlikely, since carbachol-induced IP_3 accumulation is unaffected by sodium nitroprusside (SNP), nitrergic field stimulation, or 8-bromo-cyclic GMP (Gibson et al., 1994a), and the calcium-activated chloride current stimulated by calcium release from the

SR is also resistant to SNP (Wayman et al., 1996b). However, SNP and 8-bromo-cyclic GMP do inhibit the nonselective cation current activated by store depletion, suggesting that inhibition of capacitative calcium entry might be one mechanism by which the NO/cyclic GMP pathway causes relaxation of the mouse anococcygeus. It is not yet clear whether this inhibition is due to a direct action on the SOC or to cyclic GMP-dependent refilling of the depleted SR, thus removing the primary drive for SOC opening; there is evidence that SNP can increase refilling of the SR, suggesting that the latter mechanism is more likely (Gibson et al., 1994b; Raymond and Wendt, 1996). However, it is unlikely that inhibition of capacitative calcium entry is the only mechanism underlying the relaxant effects of NO. In the mouse anococcygeus, SNP can relax tone induced by the thromboxane-mimetic U46619 in calcium-free medium (Gibson et al., 1994b), and in the rat anococcygeus SNP causes relaxation in the absence of any change in cytosolic calcium concentrations (Raymond and Wendt, 1996). It seems most likely that another component of the relaxant effects of NO/cyclic GMP involves interaction with the calcium sensitization/contractile protein elements of the smooth muscle (Gibson et al., 1994b; Raymond and Wendt, 1996).

A summary of the postsynaptic actions of the nitrergic transmitter in the mouse anococcygeus is given in Figure 2.4.

Conclusions

The anococcygeus muscle has provided an excellent model for investigation of the nitrergic neurotransmission process. The innervation pattern of the tissue is relatively simple, and the smooth muscle fibers are arranged in parallel bundles rather than the circular and longitudinal arrangement that can complicate matters in other tissues. The debate on the nature of the chemical substance actually released by the nerves has highlighted the need to adjust preconceived ideas about neurotransmission to interpret accurately new experimental findings. In his original paper on the rat anococcygeus, Gillespie (1972) expressed the hope that the tissue would prove useful in biomedical teaching and research. This hope has certainly been fulfilled.

References

Ambache, N., Killick, S.W., and Zar, M.A. 1975. Extraction from ox retractor penis of an inhibitory substance which mimics its atropine-resistant neurogenic relaxation. *Br. J. Pharmacol.* 54:409–410.

Ames, B.N., Cathcart, M., Schwiers, E., and Hochstein, P. 1981. Uric acid provides an antioxidant defense in humans against oxidant- and radical-caused aging and cancer: a hypothesis. *Proc. Natl. Acad. Sci. USA* 78:6858–6862.

Berridge, M.J. 1995. Capacitative calcium entry. *Biochem. J.* 312:1–11.
Blank, M.A., Brown, J.R., Hunter, J.C., Bloom, S.R., and Tyers, M.B. 1986. Effects of VIP and related peptides and Gila monster venom on genitourinary smooth muscle. *Eur. J. Pharmacol.* 132:155–161.
Boland, B., Himpens, B., Gillis, J.M., and Casteels, R. 1992. Relationship between force and Ca^{2+} in anococcygeal and vas deferens smooth muscle cells of the mouse. *Pflugers Arch.* 421:43–51.
Bowman, A., and Drummond, A.H. 1984. Cyclic GMP mediates neurogenic relaxation in the bovine retractor penis muscle. *Br. J. Pharmacol.* 81:665–674.
Bowman, A., and Gillespie, J.S. 1982. Block of some non-adrenergic inhibitory responses of smooth muscle by a substance from haemolysed erythrocytes. *J. Physiol.* 328:11–25.
Bowman, A., Gillespie, J.S., and Pollock, D. 1982. Oxyhaemoglobin blocks non-adrenergic inhibition in the bovine retractor penis muscle. *Eur. J. Pharmacol.* 85:221–224.
Brave, S.R., Bhat, S., Hobbs, A.J., Tucker, J.F., and Gibson, A. 1993a. The influence of L-N^G-nitro-arginine on sympathetic nerve induced contraction and noradrenaline release in the rat isolated anococcygeus muscle. *J. Auton. Pharmacol.* 13:219–225.
Brave, S.R., Gibson, A., and Tucker, J.F. 1993b. The inhibitory effects of hydroquinone on nitric oxide-induced relaxation of the mouse anococcygeus are prevented by native thiols. *Br. J. Pharmacol.* 109 (Suppl.):10P.
Brave, S.R., Tucker, J.F., Gibson, A., Bishop, A.E., Riveros-Moreno, V., Moncada, S., and Polak, J.M. 1993c. Localisation of nitric oxide synthase within non-adrenergic, non-cholinergic nerves in the mouse anococcygeus. *Neurosci. Lett.* 161:93–96.
Burnstock, G. 1986. The non-adrenergic non-cholinergic nervous system. *Arch. Int. Pharmacodyn. Ther.* 280 (Suppl.):1–15.
Burnstock, G., Cocks, T., and Crowe, R. 1978. Evidence for purinergic innervation of the anococcygeus muscle. *Br. J. Pharmacol.* 64:13–20.
Byrne, N., and Large, W.A. 1987a. Action of noradrenaline on single smooth muscle cells freshly dispersed from the rat anococcygeus muscle. *J. Physiol.* 389:513–525.
Byrne, N., and Large, W.A. 1987b. Membrane mechanism associated with muscarinic receptor activation in single cells freshly dispersed from the rat anococcygeus muscle. *Br. J. Pharmacol.* 92:371–379.
Carvajal, A., Gibson, A., and Yu, O. 1986. Evidence against VIP-involvement in neurogenic relaxations of the mouse anococcygeus. *J. Pharm. Pharmacol.* 38:767–769.
Cellek, S., Kasakov, L., and Moncada, S. 1996. Inhibition of nitrergic relaxations by a selective inhibitor of the soluble guanylate cyclase. *Br. J. Pharmacol.* 118: 137–140.
Coates, J., and Weetman, D.F. 1983. Demonstration of α_{1S}-adrenoceptors after exposure of the rat anococcygeus to benextramine. *Br. J. Pharmacol.* 80:343–346.
Creed, K.E., and Gillespie, J.S. 1977. Some electrical properties of the rabbit anococcygeus muscle and a comparison of the effects of inhibitory nerve stimulation in the rat and rabbit. *J. Physiol.* 273:137–153.
Creed, K.E., Gillespie, J.S., and McCaffrey, H. 1977. The rabbit anococcygeus muscle and its response to field stimulation and to some drugs. *J. Physiol.* 273:121–135.

Creed, K.E., Gillespie, J.S., and Muir, T.C. 1975. The electrical basis of excitation and inhibition in the rat anococcygeus muscle. *J. Physiol.* 245:33–47.

Dail, W.G., Carrillo, Y., and Walton, G. 1990. Innervation of the anococcygeus muscle of the rat. *Cell Tissue Res.* 259:139–146.

Dail, W.G., Galloway, B., and Bordegaray, J. 1993. NADPH diaphorase innervation of the rat anococcygeus and retractor penis muscles. *Neurosi. Lett.* 160:17–20.

Davies, R.E., Bashforth, P.M., and Docherty, R.J. 1998. A comparison of the effects of capsaicin with inhibitory nerve stimulation in the rat anococcygeus muscle in vitro. *Eur. J. Pharmacol.* 355:195–202.

Dehpour, A.R., Khoyi, M.A., Koutcheki, H., and Zarrindast, M.R. 1980. Pharmacological study of the anococcygeus muscle of the dog. *Br. J. Pharmacol.* 71:35–40.

De Luca, A., Li, C.G., Rand, M.J., Reid, J.J., Thaina, P., and Wong-Dusting, H.K. 1990. Effects of ω-conotoxin GVIA on autonomic neuroeffector transmission in various tissues. *Br. J. Pharmacol.* 101:437–447.

De Man, J.G., De Winter, B., Boeckxstaens, G.E., Herman, A.G., and Pelckmans, P.A. 1996. Effect of thiol modulators and Cu/Zn superoxide dismutase inhibition on nitrergic relaxations in the rat gastric fundus. *Br. J. Pharmacol.* 119: 1022–1028.

Docherty, J.R., and Starke, K. 1981. Postsynaptic alpha-adrenoceptor subtypes in rabbit blood vessels and rat anococcygeus muscle studied in vitro. *J. Cardiovasc. Pharmacol.* 3:854–866.

Doggrell, S.A. 1981. Effect of anti-muscarinic agents on the contractile responses to cholinomimetics in the rat anococcygeus muscle. *Br. J. Pharmacol.* 73:829–835.

Doggrell, S.A., and Vincent, L. 1981. The postsynaptic effects of anti-depressant drugs in the rat anococcygeus muscle. *J. Pharm. Pharmacol.* 33:720–724.

Doggrell, S.A., and Waldron, J.B. 1982a. Effect of cocaine on tone and contractile responses to acetylcholine in the rat anococcygeus muscle. *J. Pharm. Pharmacol.* 34:408–410.

Doggrell, S.A., and Waldron, J.B. 1982b. The effects of 6-hydroxydopamine or (−)-noradrenaline treatment in vitro on noradrenergic transmission in the rat anococcygeus muscle. *J. Auton. Pharmacol.* 2:231–240.

Ellis, A., Li, C.G., and Rand, M.J. 1998. Effect of xanthine oxidase inhibition on endothelium-dependent and nitrergic relaxations. *Eur. J. Pharmacol.* 356:41–47.

Fonseca, M., Uddin, N., and Gibson, A. 1998. No evidence for a significant non-nitrergic, hyperpolarising factor contribution to field stimulation-induced relaxation of the mouse anococcygeus. *Br. J. Pharmacol.* 124:524–528.

Furchgott, R.F. 1987. Studies on the relaxation of rabbit aorta by sodium nitrite: the basis for the proposal that the acid-activatable inhibitory factor from bovine retractor penis is inorganic nitrite and the endothelium-derived relaxing factor is nitric oxide. In: *Mechanisms of Vasodilatation* (ed. P.M. Vanhoutte). Raven Press, New York.

Furchgott, R.F., and Zawadski, J.V. 1980. The obligatory role of endothelial cells in the relaxation of arterial smooth muscle by acetylcholine. *Nature* 288:373–376.

Gibbins, I.L., and Haller, C.J. 1979. Ultrastructural identification of non-adrenergic, non-cholinergic nerves in the rat anococcygeus muscle. *Cell Tissue Res.* 200: 257–271.

Gibson, A., Babbedge, R., Brave, S.R., Hart, S.L., Hobbs, A.J., Tucker, J.F., Wallace, P., and Moore, P.K. 1992. An investigation of some *S*-nitrosothiols, and of hydroxy-arginine, on the mouse anococcygeus. *Br. J. Pharmacol.* 107:715–721.

Gibson, A., Bern, H.A., Ginsburg, M., and Botting, J.H. 1984. Neuropeptide-induced contraction and relaxation of the mouse anococcygeus muscle. *Proc. Natl. Acad. Sci. USA* 81:625–629.

Gibson, A., Brave, S.R., McFadzean, I., Mirzazadeh, S., Tucker, J.F., and Wayman, C. 1994a. Nitrergic stimulation does not inhibit carbachol-induced inositol phosphate generation in the rat anococcygeus. *Neurosci. Lett.* 178:35–38.

Gibson, A., Brave, S.R., McFadzean, I., Tucker, J.F., and Wayman, C. 1995. The nitrergic transmitter of the anococcygeus—NO or not? *Arch. Int. Pharmacodyn. Ther.* 329:39–51.

Gibson, A., and Gillespie, J.S. 1973. The effect of immunosympathectomy and of 6-hydroxydopamine on the responses of the rat anococcygeus to nerve stimulation and to some drugs. *Br. J. Pharmacol.* 47:261–267.

Gibson, A., and James, T.A. 1977. The nature of potassium chloride-induced relaxations of the rat anococcygeus muscle. *Br. J. Pharmacol.* 60:141–145.

Gibson, A., McFadzean, I., Tucker, J.F., and Wayman, C. 1994b. Variable potency of nitrergic-nitrovasodilator relaxations of the mouse anococcygeus against different forms of induced tone. *Br. J. Pharmacol.* 113:1494–1500.

Gibson, A., McFadzean, I., Wallace, P., and Wayman, C.P. 1998. Capacitative Ca^{2+} entry and the regulation of smooth muscle tone. *Trends. Pharmacol. Sci.* 19: 266–269.

Gibson, A., and Mirzazadeh, S. 1989. *N*-Methylhydroxylamine inhibits and M&B22948 potentiates relaxations of the mouse anococcygeus to non-adrenergic, non-cholinergic field stimulation and to nitrovasodilator drugs. *Br. J. Pharmacol.* 96:637–644.

Gibson, A., Mirzazadeh, S., Hobbs, A.J., and Moore, P.K. 1990. L-N^G-Monomethyl arginine and L-N^G-nitro arginine inhibit non-adrenergic, non-cholinergic relaxation of the mouse anococcygeus muscle. *Br. J. Pharmacol.* 99:602–606.

Gibson, A., and Pollock, D. 1973. The effects of drugs on the sensitivity of the rat anococcygeus muscle to agonists. *Br. J. Pharmacol.* 49:506–513.

Gibson, A., and Tucker, J.F. 1982. The effects of vasoactive intestinal polypeptide and of adenosine 5′-triphosphate on the isolated anococcygeus muscle of the mouse. *Br. J. Pharmacol.* 77:97–103.

Gibson, A., and Wedmore, C.V. 1981. Responses of the isolated anococcygeus muscle of the mouse to drugs and to field stimulation. *J. Auton. Pharmacol.* 1:225–233.

Gibson, A., and Yu, O. 1983. Pharmacology of postsynaptic alpha-adrenoceptors in the mouse anococcygeus muscle. *J. Auton. Pharmacol.* 3:1–6.

Gillespie, J.S. 1972. The rat anococcygeus muscle and its response to nerve stimulation and some drugs. *Br. J. Pharmacol.* 45:404–416.

Gillespie, J.S. 1980. The physiology and pharmacology of the anococcygeus muscle. *Trends Pharmacol. Sci.* 1:453–457.

Gillespie, J.S. 1997. The rat anococcygeus muscle and its response to nerve stimu-lation and to some drugs. A Commentary. *Br. J. Pharmacol.* 120 (Suppl.): 378–379.

Gillespie, J.S., Hunter, J.C., and Martin, W. 1981. Some physical and chemical properties of the smooth muscle inhibitory factor in extracts of the bovine retractor penis muscle. *J. Physiol.* 315:111–125.

Gillespie, J.S., Liu, X., and Martin, W. 1989. The effects of L-arginine and N^G-monomethyl L-arginine on the responses of the rat anococcygeus muscle to NANC nerve stimulation. *Br. J. Pharmacol.* 98:1080–1082.

Gillespie, J.S., and Lullman-Rauch, R. 1974. On the ultrastructure of the rat anococcygeus muscle. *Cell Tissue Res.* 149:91–104.

Gillespie, J.S., and Martin, W. 1980. A smooth muscle inhibitory material from the bovine retractor penis and rat anococcygeus muscles. *J. Physiol.* 309:55–64.

Gillespie, J.S., and McGrath, J.C. 1973. The spinal origin of the motor and inhibitory innervation of the rat anococcygeus muscles. *J. Physiol.* 230:659–672.

Gillespie, J.S., and McGrath, J.C. 1974. The response of the cat anococcygeus muscle to nerve or drug stimulation and a comparison with the rat. *Br. J. Pharmacol.* 50:109–118.

Gillespie, J.S., and McKnight, A.T. 1978. The actions of some vasoactive polypeptides and their antagonists on the anococcygeus muscle. *Br. J. Pharmacol.* 62: 267–274.

Gillespie, J.S., and Sheng, H. 1989. A comparison of haemoglobin and erythrocytes as inhibitors of smooth muscle relaxation by the NANC transmitter in the BRP and rat anococcygeus and by EDRF in the rabbit aortic strip. *Br. J. Pharmacol.* 98:445–450.

Gillis, J.M., Cao, M.L., and Godfrainddebecker, A. 1988. Density of myosin filaments in the rat anococcygeus muscle at rest and in contraction. *J. Muscle. Res. Cell. Mot.* 9:9–17.

Graham, A.M., and Sneddon, P. 1993. Evidence for nitric oxide as an inhibitory neurotransmitter in rabbit isolated anococcygeus. *Eur. J. Pharmacol.* 237: 93–99.

Griffith, T.M., Edwards, D.H., Lewis, M.J., and Henderson, A.H. 1985. Evidence that cyclic guanosine monophosphate (cGMP) mediates endothelium-dependent relaxation. *Eur. J. Pharmacol.* 112:195–202.

Gross, J., Mutschler, E., and Lambrecht, G. 1997. Evidence for muscarinic M-4 receptors mediating nonadrenergic noncholinergic relaxations in rabbit anococcygeus muscle. *N. S. Arch. Pharmacol.* 356:505–516.

Grunewald, R.A. 1993. Ascorbic acid in the brain. *Brain Res. Rev.* 18:123–133.

Hobbs, A.J., Tucker, J.F., and Gibson, A. 1991. Differentiation by hydroquinone of relaxations induced by exogenous and endogenous nitrates in non-vascular smooth muscle: role of superoxide anions. *Br. J. Pharmacol.* 104:645–650.

Iijima, T. 1983. Quinacrine-induced degeneration of non-adrenergic, non-cholinergic autonomic nerves in the rat anococcygeus muscle. *Cell Tissue Res.* 230:639–648.

Ishii, T., and Shimo, Y. 1980. Potassium-induced relaxation of the anococcygeus muscle. *Arch. Int. Pharmacodyn. Ther.* 243:27–36.

Kasakov, L., Belai, A., Vlaskovska, M., and Burnstock, G. 1994. Noradrenergic-nitrergic interactions in the rat anococcygeus muscle: evidence for postjunctional modulation by nitric oxide. *Br. J. Pharmacol.* 112:403–410.

Kasakov, L., Cellek, S., and Moncada, S. 1995. Characterisation of nitrergic neurotransmission during short- and long-term electrical stimulation of the rabbit anococcygeus muscle. *Br. J. Pharmacol.* 115:1149–1154.

Kerr, S.W., Buchanan, L.V., Bunting, S., and Mathews, W.R. 1993. Evidence that *S*-nitrosothiols are responsible for the smooth muscle relaxant activity of the bovine retractor penis inhibitory factor. *J. Pharmacol. Exp. Ther.* 263:285–292.

La, M.L., Paisley, K., Martin, W., and Rand, M.J. 1997. Effects of hydroxocobalamin on nitrergic transmission in rat anococcygeus and bovine retractor penis muscles: sensitivity to light. *Eur. J. Pharmacol.* 321:R5–R6.

Larsen, B.A., Gibson, A., and Bern, H.A. 1985. The effects of urotensins in tetrapods: physiology or pharmacology? In: *Neurosecretion and the Biology of Neuropeptides* (eds. H. Kobayashi, H.A. Bern, and A. Urano), pp. 486–493. Springer-Verlag, Berlin.

Lefebvre, R.A. 1996. Influence of superoxide dismutase inhibition on the discrimination between NO and the nitrergic neurotransmitter in the rat gastric fundus. *Br. J. Pharmacol.* 118:2171–2177.

Lefebvre, R.A. 1998. Influence of a selective guanylate cyclase inhibitor, and of the contraction level, on nitrergic relaxations in the gastric fundus. *Br. J. Pharmacol.* 124:1439–1448.

Li, C.G., and Rand, M.J. 1989a. Evidence for a role of nitric oxide in the neurotransmitter system mediating relaxation of the rat anococcygeus muscle. *Clin. Exp. Pharmacol. Physiol.* 16:933–938.

Li, C.G., and Rand, M.J. 1989b. Prejunctional inhibition of non-adrenergic non-cholinergic transmission in the rat anococcygeus. *Eur. J. Pharmacol.* 168:107–110.

Lilley, E., and Gibson, A. 1995. Inhibition of relaxations to nitrergic stimulation of the mouse anococcygeus by duroquinone. *Br. J. Pharmacol.* 116:3231–3236.

Lilley, E., and Gibson, A. 1996. Antioxidant protection of NO-induced relaxations of the mouse anococcygeus against inhibition by superoxide anions, hydroquinone and carboxy-PTIO. *Br. J. Pharmacol.* 119:432–438.

Lilley, E., and Gibson, A. 1997. Release of the antioxidants and urate from a nitrergically-innervated smooth muscle. *Br. J. Pharmacol.* 122:1746–1752.

Liu, X., Gillespie, J.S., and Martin, W. 1994. Non-adrenergic, non-cholinergic relaxation of the bovine retractor penis muscle: role of *S*-nitrosothiols. *Br. J. Pharmacol.* 111:1287–1295.

Liu, X., Miller, S.M., and Szurszewski, J.H. 1997. Protection of nitrergic neurotransmission by colocalisation of neural nitric oxide synthase with copper zinc superoxide dismutase. *J. Auton. Nerv. Syst.* 62:126–133.

Martin, W., Furchgott, R.F., Villani, G.M., and Jothianandan, D. 1986. Phosphodiesterase inhibitors induce endothelium-dependent relaxation of rat and rabbit aorta by potentiating the effects of spontaneously released endothelium-derived relaxing factor. *J. Pharmacol. Exp. Ther.* 237:539–547.

Martin, W., McAllister, K.H.M., and Paisley, K. 1994. NANC neurotransmission in the bovine retractor penis muscle is blocked by superoxide anion following inhibition of superoxide dismutase with diethyldithiocarbamate. *Neuropharmacology* 33:1293–1301.

Martin, W., Smith, J.A., Lewis, M.J., and Henderson, A.H. 1988. Evidence that inhibitory factor extracted from bovine retractor penis is nitrite, whose acid-activated derivative is stabilized nitric oxide. *Br. J. Pharmacol.* 93:579–586.

Martin, W., Villani, G.M., Jothianandan, D., and Furchgott, R.F. 1985. Selective blockade of endothelium-dependent and glyceryl trinitrate-induced relaxation by hemoglobin and by methylene blue in the rabbit aorta. *J. Pharmacol. Exp. Ther.* 232:708–716.

McGrath, J.C. 1984. Alpha-adrenoceptor antagonism by apoyohimbine and some observations on the pharmacology of alpha-adrenoceptors in the rat anococcygeus and vas deferens. *Br. J. Pharmacol.* 82:769–781.

Mirzazadeh, S., Hobbs, A.J., Tucker, J.F., and Gibson, A. 1981. Cyclic nucleotide content of the rat anococcygeus during relaxations induced by drugs or by non-adrenergic, non-cholinergic field stimulation. *J. Pharm. Pharmacol.* 43: 247–257.

Mitchell, J.A., Sheng, H., Forstermann, U., and Murad, F. 1991. Characterisation of nitric oxide synthases in non-adrenergic non-cholinergic nerve containing tissue from the rat anococcygeus. *Br. J. Pharmacol.* 104:289–291.

Mudumbi, R.V., and Leighton, H.J. 1994a. Analysis of bradykinin-induced relaxations in the rat isolated anococcygeus muscle. *Life Sci.* 54:813–821.

Mudumbi, R.V., and Leighton, H.J. 1994b. Effects of ω-conotoxin GVIA on electrical-field stimulation-induced and agonist-induced changes in cytosolic Ca^{2+} and tension in isolated rat anococcygeus muscle. *J. Auton. Pharmacol.* 14: 253–265.

Olson, L., and Alund, M. 1979. Quinacrine-binding nerves: presence in the mouse anococcygeus muscle, disappearance after muscle transection. *Med. Biol.* 57: 182–186.

Oriowo, M.A. 1983. Muscarinic receptor sub-type in the rat anococcygeus muscle. *J. Pharm. Pharmacol.* 35:469–470.

Putney, J.W., Jr. 1990. Capacitative calcium entry revisited. *Cell Calcium* 11: 611–624.

Rajayanagam, M.A.S., Li, C.G., and Rand, M.J. 1993. Differential effects of hydroxocobalamin on NO-mediated relaxations in rat aorta and anococcygeus muscle. *Br. J. Pharmacol.* 108:3–5.

Ramagopal, M.V., and Leighton, H.J. 1989. Effects of N^G-monomethyl-L-arginine on field stimulation-induced decreases in cytosolic Ca^{2+} levels and relaxation in the rat anococcygeus muscle. *Eur. J. Pharmacol.* 174:297–299.

Rand, M.J. 1992. Nitrergic transmission: nitric oxide as a mediator of non-adrenergic, non-cholinergic neuro-effector transmission. *Clin. Exp. Pharmacol. Physiol.* 19:147–169.

Rand, M.J., and Li, C.G. 1991. Activation of noradrenergic and nitrergic mechanisms in the rat anococcygeus muscle by nicotine. *Clin. Exp. Pharmacol. Physiol.* 19:103–111.

Rand, M.J., and Li, C.G. 1993. Differential effects of hydroxocobalamin on relaxations induced by nitrosothiols in rat aorta and anococcygeus muscle. *Eur. J. Pharmacol.* 241:249–254.

Rand, M.J., and Li, C.G. 1995a. Nitric oxide as a neurotransmitter in peripheral nerves: nature of transmitter and mechanism of transmission. *Annu. Rev. Physiol.* 57:659–682.

Rand, M.J., and Li, C.G. 1995b. Nitric oxide in the autonomic nervous system. In: *Nitric Oxide in the Nervous System* (ed. S.R. Vincent), pp. 227–279. Academic Press, London.

Rand, M.J., and Li, C.G. 1995c. Discrimination by the NO-trapping agent, carboxy-PTIO, between NO and the nitrergic transmitter but not between NO and EDRF. *Br. J. Pharmacol.* 116:1906–1910.

Raymond, G.L., and Wendt, I.R. 1996. Force and intracellular Ca^{2+} during cyclic nucleotide-mediated relaxation of rat anococcygeus muscle and the effects of cyclopiazonic acid. *Br. J. Pharmacol.* 119:1029–1037.

Schmidt, H.H.H.W., Gagne, G.D., Nakane, M., Pollock, J.S., Miller, M.F., and Murad, F. 1992. Mapping of neural nitric oxide synthase in the rat suggests frequent colocalisation with NADPH diaphorase but not with soluble guanylyl cyclase, and novel paraneural functions for nitrinergic signal transduction. *J. Histochem. Cytochem.* 40:1439–1456.

Selemidis, S., and Cocks, T.M. 1997. Evidence that both nitric oxide (NO) and a non-NO hyperpolarising factor elicit NANC nerve-mediated relaxation in the rat isolated anococcygeus. *Br. J. Pharmacol.* 120:662–666.

Sneddon, P., Westfall, D.P., and Fedan, J.S. 1982. Investigation of relaxations of the rabbit anococcygeus muscle by nerve stimulation and ATP using the ATP antagonist $ANAPP_3$. *Eur. J. Pharmacol.* 80:93–98.

Song, Z.M., Brookes, S.J.H., and Costa, M. 1993. NADPH diaphorase reactivity in nerves supplying the rat anococcygeus muscle. *Neurosci. Lett.* 158:221–224.

Walker, J.S., Wendt, I.R., and Gibbs, C.L. 1988. Heat production of rat anococcygeus muscle during isometric contraction. *Am. J. Physiol.* 255:C536–C542.

Way, K.J., and Reid, J.J. 1994. Nitric oxide-mediated neurotransmission is attenuated in the anococcygeus muscle from diabetic rats. *Diabetologia* 37:232–237.

Wayman, C.P., Gibson, A., and McFadzean, I. 1998. Depletion of either ryanodine- or IP_3-sensitive calcium stores activates capacitative calcium entry in mouse anococcygeus smooth muscle cells. *Pflugers Arch.* 435:231–239.

Wayman, C.P., McFadzean, I., Gibson, A., and Tucker, J.F. 1996a. Two distinct membrane currents activated by cyclopiazonic acid-induced calcium store depletion in single smooth muscle cells of the mouse anococcygeus. *Br. J. Pharmacol.* 117: 566–572.

Wayman, C.P., McFadzean, I., Gibson, A., and Tucker, J.F. 1996b. Inhibition by sodium nitroprusside of a calcium store depletion-operated non-selective cation current in smooth muscle cells of the mouse anococcygeus. *Br. J. Pharmacol.* 118:2001–2008.

Weiser, M., Mutschler, E., and Lambrecht, G. 1997. Characterisation of postjunctional muscarinic receptors mediating contraction in rat anococcygeus muscle. *N. S. Arch. Pharmacol.* 356:671–677.

Wood, J., and Garthwaite, J. 1994. Models of the diffusional spread of nitric oxide: implications for neural nitric oxide signalling and its pharmacological implications. *Neuropharmacology* 33:1235–1244.

3
Interactions Between Nitric Oxide and Other NANC Inhibitory Neurotransmitters in the Periphery: Clues from the Endothelium

TOM M. COCKS AND STAVROS SELEMIDIS

Introduction

From the beginning of the story of endothelium-dependent vasodilatation, another equally intriguing search was in progress for the identity of an inhibitory neurotransmitter released from certain nonadrenergic, non-cholinergic (NANC) nerves innervating smooth muscle in the lower regions of the gut. Thus, it was perhaps of no surprise that when the endothelium-derived relaxing factor (EDRF) was identified as nitric oxide (NO), the NANC inhibitory factor in the gut was also proposed to be NO. Also not surprising was the realization that not all non-prostanoid-mediated, endothelium-dependent dilatation in the vasculature was due to NO. Another mechanism was known to occur in response to many endothelium-dependent vasodilators, that being smooth muscle hyperpolarization, and it soon became apparent that this was due to the release of a factor called endothelium-derived hyperpolarizing factor (EDHF). Given the parallel histories of endothelial cells and NANC inhibitory nerves as to how they signal smooth muscle cells to relax, it was a logical step to determine whether a similar non-NO hyperpolarizing factor was released with NO from NANC nerves. Furthermore, if there was such a factor, clues as to the identity of both it and EDHF might arise from further comparison of these systems. Therefore, in this chapter we hope to provoke new thought and discussion related to mechanisms of NANC nerve-mediated relaxation of gut smooth muscle. The link with endothelial cells will remain a strong theme, given the evolving similarities between NANC neurotransmission and endothelial cell "endotransmission."

NANC Inhibitory Neurotransmission: History

Autonomic nerves were originally described as either cholinergic or adrenergic based on whether they utilized the classical transmitters acetylcholine noradrenaline, respectively. In the gastrointestinal tract these nerves are extrinsic and mediate both excitation (cholinergic) and inhibition (adrenergic) of various intestinal smooth muscles. It was not until the early 1960s, however, that a third intrinsic inhibitory nerve that was neither adrenergic nor cholinergic (i.e., NANC) was discovered in the guinea pig taenia coli (Burnstock et al., 1964), a gut smooth muscle preparation that had been used extensively for measurements and observations of changes in membrane potential and tension (Bulbring, 1955; Bulbring et al., 1958). Briefly, Burnstock and colleagues showed two different types of responses to electrical field stimulation in the guinea pig taenia coli that were likely to be generated by separate transmitters, each showing different electrical and mechanical properties and sensitivities to pharmacological agents. The first of these responses, excitation, was sensitive to atropine and thus was due to stimulation of cholinergic nerves and the release of acetylcholine (Burnstock et al., 1964; Campbell, 1966). The second response, inhibition, comprised hyperpolarizations and relaxations that were both unaffected by the adrenergic neuron blocker guanethidine (Burnstock et al., 1964, 1966; Bennett et al., 1966a,b). These findings were novel because they were the first to describe the involvement of nonadrenergic nerve-mediated inhibition of gastrointestinal smooth muscle. Ever since the discovery of NANC neurotransmission by Burnstock and colleagues, many studies have attempted to identify the chemical identities of NANC inhibitory transmitters, which in some cases remain equivocal. The main NANC inhibitory neurotransmitter candidates include the gaseous compound NO (for reviews see Rand, 1992; Sanders and Ward, 1992; Rand and Li, 1995), ATP (for reviews see Burnstock, 1972, 1990), and a number of neuropeptides, the most favored being vasoactive intestinal peptide (VIP) (for a review see Furness and Costa, 1987) and the VIP-like peptide pituitary adenylyl cyclase-activating peptide (PACAP) (Jin et al., 1994; McConalogue et al., 1995).

Although these NANC inhibitory neurotransmitter candidates have been implicated in NANC neurotransmission, their individual contributions and roles in NANC-mediated relaxation of gastrointestinal smooth muscle appear to vary between different preparations. For instance, until recently ATP appeared to be the only NANC transmitter in the guinea pig taenia coli and NO the only candidate in the rat anococcygeus. In this monograph, we review mechanisms and interactions between NO and other candidate NANC inhibitory neurotransmitters, with particular reference to gut smooth muscle, and a new mode of NANC cotransmission that involves functional transmitter redundancy.

Nitrergic Transmission

The term *nitrergic* was coined by Rand (1992) in accordance with other "ergic" nerves. This term is now commonly used to describe the transmission of NO not only from enteric NANC inhibitory nerves to gastrointestinal smooth muscle, but also from other nerves that release NO or in which the enzyme responsible for NO synthesis, neuronal nitric oxide synthase (nNOS), has been localized. The evidence that NO or a related compound was a NANC inhibitory transmitter in the gut came initially from studies of the rat anococcygeus and the related bovine retractor penis muscle, two gut-associated smooth muscles in which inhibitory NANC responses have been well characterized (Gillespie, 1972; Klinge and Sjostrand, 1974). Also, nNOS immunoreactivity was localized to myenteric nerves in many gut preparations that displayed NANC inhibitory responses, including the rat anococcygeus and guinea pig taenia coli (Bredt et al., 1991; Mitchell et al., 1991; Furness et al., 1992; Song et al., 1993; Dail et al., 1993). Furthermore, NO was shown to be released upon electrical field stimulation of NANC inhibitory nerves (Boeckxstaens et al., 1991a,b; Archer, 1993), and the ensuing inhibitory responses were blocked by inhibitors of either nNOS [for e.g., N^G-nitro-L-arginine (L-NOARG)] or NO (e.g., oxyhemoglobin) (for reviews see Rand, 1992; Sanders and Ward, 1992; Rand and Li, 1995). Finally, exogenously applied NO mimicked the response to NANC inhibitory nerve stimulation in many gastrointestinal preparations, and both responses coincided with an increase in smooth muscle cGMP.

Multiple NANC Inhibitory Cotransmitters

It is now accepted that the diverse responses to NANC inhibitory nerve stimulation of gastrointestinal smooth muscle are not mediated by a single transmitter. Part of the evidence for this diversity comes from studies using pharmacological tools, such as apamin, a neurotoxin polypeptide (Habermann and Reiz, 1965; Callewaert et al., 1968) that blocks small-conductance Ca^{2+}-activated K^+ channels (SK_{Ca}) (Banks et al., 1979). As such, apamin has the potential to inhibit hyperpolarization and any resultant relaxation evoked by NANC inhibitory nerves if the transmitters involved in these responses activate SK_{Ca}. However, apamin displays different potencies as such a blocker of NANC relaxations in different gut preparations. For example, in some tissues the effect of apamin was dramatic, whereas in others it was ineffective (Costa et al., 1986; Manzini et al., 1986). This variable effect of apamin could be taken to indicate either different transmitters or different contributions of SK_{Ca} to the NANC-induced responses. As for apamin, inhibitors of NOS and other candidate transmitters such as ATP, VIP, and PACAP have all been shown to cause variable inhibition of NANC responses throughout the gastrointestinal tract, which again suggests either different transmitters or variable contributions of these putative transmitters to NANC neurotransmission.

The interaction of NO with one or more NANC inhibitory transmitters has been demonstrated in many gut smooth muscle preparations. For example, ATP, has been claimed to act in synergy or as a cotransmitter with NO to mediate relaxation of the guinea pig isolated proximal colon (Maggi and Giuliani, 1993; Zagorodnyuk and Maggi, 1994). In this tissue, NANC responses were found to be complex, being rapid in onset and followed by rebound contractions and delayed secondary relaxations. Also, the fast component of the relaxation was reduced by either apamin or L-NOARG, suggesting the involvement, via activation of SK_{Ca}, of ATP and NO, respectively. Interestingly, the secondary slow relaxation was found to be resistant to these inhibitors, indicating the possible involvement of a third transmitter that may not represent either ATP or NO. Thus, NANC inhibitory neurotransmission in gastrointestinal smooth muscle may involve complex mechanisms arising from the interaction of NO with multiple cotransmitters. Such differentiation of NANC responses in terms of different transmitters and mechanisms points to a rich diversity for the control of gut motility.

Role of Nitric Oxide

As already mentioned, the role of NO in NANC inhibitory neurotransmission in the gut appears to be tissue-dependent (see Table 3.1). For example, in the rat anococcygeus (Gillespie et al., 1989; Li and Rand, 1989; Hobbs and Gibson, 1990; Liu et al., 1991), opossum lower oesophageal sphincter (Tøttrup et al., 1991; Knudsen et al., 1992), and canine ileocolonic sphincter (Bult et al., 1990), NOS inhibitors have been shown to abolish the relaxation to NANC nerve stimulation. In the guinea pig ileum (Crist et al., 1992; Xue and Goyal, 1993), proximal colon (Maggi and Giuliani, 1993; Zagorodnyuk and Maggi, 1994), gastric fundus (Ohno et al., 1996), internal anal sphincter (Rae and Muir, 1996), rabbit internal anal sphincter (Tøttrup et al., 1995), and human isolated colon (Boeckxstaens et al., 1993; Keef et al., 1993), however, similar NOS inhibitors only partially inhibit NANC relaxations, with the remainder being abolished by apamin, again pointing to a role for both NO and an apamin-sensitive transmitter, possibly ATP, as cotransmitters. In the guinea pig tenia coli (Rand and Li, 1990; Knudsen and Tøttrup, 1992; Piotrowski et al., 1993; Grider et al., 1994) and rat distal colon (Kishi et al., 1996), NANC relaxations are nearly completely resistant to NOS inhibitors, which indicates little if any role for NO.

Similarities Between NANC Nerves and Endothelial Cells

NANC inhibitory nerves innervating gastrointestinal tissue and vascular endothelial cells utilize similar signaling mechanisms to mediate inhibitory responses of smooth muscle cells (Figure 3.1). Both cell types release not

TABLE 3.1. Transmitters and types of neurotransmission for NANC inhibitory nerves in various gastrointestinal tract preparations.

Tissue	Effect of NO inhibitors on NANC responses	Possible transmitters	Type of neurotransmission	Reference
Bovine retractor penis	Abolished	NO	Single	Liu et al., 1991
Opossum lower esophageal sphincter	Abolished	NO	Single	Tøttrup et al., 1991; Knudsen et al., 1992
Canine ileocolonic sphincter	Abolished	NO	Single	Ward et al., 1992
Human internal anal sphincter	Abolished	NO	Single	Burleigh, 1992
Mouse anococcygeus	Abolished	NO	Single	Gibson et al., 1990; Fonseca et al., 1998
Rabbit internal anal sphincter	Partial/apamin-sensitive/insensitive	ATP, NO, and a third (?)	Cotransmission	Tøttrup et al., 1995
Guinea pig proximal colon	Partial/apamin-sensitive	ATP, NO	Cotransmission	Maggi Giuliani, 1993; Zagorodnyuk and Maggi, 1994
Guinea pig gastric fundus	Partial/apamin-sensitive	ATP, NO	Cotransmission	Ohno et al., 1996
Guinea pig internal anal sphincter	Partial/apamin-sensitive	ATP, NO	Cotransmission	Rae and Muir, 1996
Human isolated colon	Partial/apamin-sensitive	ATP, NO	Cotransmission	Keef et al., 1993
Rat distal colon	No effect/apamin-sensitive/charybdotoxin-sensitive	PACAP, VIP	Cotransmission	Suthamnatpong et al., 1993
Rat gastric fundus	Nearly abolished/VIP-antibody-sensitive (higher frequency)	NO, VIP	Cotransmission	Li and Rand, 1990; Shimamura et al., 1993.
Guinea pig taenia coli	Silent/apamin-sensitive	NDHF (ATP?, PACAP?), NO (backup)	Backup cotransmission	Selemidis et al., 1997
Rat anococcygeus	Silent/nifedipine-sensitive	NDHF, NO (backup)	Backup cotransmission	Selemidis and Cocks, 1997

PACAP denotes pituitary adenylyl cyclase-activating peptide, NDHF nerve-derived hyperpolarizing factor, and VIP vasoactive intestinal peptide.

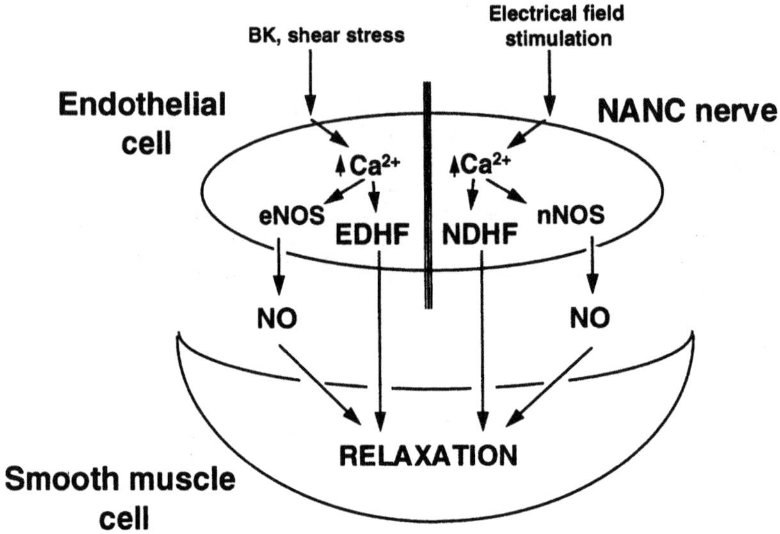

FIGURE 3.1. Schema depicting similarities between endothelial cells and NANC inhibitory nerves in terms of how they mediate relaxation of vascular and gut smooth muscle relaxation, respectively. Stimulation of endothelial cell surface B_2 receptors with bradykinin (BK) or an increase in shear stress evokes an increase in endothelial cell Ca^{2+}. Also, electrical field stimulation of NANC nerves triggers an increase in neuronal cell Ca^{2+}. In each case, the increase in Ca^{2+} generates and releases NO from endothelial cells and NANC nerves by activating either the endothelial (eNOS) or the neuronal isoforms (nNOS) of NOS. The same increase in cell Ca^{2+} causes the release of non-NO hyperpolarizing factors endothelium-derived hyperpolarizing factor (EDHF) and nerve-derived hyperpolarizing factor (NDHF).

only NO but also non-NO factors that mediate relaxation via smooth muscle cell hyperpolarization. Non-NO, nerve-derived hyperpolarizing transmitter candidates include ATP and PACAP and, for endothelial cells, a chemically undefined hyperpolarizing factor termed endothelium-derived hyperpolarizing factor (EDHF) (Taylor and Weston, 1988; Garland et al., 1995). Also, as for NANC cotransmitters, the relative roles and contributions of different endothelium-derived relaxing factors or "co-endotransmitters" appear to vary between different vascular preparations (Garland et al., 1995). For example, in the rat aorta NO appears to be the only relaxing factor evident, whereas in vessels such as the rat hepatic artery NO acts in parallel or in synergy with EDHF (Parsons et al., 1994).

Redundancy in Endothelium-Dependent Relaxing Mechanisms

In some blood vessels, such as pig and bovine coronary arteries, EDHF acts, as a functionally "silent" relaxing factor if NO is available (Kilpatrick and Cocks, 1994; Drummond and Cocks, 1996). In these tissues NO is the main relaxing factor, even though EDHF can almost completely "back up" the response to endothelium-dependent relaxing agonists following blockade of NO (Kilpatrick and Cocks, 1994; Drummond and Cocks, 1996). Interestingly, blockade of both NO and EDHF reveals a second-order backup relaxation mechanism of endothelium-dependent relaxation in some vessels, which represents the action of vasoactive cytochrome-P_{450}-derived metabolites of arachidonic acid (Drummond et al., in press). These findings indicate that the vasculature possesses redundancy in mechanisms for endothelium-dependent smooth muscle relaxation, such that when the main relaxing mechanism is unavailable, multiple backup systems are able to compensate for its loss.

Effect of Functional Antagonism on NANC Inhibitory Neurotransmission

Endothelium-dependent relaxations in blood vessels are measured in preparations contracted to set levels of active force with various vasoconstrictor agents. In these bioassays, the magnitude of contraction and thus the level of functional antagonism influence the ability of the vessels to relax. For instance, endothelium-dependent relaxations are abolished by NOS inhibitors in preparations contracted to high levels of active force (approximately 80% of their maximum contraction), whereas similar responses are virtually unaffected by the same NOS inhibitors in preparations contracted to low levels of active force (40% of their maximum contraction) (Drummond and Cocks, 1996). Therefore, the degree of functional antagonism is a crucial factor when determining the contribution and role of endothelium-derived factors in vascular smooth muscle relaxation.

As in the vasculature and EDRF, similar considerations of the degree and mechanism of contraction need to be taken into account in studies that aim to assess the contributions of multiple neurotransmitters to gut smooth muscle relaxation. For example, as mentioned earlier, the rat anococcygeus has been widely recognized as a purely nitrergically innervated gastrointestinal preparation (i.e., NO acts as the only NANC inhibitory transmitter). When we transposed our techniques of bioassaying EDRF in blood vessels to the rat anococcygeus, however, it became apparent that if preparations are contracted to low levels of active force to avoid any effects of altered or excessive functional antagonism, the contribution of NO to

NANC nerve-mediated relaxations in this tissue was poor (Selemidis and Cocks, 1997). In fact, at low frequencies of stimulation it had no role at all (Figure 3.2). By contrast, block of any hyperpolarization-mediated relaxation that may have involved closure of L-type VOCCs with nifedipine revealed relaxations that were now abolished by NO inhibition (Figure 3.2).

In another classic NANC inhibitory nerve-innervated preparation, the guinea pig taenia coli, we also found NANC nerve-mediated relaxations to be completely resistant to NO inhibition (Selemidis et al., 1997). Thus, the combined treatments of the NOS inhibitor L-NOARG with either the NO-scavenging molecule oxyhemoglobin, the free radical generator 7-ethoxyresorufin, or the guanylate cyclase inhibitor 1H[1,2,4]oxadiazolo[4,3-a]quinoxalin-1-one (ODQ) had no effect on NANC relaxations (Selemidis et al., 1997). Like those in the rat anococcygeus, block of hyperpolarization-mediated relaxation with apamin unmasked relaxations in the guinea pig taenia coli that were now abolished by L-NOARG and ODQ. Such "silent" effects of NO inhibition on NANC relaxations in the rat anococcygeus and guinea pig taenia coli led us to conclude that in both tissues NO served as a backup relaxing factor (Selemidis and Cocks, 1997; Selemidis et al., 1997) for the predominant hyperpolarization-dependent mechanism that most likely represents the actions of a non-NO transmitter. We termed this factor nerve-derived hyperpolarizing factor (NDHF) (Selemidis and Cocks, 1997; Selemidis et al., 1997). Therefore, nitrergic neurotransmission in the rat anococcygeus and guinea pig taenia coli is similar, in that it does not represent the main relaxing mechanism but rather acts as an important backup mechanism for smooth muscle relaxation in these tissues. By contrast, the contributions of NO and EDHF to endothelium-dependent relaxation in blood vessels are reversed. Here NO is the main endotransmitter, with EDHF acting as first-line backup.

Backup Cotransmission

The findings described above in the rat anococcygeus (Selemidis and Cocks, 1997) and guinea pig taenia coli (Selemidis et al., 1997) indicate that NO functions as a "backup" or "redundant" NANC neurotransmitter. Therefore, our studies have provided the basis for proposing a novel mode of cotransmission in the autonomic nervous system. Unlike the more "classical" type of cotransmission, where multiple neurotrans-mitters act to varying degrees in parallel to mediate smooth muscle relaxation, backup cotransmission, as reported here, involves a dominant transmitter able to mediate most if not all of the response to NANC inhibitory nerve stimulation and a redundant transmitter that is concomitantly released with the main neurotransmitter yet remains functionally "silent."

Our findings in the rat anococcygeus (Selemidis and Cocks, 1997) appear rather "heretical," in that they suggest NANC nerves are not only

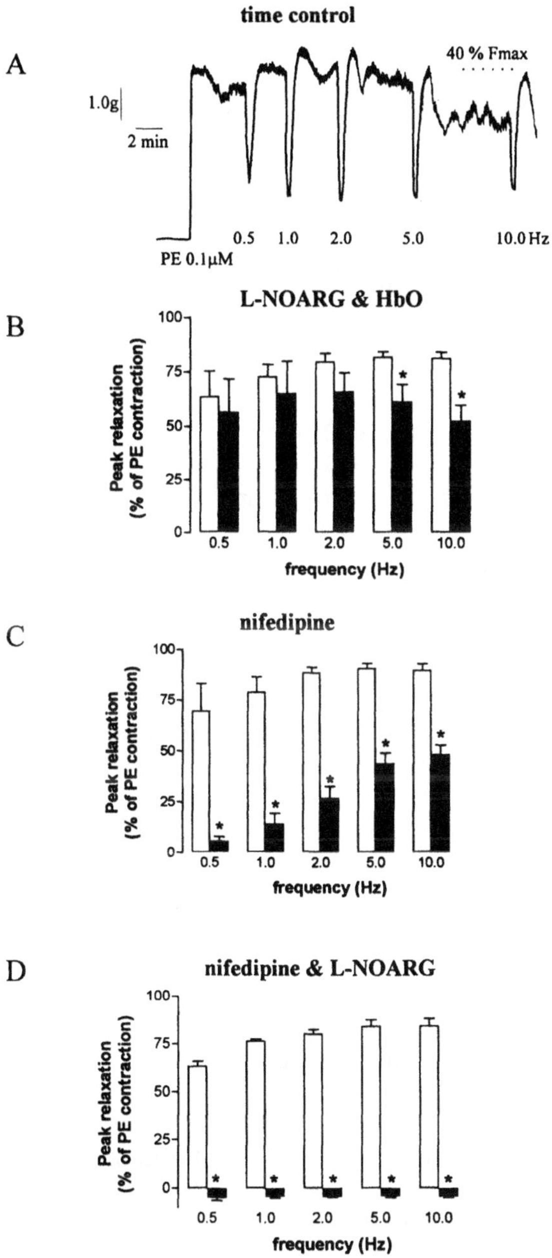

FIGURE 3.2. Demonstration that NO acts as a backup coinhibitory neurotransmitter for NDHF in the rat anococcygeus. (A) Original chart recording depicts time control relaxations to electrical field stimulation (EFS) (0.5–10 Hz, 0.2 ms duration, supramaximal voltage, 20 s) of NANC inhibitory nerves. (B-D) Group data for first control relaxations (open columns) and second responses (filled columns) to EFS of NANC inhibitory nerves obtained in the presence of (B) a combination of L-NOARG (100 μM) and oxyhemoglobin (HbO) (30 μM), (C) nifedipine (0.3 μM), or (D) a combination of nifedipine (0.3 μM) and L-NOARG (100 μM). All tissues were contracted to 40% F_{max} with phenylephrine (PE) regardless of treatment. Values are means ± SEM. Asterisks denote significant differences from controls ($P < 0.05$).

"nitrergic," as previously widely believed. Such a classification, however, relies on the assumption that under normal physiological conditions smooth muscle "tone" in the anococcygeus depends on predominantly depolarization-dependent mechanisms. In addition, our studies indicate that the mechanisms of neurotransmission in the anococcygeus are similar to those in the guinea pig taenia coli and together give a clear message that the ability to demonstrate only nitrergic responses in NANC-innervated gut preparations may indicate inappropriate bioassay conditions.

Mechanisms of Redundancy

The precise reason why NO is functionally silent in the rat anococcygeus and the guinea pig taenia coli is unknown. One explanation, however, could be that the stimulus required for NO release (i.e., frequency of stimulation or level of neuronal Ca^{2+}) is higher than that required for NDHF release. Therefore, NDHF would be able to elicit nearly all the relaxation to NANC nerve stimulation before enough neuronal Ca^{2+} was generated to activate NOS.

Another reason that NO is a silent neurotransmitter could be that the sensitivity of the smooth muscle relaxing mechanism activated by NO may be different from that activated by NDHF. NDHF mediates nearly all the relaxation to NANC nerve stimulation by activating either SK_{Ca} (guinea pig taenia coli) or a nifedipine-sensitive mechanism (rat anococcygeus). If NO was simultaneously released with NDHF, then the mechanism it activated to mediate relaxation might be secondary to the effects of NDHF and therefore would remain functionally silent. For example, in the guinea pig taenia coli, NO appears to activate a cGMP-dependent K^+ channel to evoke hyperpolarization and initiate relaxation (Selemidis et al., 1997). The apaminsensitive K^+ channel activated by NDHF, however, may have a shorter latency time for activation than the cGMP-dependent K^+ channel activated by NO, and as a result most of the relaxation would be mediated by NDHF. Alternatively, NDHF may inhibit in some way the cGMP-dependent K^+ channel. Interestingly, such a "negative feedback" type of regulation has been proposed to account for redundancy of EDHF in blood vessels, although in this case NO inhibits EDHF (Bauersachs et al., 1996).

Electrophysiology

Almost invariably, stimulation of NANC inhibitory nerves in the gastrointestinal tract causes smooth muscle hyperpolarizations that are referred to as inhibitory junction potentials (IJPs). In many cases, such hyperpolarization is biphasic. The first component of the IJP has been shown to be fast in onset and with a short duration, whereas the second component is slower and longer-lasting. Such biphasic responses have been demonstrated in the proximal colon, gastric fundus, internal anal sphincter, ileum, and taenia coli of the guinea pig, in the mouse gastric fundus, and recently in the rat anococcygeus (Selemidis et al., 1998) (for an extensive review see Bennett, 1997). In all of these preparations except for the guinea pig taenia coli, apamin abolished the fast component, whereas inhibitors of the NO pathway abolished the second, slow phase. These results indicate that SK_{Ca} channels and NO mediate the first and second components of the biphasic IJP to NANC nerve stimulation in these tissues. In the guinea pig taenia coli, however, whereas the first, fast phase of the IJP to NANC inhibitory nerve stimulation is similarly blocked by apamin, the second component was shown to be unaffected by NO inhibitors (Bridgewater et al., 1995). This lack of effect of NOS inhibitors on the biphasic IJP is paradoxical, since similar NOS inhibitors combined with apamin abolish the relaxation to NANC inhibitory nerve stimulation in the guinea pig taenia coli (Selemidis et al., 1997). Dissociation between the electrical and mechanical responses to NANC inhibitory nerve stimulation also occurs in the rat anococcygeus (Selemidis et al., 1998; Selemidis and Cocks, 1997). Here, however, the nifedipine-sensitive relaxation to NANC inhibitory nerve stimulation was unaffected by apamin, which blocked the first, fast phase of the biphasic IJP, whereas the second, slow phase was abolished by NOS inhibitors and ODQ (Selemidis et al., 1998). A similar disparity between electrical and mechanical recordings of NANC nerve-mediated inhibition has been demonstrated in the guinea pig internal anal sphincter (Rae and Muir, 1996), where as seen in the rat anococcygeus, the apamin-sensitive IJP contributed very little if anything to NANC relaxation. Therefore, the roles of electrical events associated with NANC inhibitory nerve-mediated relaxation need further evaluation.

Novel Mechanisms of NANC Neurotransmission: Ca^{2+} Sparks

We have recently demonstrated in the mouse longitudinal gastric fundus that NANC inhibitory neurotransmission is solely nitrergic, with no evidence for backup relaxation mechanisms (Selemidis and Cocks, in press; Figure 3.3). Interestingly, these nitrergic nerve-evoked responses and those to exogenously applied NO were inhibited by ryanodine, a blocker of sarcoplasmic reticulum (SR) intracellular Ca^{2+}-induced Ca^{2+} release. Also, we found that caffeine, an activator of Ca^{2+} release from intracellular ryanodine-sensitive Ca^{2+} pools, caused relaxation of the mouse gastric fundus, which was inhibited by ryanodine (Figure 3.3). These findings indicate that Ca^{2+} release from the SR is in some way involved in the relaxation to NO. This appears contrary to current ideas that intracellular Ca^{2+} release evokes contraction and that NO simply acts to reduce Ca^{2+} to cause smooth muscle relaxation. Elegant studies by Nelson and colleagues (1995), however, show this to be too simplistic. Their studies showed that Ca^{2+} released from

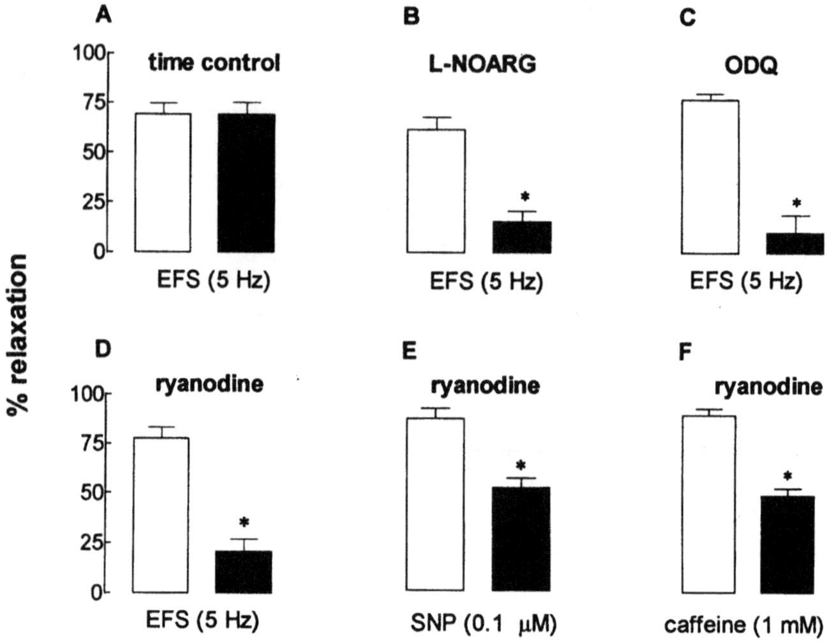

FIGURE 3.3. Evidence that NANC inhibitory innervation in mouse gastric fundus is nitrergic and that the evoked relaxations are dependent on cGMP and mediated by a ryanodine-sensitive mechanism. (A) Group data showing first control (open columns) and second time control relaxations to 5-Hz electrical field stimulation (EFS) (0.1 ms duration, 10 s, 30 V) under NANC conditions. (B-D) Similar responses as in (A), except tissues were treated with L-NOARG (100 μM), ODQ (10 μM), or ryanodine (10 μM), respectively (closed columns). (E and F) Repeated relaxations to the NO donor sodium nitroprusside (SNP) (0.1 μM) (E) and the activator of Ca^{2+} release from ryanodine-sensitive channels in the SR, caffeine (1 mM) (F) in the absence (open columns) and presence (closed) of ryanodine (F). All relaxations were obtained in tissues contracted to 40% F_{max} with U46619. Values are means ± SEM. Asterisks denote significant differences from controls ($P < 0.05$).

clusters of ryanodine-sensitive Ca^{2+} channels on the SR results in local, microdomain increases in Ca^{2+} that do not contribute to contraction. Instead these "Ca^{2+} sparks" activate nearby K_{Ca} channels on the plasma membrane to initiate hyperpolarization and thus relaxation, most likely by closure of voltage-sensitive Ca^{2+} channels. Porter et al. (1998) showed that NO increases the frequency of Ca^{2+} sparks and thus increases the activity of K^+ channels. Although ryanodine-sensitive, nitrergic responses in the mouse gastric fundus were unaffected by a variety of specific K^+ channel inhibitors and by nonselective K^+ channel inhibition with high extracellular K^+ (Selemidis and Cocks, in press), we proposed that Ca^{2+} sparks initiated by NO-mediated increases in cGMP evoke relaxation of gastrointestinal smooth muscle via other Ca^{2+}-sensitive processes.

In another study by Mironneau et al. (1996), caffeine was shown to increase Ca^{2+} spark frequency via a ryanodine-sensitive mechanism, which then activated K_{Ca} channels to evoke hyperpolarization. With higher concentrations of caffeine, however, oscillating Ca^{2+} waves, which spread within the cell, were generated at the same sites of Ca^{2+} spark release. These authors proposed that Ca^{2+} sparks initiate Ca^{2+} waves. Because Ca^{2+} waves have also been shown to travel between cells via gap junctions (Young et al., 1996), they may be important in initiating coordinated tissue responses. For instance, since it is unlikely that all smooth muscle cells are innervated by NANC inhibitory nerves and that smooth muscle in most gut tissue is composed of multiple layers of cells, it is possible that Ca^{2+} sparks generated in innervated smooth muscle cells initiate intercellular Ca^{2+} waves that in turn can evoke responses in noninnervated cells. Based on these intriguing possibilities, we investigated the effect of gap junction uncoupling on responses to NANC nerves in various gut preparations.

Role of Gap Junctions

We recently showed that the gap junction uncoupler 18-β-glycyrrhetinic acid (Goldberg et al., 1996) abolished NO-independent relaxations to NANC nerve stimulation in the rat anococcygeus muscle (Figure 3.4). Although this result suggests a possible functional role for gap junctions in NANC inhibitory neurotransmission in the gut, it also reveals an intriguing further similarity between endothelial cells and NANC nerves in how these cells signal smooth muscle to relax, since it has recently been demonstrated that gap junctions are involved in some way in vasodilatation to EDHF (Kühberger et al., 1994; Chaytor et al., 1998; Taylor et al., 1998; Hutcheson et al., 1999; Dora et al., 1999).

Like most types of nerves, peripheral NANC inhibitory nerves in the gut are believed to mediate smooth muscle relaxation by the release of humoral neurotransmitters, which then diffuse across the neuroeffector junction to activate specific receptors on postjunctional effector cells, in this case smooth muscle cells, to trigger the response. The data we have presented here (Figure 3.4), however, suggest that such classical chemical neurotransmission may require modification if gap junctions are in some way involved. If they are, then their location and role in NANC-induced responses need clarification. Our analogy with "endotransmission" in blood vessels may help here. Proposals that endothelium-dependent relaxation to EDHF involves direct communication between endothelial and smooth muscle cells via gap junctions are not new (Kühberger et al., 1994). However, Bény and coworkers (1990) showed that endothelial cells and smooth muscle cells of the pig coronary artery were not dye coupled. This finding did not support the idea that EDHF was able to travel to smooth muscle cells via gap junctions. Furthermore, Little et al. (1995) found that

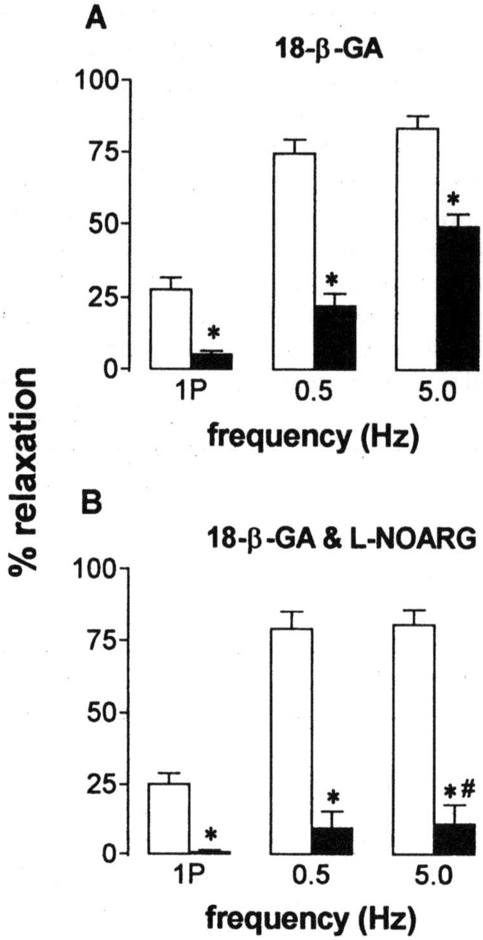

FIGURE 3.4. Demonstration of a role for gap junctions in non-NO, NANC nerve-mediated relaxation of the rat anococcygeus. (A) Group data showing first control relaxations (open columns) and second responses (filled columns) obtained in the presence of the gap junction uncoupling agent 18-β-glycyrrhetinic acid (18-β-GA) (50 μM) to electrical field stimulation (EFS) (0.1 ms duration, 10 s, supramaximal voltage) under NANC conditions. (B) Similar responses as in (A) except tissues were treated with a combination of 18-β-GA) (50 μM) and L-NOARG (100 μM). All tissues were contracted to 40% F_{max} with phenylephrine (PE) regardless of treatment. Values are means ± SEM. Asterisks denote significant differences from controls ($P < 0.05$).

gap junctions were prevalent between homologous cells in both the endothelium and smooth muscle cell layers of the artery wall, with few heterologous junctions between endothelial and smooth muscle cells. Such homologous junctions in the artery wall may function, particularly in larger arteries, to initiate a coordinated response to endothelium-dependent vascular dilatation. Because very little dye transfer has been demonstrated

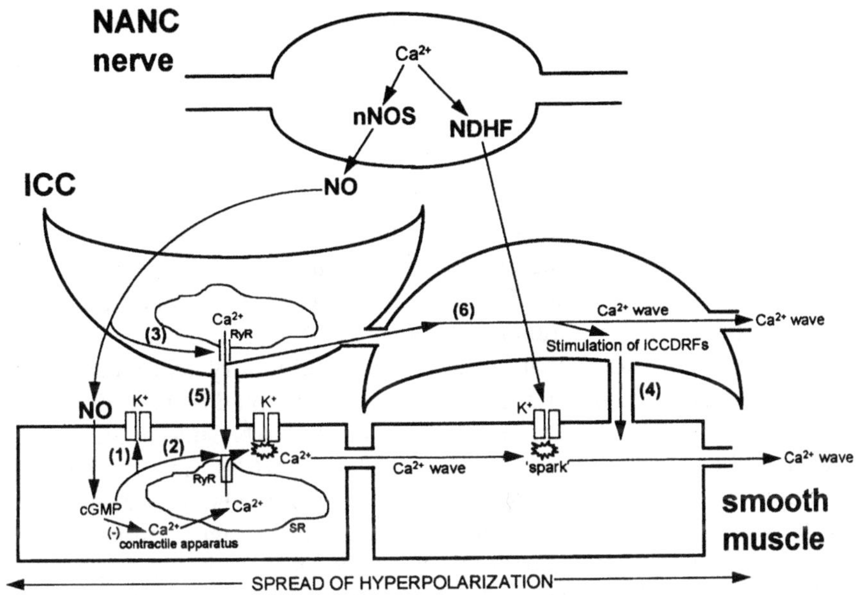

FIGURE 3.5. Schematic representation of possible mechanisms underlying gastrointestinal smooth muscle relaxation to NANC inhibitory nerve stimulation. NO activates K^+ channels via cGMP either directly (1) or indirectly (2) via cGMP-dependent Ca^{2+} sparks from the ryanodine-sensitive Ca^{2+} channel in the SR (Porter et al., 1998). A modification of these NO-dependent mechanisms is the possible amplification of Ca^{2+} release from interstitial cells of Cajal (ICC) (3) by NO which may cause either further NO release (Publicover et al., 1993) or the transfer of inhibitory messengers (interstitial cell-derived relaxing factors or ICCDRFs) to smooth muscle cells via gap junctions (4) (Berezin et al., 1988; for a review see also Sanders, 1996). Also, this ICC-dependent Ca^{2+} release may travel as Ca^{2+} waves to smooth muscle cells (5) or ICC (6). Other non-NO, nerve-derived hyperpolarizing factors such as NDHF, as shown here, directly activate smooth muscle cell K^+ channels to initiate hyperpolarisation. Ryr denotes ryanodine receptor, ICCDRF (interstitial cell-) and NDHF nerve-derived hyperpolarizing factor.

between the endothelium and the smooth muscle of large arteries, which suggests sparse electrical coupling between the two cell layers, it remains possible, however, that the few myoendothelial junctions are sufficient to transfer EDHF, whose effect then spreads and is amplified between smooth muscle cells within the artery wall via gap junctions. For example, such a mechanism could involve Ca^{2+} waves, since these are known to spread between smooth muscle cells via gap junctions (Young et al., 1996). Regardless of their exact location and role in vasorelaxation to EDHF, if, as our data suggest, gap junctions are similarly involved in NANC inhibitory neurotransmission in the rat anococcygeus, as they appear to be in blood vessels, then they may be located either between smooth muscle cells or as inferred by Sanders (1996), between interstitial cells of Cajal and smooth muscle cells (Figure 3.5). Finally, direct heterologous communication via specialized gap junctions between NANC nerves and smooth muscle cells remains an exciting possibility.

Conclusions

In conclusion, studies on NANC inhibitory nerve transmission in the gastrointestinal tract continue to reveal fascinating new mechanisms of heterocellular communication. Furthermore, NANC inhibitory neurotransmission continues to parallel mechanisms of endotransmission in the vasculature. Both endothelial cells and NANC inhibitory nerves act as automatic controllers of smooth muscle tone and thus could be regarded as different "phenotypes" of a common cell, i.e., either NANC nerves are "linear" endothelial cells or endothelial cells are "planar" NANC nerves. Therefore, although the two cell types are not embryologically related, they nevertheless have evolved remarkably similar mechanisms to signal smooth muscle to relax. Finally, EDHF has recently been shown to be K^+—a simpler molecule than NO—which causes relaxation of vascular smooth muscle via activation of inward rectifying K^+ channels and Na^+-K^+-ATPase (Edwards et al., 1998). With respect to NANC inhibitory nerves, the question must now be raised: is NDHF neurally released K^+?

References

Archer, S. 1993. Measurement of nitric oxide in biological models. *FASEB J.* 7:349–360.

Banks, B.E.C., Brown, C., Burgess, G.M., Burnstock, G., Claret, M., Cocks, T.M., and Jenkinson, D.H. 1979. Apamin blocks certain neurotransmitter-induced increases in potassium permeability. *Nature* 282:415–417.

Bauersachs, J., Popp, R., Hecker, M., Sauer, E., Fleming, I., and Busse, R. 1996. Nitric oxide attenuates the release of endothelium-derived hyperpolarising factor. *Circulation* 94:3341–3347.

Bény, J.L. 1990. Endothelial and smooth muscle cells hyperpolarized by bradykinin are not dye coupled. *Am. J. Physiol.* 258 (*Heart Circ. Physiol.* 27):H836–H841.

Bennett, M.R., Burnstock, G., and Holman, M.E. 1966a. Transmission from intramural inhibitory nerves to the smooth muscle of the guinea-pig taenia coli. *J. Physiol.* 182:541–558.

Bennett, M.R., Burnstock, G., and Holman, M.E. 1966b. Transmission from perivascular inhibitory nerves to the smooth muscle of the guinea-pig taenia coli. *J. Physiol.* 182:527–540.

Bennett, M.A. 1997. Non-adrenergic non-cholinergic (NANC) transmission to smooth muscle: 35 years on. *Prog. Neurobiol.* 52:159–195.

Berezin, I., Huizinga, J.D., and Daniel, E.E. 1988. Interstitial cells of Cajal in the canine colon: a special communication network at the inner border of the circular muscle. *J. Comp. Neurol.* 273(1):42–51.

Boeckxstaens, G.E., Bult, H., and Pelckmans, P.A. 1991a. Nitric oxide release in response to stimulation of nonadrenergic, noncholinergic nerves. *J. Cardiovasc. Pharmacol.* 17(Suppl. 3):S238–S242.

Boeckxstaens, G.E., Pelckmans, P.A., Bogers, J.J., Bult, H., and De Man, J.G. 1991b. Release of nitric oxide upon stimulation of non-adrenergic, non-cholinergic nerves in the rat gastric fundus. *J. Pharmacol. Exp. Ther.* 256:441–447.

Boeckxstaens, G.Y., Pelckmans, P.A., Herman, A.G., and Van Maercke, Y.M. 1993. Involvement of nitric oxide in the inhibitory innervation of the human isolated colon. *Gastroenterology* 104:690–697.

Bredt, D.S., Hwang, P.M., and Snyder, S.H. 1991. Localization of nitric oxide synthase indicating a neural role for nitric oxide. *Nature* 347:768–770.

Bridgewater, M., Cunnane, T.C., and Brading, A.F. 1995. Characteristic features of inhibitory junction potentials evoked by single stimuli in the guinea-pig isolated taenia coli. *J. Physiol.* 485:145–155.

Bulbring, E. 1955. Correlation between membrane potential, spike discharge and tension in smooth muscle. *J. Physiol.* 128:200–221.

Bulbring, E., Burnstock, G., and Holman, M.E. 1958. Excitation and conduction in the smooth muscle of the isolated taenia coli of the guinea-pig. *J. Physiol.* 142:420–437.

Bult, H., Boeckxstaens, G.E., Pelckmans, P.A., Jordaens, F.H., Van Maercke, Y.M., and Herman, A.G. 1990. Nitric oxide as an inhibitory non-adrenergic non-cholinergic neurotransmitter. *Nature* 345:346–347.

Burleigh, D.E. 1992. N^{g}-nitro-L-arginine reduces nonadrenergic, noncholinergic relaxations of human gut. *Gastroenterology* 102:679–683.

Burnstock, G. 1972. Purinergic nerves. *Pharmacol. Rev.* 24:509–581.

Burnstock, G. 1990. Overview: purinergic mechanisms. *Ann NY Acad Sci.* 603:1–18.

Burnstock, G., Campbell, G., Bennett, M., and Holman, M.E. 1964. Innervation of the guinea-pig taenia coli: Are there intrinsic inhibitory nerves which are distinct from sympathetic nerves? *Int. J. Neuropharmacol.* 3:163–166.

Burnstock, G.,Campbell, G., and Rand, M.J. 1966. The innervation of the taenia of the guinea-pig caecum. *J. Physiol.* 182:504–526.

Callewaert, G.L., Shipolini, R., and Vernon, C.A. 1968. The disulphide bridges of apamin. *FEBS Lett.* 1:111–113.

Campbell, G. 1966. Nerve-mediated excitation of the taenia of the guinea-pig caecum. *J. Physiol.* 185:148–159.

Chaytor, A.T., Evans, W.H., and Griffith, T.M. 1998. Central role of heterocellular gap junctional communication in endothelium-dependent relaxations of rabbit arteries. *J. Physiol.* 508:561–573.

Costa, M., Furness, J.B., and Humphreys, C.M.S. 1986. Apamin distinguishes two types of relaxation mediated by enteric nerves in the guinea-pig gastrointestinal tract. *Naunyn. Schmied. Arch. Pharmacol.* 332:79–88.

Crist, J.R., He, X.D., and Goyal, R.K. 1992. Both ATP and the peptide VIP are inhibitory neurotransmitters in guinea-pig ileum circular muscle. *J. Physiol.* 447:119–131.

Dail, W.G., Galloway, B., and Bordegaray, J. 1993. NADPH diaphorase innervation of the rat anococcygeus and retractor penis muscles. *Neurosci. Lett.* 160:17–20.

Dora, K.A., Martin, P.E.M., Chaytor, A.T., Evans, W.H., Garland, C.J., and Griffith, T.M. 1999. Role of heterocellular gap junctional communication in endothelium-dependent smooth muscle hyperpolarisation: inhibition by a connexin-mimetic peptide. *Biochem. Biophys. Res. Commun.* 254:27–31.

Drummond, G.R., and Cocks, T.M. 1996. Evidence for mediation by endothelium-derived hyperpolarizing factor of relaxation to bradykinin in the bovine isolated coronary artery independently of voltage-operated Ca^{2+} channels. *Br. J. Pharmacol.* 117:1035–1040.

Drummond, G.R., Selemidis, S., and Cocks, T.M. In press. Apamin-sensitive non-nitric oxide (NO) endothelium-dependent relaxations to bradykinin in the bovine isolated coronary artery: no role for cytochrome P_{450} and K^+. *Br. J. Pharmacol.*

Edwards, G., Dora, K.A., Gardener, M.J., Garland, C.J., and Weston, A.H. 1998. K^+ is an endothelium-derived hyperpolarising factor in rat arteries. *Nature* 396:269–272.

Fonseca, M., Uddin, N., and Gibson, A. 1998. No evidence for a significant non-nitrergic, hyperpolarising factor contribution to field stimulation-induced relaxation of the mouse anococcygeus. *Br. J. Pharmacol.* 124:524–528.

Furness, J.B., and Costa, M. 1987. *The Enteric Nervous System.* Churchill Livingstone, Edinburgh, London, Melbourne.

Furness, J.B., Pompolo, S., Shuttleworth, C.W.R., and Burleigh, D.E. 1992. Light- and electron-microscopic immunochemical analysis of nerve fibre types innervating the taenia of the guinea-pig caecum. *Cell. Tiss. Res.* 270:125–137.

Garland, C.J., Plane, F., Kemp, B., and Cocks, T.M. 1995. Endothelium-dependent hyperpolarisation: a role in the control of vascular tone. *Trends Pharmacol. Sci.* 16:23–30.

Gibson, A., Mirzazadeh, S., Hobbs, A.J., and Moore, P.K. 1990. L-N^G-nonomethyl arginine and L-N^G-nitro arginine inhibit non-adrenergic, non-cholinergic relaxation of the mouse anococcygeus muscle. *Br. J. Pharmacol.* 99:602–606.

Gillespie, J.S. 1972. The rat anococcygeus muscle and its response to nerve stimulation and to some drugs. *Br. J. Pharmacol.* 45:404–416.

Gillespie, J.S., Liu, X., and Martin, W. 1989. The effects of L-arginine and N^G-monomethyl L-arginine on the response of the rat anococcygeus muscle to NANC nerve stimulation. *Br. J. Pharmacol.* 98:1080–1082.

Goldberg, G.S., Moreno, A.P., Bechberger, J.F., Hearn, S.S., Shivers, R.R., Macphee, D.J., Zhang, Y.C., and Naus, C.C.G. 1996. Evidence that disruption of connexon particle arrangements in gap junction plaques is associated with inhibition of gap junctional communication by a glycyrrhetinic acid derivative. *Exp. Cell Res.* 222:48–53.

Grider, J.R., Katsoulis, S., Schmidt, W.E., and Jin, J.G. 1994. Regulation of the descending relaxation phase of intestinal peristalsis by PACAP. *J. Auton. Nerv. Syst.* 50:151–159.

Habermann, E., and Reiz, K.G. 1965. On the biochemistry of bee venom peptides, melittin and apamin. *Biochem. Z.* 341:451–466.

Hobbs, A.J., and Gibson, A. 1990. L-N^G-nitroarginine and its methyl ester are potent inhibitors of non-adrenergic, non-cholinergic transmission in the rat anococcygeus. *Br. J. Pharmacol.* 100:749–752.

Hutcheson, I.A., Chaytor, A.T., Evans, W.H., and Griffith, T.M. 1999. Nitric oxide-independent relaxations to acetylcholine and A23187 involve different routes of heterocellular communication: role for gap junctions and phospholipase A_2. *Circ. Res.* 84:53–63.

Jin, J.G., Katsoulis, S., Schmidt, W.E., and Grider, J.R. 1994. Inhibitory transmission in tenia coli mediated by distinct vasoactive intestinal peptide and apamin-sensitive pituitary adenylate cyclase activating peptide receptors. *J. Pharmacol. Exp. Ther.* 270:433–439.

Keef, K.D., Du, C., Ward, S.M., McGregor, and Sanders, K.M. 1993. Enteric inhibitory neural regulation of human colonic circular muscle: role of nitric oxide. *Gastroenterology* 105:1009–1016.

Kilpatrick, E.V., and Cocks, T.M. 1994. Evidence for differential roles of nitric oxide (NO) and hyperpolarization in endothelium-dependent relaxation of pig isolated coronary artery. *Br. J. Pharmacol.* 112:557–565.

Kishi, M., Takeuchi, T., Suthamnatpong, N., Ishii, T., Nishio, H., Hata, F., and Takewaki, T. 1996. VIP- and PACAP-mediated nonadrenergic, noncholinergic inhibition in longitudinal muscle of the rat distal colon: involvement of activation of charybdotoxin- and apamin-sensitive K^+ channels. *Br. J. Pharmacol.* 119: 623–630.

Klinge, E., and Sjostrand, N.O. 1974. Contraction and relaxation of the retractor penis muscle and the penile artery of the bull. *Acta. Physiol. Scand. Suppl.* 420:1–88.

Knudsen, M.A., Svane, D., and Tottrup, A. 1992. Action profiles of nitric oxide, *S*-nitroso-L-cysteine, SNP, and NANC responses in opossum lower esophageal sphincter. *Am. J. Physiol.* 262:G840–G846.

Knudsen, M.A., and Tottrup, A. 1992. A possible role of the L-arginine-nitric oxide pathway in the modulation of cholinergic transmission in the guinea-pig taenia coli. *Br. J. Pharmacol.* 107:837–841.

Kühberger, E., Groschner, K., Kukovetz, W.R., and Brunner, F. 1994. The role of myoendothelial cell contact in non-nitric oxide-, non-prostanoid-mediated endothelium-dependent relaxation of porcine coronary artery. *Br. J. Pharmacol.* 113:1289–1294.

Li, C.G., and Rand, M.J. 1989. Evidence for a role of nitric oxide in the neurotransmitter system mediating relaxation of the rat anococcygeus muscle. *Clin. Exp. Pharm. Physiol.* 16:933–938.

Li, C.G., and Rand, M.J. 1990. Nitric oxide and vasoactive intestinal polypeptide mediate non-adrenergic, non-cholinergic inhibitory transmission to smooth muscle of the rat gastric fundus. *Eur. J. Pharmacol.* 191:303–309.

Little, T.L., Xia, J., and Duling, B.R. 1995. Dye tracers define differential endothelial and smooth muscle coupling patterns within the arteriolar wall. *Circ. Res.* 76:498–504.

Liu, X., Gillespie, J.S., Gibson, I.F., and Martin, W. 1991. Effects of N^G-substituted analogues of L-arginine on NANC relaxation of the rat anococcygeus and bovine retractor penis muscles and bovine penile artery. *Br. J. Pharmacol.* 104:53–58.

Maggi, C.A., and Giuliani, S. 1993. Multiple inhibitory mechanisms mediate non-adrenergic non-cholinergic relaxation in the circular smooth muscle of the guinea-pig colon. *Naunyn-Schmied. Arch. Pharmacol.* 347:630–634.

Manzini, S., Maggi, C.A., and Meli, A. 1986. Pharmacological evidence that at least two different NANC inhibitory systems are present in the rat small intestine. *Eur. J. Pharmacol.* 123:229–236.

McConalogue, K., Lyster, D.J.K., and Furness, J.B. 1995. Electrophysiological analysis of the actions of pituitary adenylyl cyclase activating peptide in the taenia of the guinea-pig caecum. *Naunyn-Schmied. Arch. Pharmacol.* 352:538–544.

Mironneau, J., Arnaudeau, S., Macrez-Lepretre, N., and Boittin, F.X. 1996. Ca^{2+} sparks and Ca^{2+} waves activate different Ca^{2+}-dependent ion channels in single myocytes from rat portal vein. *Cell Calcium* 20:153–160.

Mitchell, J.A., Sheng, H., Forstermann, U., and Murad, F. 1991. Characterisation of nitric oxide synthases in non-adrenergic non-cholinergic nerve containing tissue from the rat anococcygeus muscle. *Br. J. Pharmacol.* 104:289–291.

Nelson, M.T., Cheng, H., Rubart, M., Santana, L.F., Bonev, A.D., Knot, H.J., and Lederer, W.J. 1995. Relaxation of arterial smooth muscle by calcium sparks. *Science* 270:633–637.

Ohno, N., Xue, L., Yamamoto, Y., and Suzuli, H. 1996. Properties of the inhibitory junction potential in smooth muscle of the guinea-pig gastric fundus. *Br. J. Pharmacol.* 117:874–978.

Parsons, S.J.W., Hill, A., Waldron, G.J., Plane, F., and Garland, C.J. 1994. The relative importance of nitric oxide and nitric oxide-independent mechanisms in acetylcholine-evoked dilatation of the rat mesenteric bed. *Br. J. Pharmacol.* 113:1275.

Piotrowski, W., Simon, M.C., and Brennan, L. 1993. Effects of N^G-nitro-L-arginine and methylene blue on non-adrenergic, non-cholinergic responses of isolated guinea-pig taenia caeci. *Br. J. Pharmacol.* 157P.

Porter, V.A., Bonev, A.D., Knot, H.J., Heppner, T.J., Stevenson, A.S., Kleppisch, T., and Nelson, M.T. 1998. Frequency modulation of Ca^{2+} sparks is involved in regulation of arterial diameter by cyclic nucleotides. *Am. J. Physiol.* 274 (*Cell Physiol.* 43):C1346–C1355.

Publicover, N.G., Hammond, E.M., and Sanders, K.M. 1993. Amplification of nitric oxide signaling by interstitial cells isolated from canine colon. *Proc. Natl. Acad. Sci.* 90(5):2087–2091.

Rae, M.G., and Muir, T.C. 1996. Neuronal mediators of inhibitory junction potentials and relaxation in the guinea-pig internal anal sphincter. *J. Physiol.* 493:517–527.

Rand, M.J. 1992. Nitrergic transmission: nitric oxide as a mediator of non-adrenergic, non-cholinergic neuro-effector transmission. *Clin. Exp. Pharm. Physiol.* 19:147–169.

Rand, M.J., and Li, C.G. 1990. Nitric oxide mediates non-adrenergic, non-cholinergic relaxation in some neuro-effector systems: examples of nitrergic transmission. Xth International Congress of Pharmacology, Amsterdam. *Eur. J. Pharmacol.* 183:1144.

Rand, M.J., and Li, C.G. 1995. Nitric oxide as a neurotransmitter in peripheral nerves: nature of transmitter and mechanism of transmission. *Annu. Rev. Physiol.* 57:659–682.

Sanders, K.M. 1996. A case for interstitial cells of Cajal as pacemakers and mediators of neurotransmission in the gastrointestinal tract. *Gastroenterology* 111:492–515.

Sanders, K.M., and Ward, S.M. 1992. Nitric oxide as a mediator of nonadrenergic, noncholinergic neurotransmission. *Am. J. Physiol.* 262:G379–G392.

Selemidis, S., and Cocks, T.M. 1997. Evidence that both nitric oxide (NO) and a non-NO hyperpolarizing factor elicit NANC nerve-mediated relaxation in the rat isolated anococcygeus. *Br. J. Pharmacol.* 120:662–666.

Selemidis, S., and Cocks, T.M., In press. Nitrergic relaxation of the mouse gastric fundus is mediated by cGMP-dependent and ryanodine-sensitive mechanisms. *Br. J. Pharmacol.*

Selemidis, S., Satchell, D.G., and Cocks, T.M. 1997. Evidence that NO acts as a redundant NANC inhibitory neurotransmitter in the guinea-pig isolated taenia coli. *Br. J. Pharmacol.* 121:604–611.

Selemidis, S., Ziogas, J., and Cocks, T.M. 1998. Apamin- and nitric oxide-sensitive biphasic non-adrenergic non-cholinergic inhibitory junction potentials in the rat anococcygeus muscle. *J. Physiol.* 513:835–844.

Shimamura, K., Fujisawa, A., Toda, N., and Sunano, S. 1993. Effects of N^G-nitro-L-arginine on electrical and mechanical responses to stimulation of non-adrenergic, non-cholinergic inhibitory nerves in circular muscle of the rat gastric fundus. *Eur. J. Pharmacol.* 231:103–109.

Song, Z.M., Brookes, S.J.H., and Costa, M. 1993. NADPH-diaphorase reactivity in nerves supplying the rat anococcygeus muscle. *Neurosci. Lett.* 158:221–224.

Suthamnatpong, N., Hata, F., Kanada, A., Takeuchi, T., and Yagasaki, O. 1993. Mediators of nonadrenergic, noncholinergic inhibition in the proximal, middle and distal regions of rat colon. *Br. J. Pharmacol.* 108:348–355.

Taylor, H.J., Chaytor, A.T., Evans, W.H., and Griffith, T.M. 1998. Inhibition of the gap junctional component of endothelium-dependent relaxations in rabbit iliac artery by 18-alpha glycyrrhetinic acid. *Br. J. Pharmacol.* 125:1–3.

Taylor, S.G., and Weston, A.H. 1988. Endothelium-derived hyperpolarizing factor: a new endogenous inhibitor from the vascular endothelium. *Trends Pharmacol. Sci.* 9:272–274.

Tøttrup, A., Knudsen, M.A., Hanberg Sorensen, F., and Glavind, E.B. 1995. Pharmacological identification of different inhibitory mediators involved in the innervation of the internal anal sphincter. *Br. J. Pharmacol.* 115:158–162.

Tøttrup, A., Svane, D., and Froman, A. 1991. Nitric oxide as a mediator of NANC inhibition of opossum lower esophageal sphincter. *Am. J. Physiol.* 260:G385–G389.

Ward, S.M., McKeen, E.S., and Sanders, K.M. 1992. Role of nitric oxide in non-adrenergic, non-cholinergic inhibitory junctional potentials in canine ileocolonic sphincter. *Br. J. Pharmacol.* 105:776–782.

Xue, D.H., and Goyal, R.K. 1993. Nitric oxide involvement in the peptide VIP-associated inhibitory junction potential in the guinea-pig ileum. *J. Physiol.* 461:485–499.

Young, S.H., Ennes, H.S., and Mayer, E.A. 1996. Propagation of calcium waves between colonic smooth muscle cells in culture. *Cell Calcium* 20:257–271.

Zagorodnyuk, V., and Maggi, C.A. 1994. Electrophysiological evidence for different release mechanism of ATP and NO as inhibitory NANC transmitters in guinea-pig colon. *Br. J. Pharmacol.* 112:1077–1082.

4
Nitrergic Neurotransmission in the Lower Urinary Tract and Penile Erectile Tissues

KARL-ERIK ANDERSSON

Introduction

Nonadrenergic, noncholinergic (NANC) nerves and neurotransmission can be demonstrated both in the lower urinary tract and in penile erectile tissues (Andersson, 1993; Andersson and Wagner, 1995). Many of the inhibitory NANC-mediated responses have been reported to involve nitric oxide (NO), but with the exception of the role of NO in penile erection, their functional importance has not been clarified. However, the demonstration of inhibitory NANC neurotransmission in normal, as well as pathophysiologically changed, lower urinary tract smooth muscles, suggests that NO may have a role physiologically and in different disorders of the lower urinary tract.

Nitric Oxide and the Lower Urinary Tract

Distribution of NOS-Containing Nerves

Both immunohistochemistry and NADPH-diaphorase (NADPH-d) have been used to localize nitric oxide synthase (NOS) in the lower urinary tract. Neuronal NOS and neuronal NADPH-d seem to be identical in brain and peripheral tissues, and NADPH-d histochemistry should therefore provide an appropriate marker for neurons producing NO (Dawson et al., 1991; Hope et al., 1991). In addition, NADPH-d may provide information on non-neuronal forms of NOS.

In the pig and lamb detrusor, trigone, and urethra, NADPH-d-positive fibers and thick nerve branches were found in or around the muscular bundles (Larsson et al., 1992; Persson et al., 1993; Triguero et al., 1993). Thin nerve fibers within the muscle bundles were often found in the urethral/trigonal area, whereas they were less common in the detrusor. NADPH-d-positive nerve fibers were frequently seen around arteries but not around veins. A positive staining was found in the endothelium of some arteries. There was also distinct staining of the urothelium (Larsson et al., 1992; Persson et al., 1993). Numerous NADPH-d-positive, fine varicose nerve fibers were demonstrated around arteries and in and around smooth muscle fibers in the female rabbit lamina propria (Zygmunt et al., 1993a).

In the lateral wall and trigone regions of the human bladder, plexuses of NADPH-d-containing nerve fibers were found (Smet et al., 1994). Biopsy specimens from the lateral bladder wall contained many NADPH-d-positive nerve terminals, particularly in the subepithelial region immediately beneath the urothelium; occasionally they penetrated into the epithelial layer. Fewer NADPH-d-positive nerves were observed in the trigone region than in the bladder wall, which contrasts to the findings in the pig and sheep. NADPH-d activity was also detectable in the urothelium and in intramural ganglia.

There are conflicting results concerning the NADPH-d staining of the detrusor of other species. Keast (1992) found no NADPH-d staining in the rat detrusor, whereas McNeill et al. (1992b) and Grozdanovic et al. (1992) found staining in the rat and mouse detrusor, respectively.

The difference observed in the distribution of NADPH-d-positive nerves between pig detrusor, trigone, and urethra could be confirmed by immunohistochemistry, using antiserum produced in rabbits against a C-terminal fragment (Alm et al., 1993) of a cloned NOS from rat cerebellum (Bredt et al., 1990, 1991). The density of NOS immunoreactivity was distinctly higher in trigonal and urethral tissue than in the detrusor (Persson et al., 1993; Persson and Andersson, 1994). There was no NOS staining of the urothelium, which contrasts to the results with NADPH-d. Smet et al. (1996a) found that NOS immunoreactivity was widely distributed in the lower urinary tract of guinea pigs and humans. In guinea pigs 50%–68% of all intrinsic vesical neurons expressed NOS immunoreactivity, whereas in humans 72%–96% of neurons in the wall of the bladder contained NOS. In both humans and guinea pigs, varicose NOS-immunoreactive (NOS-IR) nerve terminals provided a moderate innervation to the detrusor muscle of the bladder body and a denser innervation to the urethral muscle. NOS-IR nerves also projected to the subepithelium and around blood vessels, but were rarely observed encircling intramural vesical ganglia (Smet et al., 1996b).

NOS immunoreactivity was also demonstrated in the bladder neck and membranous urethra of rats and was seen in nerves of the mucosal stroma, to a large extent encircling small arteries (Burnett et al., 1992; Alm et al., 1993, 1995). Some nerves were also found near the smooth musculature, and some were running in close proximity to the urothelium (Alm et al., 1993). In the rat detrusor, the number of NOS-IR nerves was low, except around the ureteral orifices (Alm et al., 1995).

In the pig, colocalization studies revealed that some NOS-IR nerves had profiles that were similar to those of nerves stained for neuropeptide

Y (NPY), vasoactive intestinal peptide (VIP), and acetylcholinesterase (AChE). Nitric oxide-containing nerves were present in a density lower than that of the cholinergic, but higher than the density of any peptidergic nerve (Persson et al., 1995). In the rat and human bladder, colocalization studies using antibodies to the vesicular acetylcholine transporter protein (VAChT), which is specific for acetylcholine-containing nerves (Arvidsson et al., 1997), confirmed that NOS and VIP were contained within a population of cholinergic (acetylcholine-containing) nerves (Alm et al., unpublished results).

Origin of NOS-Containing Nerves

Efferent Pathways

Electrical stimulation of sacral nerve roots lowers urethral pressure (McGuire and Herlihy, 1978; Torrens, 1978; Slack and Downie, 1983). It may therefore be expected that parasympathetic pathways in the pelvic nerve should be involved in relaxation of the bladder outlet. The region of the sacral parasympathetic nucleus, containing cholinergic preganglionic neurons projecting toward the urinary tract, exhibits NADPH-d activity in the rat (Vizzard et al., 1993a,b), but not in the cat (Vizzard et al., 1994a). In the dog, NADPH-d activity could be demonstrated in sympathetic preganglionic neurons in the rostral lumbar segments, whereas parasympathetic preganglionic neurons in the sacral segments were not stained (Vizzard et al., 1997). However, the fact that removal of the major pelvic ganglia (Parlani et al., 1993b), but not hypogastric nerve transection (Noto et al., 1993), inhibits urethral relaxation supports a parasympathetic origin of NO and other possible transmitters involved in urethral relaxation. This is also supported by the fact that chemical sympathectomy with 6-hydroxydopamine did not change the relaxation induced by electrical stimulation in sheep urethral smooth muscle (Garcia-Pascual et al., 1991). Furthermore, NADPH-d-reactive neurons are not present in the sympathetic inferior mesenteric ganglia (McNeill et al., 1992b). To establish which type of nerve (parasympathetic, sympathetic, or sensory) produces NO in the rat lower urinary tract, Persson et al. (1997) performed chemical denervation of primary afferents and sympathetic nerves by systemic treatment with capsaicin and 6-hydroxydopamine, respectively, followed by identification of neuronal NOS (nNOS) immunoreactivity. Nerve fibers immunoreactive for substance P and calcitonin gene-related peptide (CGRP) were found in control tissue, but could not be detected following capsaicin treatment. NOS-IR fibers appeared to be unaffected by capsaicin treatment. Administration of 6-hydroxydopamine resulted in a complete disappearance of tyrosine hydroxylase-immunoreactive (TH-IR) nerves, whereas NOS-containing nerve fibers did not appear to be affected by the treatment. Systemic treatment with capsaicin or 6-hydroxydopamine had no effect on nerve-evoked, NO-mediated relaxations. Thus, NOS-containing nerves in the rat lower urinary tract do not belong to nerve populations sensitive to either the sympathetic neurotoxin, 6-hydroxydopamine, or the sensory neurotoxin, capsaicin.

Retrograde axonal tracing of the bladder revealed that only 4%–18% of the fluorogold-labeled cells in the major pelvic ganglia were stained for NADPH-d (McNeill et al., 1992b). However, 44% of the cells in the major pelvic ganglia were NADPH-d-positive following fluorogold injection into the urethra (Bennett et al., 1995). With NOS immunohistochemistry, the percentage of neurons stained in the major pelvic ganglia was lower than with NADPH-d histochemistry, but the difference in staining between the urethra (11%) and bladder (0%) persisted (Vizzard et al., 1994b). Thus, in line with several functional observations (Andersson and Persson, 1993), postganglionic NO-containing efferent nerves in the major pelvic ganglia seem to project preferably to the urethra. Persson et al. (1998) studied the possible coexistence of NO and acetylcholine in the rat major pelvic ganglion (MPG) using double immunohistochemistry with antisera raised against NOS and choline acetyltransferase (ChAT). The results suggested that the NOS-IR nerves that mediate inhibition of rat urethral smooth muscle tone originate from the MPG and contain ChAT.

Afferent Pathways

Afferent neurons in the lumbosacral dorsal root ganglia and within the dorsal horn of the spinal cord show NADPH-d activity (Dun et al., 1993; McNeill et al., 1992b; Vizzard et al., 1993a,b), suggesting a role for NO in afferent transmission from the urinary tract. In fact, as many as 80% of the neurons in the dorsal root ganglia exhibit NADPH-d activity following retrograde axonal tracing from the bladder (Vizzard et al., 1993b). In contrast, only a small number of neurons in the lumbosacral dorsal root ganglia show NOS immunoreactivity (Vizzard et al., 1994b), raising the possibility that NADPH-d activity is not a useful marker for NO in afferent pathways in the lower urinary tract. The nitrergic nerves observed in the detrusor (Persson et al., 1993, 1995; Smet et al., 1994), and particularly within and beneath the urothelium (Smet et al., 1994), may be afferent terminals. If this is the case, it is not surprising that NO has little "efferent" effect on the detrusor smooth muscle (Persson and Andersson, 1992; Persson et al., 1992). Nitric oxide may be released from afferent nerves by, for example, dimethylsulfoxide (DMSO) and capsaicin (Birder et al., 1997). This may explain why capsaicin treatment produced an NO-dependent relaxation of the isolated dog urethra (Nishizawa et al., 1997).

Cyclic GMP—Immunoreactive Cells in the Lower Urinary Tract

Smet et al. (1996b), using an antibody to cGMP, identified a population of cells in the bladder wall of humans and guinea pigs that may be the physiological targets of neuronally released NO. The detrusor muscle fibers did not express cGMP, but numerous interstitial cells throughout the bladder body demonstrated an intense cGMP immunoreactivity after stimulation with sodium nitroprusside. These cells had long dendritic processes extending parallel to the smooth muscle fibers and contained vimentin, an intermediate filament expressed by cells of mesenchymal origin. Similar cGMP-IR cells were found in the rabbit urethra (Waldeck et al., 1998). Smooth muscle cells of the human and guinea pig urethra expressed strong cGMP immunoreactivity (Smet et al., 1996b). Smet et al. (1996b) suggested that if the NO-mediated induction of cGMP is a reliable indicator of the physiological responsiveness of a cell to NO, then smooth muscle cells appear to be the predominant targets of NO in the urethra, whereas in the bladder body, interstitial cells may serve this role.

NOS Activity in the Lower Urinary Tract

In the lower urinary tract, as in other tissues, NO is synthetized from L-arginine by NOS. All three major isoforms of NOS—neuronal (nNOS), endothelial (eNOS), and inducible (iNOS) (Förstermann et al., 1991; Griffith and Stuehr, 1995)—may be involved in NO formation, normally or pathophysiologically.

NOS activity, as revealed by the ability to convert [^{3}H]arginine and [^{14}C]arginine to [^{3}H]citrulline and [^{14}C]citrulline, respectively, was found to be highest in the human prostatic urethra, intermediate in the bladder neck, and lowest in the detrusor (Ehrén et al., 1994). High NOS activity was also found in the urethra of rats (Burnett et al., 1992) and rabbits (Dokita et al., 1994). Both the soluble and particulate fractions from the rabbit urethra converted [^{14}C]arginine to [^{14}C]citrulline, the soluble activity being Ca^{2+}-dependent, but not the particulate (Dokita et al., 1994). As emphasized by Dokita et al. (1994), at least three sources of NOS activity are possible in the urethra: the NANC neurons, the urethral epithelial lining and its vascular supply, and smooth muscle cells. They suggested that the NOS of the soluble fraction most probably is localized in urethral neurons, since the enzyme had the characteristics of the neurogenic enzyme. However, a NOS enzyme that can be induced by lipopolysaccharide seems to be expressed by bladder smooth muscle cells (Weiss et al., 1994). Birder et al. (1998), using a microsensor to directly measure endogenous NO production, found that noradrenaline as well as capsaicin could release NO from the bladder urothelium. This may provide a novel mechanism for modulation of both bladder and urothelial function.

Functional Effects of Nitric Oxide

Detrusor

Published information on the effects of NO on the detrusor smooth muscle is controversial. In pig detrusor strips, a NANC-mediated relaxation, preceded by a contraction, was found in response to electrical stimulation of nerves (Klarskov, 1987). In human detrusor, such a relaxation was seen only occasionally, and was short-lasting and fading. In small biopsy preparations of the human detrusor, James et al. (1993a,b) found that electrical stimulation evoked relaxations sensitive to N^{G}-nitro-L-arginine (L-NNA), but insensitive to tetrodotoxin. They suggested that NO might be generated from the detrusor muscle and act as an important factor for bladder relaxation during the filling phase. However, Elliott and Castleden (1993) were unable to demonstrate nerve-mediated relaxation in human detrusor muscle. Furthermore, Moon et al. (1997) reported that both nitroprusside and L-arginine produced contractions in strips of human bladder. On the other hand, rat detrusor strips were relaxed by NO generated by the photoinduced adequate NO (PIANO) system (Chung et al., 1996), and it was concluded that the rat detrusor was capable of responding to NO.

If NO has an important role in detrusor relaxation, the detrusor muscle would be expected to have a high sensitivity to agents acting by increasing the intracellular concentrations of cyclic GMP. In the pig detrusor, the NO donor morpholinosydnonimine (SIN-1) and NO relaxed carbachol- and endothelin-1-contracted preparations by approximately 60%. However, isoprenaline was about 1000 times more potent than SIN-1 and NO and caused complete relaxation (Persson and Andersson, 1992). Nitroprusside, SIN-1, and NO were only moderately effective in relaxing isolated rat, pig, and rabbit detrusor muscle, compared with their effects on the urethral muscle (Persson and Andersson, 1992; Persson et al., 1993, 1995). These results agree well with those of Morita et al. (1992), who found that in rabbits cGMP is related mainly to urethral relaxation and cAMP to urinary bladder relaxation.

The possible role of the L-arginine/NO pathway as a neuromodulator of excitatory responses in the bladder has been studied in vitro (Persson et al., 1993; Ehrén et al., 1994; Liu and Lin-Shiau, 1997). In the pig bladder, Persson et al. (1993) found that L-NNA caused an enhancement of the contractile response to electrical field stimulation. This was also found in human detrusor (Ehrén et al., 1994). L-Arginine, but not D-Arginine, decreased the electrically evoked contractions by 25%–30% in the pig detrusor. This effect of L-arginine was reversed by L-NNA. L-Arginine had no effect on NANC contractions in the presence of scopolamine, indicating that the inhibitory response was associated with the cholinergic component of the contraction (Persson et al., 1993). Whether the effect of L-arginine on nerve-evoked cholinergic contractions is due to functional antagonism at the smooth muscle level or to prejunctional effects on nerves is unknown. In the mouse bladder, Liu and Lin-Shiau (1997) found that NO and NO donors enhanced contractions elicited by electrical field stimulation.

Trigone

NANC-mediated relaxant responses to electrical stimulation have been demonstrated in human and pig trigonal muscle (Klarskov et al., 1983). A similar relaxant response to electrical stimulation of isolated superficial trigone muscle preparations from the human bladder was found by Speakman et al. (1988). The relaxations produced in isolated pig trigonal strips were concentration-dependently reduced by L-NNA (Persson and Andersson, 1992); high concentrations of L-NNA abolished all relaxation and unmasked a contractile component. In addition, administration of exogenous NO induced concentration-dependent relaxations of the pig trigone. It was concluded that NO was the transmitter responsible for the relaxant response to electrical stimulation in the trigone (Persson and Andersson, 1992).

Bladder Neck and Urethra

Andersson et al. (1991, 1992) demonstrated that the NANC nerve-mediated relaxation found previously (Andersson et al., 1983) in the isolated female rabbit urethra could be inhibited concentration-dependently by L-NNA. D-NNA had no effect. Maximal relaxation was increased after pretreatment with L-arginine, and the inhibitory effect of L-NNA was counteracted. Administration of NO also induced concentration-dependent relaxations in preparations contracted by noradrenaline. These results were confirmed by Dokita et al. (1991), who also showed that a selective cGMP-phosphodiesterase inhibitor potentiated the relaxation and methylene blue reduced the relaxation. The involvement of the L-arginine/NO pathway in the relaxation of isolated urethral and bladder neck smooth muscle has now been demonstrated by several investigators in various species, including sheep (Garcia-Pascual et al., 1991; Thornbury et al., 1992; Triguero et al., 1993), rat (Persson et al., 1992), pig (Persson and Andersson, 1992; Bridgewater et al., 1993), dog (Hashimoto et al., 1993), and man (Andersson et al., 1992).

Nerve-induced relaxation of the rabbit urethra increases the smooth muscle content of cGMP but not cAMP (Persson and Andersson, 1994). L-NA prevented both the urethral relaxation and the increase in cGMP content. Furthermore, in the presence of the cGMP phosphodiesterase inhibitor zaprinast, the increase in cGMP levels was more pronounced (Persson and Andersson, 1994). Thus, it seems that cGMP has a role as a second messenger in the rabbit urethra and that this system is activated during NANC nerve-mediated relaxation.

To investigate whether the NO-mediated relaxation of rabbit urethral smooth muscle is associated with a hyperpolarization of the membrane potential, electrophysiological and mechanical experiments were performed (Waldeck et al., 1998). Intracellular recordings of the membrane potential did not reveal any changes in the membrane potential after application of NO. Thus, in contrast to what has been observed in several other tissues, the relaxant effect of NO in the rabbit urethra does not seem to be mediated by hyperpolarization. The role of the network of cGMP-IR cells that may constitute target cells for NO (Smet et al., 1996b; Waldeck et al., 1998) remains to be established.

Considering the colocalization of NOS with various peptides (Persson et al., 1995), an interaction between NO and nerve-released peptides contributing to relaxation may be expected. In the dog (Hashimoto et al., 1992) and pig (Bridgewater et al., 1993; Werkström et al., 1995), the relaxant response of urethral smooth muscle preparations to electrical stimulation of nerves has more than one component. Hashimoto et al. (1992) found that the relaxation was frequency-dependent and seemed to consist of a transient and a slow component, suggesting that at least two neurogenic factors were involved. The transient component of the relaxation could be inhibited by L-NNA, but not the slow component (Hashimoto et al., 1993).

In vivo studies using inhibitors of NO synthesis have suggested that NO has a role in outlet relaxation and that lack of NO can produce bladder overactivity (Andersson and Persson, 1993). However, NO synthesis inhibitors are not ideal tools for studying NO effects in vivo and may have nonspecific actions. Therefore, mice with targeted deletion of neuronal NOS (nNOS–) should be an interesting alternative approach to elucidate the role of NO in lower urinary tract function and dysfunction.

In male nNOS–/– mice, Burnett et al. (1997) found markedly dilated bladders with muscular hypertrophy, findings compatible with deficient urethral relaxation and increased outflow resistance. In support of the view that the bladder changes were caused by disturbances in outflow relaxation, the decrease in tension produced by low-frequency stimulation of nerves of isolated urethral preparations from wild-type controls was absent in preparations from nNOS– mice. In contrast to these findings, Sutherland et al. (1997), investigating female nNOS– mice with voiding, urodynamic, and muscle strip testing, as well as histologic examination, found no marked differences between these animals and normal controls. Detrusor strips from nNOS– mice demonstrated the same responses as those from controls (the relaxant capacity of urethral preparations was not investigated).

It was shown, however, that these mice, carrying a mutation in the nNOS gene, are still able to express an alternatively spliced mRNA of nNOS, which could be the source of NO in nNOS mutant mice (Eliasson et al., 1997). Since NO is produced, and the NO/cGMP pathway apparently is intact in these animals, they are not suitable for testing the importance of the NO/cGMP pathway for lower urinary tract function. The availability of mice lacking cGMP-dependent kinase type I (Pfeifer et al., 1998a) provides an opportunity to answer this question.

Nitric oxide activates soluble guanylate cyclase and increases the production of cGMP, but the mechanisms by which cGMP relaxes penile smooth muscles have not been definitely established. Cyclic GMP may have different targets within the cell: ion channels, phosphodiesterases, or protein kinases (Lincoln and Cornwell, 1993). Two different cGMP-dependent protein kinases (cGKI and cGKII) have been identified (Lohmann et al., 1997; Pfeifer et al., 1998a). cGKII is expressed in the small intestine, brain, and cartilage (Uhler, 1993; Jarchau et al., 1994; Pfeifer et al., 1996), whereas the highest concentrations of cGKI are found in smooth muscle, platelets, and Purkinje cells of the cerebellum (Waldmann et al., 1986; Keilbach et al., 1992). Inactivation of cGKI in mice abolished NO/cGMP-dependent relaxation of vascular and intestinal smooth muscle, causing hypertension and intestinal dysmotility (Pfeifer et al., 1998b). The lower urinary tract function of these mice was investigated (Persson et al., unpublished results). In contrast to the findings in wild-type controls, which exhibited an intense cGKI immunoreactivity, no cGKI immunoreactivity was found in urethral and bladder smooth muscle of cGKI–/– mice. However, the supply of NOS-IR nerves was similar in both genotypes. The bladder weight of cGKI–/– mice (male and female) was not different from the bladder weight of wild-type animals, which is in line with the findings of Sutherland et al. (1997) that female mice lacking nNOS did not show any signs of outflow obstruction. There were no differences between the genotypes in bladder contractions stimulated by K^+, carbachol, or electrical fields. Urodynamic investigations showed that the cGKI–/– mice, in contrast to the wild-types control, had bladder overactivity.

Urethral Lamina Propria

Isolated preparations of the rabbit urethral lamina propria, contracted by noradrenaline, produce frequency-dependent NANC relaxations in response to electrical field stimulation and to acetylcholine (Mattiasson et al., 1985; Zygmunt et al., 1993a,b). Both relaxations induced by electrical stimulation and those induced by acetylcholine could be abolished by L-NNA but not D-NNA. It was shown (Zygmunt et al., 1993a) that electrically induced relaxations could be evoked even in preparations in which acetylcholine-induced relaxation was poor or absent. The NANC neurotransmission was less sensitive to inhibition by ω-conotoxin than transmission mediated by adrenergic and cholinergic nerves (Zygmunt et al., 1993b) The NO donor SIN-1 relaxed lamina propria preparations contracted by noradrenaline in a concentration-dependent way, an effect not influenced by L-NNA (Zygmunt et al., 1993a).

Physiological Role of Nitric Oxide

An inhibitory, relaxation-mediating system not only may serve each of the components of the lower urinary tract—the detrusor, trigone, and bladder neck/urethra—but also may be of importance for their integrated function. Such a system may be involved in several important events.

Accommodation of the Bladder During Filling

The normal bladder responds to filling at a physiological rate, and can accommodate large volumes of urine, with a minimal increase in intravesical pressure (Coolsaet, 1985). There have been many suggestions regarding the underlying mechanism. The phenomenon has been attributed not only to the physical properties of the bladder (Nesbit et al., 1947; Tang and Ruch, 1955), but also to the existence of an inhibitory neural mechanism operative during filling and storage. Such a mechanism may include inhibition of parasympathetic nervous activity (Klevmark, 1977) or an increase in sympathetic nervous activity (Edvardsen, 1968; de Groat, 1975; Maggi et al., 1985). Since there is a predominance of β- over α-adrenoceptors in the normal detrusor (Andersson, 1993), such an increase in sympathetic activity would lead to detrusor relaxation.

Theoretically, an increased activity of NO-releasing inhibitory nerves could be one factor keeping the bladder relaxed during filling. In fact, NO has been suggested to have such a function in the stomach, i.e., as a mediator of adaptive relaxation to accommodate food or fluid (Desai et al., 1991). Inhibition of NOS by N^G-nitro-L-arginine methylester (L-NAME) had no effect on the length–tension relationship of rat detrusor muscle strips (Yanni and Williams, 1996), suggesting that distension of detrusor muscle does not release NO. However, length changes in vitro may not reflect the situation in vivo. Nevertheless, available results show that some of the NOS-containing nerves of the detrusor, as judged from their localization, are sensory afferents. Nitric oxide may be released from intramural nerves (or some other source, such as the urothelium) as a response to distension of the bladder, but it is not probable that NO produces direct relaxation of the detrusor smooth muscle, since the detrusor sensitivity to NO and agents acting via the cyclic GMP system seems to be low. This does not exclude that NO may modulate the effects of other transmitters.

Relaxation of the Outlet Region Associated with Micturition

It is now well documented that the normal pattern of voiding in humans is characterized by an initial drop in urethral pressure followed 5 to 15 seconds later by an increase in intravesical pressure (Tanagho and Miller, 1970; Asmussen and Ulmsten, 1976; Low, 1977; McGuire, 1978; Rud et al., 1978; Low et al., 1989). One explanation for the fall in intraurethral pressure is that increased parasympathetic activity, resulting in stimulation of muscarinic receptors on noradrenergic nerves, diminishes noradrenaline release and thereby tone in the proximal urethra (Mattiasson et al., 1987). Another possibility is that a NANC mechanism mediates this response (Andersson et al., 1983; Hills et al., 1984; Ito and Kimoto, 1985; Klarskov et al., 1983; Klarskov, 1987). Sacral ventral root stimulation was found to produce an atropine-sensitive urethral constriction when basal urethral resistance was low, but dilatation when resistance was high (Slack and Downie, 1983). The latter response was reduced, but not abolished, by atropine. When urethral constriction had been produced by phenylephrine, injection of acetylcholine produced a consistent decrease in urethral resistance, which was not reduced by atropine. It was suggested that parasympathetic dilatation of the urethra may be mediated by an unknown NANC transmitter released from postganglionic neurons. One possibility is that this transmitter is NO.

The Sealing Mechanism of the Urethral Lamina Propria

Nitric oxide released from nerves or urothelium (Birder et al., 1998) or from the endothelium of the rich suburothelial vasculature can cause relaxation of the urethral lamina propria. Whether such NO-mediated relaxation can influence the inner urethral softness (Zinner et al., 1980) and contribute to the maintenance of continence is speculative.

Nitric Oxide and Functional Disturbances of the Lower Urinary Tract

Excess NOS Activity

Functional disturbances in the outflow region include excessive urethral pressure variations ("unstable urethra"), which may be associated with certain voiding disturbances in women (Low, 1977; Ulmsten et al., 1977, 1982; Kulseng-Hanssen, 1983, 1987; Low et al., 1989). It has been suggested that these pressure variations are caused primarily by relaxation of the urethral smooth muscle (Kulseng-Hanssen, 1987), but it is unclear whether the mechanism underlying this effect is different from that occurring during the initiation of normal voiding.

To address the possibility that excess NO may cause symptoms related to the unstable urethra, NO donors were administered to rats undergoing cystometry. The NO donors decreased bladder capacity and micturition volume, and initiated spontaneous bladder contractions (Persson et al., 1992). Urethral instability is often combined with unstable bladder contractions, suggesting that bladder instability may be initiated from the bladder outlet region (Hindmarsh et al., 1983). The observation of spontaneous bladder contractions in the rat following administration of NO donors may be related to the existence of a urethrovesical reflex that triggers involuntary detrusor contraction when urine flows through the urethra. Stimulation of NOS by L-arginine perfusion to fetal lambs resulted in a persistently open sphincter (Mevorach et al., 1994), demonstrating that sphincter activity can be reduced by NO.

Reduced NOS Activity

If the activity of the NO system in the lower urinary tract for some reason is reduced, the ability of the bladder outlet to relax may be impaired and micturition disorders may develop (Burnett et al., 1997). Detrusor contractions against a "partly closed" urethra may result in incomplete emptying of the bladder and bladder instability due to lack of coordination between the bladder and its outlet. To investigate this hypothesis in vivo, NOS inhibitors were administered to rats undergoing cystometry. In the unanesthetized rat, inhibition of the L-arginine/NO pathway by L-NAME given intraarterially near the bladder led to bladder hyperactivity and decreased bladder capacity (Persson et al., 1992). Although it could not be assessed directly whether the effect of NOS inhibition was due to impaired urethral relaxation, information from in vitro studies may allow such an assumption. Theobald (1996), investigating the effect of NOS inhibition on bladder function in cats, found a decrease in the micturition volume threshold and a modest increase in micturition contraction, which is in line with previous observations.

Direct measurement of urethral pressure in rats has confirmed that NOS inhibition actually inhibits urethral relaxation in vivo (Bennett et al., 1995). Data obtained by Mevorach et al. (1994) in fetal lamb showed increased bladder capacity, increased residual volume, and increase in low-level bladder contractions following infusion of a NOS inhibitor. Inadequate sphincter relaxation was suggested to be the underlying cause of the observed disorders. The reason that bladder capacity decreased in the rat and increased in the fetal lamb following NOS inhibition is not clear. It may be speculated that NO is more important for detrusor function in younger individuals. In fact, in an ontogenetic study, nerve-induced NO-mediated relaxations of the fetal calf detrusor were identified (Lee et al., 1994). As mentioned above, such relaxations cannot always be demonstrated in adults. Evidence against an involvement of the NO system in normal bladder function in mini-pigs was reported by Turner (1994), using concentrations of the NOS inhibitor that apparently affected blood pressure.

The L-arginine/NO pathway in the bladder outflow region would provide a possible therapeutic target in clinical situations related to sphincter relaxation failure. In patients with bladder neck dyssynergia, the density of excitatory nerves was increased compared with control patients, whereas the density of NOS nerves appeared to be unaltered (Noble et al., 1993). Thus, in these patients increased excitatory rather than decreased inhibitory innervation seems to account for the inability of the bladder neck to relax. Whether bladder outlet obstruction causes an alteration in pattern and density of NOS-expressing nerves and subsequent functional alterations remains to be investigated.

In postmenopausal women, the lamina propria of the urethra undergoes atrophic changes that may be improved by estrogen treatment (Miodrag et al., 1988). Estrogen has been reported to increase NOS activity in various tissues (Weiner et al., 1994). However, the relation between lamina propria, estrogens, and NO awaits further exploration.

Afferent Activity and "Unstable" Bladder Contractions

Dysfunction of the afferent nervous system might be involved in lower urinary tract disorders, such as neurogenic inflammation, incontinence, and bladder instability. Activation of capsaicin-sensitive nerves produces relaxation of the rat external urethral sphincter through release of CGRP (Parlani et al., 1993a). Whether NO is released from sensory nerves and participates in the relaxation of the urethra is not known. Ho et al. (1998) demonstrated nitrergic nerve fibers in the striated urethral sphincter muscle preparations obtained from human males. The presence of NOS immunoreactivity in the sarcolemma was suggested to indicate a role of NO in the regulation of urethral striated muscle metabolism and contraction.

It has been suggested that unstable detrusor contractions may be initiated from the bladder outlet region, rather than from the detrusor itself (Hindmarsh et al., 1983). Morphological changes, such as those seen in outflow obstruction caused by benign prostatic hyperplasia, are associated with unstable bladder contractions in 50% to 70% of the cases (McGuire, 1984). The mechanisms behind this hyperactivity are not known, but increased afferent activity, which may be produced by lack of an inhibitory substance in either the detrusor or the outlet region, has been suggested as one of the mechanisms for initiation of these contractions. If the nitrergic nerves observed in the detrusor, and particularly within and beneath the urothelium, are afferent terminals, NO may be involved in the regulation of the threshold for bladder afferent firing. If so, inhibitors of NO synthesis could be expected to cause unstable bladder contractions. A rat model with continuous cystometry was used to study whether this was the case (Persson et al., 1991, 1992). Inhibition of the L-arginine/NO pathway led to bladder hyperactivity and decreased bladder capacity (Persson et al., 1991, 1992). Mice lacking cGKI, an important target for cGMP, also exhibited bladder overactivity (see above). In addition, evidence that baseline NOS activity prevents uninhibited bladder contractions exists from urodynamic studies performed in a fetal lamb model (Mevorach et al., 1994). The symptoms of detrusor instability were improved in humans by glyceryl trinitrate (James and Iacovou, 1993). Thus, bladder hyperactivity may involve a loss of NO-mediated regulation of the threshold for bladder afferent firing.

NOS Activity and Bladder Inflammation

Vizzard et al. (1996) found that the expression of nNOS increased in bladder afferent pathways following chronic bladder irritation (cyclophosphamide-induced cystitis). Urine fractions from patients with urinary tract infections showed increased NOS activity, which was calcium-independent and inhibited by L-canavanine, implying involvement of the inducible form of NOS (Smith et al., 1994). Nitric oxide was measured directly in the urinary bladder in patients with infectious cystitis, interstitial cystitis, irradiation cystitis, and cystitis induced by antitumor treatment (Lundberg et al., 1996). The investigators found that NO levels were 30–50 times higher in all varieties of cystitis than in controls, and they suggested that direct measurement of NO in the urinary bladder could be an attractive diagnostic method for detection of mucosal inflammation. NOS activity in bladder biopsy and urine samples was lower in patients with interstitial cystitis than in control patients (Smith et al., 1996). The underlying mechanism of these somewhat unexpected results was explained by the presence of mast cells with a presumed pathogenic role in interstitial cystitis, mediating inhibition of NOS activity. Wheeler et al. (1997) gave eight patients with interstitial cystitis L-arginine orally (1.5 g daily) for 6 months. There urinary levels of NOS-related enzymes and metabolites were increased, and there was a decrease in interstitial cystitis-related symptoms (Smith et al., 1997). These results were not confirmed by Ehrén et al. (1998), who treated nine women with interstitial cystitis with 3–10 g of L-arginine daily for 5 weeks. They concluded that L-arginine treatment in the doses used did not change NO production in the urinary bladder in patients with interstitial cystitis, and they did not notice any relief of the patients' symptoms. Controlled clinical trials have to be performed to assess the potential value of treatment of interstitial cystitis with L-arginine.

Nitric Oxide and Penile Erectile Tissues

Penile erection is produced by an increased blood flow to the corpora cavernosa that is made possible by opening of penile resistance vessels (helicine arteries), relaxation of the corpora cavernosa smooth muscle, and occlusion of the venous outflow (Andersson and Wagner, 1995). The erectile response in several animal models has been shown to depend upon relaxation of both the vasculature and the corpora cavernosa tissue by NO. In vitro, several investigators have shown that both acetylcholine- and neuronally mediated relaxation of animal and human corpora cavernosa involves release of NO or a NO-like substance (Andersson and Wagner, 1995; Burnett, 1997).

NOS in the Penis

Many studies using specific antibodies have demonstrated nNOS and eNOS, but not iNOS, in human and animal penile erectile tissues. However, rat corpora cavernosa tissue treated with lymphokines and/or bacterial lipopolysaccharide was shown to have the capacity to express iNOS (Hung et al., 1995). Garban et al. (1997) delivered a mix of iNOS inducers continuously for 3–6 days to old rats and found an improvement of their erectile dysfunction. The authors suggested that their results may open the way for the possible use of NOS isoforms in the management of erectile dysfunction. Rajasekeran et al. (1998), studying primary cultures of human corpora cavernosa cells, found that these cells were able to express both eNOS and iNOS, and they suggested that both forms may significantly contribute to NO production during the erectile process.

Alternative splicing of the nNOS gene is prominent, with more than 10 differently spliced nNOS transcripts reported (see Eliasson et al., 1997). Magee et al. (1996) cloned from rat penile corpora cavernosa a cDNA coding for a nNOS differing from the cerebellar NOS. The penile NOS (also expressed in urethra, prostate, and skeletal muscle) was suggested to be regulated differently from the cerebellar nNOS. It cannot be excluded that this form of NOS is responsible for the synthesis of NO during penile erection.

Sources of Nitric Oxide

Theoretically, not only the nerves innervating the corpora cavernosa, but also the sinusoidal endothelium and the corpora cavernosa smooth muscle itself, could be the source of the NO involved in erection.

Penile Nerves

There seems to be no doubt about the presence of NOS in the cavernous nerves and their terminal endings within the corpora cavernosa, in the branches of the dorsal penile nerves, and in the nerve plexuses in the adventitia of the deep cavernous and helicine arteries (Burnett et al., 1992, 1993, 1996; Keast, 1992; Alm et al., 1993; Ding et al., 1993; Dail et al., 1995; Hedlund et al., 1995a; Bloch et al., 1998). Using an antiserum to the constitutive NOS found in rat cerebellum, Burnett et al. (1992) demonstrated staining of the pelvic plexus and the axonal processes forming the cavernous nerve. The nerve plexus of the adventitia of the deep cavernosal arteries and the neuronal processes in the sinusoids and the periphery of the corpora cavernosa stained prominently. Dorsal penile and cavernosal arteries stained for NOS in both the adventitial and the endothelial layers, although endothelial staining was faint in the cavernosal vessels. Alm et al. (1993), using antiserum produced in rabbits against a C-terminal fragment of rat cerebellar NOS, confirmed these findings. By means of NADPH-d histochemistry, Ding et al. (1993) found that the thick-walled helicine arteries of the rat had a denser NOS innervation than the deep penile arteries.

Burnett et al. (1993) demonstrated nNOS also in the human penis. Neuronal NOS was identified in the cavernous nerves and their terminal endings within the corpora cavernosa, in the branches of the dorsal penile nerves, and in nerve plexuses in the adventitia of the deep cavernous arteries. In agreement with previous findings in the rat, Bloch et al. (1998) found that the small intracavernosal helicine arteries of the human penis had a particularly dense nNOS innervation.

By retrograde labeling, several investigators have found that the majority of penile nerves containing immunoreactivity for NOS are derived from the pelvic plexus (Ding et al., 1993, 1995; Domoto and Tsumori, 1994; Dail et al., 1995; Vanhatalo and Soinila, 1995; Vanhatalo et al., 1996). It was suggested (Dail et al., 1995) that most neurons derived from the pelvic plexus and destined for the penis comprise a uniform population. These neurons are capable of synthesizing NO and exhibit cholinergic characteristics. Hedlund et al. (1999) found by double immunolabeling that VAChT-IR and NOS-IR terminals (Figure 4.1), VAChT-IR and VIP-IR terminals, and NOS-IR and VIP-IR terminals generally showed coinciding profiles, whereas the profiles of VAChT-IR and TH-IR terminals and NOS-IR and TH-IR terminals ran closely together, although the profiles did not coincide. In general, there were fewer VAChT-IR nerves than AChE-IR nerves, which is not surprising, since AChE staining is not specific for cholinergic nerves (Lincoln and Burnstock, 1993). Since VAChT is specific for cholinergic neurons, the findings provide direct evidence that the NOS-containing penile neurons are cholinergic. However, the VAChT-containing nerves were fewer than those containing NOS, and it cannot be excluded that NOS may occur also in non-VAChT neurons, including a population of nerves containing only NOS (Elfvin et al., 1997). In a subpopulation of afferent nerves, McNeill et al. (1992a) demonstrated that CGRP and NOS (NADPH-d staining) occurred together, and suggested that these afferents may be an important source of NO for penile tumescence. Colocalization of NOS and VIP within nerves innervating the penis of both animals and humans has been demonstrated by many investigators (Ehmke et al., 1995; Hedlund et al., 1995a,b; Tamura et al., 1995, 1997; Vanhatalo et al., 1996; Dail et al., 1997; Schirar et al., 1997), but the functional importance of this finding has not been established (see below)

Jen et al. (1996) reported that NOS and TH were colocalized in nerves supplying the postnatal human penis, and Tamura et al. (1995) reported that also in the adult human penis, NOS could be found in nerves containing TH, suggesting that NO may be generated by adrenergic nerves. The physiological significance of such a localization of NOS is presently unclear, and the finding has to be confirmed.

Chemical sympathectomy abolished TH-IR nerve structures in the rat corpora cavernosa and reduced the number of NPY-IR nerves, whereas

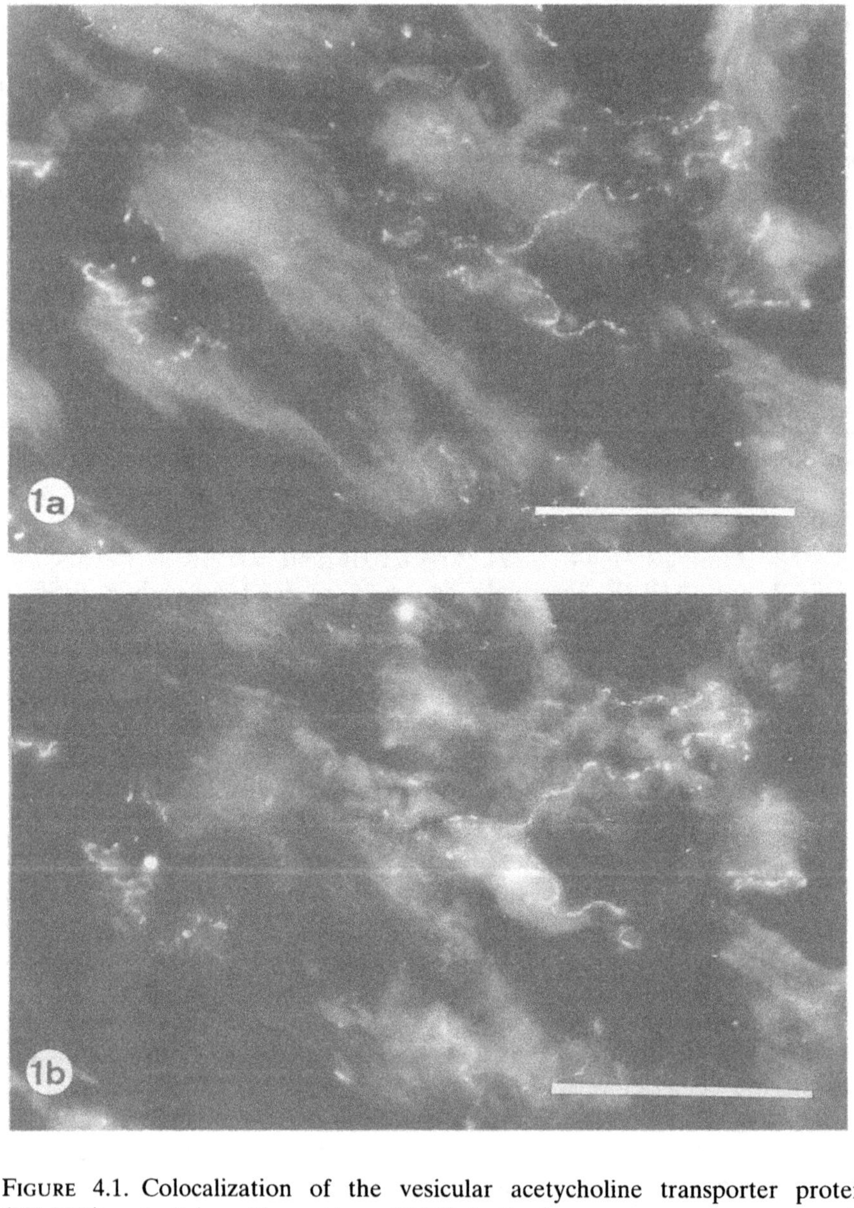

FIGURE 4.1. Colocalization of the vesicular acetycholine transporter protein (VAChT) and nitric oxide synthase (NOS) in the human corpus cavernosum. (a) VAChT-immunoreactive terminals along smooth muscle bundles. Texas Red-immunofluorescence. (b) the same section as in (a): NOS-IR terminals displaying profiles coinciding with VAChT-IR terminals. Fluorescein 5′-Iso Thio Cyanate (FITC)-immunofluorescence. Bar 100 μm.

the amount and distribution of nerves containing immunoreactivities for NOS, VAChT, and VIP were unaffected (Hedlund et al., 1999). Thus, NOS-IR, VAChT-IR, and VIP-IR nerve fibers may represent a population of parasympathetic cholinergic neurons innervating the erectile compartments and mediating erection by relaxing corpora cavernosa and vascular tissue. It seems as if NOS can also occur in other nerve populations, but the functional importance of these nerves has so far not been established.

Sinusoidal and Vascular Endothelium

The endothelium may provide a complementary source of NO during penile erection, and acetylcholine-induced relaxation of erectile tissue has been shown to involve the release of NO from endothelial cells (Andersson and Wagner, 1995). Although a positive NADPH-d reaction was described in some parts of the sinusoidal endothelium of the rat penis by Keast (1992), this has not been found by later investigators (Schirar et al., 1994; Dail et al., 1995). Dail et al. (1995) found that the endothelium of penile vessels stained for both eNOS and NADPH-d. However, the endothelium of the cavernous sinuses did not contain immunoreactivity for eNOS and did not stain for NADPH-d. Hedlund et al. (1999) found no immunoreactivity for eNOS in endothelial cells of the sinusoidal endothelium of the rat corpora cavernosa, a finding supporting the observations of Schirar et al. (1994) and Dail et al. (1995). This is in contrast to findings in both normal mice and mice with targeted deletion of nNOS (Burnett et al., 1996). In both types, eNOS was present in the endothelium of the penile vasculature and in sinusoidal endothelium within the corpora cavernosa. In human corpora cavernosa the endothelium of the cavernous sinuses stained only faintly for eNOS (Bloch et al., 1998). Since pronounced relaxant effects of carbachol or acetylcholine have been demonstrated in precontracted corpora cavernosa tissue from humans and several other mammals (Andersson and Wagner, 1995), but only minor relaxations can be produced in the rat (Hedlund et al., 1999), species differences concerning the role of endothelially derived NO seem to exist, which should be taken into consideration when studying penile erection in different models.

Stromal Smooth Muscle

Different isoforms of NOS may be expressed in penile stromal smooth muscle. Smooth muscle vascular and trabecular tissue from rat corpora cavernosa treated with lymphokines and/or bacterial lipopolysaccharide was shown to express iNOS (Hung et al., 1995). Bloch et al. (1998) found distinct NADPH-d staining of human penile fibromuscular stroma. The expression of NADPH-d staining in the smooth muscle cells was confirmed by positive immunohistochemistry with eNOS but not nNOS antibodies.

Bloch et al. (1998) suggested that eNOS may be a main source of NO along with nNOS.

Regulation of NOS Activity

Kim et al. (1993), investigating isolated corpora cavernosa tissue from humans and rabbits, found that relaxant responses to acetylcholine and electrical stimulation of nerves were progressively inhibited as a function of decreasing oxygen tension. They concluded that physiological concentrations of oxygen modulate penile erection by modulating NO synthesis in corpora cavernosa tissue.

Results from several investigators support the view that a primary action of androgens is to stimulate the synthesis of NO (Mills et al., 1992, 1994; Chamness et al., 1995; Zvara et al., 1995; Lugg et al., 1996; Reilly et al., 1997). Chamness et al. (1995) demonstrated a 45% decrease in penile NOS activity after castration and a 57% decrease in NOS protein. The decrease in penile NOS activity was confirmed by Lugg et al. (1996), who also found that the enzyme inhibition was reversed by cavernosal nerve stimulation. However, they showed that androgen depletion did not reduce the NOS content.

In a preliminary study, Zorgniotti and Lizza (1994) found that oral L-arginine, but not placebo, produced an improvement of erectile function in men with erectile dysfunction. In aging rats, Moody et al. (1997) found that long-term administration of high doses of L-arginine improved the erectile response. It was speculated that L-arginine in the penis may be a substrate-limiting factor for NOS activity and that L-arginine may up-regulate penile NOS activity, but not its expression. On the other hand, Reilly et al. (1997) found that L-arginine administration did not augment the erectile response in rats after castration.

Nitric Oxide-Mediated Penile Relaxation

An important role for NO in the relaxation of corpus cavernosum smooth muscle and vasculature is widely accepted (Andersson and Wagner, 1995; Burnett, 1997). However, the precise mechanisms of action have not been clarified and may vary with species. Thus, Cellek and Moncada (1997) suggested that the human and rabbit genitourinary organs have a powerful nitrergic innervation that does not merely modulate, but actually controls, the sympathetic response. Further experimental evidence supporting this has been presented by several investigators (Carrier et al., 1997; Dahiya et al., 1997). Fewer NO-containing neurons and less expression of NOS mRNA were found in the corpora cavernosa of old than of young rats, whereas the expression of α_1-adrenoceptor mRNA was unchanged. Such an alteration in the balance between factors contributing to corporeal relaxation (e.g., NO) and contraction (e.g., α_1-adrenoceptors), favoring contraction, may result in erectile dysfunction.

Cavernous Smooth Muscle

In vitro, several investigators have shown that both acetylcholine- and neuronally mediated relaxation in animal and human corpus cavernosum involves release of NO or a NO-like substance (Andersson and Wagner, 1995). Both the endothelium and the nerves innervating the corpus cavernosum are likely sources of NO, and thus more than one isoform of NOS may be involved. Several investigators have shown that isolated human cavernous tissue responds with relaxation to electrical stimulation of nerves after destruction of the endothelium, while responses to acetylcholine, bradykinin, and substance P (SP) are abolished (Kimoto et al., 1990). In anesthetized dogs, intracavernous injection of 3-[(3-cholamidopropyl)-dimethylammonio]-1-propanesulfonate (CHAPS) to destroy the sinusoidal endothelium, abolished the erectile response to acetylcholine, but only partially inhibited the response to electrostimulation (Trigo-Rocha et al., 1993) The NOS demonstrated in rabbit corpus cavernosum was shown to be a cytosolic, constitutive isoform of NOS (Bush et al., 1992), whereas the endothelium-derived NOS is known to be primarily membrane-bound (Förstermann et al., 1991). If this is the case also in humans, it would suggest that the most important source of NO in penile tissue is neuronal. On the other hand, in nNOS-deleted mice (see below), there was a consistent 20% increase in eNOS compared with findings in wild-type animals (Burnett et al., 1996). This raises the questions whether eNOS can preserve genital smooth muscle function or whether these results in nNOS-deleted mice suggest that NO is not the sole relaxation-producing transmitter. Is the NO–cGMP–cGKI pathway necessary for penile erection?

In rats, intracavernous injection of different drugs known to act at different levels of the cAMP and cGMP pathways revealed that neither cAMP nor drugs that stimulate adenylate cyclase activity induced any change in intracavernosal pressure (Martinez-Pineiro et al., 1993). Intracavernosal injection of cGMP and nitroprusside, on the other hand, caused dose-dependent changes in intracavernous pressure that could be inhibited by methylene blue, and it was concluded that cavernous smooth muscle relaxation in the rat is mediated by activation of guanylate cyclase, and that the cAMP system apparently has no important role (Martinez-Pineiro et al., 1993). In monkeys, on the other hand, both guanylate cyclase and adenylate cyclase may be involved in cavernous smooth muscle relaxation, with cyclic GMP being the predominant intracellular second messenger (Trigo-Rocha et al., 1994). This may also be the case in humans.

Mice lacking nNOS (Huang et al., 1993) have erections, show normal mating behavior, and respond with erection to electrical stimulation of the cavernous nerves (Burnett et al., 1996). Surprisingly, isolated corporeal tissue from both wild-type and nNOS-deleted animals showed similar responses to electrical stimulation (Burnett et al., 1996). It was suggested that eNOS is essential for erection not only in nNOS-deleted but also in normal mice. The importance of eNOS was stressed by Bloch et al. (1998). They demonstrated that eNOS is expressed in cavernosal smooth muscle and suggested that eNOS could be a main source of NO along with nNOS. This may be the case. However, it was shown that mice carrying a mutation in the nNOS gene are still able to express an alternatively spliced mRNA of nNOS, which could be the source of NO in nNOS mutant mice (Eliasson et al., 1997). Since NO is produced and the guanylyl–cGMP–cGKI pathway apparently is intact in the corpora cavernosa of these animals, they are not suitable for testing whether or not the NO–cGMP pathway is a prerequisite for normal penile erection.

cGKI-deficient (cGKI–/–) mice (see above) have a very low ability to reproduce. Corpus cavernosum tissue from these mice has an inability or markedly reduced ability to relax in response to neuronally or endothelially released or exogenously administered NO (Hedlund et al., unpublished results). The expression of cGKI in penile tissue fom cGKI+/+ mice, as revealed by immunohistochemistry, was confined to the smooth muscle of the walls of the central and helicine arteries, and to the smooth muscle of the trabecular septa surrounding the cavernous spaces. This is in line with its presumed role in the erectile events. The total innervation (Protein game product 9.5 PGP immunoreactivity) and distribution of nerve populations containing transmitters or transmitter-forming enzymes (VAChT, VIP, TH, and NOS) believed to be important in the regulation of tone in corpus cavernosum tissue (Andersson and Wagner, 1995) were similar in normal and cGKI-null mice. Analysis of the NO/cGMP-induced relaxation clearly showed that cGKI is the major mediator of the cGMP signaling cascade in corpus cavernosum tissue. Its absence cannot be compensated by the cAMP signaling cascade that relaxes normal and cGKI-null penile erectile tissue to a similar extent. Taken together, these findings suggest that activation of cGKI is a key step in the signal cascade leading to penile erection.

Cyclic AMP and cGMP are synthetized from the corresponding nucleoside triphosphates by their respective membrane-bound or soluble adenylate or guanylate cyclases. Cyclic AMP and cGMP are inactivated by phosphodiesterases (PDEs) by hydrolytic cleavage of the 3′-ribose-phosphate bond. Presently, more than seven different families of PDE are known, which differ in their specificity for cAMP and cGMP, cofactor requirements, and kinetic properties (Beavo, 1995; Polson and Strada, 1996). Each family can again be divided into subfamilies. Because of their central role in the regulation of smooth muscle tone and the considerable variation of PDE isoenzymes with respect to species and tissues, PDEs have become an attractive target for drug development. PDE-2, PDE-3, PDE-4, and PDE-5 isoenzymes have been identified in human cavernous tissue (Stief et al., 1995; Ballard et al., 1996, 1998). Functionally, PDE-3 and PDE-5 seem to be the most important (Stief et al., 1998; Ballard et al., 1998).

The identification of PDE families has been paralleled by the synthesis of selective or partially selective inhibitors. Sildenafil is a highly selective inhibitor of PDE-5 (Boolell et al., 1996a,b). It enhances NO-mediated relaxation of rabbit and human corpus cavernosum in vitro (Ballard et al., 1996, 1998; Tang et al., 1996; Chuang et al., 1998; Stief et al., 1998) and dose-dependently increases the intracavernous pressure in anesthetized dogs (Carter et al., 1998). Sildenafil increases the intracellular concentrations of cGMP (Chuang et al., 1998; Jeremy et al., 1997). The effect of sildenafil may thus be due to an amplification of the endogenous NO–cGMP pathway. This seems to involve a novel cellular signal transduction pathway in which force is dissociated from myosin light chain phosphorylation (Chuang et al., 1998).

As mentioned previously, many penile nerves contain NO, VIP, and acetylcholine, and the possible interactions among these agents should be of particular interest. The effects of NO and the NO donor linsidomine (SIN-1) on human isolated cavernous artery and corpus cavernosum were studied by Hempelmann et al. (1995). They found nonsynergistic, independent relaxant effects in both types of preparation. Suh et al. (1995) investigated the effect of VIP and VIP combined with acetylcholine given intracavernously in rats. The found that VIP and acetylcholine, individually or in combination, did not produce full erection and concluded that neither VIP nor acetylcholine was likely to be a principal transmitter.

Penile Vascular Smooth Muscle

Dilatation of the cavernosal and helicine arteries with resultant increase in penile blood flow is a primary hemodynamic event in erection. However, information is scarce concerning NO-dependent mechanisms in the penile small (helicine) arteries, which receive a dense NOS innervation and also exhibit a distinct eNOS immunoreactivity. Simonsen et al. (1997) investigated in vitro small penile arteries from patients undergoing penile surgery or from organ donors. They found that in phenylephrine-contracted preparations, electrical field stimulation induced frequency-dependent relaxations that were independent of the endothelium. Inhibition of NOS synthase by L-NNA abolished the responses to low-frequency stimulation, but tetrodotoxin-sensitive relaxations could still be elicited at frequencies of 16 Hz and above. Acetylcholine-induced relaxations were only partially reduced in the presence of L-NNA. The authors concluded that the relaxations induced by nerve stimulation or by release of mediator from the endothelium partially involved NO and also another transmitter. Non-NO mediators released from penile nerves and from the endothelium have also been described in horse penile arteries (Simonsen et al., 1995; Prieto et al., 1998). Release of NO from the nerves and endothelium of the human penile deep dorsal vein has been reported, and it was suggested that this vessel may be an active component in the regulation of penile vascular resistance (Segarra et al., 1998).

Nitric Oxide and Erectile Dysfunction

Aging is an important risk factor for erectile dysfunction, and it has been estimated that 55% of men are impotent at the age of 75 (Kaiser, 1991; Melman and Gingell, 1999). Garban et al. (1995) found that the soluble NOS activity decreased significantly in penile tissue from senescent rats. The expression of NOS mRNA is lower in old than in young rats (Dahiya et al., 1997). In another rat model of aging, the number of NOS-containing nerve fibers in the penis decreased significantly, and the erectile response to both central and peripheral stimulation decreased (Carrier et al., 1997). As mentioned previously, long-term administration of high doses of L-arginine to aging rats improved the erectile responses (Moody et al., 1997). In the aging rabbit, endothelium-dependent relaxation of the corpora cavernosum was attenuated; however, eNOS was up-regulated in both vascular endothelium and corporal smooth muscle (Haas et al., 1998).

Diabetes mellitus is often associated with erectile dysfunction (Saenz de Tejada and Goldstein, 1988; Melman and Gingell, 1999) and with impaired NOS-dependent erectile mechanisms. In isolated corpus cavernosum from diabetic patients with impotence, both neurogenic and endothelium-dependent relaxation was impaired (Sanz de Tejada et al., 1989). This was also found in rabbits with diabetes induced by alloxan (Azadzoi and Saenz de Tejada, 1992). Penile NOS activity and penile NOS content were reduced in rat models of both type I and type II diabetes with erectile dysfunction (Vernet et al., 1995). However, in streptozotocin-induced diabetic rats, NOS binding increased (Sullivan et al., 1996), and NOS activity in penile tissue was significantly higher than in controls, despite a significant degradation of mating behavior and indications of defective erectile potency (Elabbady et al., 1995). In humans diabetic erectile dysfunction was suggested to be related to the effects of advanced glycation end products on NO formation (Seftel et al., 1997).

Atherosclerosis and hypercholesterolemia are significant risk factors involved in the development of vasculogenic erectile dysfunction. Hypercholesterolemia was also found to impair endothelium-mediated relaxation of rabbit corpus cavernosum smooth muscle (Azadzoi and Saenz de Tejada, 1991; Azadzoi et al., 1998). Hypercholesterolemia did not affect NOS activity but impaired the endothelium-dependent, but not the neurogenic, relaxation of rabbit corpus cavernosum tissue. Since the endothelium-dependent relaxation was improved after treatment with L-arginine, it was speculated that NO formation was deficient due to a lack of availability of L-arginine in the hypercholesterolemic animals.

In a rabbit model of atherosclerotic erectile dysfunction (Azadzoi and Goldstein, 1992; Azadzoi et al., 1996), it was shown that chronic cavernosal ischemia impaired not only endothelium-dependent but also neurogenic corpus cavernosum relaxation and NOS activity (Azadzoi et al., 1998). There was also an increased output of constrictor eicosanoids in the corpora cavernosa. L-Arginine administration failed to improve corpus cavernosum relaxation, which was suggested to be due to impairment of NOS activity and reduction of NO formation.

Smoking is a major risk factor in the development of impotence (Mannino et al., 1994). In rats, passive chronic smoking caused age-independent moderate systemic hypertension and marked decreases in penile NOS activity and nNOS content (Xie et al., 1997). This was not reflected in a reduction of the erectile response to electrical nerve stimulation or by a decrease in penile eNOS.

Acknowledgments. This work was supported by the Swedish Medical Research Council (grant no 6837) and by the Medical Faculty, University of Lund, Sweden.

References

Alm, P., Larsson, B., Ekblad, E., Sundler, F., and Andersson, K.-E. 1993. Immunohistochemical localization of peripheral nitric oxide synthase containing nerves using antibodies raised against synthetized C- and N-terminal fragments of a cloned enzyme from rat brain. *Acta Physiol. Scand.* 148:421–429.

Alm, P., Zygmunt, P.K.E., Iselin, C., Larsson, B., Uvelius, B., Werner, S., and Andersson, K.-E. 1995. Nitric oxide synthase-immunoreactive, adrenergic, cholinergic, and peptidergic nerves of the female rat urinary tract: a comparative study. *J. Auton. Nerv. Syst.* 56:105–114.

Andersson, K.-E. 1993. The pharmacology of lower urinary tract smooth muscles and penile erectile tissues. *Pharmacol. Rev.* 45:253–308.

Andersson, K.-E., Garcia-Pascual, A., Forman, A., and Tøttrup, A. 1991. Non-adrenergic, non-cholinergic nerve-mediated relaxation of rabbit urethra is caused by nitric oxide. *Acta Physiol. Scand.* 141:133–134.

Andersson, K.-E., Garcia-Pascual, A., Persson, K., Forman, A., and Tøttrup, A. 1992. Electrically-induced, nerve-mediated relaxation of rabbit urethra involves nitric oxide. *J. Urol.* 147:253–259.

Andersson, K.-E., Mattiasson, A., and Sjögren, C. 1983. Electrically induced relaxation of the noradrenaline contracted isolated urethra from rabbit and man. *J. Urol.* 129:210–213.

Andersson, K.-E., and Persson, K. 1993. The L-arginine/nitric oxide pathway and non-adrenergic, non-cholinergic relaxation of the lower urinary tract. *Gen. Pharmacol.* 24:833–839.

Andersson, K.-E., and Wagner, G. 1995. Physiology of penile erection. *Physiol. Rev.* 75:191–236.

Arvidsson, U., Reidl, M., Elde, R., and Meister, B. 1997. Vesicular acetylcholine transporter (VAChT) protein: a novel and unique marker for cholinergic neurons in the central and peripheral nervous systems. *J. Comp. Neurol.* 378:454–467.

Asmussen, M., and Ulmsten, U. 1976. Simultaneous urethro-cystometry with a new technique. *Scand. J. Urol. Nephrol.* 10:7–11.

Azadzoi, K.M., and Goldstein, I. 1992. Erectile dysfunction due to atherosclerotic vascular disease: the development of an animal model. *J. Urol.* 147:1675–1681.

Azadzoi, K.M., Goldstein, I., Siroky, M.B., Traish, A.B., Krane, R.J., and Saenz de-Tejada, I. 1998. Mechanisms of ischemia-induced cavernosal smooth muscle relaxation impairment in a rabbit model of vasculogenic erectile dysfunction. *J. Urol.* 160:2216–2222.

Azadzoi, K.M., and Saenz de Tejada, I. 1991. Hypercholesterolemia impairs endothelium-dependent relaxation of rabbit corpus cavernosum smooth muscle. *J. Urol.* 146:238–240.

Azadzoi, K.M., and Saenz de Tejada, I. 1992. Diabetes mellitus imparis neurogenic and endothelium-dependent relaxation of rabbit corpus cavernosum smooth muscle. *J. Urol.* 148:1587–1591.

Azadzoi, K.M., Siroky, M.B., and Goldstein, I. 1996. Study of etiologic relationship of arterial atherosclerosis to corporal veno-occlusive dysfunction in the rabbit. *J. Urol.* 155:1795–800.

Ballard, S.A., Gingell, C.J., Tang, K., Turner, L.A., Price, M.E., and Naylor, A.M. 1998. Effects of sildenafil on the relaxation of human corpus cavernosum tissue in vitro and on the activities of cyclic nucleotide phosphodiesterase isozymes. *J. Urol.* 159:2164–2171.

Ballard, S.A., Turner, L.A., and Naylor, A.M. 1996. Sildenafil, a potent selective inhibitor of type 5 phosphodiesterase, enhances nitric oxide-dependent relaxation of rabbit corpus cavernosum (abstr). *Br. J. Pharmacol.* 118:153P.

Beavo, J.A. 1995. Cyclic nucleotide phosphodiesterases: functional implications of mutiple isoforms. *Physiol. Rev.* 75:725–748.

Bennett, B.C., Kruse, M.N., Roppolo, J.R., Flood, H.D., Fraser, M., and de Groat, W.C. 1995. Neural control of urethral outlet activity in vivo: role of nitric oxide. *J. Urol.* 153:2004–2009.

Birder, L.A., Apodaca, G., de Groat, W.C., and Kanai, A.J. 1998. Adrenergic- and capsaicin-evoked nitric oxide release from urothelium and afferent nerves in urinary bladder. *Am. J. Physiol.* 275:F226–F229.

Birder, L.A., Kanai, A.J., and de Groat, W.C. 1997. DMSO: effect on bladder afferent neurons and nitric oxide release. *J. Urol.* 158:1989–1995.

Bloch, W., Klotz, T., Sedlaczek, P., Zumbé, J., Engelmann, U., and Addicks, K. 1998. Evidence for the involvement of endothelial nitric oxide synthase from smooth muscle cells in the erectile function of the human corpus cavernosum. *Urol. Res.* 26:129–135.

Boolell, M., Allen, M.J., Ballard, S.A., Gepi-Attee, S., Muirhead, G.J., Naylor, A.M., Osterloh, I.H., and Gingell, C. 1996a. Sildenafil: an orally active type 5 cyclic GMP-specific phosphodiesterase inhibitor for the treatment of penile erectile dysfunction. *Int. J. Impotence Res.* 8:47–52.

Boolell, M., Gepi-Attee, S., Gingell, J.C., and Allen, M.J. 1996b. Sildenafil, a novel oral therapy for male erectile dysfunction. *Br. J. Urol.* 78:257–261.

Bredt, D.S., Hwang, P.M., Glatt, C.E., Lowenstein, C., Reed, R.R., and Snyder, S.H. 1991. Cloned and expressed nitric oxide synthase structurally resembles cytochrome P-450 reductase. *Nature* 351:714–718.

Bredt, D.S., Hwang, P.M., and Snyder, S.H. 1990. Localization of nitric oxide synthase indicating a neural role for nitric oxide. *Nature* 347:768–770.

Bridgewater, M., MacNeil, H.F., and Brading, A.F. 1993. Regulation of tone in pig urethral smooth muscle. *J. Urol.* 150:223–228.

Burnett, A.L. 1997. Nitric oxide in the penis: physiology and pathology. *J. Urol.* 157:320–324.

Burnett, A.L., Calvin, D.C., Chamness, S.L., Liu, J.-X., Nelson, R.J., Klein, S.L., Dawson, V.L., Dawson, T.M., and Snyder, S.H. 1997. Urinary bladder-urethral sphincter dysfunction in mice with targeted disruption of neuronal nitric oxide synthase models idiopathic voiding disorders in humans. *Nature Med.* 3:571–574.

Burnett, A.L., Lowenstein, C.J., Bredt, D.S., Chang, T.S.K., and Snyder, S.H. 1992. Nitric oxide: a physiologic mediator of penile erection. *Science* 257:401–403.

Burnett, A.L., Nelson, R.J., Calvin, D.C., Liu, J-X., Demas, G.E., Klein, S.L., Kriegsfeld, L.J., Dawson, V.L., Dawson, T.M., and Snyder, S.H. 1996. Nitric oxide-dependent penile erection in mice lacking neuronal nitric oxide synthase. *Mol. Med.* 2:288–296.

Burnett, A.L., Tillman, S.L., Chang, T.S., Epstein, J.I., Lowenstein, C.J., Bredt, D.S., Snyder, S.H., and Walsh, P.C. 1993. Immunohistochemical localization of nitric oxide synthase in the autonomic innervation of the human penis. *J. Urol.* 150:73–76.

Bush, P.A., Gonzalez, N.E., and Ignarro, L.J. 1992. Biosynthesis of nitric oxide and citrulline from L-arginine by constitutive nitric oxide synthase present in rabbit corpus cavernosum. *Biochem. Biophys. Res. Commun.* 186:308–314.

Carrier, S., Nagaraju, P., Morgan, D.M., Baba, K., Nunes, L., and Lue, T. 1997. Age decreases nitric oxide synthase-containing nerve fibers in the rat penis. *J. Urol.* 157:1088–1092.

Carter, A.J., Ballard, S.A., and Naylor, A.M. 1998. Effect of the selective phosphodiesterase type 5 inhibitor sildenafil on erectile function in the anesthetized dog. *J. Urol.* 160:242–246.

Cellek, S., and Moncada, S. 1997. Nitrergic control of peripheral sympathetic responses in the human corpus cavernosum. *Proc. Natl. Acad. Sci. USA* 94:8226–8231.

Chamness, S.L., Ricker, D.D., Crone, J.K., Dembeck, C.L., Maguire, M.P., Burnett, A.L., and Chang, T.S.K. 1995. The effect of androgen on nitric oxide synthase in the male reproductive tract of the rat. *Fertil. Steril.* 63:1101–1107.

Chuang, A.T., Strauss, J.D., Murphy, R.A., and Steers, W.D. 1998. Sildenafil, a type-5 cGMP phosphodiesterase inhibitor, specifically amplifies endogenous cGMP-dependent relaxation in rabbit corpus cavernosum smooth muscle in vitro. *J. Urol.*160:257–261.

Chung, B.H., Choi, S.K., and Chang, K.C. 1996. Effects of nitric oxide on detrusor relaxation. *J. Urol.* 155:2090–2093.

Coolsaet, B. 1985. Bladder compliance and detrusor activity during the collection phase. *Neurourol. Urodyn.* 4:263–273.

Dahiya, R., Lin, A., Bakicioglu, M.E., Huang, S.T., and Lue, T.F. 1997. mRNA and protein expression of nitric oxide synthase and adrenoceptor alpha 1 in young and old rat penile tissues. *Br. J. Urol.* 80:300–306.

Dail, W.G., Barba, V., Leyba, L., and Galindo, R. 1995. Neural and endothelial nitric oxide synthase activity in rat penile erectile tissue. *Cell Tissue Res.* 282:109–116.

Dail, W.G., Galindo, R., Leyba, L., and Barba, V. 1997. Denervation-induced changes in perineuronal plexuses in the major pelvic ganglion of the rat: immunohistochemistry for vasoactive intestinal polypeptide and tyrosine hydroxylase and histochemistry for NADPH-diaphorase. *Cell Tissue Res.* 287:315–324.

Dawson, T.M., Bredt, D.S., Fotuhi, M., Hwang, P.M., and Snyder, S.H. 1991. Nitric oxide synthase and neuronal NADPH diaphorase are identical in brain and peripheral tissues. *Proc. Natl. Acad. Sci. USA* 88:7797–7801.

de Groat, W.C. 1975. Nervous control of the urinary bladder of the cat. *Brain Res.* 87:201–211.

Desai, K.M., Sessa, W.C., and Vane, J.R. 1991. Involvement of nitric oxide in the reflex relaxation of the stomach to accommodate food or fluid. *Nature* 351:477–479.

Ding, Y.Q., Takada, M., Kaneko, T., and Mizuno, N . 1995. Colocalization of vasoactive intestinal polypeptide and nitric oxide in penis-innervating neurons in the major pelvic ganglion of the rat. *Neurosci. Res.* 22:129–131.

Ding, Y., Wang, Y., Qin, B., and Li, J. 1993. The major pelvic ganglion is the main source of nitric oxide synthase-containing nerve fibers in penile erectile tissue of the rat. *Neurosci. Lett.* 164:187–189.

Dokita, S., Morgan, W.R., Wheeler, M.A., Yoshida, M., Latifpour, J., and Weiss, R.M. 1991. N^G-Nitro-L-arginine inhibits non-adrenergic, non-cholinergic relaxation in rabbit urethral smooth muscle. *Life Sci.* 4:2429–2436.

Dokita, S., Smith, S.D., Nishimoto, T., Wheeler, M.A., and Weiss, R.M. 1994. Involvement of nitric oxide and cyclic GMP in rabbit urethral relaxation. *Eur. J. Pharmacol.* 269:269–275.

Domoto, T., and Tsumori, T. 1994. Co-localization of nitric oxide synthase and vasoactive intestinal peptide immunoreactivity in neurons of the major pelvic ganglion projecting to the rat rectum and penis. *Cell Tissue Res.* 278:273–278.

Dun, N.J., Dun, S.L., Wu, S.Y., Förstermann, U., Schmidt, H.H.H.W., and Tseng, L.F. 1993. Nitric oxide synthase immunoreactivity in the rat, mouse, cat and squirrel monkey spinal cord. *Neuroscience* 54:845–857.

Edvardsen, P. 1968. Nervous control of urinary bladder in cats. III. Effects of autonomic blocking agents in the intact animal. *Acta Physiol. Scand.* 72: 183–193.

Ehmke, H., Junemann, K.P., Mayer, B., and Kummer, W. 1995. Nitric oxide synthase and vasoactive intestinal polypeptide colocalization in neurons innervating the human penile circulation. *Int. J. Impotence Res.* 7:147–156.

Ehrén, I., Iversen, H., Jansson, O., Adolfsson, J., and Wiklund, N.P. 1994. Localization of nitric oxide synthase activity in the human lower urinary tract and its correlation with neuroeffector responses. *Urology* 44:683–687.

Ehrén, I., Lundberg, J.O., Adolfsson, J., and Wiklund, N.P. 1998. Effects of L-arginine treatment on symptoms and bladder nitric oxide levels in patients with interstitial cystitis. *Urology* 52:1026–1029.

Elabbady, A.A., Gagnon, C., Hassouna, M.M., Bégin, L.R., and Elhilali, M.M. 1995. Diabetes mellitus increases nitric oxide synthase in penises but not in major pelvic ganglia of the rat. *Br. J. Urol.* 76:196–202.

Elfvin, L.G., Holmberg, K., Emson, P., Schemann, M., and Hokfelt, T. 1997. Nitric oxide synthase, choline acetyltransferase, catecholamine enzymes and neuropeptides and their colocalization in the anterior pelvic ganglion, the inferior mesenteric ganglion and the hypogastric nerve of the male guinea pig. *J. Chem. Neuroanat.* 14:33–49.

Eliasson, M.J.L., Blackshaw, S., Schell, M.J., and Snyder, S.H. 1997. Neuronal nitric oxide synthase alternatively spliced forms: prominent functional localizations in the brain. *Proc. Natl. Acad. Sci. USA* 94:3396–3340.

Elliott, R.A., and Castleden, C.M. 1993. Nerve mediated relaxation in human detrusor muscle *Br. J. Clin. Pharmacol.* 36:479.

Förstermann, U., Schmidt, H.H.H.W., Pollock, J.S., Sheng, H., Mitchell, J.A., Warner, T.D., Nakane, M., and Murad, F. 1991. Isoforms of nitric oxide synthase. Characterization and purification from different cell types. *Biochem. Pharmacol.* 42:1849–1857.

Garban, H., Marquez, D., Magee, T., Moody, J., Rajavashisth, T., Rodriguez, J.A., Hung, A., Vernet, D., Rajfer, J., and Gonzalez-Cadavid, N.F. 1997. Cloning of rat and human inducible penile nitric oxide synthase. Application for gene therapy of erectile dysfunction. *Biol. Reprod.* 56:954–963.

Garban, H., Vernet, D., Freedman, A., Rajfer, J., and Gonzalez-Cavidad, N. 1995. Effect of aging on nitric oxide-mediated penile erection in rats. *Am J. Physiol.* 268:H467–H475.

Garcia-Pascual, A., Costa, G., Garcia-Sacristan, A., and Andersson, K.-E. 1991. Relaxation of sheep urethral smooth muscle induced by electrical stimulation of nerves: involvement of nitric oxide. *Acta Physiol. Scand.* 141:531–539.

Griffith, O.W., and Stuehr, D.J. 1995. Nitric oxide synthases: properties and catalytic mechanism. *Annu. Rev. Physiol.* 57:707–736.

Grozdanovic, Z., Baumgarten, H.G., and Brüning, G. 1992. Histochemistry of NADPH-diaphorase, a marker for neuronal nitric oxide synthase, in the peripheral autonomic nervous system of the mouse. *Neuroscience* 48:225–235.

Haas, C.A., Seftel, A.D., Razmjouei, K., Ganz, M.B., Hampel, N., and Ferguson, K. 1998. Erectile dysfunction in aging: upregulation of endothelial nitric oxide synthase. *Urology* 51:516–522.

Hashimoto, S., Kigoshi, S., and Muramatsu, I. 1992. Neurogenic responses of urethra isolated from the dog. *Eur. J. Pharmacol.* 213:117–123.

Hashimoto, S., Kigoshi, S., and Muramatsu, I. 1993. Nitric oxide-dependent and -independent neurogenic relaxations of isolated dog urethra. *Eur. J. Pharmacol.* 231:209–214.

Hedlund, P., Alm, P., and Andersson, K.-E. 1995a. Distribution and function of nitric oxide-containing nerves in canine corpus cavernosum and spongiosum. *Acta Physiol. Scand.* 155:445–455.

Hedlund, P., Alm, P., and Andersson, K.-E. 1999. NO synthase in cholinergic nerves and NO-induced relaxation in the rat isolated corpus cavernosum. *Br. J. Pharmacol.* 127:349–360.

Hedlund, P., Alm, P., Ekström, P., Fahrenkrug, J., Hannibal, J., Hedlund, H., Larsson, B., and Andersson, K.-E. 1995b. Pituitary adenylate cyclase-activating polypeptide, helospectin, and vasoactive intestinal polypeptide in human corpus cavernosum. *Br. J. Pharmacol.* 116:2258–2266.

Hempelmann, R.G., Papadopoulos, I., and Herzig, S. 1995. Non-synergistic relaxant effects of vasoactive intestinal polypeptide and SIN-1 in human isolated cavernous artery and corpus cavernosum. *Eur. J. Pharmacol.* 276:277–280.

Hills, J., Meldrum, L., Klarskov, P., and Burnstock, G. 1984. A novel non-adrenergic, non-cholinergic nerve mediated relaxation of the pig bladder neck; an examination of possible neuro-transmitter candidates. *Eur. J. Pharmacol.* 99:287–293.

Hindmarsh, J.R., Gosling, P.T., and Deane, A.M. 1983. Bladder instability. Is the primary defect in the urethra? *Br. J. Urol.* 55:648–651.

Ho, K.M.T., McMurray, G., Brading, A.F., Noble, J.G., Ny, L., and Andersson, K.E. 1998. Nitric oxide synthase in the heterogeneous population of intramural striated muscle fibres of the human membranous urethral sphincter. *J. Urol.* 159:1091–1096.

Hope, B.T., Michael, G.J., Knigge, K.M., and Vincent, S.R. 1991. Neuronal NADPH diaphorase is a nitric oxide synthase. *Proc. Natl. Acad. Sci. USA* 88:2811–2814.

Huang, P.L., Dawson, T.M., Bredt, D.S., Snyder, S.H., and Fishman, M.C. 1993. Targeted disruption of the neuronal nitric oxide synthase gene. *Cell* 75: 1273–1286.

Hung, A., Vernet, D., Xie, Y., Rajavashisth, T., Rodriguez, J.A., Rajfer, J., and Gonzalez-Cavidad, N.F. 1995. Expression of inducible nitric oxide synthase in smooth muscle cells from rat penile corpora cavernosa. *J. Androl.* 16:469–481.

Ito, Y., and Kimoto, Y. 1985. The neural and non-neural mechanisms involved in urethral activity in rabbits. *J. Physiol. (Lond.)* 367:57–72.

James, M.J., Birmingham, A.T., and Hill, S.J. 1993a. Partial mediation by nitric oxide of the relaxation of human isolated detrusor strips in response to electrical field stimulation. *Br. J. Clin. Pharmacol.* 35:366–372.

James, M.J., Harriss, D.R., Birmingham, A.T., and Hill, S.J. 1993b. Significance of atropine concentration for the incidence of tetrodotoxin-resistant relaxations of human detrusor strips. *Br. J. Clin. Pharmacol.* 36:480–481.

James, M.J., and Iacovou, J.W. 1993. The use of GTN patches in detrusor instability. A pilot study. *Neurourol. Urodyn.* 12:399–400.

Jarchau, T., Hausler, C., Markert, T., Pohler, D., Vanderkerckhove, J., De Jonge, H.R., Lohmann, S.M., and Walter, U. 1994. Cloning, expression, and in situ localization of rat intestinal cGMP-dependent protein kinase II. *Proc. Natl. Acad. Sci. USA* 91:9426–9430.

Jen, P.Y.P., Dixon, J.S., Gearhart, J.P., and Gosling, J.A. 1996. Nitric oxide synthase and tyrosine hydroxylase are colocalized in nerves supplying the postnatal human male genitourinary organs. *J. Urol.* 155:1117–1121.

Jeremy, J.Y., Ballard, S.A., Naylor, A.M., Miller, M.A.W., and Angelini, G.D. 1997. Effects of sildenafil, a type-5 cGMP phosphodiesterase inhibitor, and papaverine on cyclic GMP and cyclic AMP levels in the rabbit corpus cavernosum in vitro. *Br. J. Urol.* 79:958–963.

Kaiser, F.E. 1991. Sexuality and impotence in the aging man. *Clin. Geriatr. Med.* 7:63–72.

Keast, J.R. 1992. A possible neural source of nitric oxide in the rat penis. *Neurosci. Lett.* 143:69–73.

Keilbach, A., Ruth, P., and Hofmann, F. 1992. Detection of cGMP-dependent protein kinase isozymes by specific antibodies. *Eur. J. Biochem.* 208:467–473.

Kim, N., Vardi, Y., Padma-Nathan, H., Daley, J., Goldstein, I., and Saenz de Tejada, I. 1993. Oxygen tension regulates the nitric oxide pathway. Physiological role in penile erection. *J. Clin. Invest.* 91:437–442.

Kimoto, Y., Kessler, R., and Constantinou, C.E. 1990. Endothelium dependent relaxation of human corpus cavernosum by bradykinin. *J. Urol.* 144:1015–1017.

Klarskov, P. 1987. Non-cholinergic, non-adrenergic inhibitory nerve responses of bladder outlet smooth muscle in vitro. *Br. J. Urol.* 60:337–342.

Klarskov, P., Gerstenberg, T., Ramirez, D., and Hald, T. 1983. Non-cholinergic, non-adrenergic nerve mediated relaxation of trigone, bladder neck and urethral smooth muscle in vitro. *J. Urol.* 129:848–850.

Klevmark, B. 1977. Motility of the urinary bladder in cats during filling at physiological rates. II. Effects of extrinsic bladder denervation on intramural tension and intravesical pressure patterns. *Acta Physiol. Scand.* 101:176–184.

Kulseng-Hanssen, S. 1983. Prevalence and pattern of unstable urethral pressure in one hundred seventy-four gynecologic patients referred for urodynamic investigation. *Am. J. Obstet. Gynecol.* 146:895–900.

Kulseng-Hanssen, S. 1987. Urethral pressure variations in women with neurourological symptoms: II. Relationship to urethral smooth muscle. *Neurourol. Urodyn.* 6:79–85.

Larsson, B., Alm, P., Persson, K., and Andersson, K.-E. 1992. Studies on the localization of some neurotransmitters, nerve markers, and NADPH diaphorase activity in tissues from the pig lower urinary tract. *Neurourol. Urodyn.* 11:444–445.

Lee, J.G., Coplen, D., Macarak, E., Wein, A.J., and Levin, R.M. 1994. Comparative studies on the ontogeny and autonomic responses of the fetal calf bladder at different stages of development: involvement of nitric oxide on field stimulated relaxation. *J. Urol.* 151:1096–1101.

Lincoln, J., and Burnstock, G. 1993. Autonomic innervation of the urinary bladder and urethra. In: The Autonomic Nervous System. Vol. 6, Chapter 8: Nervous Control of the Urogenital System (ed. by CA Maggi). Harwood Academic Publishers, London, pp. 33–68.

Lincoln, T.M., and Cornwell, T.L. 1993. Intracellular cyclic GMP receptor proteins. *FASEB J* 7:328–338.

Liu, S.H., and Lin-Shiau, S.Y. 1997. Enhancement by nitric oxide of neurogenic contraction in the mouse urinary bladder. *Naunyn-Schmiedeberg's Arch. Pharmacol.* 356:850–852.

Lohmann, S.M., Vaandrager, A.B., Smolenski, A., Walter, U., and De Jonge, R. 1997. Distinct and specific functions of cGMP-dependent protein kinases. *Trends Biochem. Sci.* 22:307–312.

Low, J.A. 1977. Urethral behaviour during the involuntary detrusor contraction. *Am. J. Obstet. Gynecol.* 128:32–42.

Low, J.A., Armstrong, J.B., and Mauger, G.M. 1989. The unstable urethra in the female. *Obstet. Gynecol.* 74:69–74.

Lugg, J., Ng, C., Rajfer, J., and Gonzalez-Cavidad, N. 1996. Cavernosal nerve stimulation in the rat reverses castration-induced decrease in penile NOS activity. *Am. J. Physiol.* 271:E354-E361.

Lundberg, J.O., Ehren, I., Jansson, O., Adolfsson, J., Lundberg, J.M., Weitzberg, E., Alving, K., and Wiklund, N.P. 1996. Elevated nitric oxide in the urinary bladder in infectious and noninfectious cystitis. *Urology* 48:700–702.

Magee, T., Fuentes, A.M., Garban, H., Rajavashisth, T., Marquez, D., Rodriguez, J.A., Rajfer, J., and Gonzalez-Cavidad, N.F. 1996. Cloning of a novel nitric oxide synthase expressed in penis and lower urinary tract. *Biochem. Biophys. Res. Commun.* 226:145–151.

Maggi, C.A., Santicioli, P., and Meli, A. 1985. Sympathetic inhibition of reflex activation of bladder motility during filling at a physiological-like rate in urethane anaesthetized rats. *Neurourol. Urodyn.* 4:37–45.

Mannino, D.M., Kievens, R.M., and Flanders, W.D. 1994. Cigarette smoking: an independent risk factor for impotence? *Am. J. Epidemiol.* 140:1003–1008.

Martinez-Pineiro, L., Trigo-Rocha, F., Hsu, L., von Heyden, B., Lue, T.F., and Tanagho, E.A. 1993. Cyclic guanosine monophosphate mediates penile erection in the rat. *Eur. Urol.* 24:492–499.

Mattiasson, A., Andersson, K.-E., Elbadawi, A., Morgan, E., and Sjögren, C. 1987. Interaction between adrenergic and cholinergic nerve terminals in the urinary bladder of rabbit, cat and man. *J. Urol.* 137:1017–1019.

Mattiasson, A., Andersson, K.-E., and Sjögren, C. 1985. Contractant and relaxant properties of the female rabbit submucosa. *J. Urol.* 133:304–310.

McGuire, E.J. 1978. Reflex urethral instability. *Br. J. Urol.* 50:200–204.

McGuire, E.J. 1984. Detrusor response to outlet obstruction. *World J. Urol.* 2:208–210.

McGuire, E.J., and Herlihy, E. 1978. Bladder and urethral responses to isolated sacral motor root stimulation. *Invest. Urol.* 16:219–223.

McNeill, D.L., Papka, R.E., and Harris, C.H. 1992a. CGRP immunoreactivity and NADPH-diaphorase in afferent nerves of the rat penis. *Peptides* 13:1239–1246.

McNeill, D.L., Traugh, N.E., Jr., Vaidya, A.M., Hua, H.T., and Papka, R.E. 1992b. Origin and distribution of NADPH-diaphorase-positive neurons and fibers innervating the urinary bladder of the rat. *Neurosci. Lett.* 147:33–36.

Melman, A., and Gingell, J.C. 1999. The epidemiology and pathophysiology of erectile dysfunction. *J. Urol.* 161:5–11.

Mevorach, R.A., Bogaert, G.A., and Kogan, B.A. 1994. Role of nitric oxide in fetal lower urinary tract function. *J. Urol.* 152:510–514.

Mills, T.M., Stopper, V.S., and Wiedmeier, V.T. 1994. Effects of castration and androgen replacement on the hemodynamics of penile erection in the rat. *Biol. Reprod.* 51:234–238.

Mills, T.M., Wiedmeier, V.T., and Stopper, V.S. 1992. Androgen maintenance of erectile function in the rat penis. *Biol. Reprod.* 46:342–348.

Miodrag, A., Castleden, C.M., and Vallance, T.R. 1988. Sex hormones and the female urinary tract. *Drugs* 36:491–504.

Moody, J.A., Vernet, D., Laidlaw, S., Rajfer, J., and Gonzalez-Cavidad, N.F. 1997. Effects of long-term oral administration of L-arginine on the rat erectile response. *J. Urol.* 158:942–947.

Moon, A., Pickard, R.S., Gillespie, J.I., and Neal, D.E. 1997. Contractile responses to sodium nitroprusside and L-arginine in isolated human detrusor. *J. Urol.* 157:258 (abstract 1012).

Morita, T., Tsujii, T., and Dokita, S. 1992. Regional difference in functional roles of cAMP and cGMP in lower urinary tract smooth muscle contractility. *Urol. Int.* 49:191–195.

Nesbit, R.M., Lapides, J., Valk, W.W., Sutler, M., Berry, R.L., Lyons, R.H., Campbell, K.N., and Moe, G.K. 1947. The effects of blockade of the autonomic ganglia on the urinary bladder in man. *J. Urol.* 57:242–250.

Nishizawa, S., Igawa, Y., Okada, N., and Ohhashi, T. 1997. Capsaicin-induced nitric-oxide-dependent relaxation in isolated dog urethra. *Eur. J. Pharmacol.* 335: 211–219.

Noble, J.G., Crowe, R., Robson, T., Milroy, E.J.G., and Burnstock, G. 1993. Neuropeptide-Y immunoreactive nerve density within the male bladder neck. A clue to the aetiology of bladder neck dyssynergia. *Neurourol. Urodyn.* 12:367–368.

Noto, H., Kawahara, T., Shimoda, N., Suzuki, T., Nishizawa, O., Harada, T., and Tsuchida, S. 1993. Internal urethral sphincter relaxation mediated by a non-adrenergic and non-cholinergic mechanism in decerebrate cats. *J. Urol.* 149:341A (abstract 513).

Parlani, M., Conte, B., Goso, C., Szallasi, A., and Manzini, S. 1993a. Capsaicin-induced relaxation in the rat isolated external urethral sphincter: characterization of the vanilloid receptor and mediation by CGRP. *Br. J. Pharmacol.* 110:989–994.

Parlani, M., Conte, B., and Manzini, S. 1993b. Nonadrenergic, noncholinergic inhibitory control of the rat external urethral sphincter: involvement of nitric oxide. *J. Pharmacol. Exp. Ther.* 265:713–719.

Persson, K., Alm, P., Johansson, K., Larsson, B., and Andersson, K.-E. 1993. Nitric oxide synthase in pig lower urinary tract: immunohistochemistry, NADPH diaphorase histochemistry, and functional effects. *Br. J. Pharmacol.* 110:521–530.

Persson, K., Alm, P., Johansson, K., Larsson, B., and Andersson, K.-E. 1995. Co-existence of nitrergic, peptidergic and acetylcholine esterase-positive nerves in the pig lower urinary tract. *J. Auton. Nerv. Syst.* 52:115–236.

Persson, K., Alm, P., Uvelius, B., and Andersson, K.-E. 1998. Nitrergic and cholinergic innervation of the rat lower urinary tract after pelvic ganglionectomy. *Am. J. Physiol.* 274:R389–R397.

Persson, K., and Andersson, K.-E. 1992. Nitric oxide and relaxation of pig lower urinary tract. *Br. J. Pharmacol.* 106:416–422.

Persson, K., and Andersson, K.-E. 1994. Non-adrenergic, non-cholinergic relaxation and levels of cyclic nucleotides in rabbit lower urinary tract. *Eur. J. Pharmacol.* 268:159–167.

Persson, K., Igawa, Y., Mattiasson, A., and Andersson, K.-E. 1991. Inhibition of the L-arginine/nitric oxide pathway causes bladder hyperactivity in the rat. *Acta Physiol. Scand.* 144:107–108.

Persson, K., Igawa, Y., Mattiasson, A., and Andersson, K.-E. 1992. Effects of inhibition of the L-arginine/nitric oxide pathway in the rat lower urinary tract in vivo and in vitro. *Br. J. Pharmacol.* 107:178–184.

Persson, K., Johansson, K., Alm, P., Larsson, B., and Andersson, K.-E. 1997. Morphological and functional evidence against a sensory and sympathetic origin of nitric oxide synthase-containing nerves in the rat lower urinary tract. *Neuroscience* 77:271–281.

Pfeifer, A., Aszódi, A., Seidler, U., Ruth, P., Hofmann, F., and Fässler, R. 1996. Intestinal secretory defects and dwarfism in mice lacking cGMP-dependent protein kinase II. *Science* 274:2082–2086.

Pfeifer, A., Klatt, P., Massberg, S., Ny, L., Sausbier, M., Hirneiss, C., Wang, G., Korth, M., Aszódi, A., Andersson, K.-E., Krombach, F., Mayerhofer, A., Ruth, P., Fässler, R., and Hofmann, F. 1998a. Defective smooth muscle regulation in cGMP-kinase I-deficient mice. *EMBO J.* 17:3045–3051.

Pfeifer, A., Ruth, P., Dostmann, W., Sausbier, M., Klatt, P., and Hofmann, F. 1998b. Structure and function of cGMP-dependent protein kinases. *Rev. Physiol. Biochem. Pharmacol.* 150:105–149.

Polson, J.B., and Strada, S.J. 1996. Cyclic nucleotide phosphodiesterases and vascular smooth muscle. *Annu. Rev. Pharmacol. Toxicol.* 36:403–427.

Prieto, D., Simonsen, U., Hernandez, M., and Garcia-Sacristan. 1998. Contribution of K^+ channels and ouabain-sensitive mechanisms to the endothelim-dependent relaxations of horse penile small arteries. *Br. J. Pharmacol.* 123:1609–1620.

Rajasekaran, M., Mondal, D., Agrawal, K., Chen, I., Hellstrom, W., and Sikka, S. 1998. Ex vivo expression of nitric oxide synthase isoforms (eNOS/iNOS) and calmodulin in human penile cavernosal cells. *J. Urol.* 160:2210–2215.

Reilly, C.M., Zamorano, P., Stopper, V.S., and Mills, T.M. 1997. Androgenic regulation of NO availability in rat penile erection. *J. Androl.* 18:110–115.

Rud, T., Ulmsten, U., and Andersson, K.-E. 1978. Initiation of voiding in healthy women and those with stress incontinence. *Acta Obstet. Gynecol. Scand.* 57:457–461.

Saenz de Tejada, I., and Goldstein, I. 1988. Diabetic penile neuropathy. *Urol. Clin. North Am.* 15:17–22.

Saenz de Tejada, I., Goldstein, I., Azadzoi, K., Krane, R.J., and Cohen, R.A. 1989. Impaired neurogenic and endothelium-mediated relaxation of penile smooth muscle from diabetic men with impotence. *N. Engl. J. Med.* 320:1025–1030.

Schirar, A., Chang, C., and Rousseau, J.P. 1997. Localization of androgen receptor in nitric oxide synthase- and vasoactive intestinal peptide-containing neurons of the major pelvic ganglion innervating the rat penis. *J. Neuroendocrinol.* 9:141–150.

Seftel, A.D., Vaziri, N.D., Ni, Z., Razmjouel, K., Fogarty, J., Hampel, H., Polak, J., Wang, R.-Z., Ferguson, K., Block, C., and Haas, C. 1997. Advanced glycation end products in human penis: elevation in diabetic tissue, site of deposition, and possible effect through iNOS or eNOS. *Urology* 50:1016–1026.

Segarra, G., Medina, P., Domenech, C., Martinez Léon, J.B., Vila, J.M., Aldasoro, M., and Lluch, S. 1998. Neurogenic contraction and relaxation of human penile deep dorsal vein. *Br. J. Pharmacol.* 124:788–794.

Simonsen, U., Prieto, D., Delgado, J.A., Hernandez, M., Resel, L., Saenz de Tejada, I., and Garcia-Sacristan, A. 1997. Nitric oxide is involved in the inhibitory neurotransmission and endothelium-dependent relaxations of human small penile arteries. *Clin. Sci.* 92:269–275.

Simonsen, U., Prieto, D., Saenz de Tejada, I., and Garcia-Sacristan, A. 1995. Involvement of nitric oxide in the non-adrenergic non-cholinergic neurotransmission of horse deep penile arteries: role of charybdotoxin-sensitive K^+-channels. *Br. J. Pharmacol.* 116:2582–2590.

Slack, B., and Downie, J.W. 1983. Pharmacological analysis of the responses of the feline urethra to autonomic nerve stimulation. *J. Auton. Nerv. Syst.* 8:141–160.

Smet, P.J., Edyvane, K.A., Jonavicius, J., and Marshall, V.R. 1994. Distribution of NADPH-diaphorase-positive nerves supplying the human urinary bladder. *J. Auton. Nerv. Syst.* 47:109–113.

Smet, P.J., Edyvane, K.A., Jonavicius, J., and Marshall, V.R. 1996a. Neuropeptides and neurotransmitter-synthesizing enzymes in intrinsic neurons of the human urinary bladder. *J. Neurocytol.* 25:112–124.

Smet, P.J., Jonavicius, J., Marshall, V.R., and de Vente, J. 1996b. Distribution of nitric oxide synthase-immunoreactive nerves and identification of the cellular targets of nitric oxide in guinea-pig and human urinary bladder by cGMP immunohistochemistry. *Neuroscience* 71:337–348.

Smith, S.D., Wheeler, M.A., Foster, H.E., Jr., and Weiss, R.M. 1996. Urinary nitric oxide synthase activity and cyclic GMP levels are decreased with interstitial cystitis and increased with urinary tract infections. *J. Urol.* 155: 1432–1435.

Smith, S.D., Wheeler, M.A., Foster, H.E., Jr., and Weiss, R.M. 1997. Improvement in interstitial cystitis symptom scores during treatment with oral L-arginine. *J. Urol.* 158:703–708.

Smith, S.D., Wheeler, M.A., and Weiss, R.M. 1994. Nitric oxide synthase: an endogenous source of elevated nitrite in infected urine. *Kidney Int.* 45:586–591.

Speakman, M.J., Walmsley, D., and Brading, A.F. 1988. An in vitro pharmacological study of the human trigone—a site of non-adrenergic, non-cholinergic neurotransmission. *Br. J. Urol.* 61:304–309.

Stief, C.G., Taher, A., Truss, M., Ückert, S., Meyer, M., Schulz-Knappe, P., Forssmann, W.F., and Jonas, U. 1995. Die Phosphodiesterase-Isoenzyme des humanen Corpus cavernosum penis und deren funktionelle Bedeutung. *Akt. Urol.* 26 (Suppl. I):58–61.

Stief, C.G., Ückert, S., Becker, A.J., Truss, M.C., and Jonas, U. 1998. The effect of the specific phosphodiesterase (PDE) inhibitors on human and rabbit cavernous tissue in vitro and in vivo. *J. Urol.* 159:1390–1394.

Suh, J.K., Mun, K.H., Cho, C.K., Shin, H.C., Kim, Y.S., and Park, T.C. 1995. Effect of vasoactive intestinal peptide and acetylcholine on penile erection in the rat in vivo. *Int. J. Impotence Res.* 7:111–118.

Sullivan, M.E., Bell, C.R., Dashwood, M.R., Miller, M.A., Thompson, C.S., Mikhailidis, D.P., and Morgan, R.J. 1996. Autoradiographic localization of nitric oxide synthase binding sites in normal and diabetic rat corpus cavernosum. *Eur. Urol.* 30:506–511.

Sutherland, R.S., Kogan, B.A., Piechota, H.J., and Bredt, D.S. 1997. Vesicourethral function in mice with genetic disruption of neuronal nitric oxide synthase. *J. Urol.* 157:1109–1116.

Tamura, M., Kagawa, S., Kimura, K., Kawanishi, Y., Tsuruo, Y., and Ishimura, K. 1995. Coexistence of nitric oxide synthase, tyrosine hydroxylase and vasoactive intestinal polypeptide in human penile tissue—a triple histochemical and immunohistochemical study. *J. Urol.* 153:530–534.

Tamura, M., Kagawa, S., Tsuruo, Y., Ishimura, K., Kimura, K., and Kawanishi, Y. 1997. Localization of NADPH diaphorase and vasoactive intestinal polypeptide-containing neurons in the efferent pathway to the rat corpus cavernosum. *Eur. Urol.* 32:100–104.

Tanagho, E.A., and Miller, E.R. 1970. Initiation of voiding. *Br. J. Urol.* 42: 175–183.

Tang, P.C., and Ruch, T.C. 1955. Non-neurogenic basis of bladder tonus. *Am. J. Physiol.* 181:249–257.

Tang, K., Turner, L.A., Ballard, S.A., and Naylor, A.M. 1996. Effects of a novel phosphodiesterase type 5 inhibitor, sildenafil, on methacholine induced relaxation of isolated rabbit corpus cavernosum (abstr). *Br. J. Pharmacol.* 118:154P.

Theobald, R.J., Jr. 1996. The effect of N^{G}-monomethyl L-arginine on bladder function. *Eur. J. Pharmacol.* 311:73–76.

Thornbury, K.D., Hollywood, M.A., and McHale, N.G. 1992. Mediation by nitric oxide of neurogenic relaxation of the urinary bladder neck muscle in sheep. *J. Physiol. (Lond.)* 451:133–144.

Torrens, M.J. 1978. Urethral sphincteric responses to stimulation of the sacral nerves in the human female. *Urol. Int.* 33:22–26.

Trigo-Rocha, F., Hsu, G.L., Donatucci, C.F., and Lue, T.F. 1993. The role of cyclic adenosine monophosphate, cyclic guanosine monophosphate, endothelium and nonadrenergic, noncholinergic neurotransmission in canine penile erection. *J. Urol.* 149:872–877.

Trigo-Rocha, F., Hsu, G.L., Donatucci, C.F., Martinez-Pineiro, L., Lue, T.F., and Tanagho, E.A. 1994. Intracellular mechanism of penile erection in monkeys. *Neurourol. Urodyn.* 13:71–80.

Triguero, D., Prieto, D., and Garcia-Pascual, A. 1993. NADPH-diaphorase and NANC relaxations are correlated in the sheep urinary tract. *Neurosci. Lett.* 163:93–96.

Turner, W.H. 1994. Evidence against regulation of bladder compliance by nitric oxide. *J. Urol.* 151:377A (abstract 598).

Uhler, M.D. 1993. Cloning and expression of a novel cyclic GMP-dependent protein kinase from mouse brain. *J. Biol. Chem.* 268:13586–13591.

Ulmsten, U., Andersson, K.-E., and Persson, C.G.A. 1977. Diagnostic and therapeutic aspects of *urge* urinary incontinence in women. *Urol. Int.* 32:88–96.

Ulmsten, U., Henriksson, L., and Iosif, S. 1982. The unstable female urethra. *Am. J. Obstet. Gynecol.* 144:93–96.

Vanhatalo, S., Klinge, E., Sjöstrand, N.O., and Soinila, S. 1996. Nitric oxide-synthesizing neurons originating at several different levels innervate rat penis. *Neuroscience* 75:891–899.

Vanhatalo, S., and Soinila, S. 1995. Direct nitric oxide-containing innervation from the rat spinal cord to the penis. *Neurosci. Lett.* 199:45–48.

Vernet, D., Cai, L., Garban, H., Babbitt, M.L., Murray, F.T., Rajfer, J., and Gonza lezCavidad, N.F. 1995. Reduction of penile nitric oxide synthase in diabetic BB/WORdp (type I) and BBZ/WORdp (type II) rats with erectile dysfunction. *Endocrinology* 136:5709–5717.

Vizzard, M.A., Erdman, S.L., and de Groat, W.C. 1993a. Localization of NADPH-diaphorase in pelvic afferent and efferent pathways of the rat. *Neurosci. Lett.* 152:72–76.

Vizzard, M.A., Erdman, S.L., and de Groat, W.C. 1993b. Localization of NADPH diaphorase in bladder afferent and postganglionic efferent neurons in the rat. *J. Auton. Nerv. Syst.* 44:85–90.

Vizzard, M.A., Erdman, S.L., and de Groat, W.C. 1996. Increased expression of neuronal nitric oxide synthase in bladder afferent pathways following chronic bladder irritation. *J. Comp. Neurol.* 370:191–202.

Vizzard, M.A., Erdman, S.L., Erickson, V.L., Stewart, R.J., Roppolo, J.R., and de Groat, W.C. 1994a. Localization of NADPH diaphorase in the lumbosacral spinal cord and dorsal root ganglia of the cat. *J. Comp. Neurol.* 339:62–75.

Vizzard, M.A., Erdman, S.L., Förstermann, U., and de Groat, W.C. 1994b. Differential distribution of nitric oxide synthase in neural pathways to the urogenital organs (urethra, penis, urinary bladder) of the rat. *Brain Res.* 646:279–291.

Vizzard, M.A., Erickson, K., and de Groat, W.C. 1997. Localization of NADPH diaphorase in the thoracolumbar and sacrococcygeal spinal cord of the dog. *J. Auton. Nerv. Syst.* 64:128–142.

Waldeck, K., Ny, L., Persson, K., and Andersson, K.-E. 1998. Mediators and mechanisms of relaxation in rabbit urethral smooth muscle. *Br. J. Pharmacol.* 123:617–624.

Waldmann, R., Bauer, S., Gobel, C., Hofmann, F., Jakobs, K.H., and Walter, U. 1986. Demonstration of cGMP-dependent protein kinase and cGMPdepen-dent phosphorylation in cell-free extracts of platelets. *Eur. J. Biochem.* 158:203–210.

Weiner, C.P., Lizasoain, I., Baylis, S.A., Knowles, R.G., Charles, I.G., and Moncada, S. 1994. Induction of calcium-dependent nitric oxide synthase by sex hormones. *Proc. Natl. Acad. Sci USA* 91:5212–5216.

Weiss, R.M., Nangia, A.K., Smith, S.D., and Wheeler, M.A. 1994. Nitric oxide synthase (NOS) activity in urethra, bladder and bladder smooth muscle cells. *Neurourol. Urodyn.* 13:397–398.

Werkström, V., Persson, K., Ny, L., Bridgewater, M., Brading, A.F., and Andersson, K.-E. 1995. Factors involved in the relaxation of female pig urethra. *Br. J. Pharmacol.* 116:1599–1604.

Wheeler, M.A., Smith, S.D., Saito, N., Foster, H.E., Jr., and Weiss, R.M. 1997. Effect of long-term oral L-arginine on the nitric oxide synthase pathway in the urine from patients with interstitial cystitis. *J. Urol.* 158:2045–2050.

Xie, Y., Garban, H., Ng, C., Rajfer, J., and Gonzalez-Cavidad, N.F. 1997. Effect of long-term passive smoking on erectile function and penile nitric oxide synthase in the rat. *J. Urol.* 157:1121–1126.

Yanni, N.C., and Williams, R.G. 1996. L-NAME has no effect on the length-tension relationship of rat detrusor muscle in vitro. *Br. J. Pharmacol.* 117:205P.

Zinner, N.R., Sterling, A.M., and Ritter, C.R. 1980. Role of inner urethral softness in urinary continence. *Urology* 16:115–121.

Zorgniotti, A.W., and Lizza, E.F. 1994. Effect of large doses of the nitric oxide precursor, L-arginine, on erectile function. *Int. J. Impotence Res.* 6:33–36.

Zvara, P., Sioufi, R., Schipper, H.M., Begin, L.R., and Brock, G.B. 1995. Nitric oxide mediated erectile activity is a testosterone dependent event: a rat erection model. *Int. J. Impotence Res.* 7:209–219.

Zygmunt, P.K.E., Persson, K., Alm, P., Larsson, B., and Andersson, K.-E. 1993a. The L-arginine/nitric oxide pathway in the rabbit urethral lamina propria. *Acta Physiol. Scand.* 148:431–439.

Zygmunt, P.M., Zygmunt, P.K.E., Högestätt, E.D., and Andersson, K.-E. 1993b. Effects of ω-conotoxin on adrenergic, cholinergic and NANC neurotransmission in the rabbit urethra and detrusor. *Br. J. Pharmacol.* 110:1285–1290.

5
Nitrergic Neurotransmission in the Enteric Nervous System

Kenton M. Sanders and Kathleen D. Keef

Introduction

Coordinated contraction of the gastrointestinal tract involves reflexes composed of intrinsic and extrinsic neurons. Intrinsic neurons are primarily located in the myenteric plexus, although submucosal neurons may provide part of the motor input to the distal gastrointestinal tract. Sensory afferent neurons with mechanosensitive and chemosensitive receptors transduce information regarding luminal dimensions and contents and convey this information centrally and to the enteric ganglia. Within enteric ganglia, interneurons integrate and propagate the processed sensory information orally and aborally and activate muscle motor neurons. There are two functional classes of muscle motor neurons, inhibitory neurons and excitatory neurons. In the guinea pig, which is the model most often used for studies of the enteric nervous system, motor neurons account for nearly half of the total population of neurons (Furness et al., 1994). A thorough discussion of enteric neurobiology is beyond the scope of this chapter, and the reader is directed to recent reviews for general information about the enteric nervous system (Furness et al., 1994, 1995; Costa and Brookes, 1994; Gershon, 1997; Wood, 1994). This chapter will focus on the characteristics of nitrergic neurotransmission. It must be clearly stated, however, that although nitric oxide (NO) is an important enteric neurotransmitter, it does not function alone. There is considerable evidence that ATP and peptides, such as vasoactive intestinal peptide (VIP), are cotransmitters released from enteric inhibitory neurons and responsible for specific aspects of inhibitory responses.

Evidence that Nitric Oxide Might Serve in Enteric Inhibitory Neurotransmission

Garthwaite and co-workers (Garthwaite et al., 1988) demonstrated that stimulation of the cerebellum with glutamate caused the release of a previously undescribed transmitter substance, NO. These findings suggested a role for NO in the nervous system. The following year several papers reported that NO was responsible for inhibitory neural responses in autonomically innervated peripheral tissues (Gillespie et al., 1989; Li and Rand, 1989; Ramagopal and Leighton, 1989). Identification of NO as a possible neurotransmitter in the gastrointestinal tract followed shortly after in 1990 in a study demonstrating that the relaxation response of the canine ileocolonic sphincter was dependent upon NO synthesis and release (Bult et al., 1990; Boeckxstaens et al., 1990). This story has developed significantly during the past decade, and many studies have now demonstrated a role for NO in enteric neurotransmission in numerous species and in most parts of the gastrointestinal tract. Acceptance of NO as a neurotransmitter has followed successful tests of standard criteria, such as neural synthesis, release during nerve stimulation, block by antagonists, mimicry by agonists, and a means of degradation.

Synthesis of Nitric Oxide by Enteric Motor Neurons

The neural isoform of nitric oxide synthase (nNOS) was first isolated and purified from rat cerebellum (Bredt and Snyder, 1990). Neuronal NOS was found to be a Ca^{2+}/calmodulin-dependent enzyme that converts L-arginine to L-citrulline. After nNOS was purified, antibodies were generated that allowed localization of the enzyme in tissues with immunohistochemical techniques. Widespread expression of nNOS was found in central and peripheral neurons (Bredt et al., 1990), and subsequent studies with the original polyclonal antiserum and more recent monoclonal antibodies have clearly documented nNOS-like immunoreactivity in a subpopulation of enteric neurons in many species (e.g., Belai et al., 1992; Ward et al., 1992b; Young et al., 1992; Costa et al., 1992). Cell bodies and varicose processes coursing through the circular and longitudinal muscle layers are positive for nNOS. The same neurons and processes are colabeled by the NADPH diaphorase histochemical method (e.g., Young et al., 1992; Ward et al., 1992b), and this technique has been widely exploited to characterize the distribution of nNOS.

Release from Nerves

Many investigators believe that NO is not stored by enteric neurons but is synthesized on demand (see Sanders and Ward, 1992). Once synthesized, NO, which is freely and rapidly diffusible through cell membranes, does not appear to be released in a quantal, vesicular fashion, but instead it reaches postjunctional targets by diffusion. Since nNOS is activated by enhanced Ca^{2+}, the signal for synthesis and release may be entry of Ca^{2+} through specific populations of prejunctional Ca^{2+} channels. Neuronal NOS is expressed in cell bodies and all along processes of enteric motor neurons, so the site of NO release may be defined more by the location of voltage-dependent Ca^{2+} channels (and proximity of Ca^{2+} channels to nNOS) than by more traditional synaptic specializations.

The release of an NO-like substance during field stimulation of enteric neurons was first demonstrated with bioassays (Boeckxstaens et al., 1991). Relaxation responses were blocked by either tetrodotoxin (TTX) or N^G-nitro-L-arginine (L-NA) in parallel with the release of the NO-like substance. A similar NO-like substance was released when intrinsic nerves were stimulated with the nicotinic receptor agonist 1,1-dimethyl-4-phenylpiperazinium (DMPP) (Boeckxstaens et al., 1991). Chemoluminescence assays confirmed conclusions drawn from bioassays and demonstrated NO release from the internal anal sphincter (Chakder and Rattan, 1993), lower esophageal sphincter (Murray et al., 1994), and canine colon (Shuttleworth et al., 1995b).

Inhibitors of NOS Reduce Enteric Inhibitory Nerve Responses

Inhibitors of NOS have been vital tools for determining the importance of NO in the enteric nervous system. Several arginine analogues bind to the same site in NOS as L-arginine and serve as competitive inhibitors of NO synthesis (for review see Moncada et al., 1991). The competitiveness of the block is also useful, because it can be overcome by excess L-arginine as a test for specificity. The first experiments using arginine analogues to test the significance of nitrergic neurotransmission utilized N^G-monomethyl-L-arginine (L-NMMA) and showed that this compound blocked inhibitory neural responses in the rat annococcygeus muscle (Gillespie et al., 1989; Li and Rand, 1989; Ramagopal and Leighton, 1989). The effect of L-NMMA was reversed by application of excess L-arginine, but D-arginine was ineffective. L-NMMA had no effect on inhibitory responses to NO or to NO donors, showing that the compound acted via blockade of synthesis rather than a nonspecific block of the postjunctional actions of NO. Later studies showed that intrinsic neurons of the gastrointestinal tract were also dependent upon nitrergic mechanisms, and these conclusions were largely based on use of arginine analogues and oxyhemoglobin, a scavenger of NO (e.g., Bult et al., 1990; Boeckxstaens et al., 1990; for reviews see Sanders and Ward, 1992; Stark and Szurszewski, 1992; Sneddon and Graham, 1992; Rand and Li, 1995; Bennett, 1997; Brookes, 1993; Lefebvre, 1995).

Part of the enteric inhibitory response involves hyperpolarization of the membrane potential in postjunctional cells, which can be monitored with intracellular microelectrode recording techniques. This important aspect of enteric inhibitory responses was also shown to depend upon NO synthesis and release from intrinsic neurons (Dalziel et al., 1991; Stark et al., 1991). The initial studies on postjunctional responses were performed in canine tissues in which NO neurotransmission is the dominant mechanism in enteric inhibitory effects. Many other preparations, including the human (see Keef et al., 1993; Stark et al., 1993), display more complicated postjunctional responses that are likely to be due to parallel release of multiple transmitters. In these tissues the slower component of the inhibitory junction potential (IJP) response is usually attributed to NO.

Exogenous Nitric Oxide Mimics Responses to Enteric Inhibitory Neurotransmission

The ability to mimic neural responses by application of a proposed transmitter is of fundamental importance in establishing the role of a substance as a neurotransmitter. Exogenous application of NO mimics, in detail, the hyperpolarization and the inhibition of contraction caused by enteric inhibitory nerve stimulation (Thornbury et al., 1991; Stark et al., 1991). In addition, oxyhemoglobin reduces or abolishes responses to exogenous NO and enteric inhibitory neurotransmission (Thornbury et al., 1991). Similar electrical and contractile effects of NO have been demonstrated in a wide variety of gastrointestinal muscles (for review see Sanders and Ward, 1992).

Stimulation of enteric inhibitory neurons also leads to enhanced levels of cGMP (Barnette et al., 1989), and inhibitors of type V phosphodiesterase, such as zaprinast, prolong enteric inhibitory responses (Ward et al., 1992a). Nitric oxide and NO donors such as nitroglycerine have long been known to raise cGMP levels in smooth muscles (Greutter et al., 1979; Arnold et al., 1977), and a similar effect has been reported in gastrointestinal muscles (Murray et al., 1992; Ward et al., 1992a). Thus, exogenous NO also mimics neural responses in terms of utilization of second messenger systems.

Mechanism for Degradation

After stimulation of postjunctional sites, a mechanism must exist to remove the active transmitter. This often requires enzymatic degradation and/or uptake of the transmitter. In the case of NO, disposal is the natural result of the short half-life of this reactive substance in biological fluids (Taha et al., 1992). In fact, it is preservation of NO, not degradation, that is likely to require specialized mechanisms, such as junctional expression of superoxide dismutase (Rand and Li, 1995), to conserve enough NO to elicit postjunctional effects.

Prejunctional Regulation of Synthesis and Release of Nitric Oxide

As described above, nNOS is expressed in varicose nerve terminals of enteric motor neurons. The action of nNOS is to convert molecular oxygen (O_2) and L-arginine to L-citrulline (see Moncada et al., 1991). Most investigators agree that NO is not stored, but rather NO is produced "on demand" upon stimulation of inhibitory neurons. Neuronal NOS is a constitutive, Ca^{2+}/calmodulin-dependent isoform that appears to be activated by local increases in cytoplasmic Ca^{2+} (Mayer et al., 1989; Busse and Mulsch, 1990; Bredt and Snyder, 1990). Depolarization of nerve terminals leads to opening of specific classes of Ca^{2+} channels that are linked to activation of nNOS (Boeckxstaens et al., 1993).

When NO is synthesized from arginine, stoichiometric amounts of citrulline are produced. Several studies have suggested that a subpopulation of enteric neurons have the capacity to recycle citrulline to arginine and thereby sustain nitrergic neurotransmission (Shuttleworth et al., 1995a, 1997; Chakder and Rattan, 1997; Rattan and Chakder, 1997). The enzymes necessary for citrulline recycling, argininosuccinate synthetase (ASS) and argininosuccinate lyase (ASL), are expressed by enteric neurons and varicose fibers within the tunica muscularis (Shuttleworth et al., 1995a). The same neurons express nNOS. This pathway is functional during enteric inhibitory nerve stimulation; this was demonstrated by showing that L-citrulline, which has no effect on nitrergic neurotransmission by itself, was able to restore nitrergic neurotransmission in tissues pretreated with arginine analogues. The mechanism responsible for this action was thought to result from conversion of L-citrulline to L-arginine. By generating excess L-arginine, inhibition of NOS by arginine analogue was reversed, leading to a restoration of NO synthesis. In another study, sustained electrical field stimulation or pretreatment of the internal anal sphincter with arginase was shown to reduce tissue levels of arginine, and this was accompanied by reduction in nitrergic relaxation (Chakder and Rattan, 1997). Addition of either L-arginine or L-citrulline reversed these effects, but the latter was ineffective if the uptake mechanism for L-citrulline was also blocked.

Nitric Oxide Participates with Other Neurotransmitters in Enteric Inhibitory Neurotransmission

Enteric inhibitory motor neurons coexpress several biologically active substances that could contribute to inhibitory neurotransmission (see Furness et al., 1989). The colocalization of transmitter substances in the same neurons has led to the "plurichemical transmission hypothesis," which suggests that enteric neurons release a combination of transmitters and that the postjunctional response represents a complex transduction of several signals (Furness et al., 1989, 1995). In the case of enteric inhibitory neurotransmission, VIP and ATP have most frequently been suggested to act in concert with NO to produce physiological responses.

The first electrophysiological evidence suggesting that more than a single neurotransmitter contributes to enteric inhibitory responses showed that IJPs in the guinea pig ileum were composed of a fast (apamin-sensitive) hyperpolarization, followed by a slower (apamin-resistant) phase of hyperpolarization (Niel et al., 1983). Many investigators have since suggested that the fast IJP is due to ATP (Burnstock, 1980; Crist et al., 1992), whereas the second phase is due to NO (Keef et al., 1993; Lyster et al., 1992; He and Goyal, 1993). A major obstacle in clarifying the role of ATP as a transmitter is the lack of specific antagonists for the P_{2Y} class of purinergic receptors, which are thought to mediate inhibitory responses (Burnstock and Kennedy, 1985). A number of studies have also suggested that VIP contributes to inhibitory neural responses as well (for review see Lefebvre, 1993; Grider and Makhlouf, 1988). Some investigators have also reported that VIP contributes to generation of enteric inhibitory IJPs, since the VIP antagonist VIP_{10-28} reduced IJPs in the guinea pig ileum (see Crist et al., 1992; He and Goyal, 1993).

Serial Models of Nitric Oxide-Dependent Neurotransmission

Some investigators have suggested that a significant amount of the NO released in response to enteric inhibitory nerve stimulation is due to postjunctional synthesis of NO (i.e., 48% and 82% in rat and rabbit tissues, respectively) that is activated by VIP released from neurons (see Grider et al., 1992; Jin et al., 1996). These authors reported that mechanical responses to VIP were greatly attenuated by blockade of NO synthesis (Grider et al., 1992), and consistent with these data, VIP stimulated production of cAMP and cGMP and activated cGMP-dependent protein kinase (PKG) and cAMP-dependent protein kinase (PKA) (Murthy and Makhlouf, 1995). A "serial cascade" model of enteric inhibitory neurotransmission was proposed to explain these results (e.g., Murthy et al., 1995).

Studies from several other laboratories (e.g., Barbier and Lefebvre, 1993; McLaren et al., 1993; Desai et al., 1994; Lefebvre et al., 1992, 1995; Bayguinov et al., 1999; Keef et al., 1994) have not supported the "serial cascade" model. In these studies, relaxation responses to VIP were not significantly affected by arginine analogues. This is a critical test, because in the serial cascade concept, stimulation by VIP causes NO synthesis and amplification of the inhibitory response. If blockade of NO synthesis does not affect responses to VIP, then this tends to negate the postjunctional synthesis of meaningful levels of NO in response to VIP. Proponents of the serial cascade model have argued that most studies testing the effects of arginine analogues on VIP responses in intact muscle strips have used excitatory agonists to raise tone (see Murthy et al., 1994). Excitatory agonists activate protein kinase C (PKC), which inhibits the isoform of NOS thought to be present in gastrointestinal muscle cells. These authors also argue that any agonist that enhances intracellular Ca^{2+} runs the risk of activating PKC and deactivating NOS. The significance of an inhibitory mechanism that is inactivated by elevated levels of Ca^{2+} in smooth muscles has never been made clear. Other studies showed that spontaneous tone in the canine fundus (Bayguinov et al., 1999) or non-agonist-dependent phasic activity in the canine colon (Keef et al., 1994) were inhibited by VIP and that these responses were not affected by arginine analogues. It is also unclear why NOS in isolated muscle cells (which appear to generate spontaneous tone because they relax in response to VIP; see Grider et al., 1992; Murthy and Makhlouf, 1995) can be activated by VIP, but this mechanism is suppressed or not apparent in intact muscles with spontaneous tone.

Another serial model of enteric inhibitory neurotransmission involves a more structurally complicated neuromuscular junction than is normally considered. Investigators have long realized that interstitial cells of Cajal (ICC) are frequently intimately associated with varicose nerve terminals in the gastrointestinal tract, and have suggested a role for these cells in mediation or modulation of neurotransmission (see Daniel and Posey-Daniel, 1984). This hypothesis has not received intensive investigation, because ICC are difficult to study with specificity in intact muscles. Recently, a mouse model has been found in which ICC fail to develop in some parts of the GI tract (see Sanders, 1996), and tissues from these animals have provided a means of testing the role of ICC in enteric inhibitory neurotransmission. Gastrointestinal muscles that lack intramuscular ICC (IC-IM) have greatly reduced NO-dependent inhibitory responses without significant reduction in the number or distribution of nitrergic nerves (Burns et al., 1996; Ward et al., 1998). Loss of IC-IM also results in reduced electrical responsiveness to exogenous NO. These data suggest that part of the apparatus needed to transduce NO signals is expressed by ICC (Figure 5.1). There may be parallel innervation of smooth muscle cells and ICC in some regions of the gastrointestinal tract, but in the murine lower esophageal sphincter and stomach, a substantial amount of the NO-dependent inhibitory response depends upon transduction by ICC (Burns et al., 1996; Ward et al., 1998).

Interstitial cells of Cajal express second messenger systems that are responsive to NO. Immunohistochemical studies have shown that cGMP levels rise in ICC when tissues are stimulated by enteric inhibitory neurotransmission or by NO donors (Shuttleworth et al., 1993; Young et al., 1993), and it is possible that these cells participate in transduction of NO signals. It is also possible that the role of ICC in inhibitory neurotransmission is

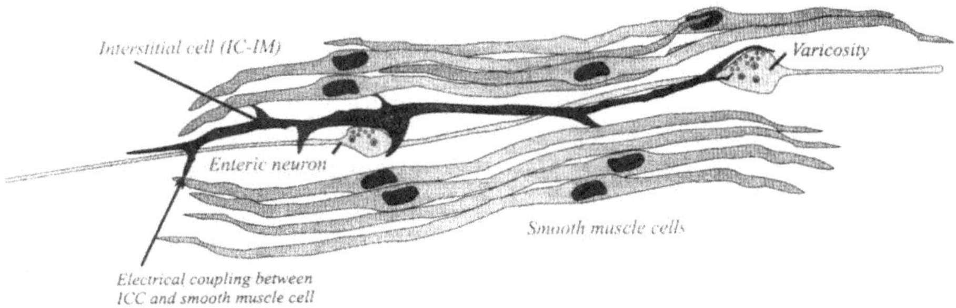

FIGURE 5.1. Neuromuscular junction that mediates NO-dependent neurotransmission in gastrointestinal muscles. The classical view of peripheral neurotransmission includes nerve endings forming close synaptic contact with effector cells. In gastrointestinal neuromuscular transmission, the effector has been typically assumed to be smooth muscle cells. But recent evidence suggests that a subpopulation of interstitial cells of Cajal (ICC) are interposed between nerve terminals and smooth muscle cells (e.g., Burns et al., 1996; Wang et al., 1999). ICC within the muscle layers (IC-IM in the stomach and IC-DMP in the small intestine) are closely associated with varicose nerve processes and form frequent close contacts and synaptic-like membrane specializations. IC-IM and IC-DMP appear to mediate nitrergic neurotransmission and conduct hyperpolarization responses to electrically coupled smooth muscle cells. Parallel transmission directly to smooth muscle cells may also occur, but this would appear to require a significantly longer diffusion pathway for NO. IC-IM may express specialized signal transduction mechanisms, a unique complement of ion channels, an amplification capacity to boost NO levels, and protective enzymes such as superoxide dismutase to help conserve the NO signal.

greater than simple transduction. Some ICC have been shown to express NOS-like immunoreactivity (Xue et al., 1994), and stimulation of these ICC with NO leads to further production of NO (Publicover et al., 1993). Thus it is possible that ICC that express NOS may serve as amplifiers of NO-dependent neurotransmission.

Postjunctional Mechanisms That Mediate Nitric Oxide Effects in Gastrointestinal Smooth Muscles

Diffusion of NO to postjunctional cells initiates a variety of responses, including activation or suppression of ion channels, changes in intracellular or subsarcolemmal Ca^{2+} concentration, and alterations in the sensitivity of the contractile apparatus to Ca^{2+}. As in vascular smooth muscle, responses of gastrointestinal muscles to NO appear to be largely dependent upon synthesis of cGMP and activation of PKG. Several reviews have addressed NO-dependent mechanisms in smooth muscle (e.g., Clementi, 1998; Lincoln et al., 1994; Hirata and Murad, 1994; Garthwaite and Boulton, 1995). In gastrointestinal muscles it is unlikely that all of the postjunctional targets for NO and PKG phosphorylation have been identified, and thus the full impact of NO in postjunctional cells is not fully understood. For example, little is known about the postjunctional responses elicited in ICC, yet mechanisms in these cells may mediate a significant portion of the inhibitory response (see previous section).

Cyclic GMP and PKG as Mediators of Nitric Oxide Effects

The heme group, which is a component of various proteins, has long been recognized as a preferred binding site for NO. When NO binds to the heme moiety of guanylyl cyclase, a conformational change occurs, leading to activation of the enzyme and production of cGMP (Arnold et al., 1977). Since many of the effects of NO appear to be mediated by events downstream from cGMP production, the heme group of guanylyl cyclase may be considered the major receptor for NO. Both the relaxation and the hyperpolarization responses to enteric inhibitory neurotransmission can typically be mimicked by addition of the membrane-permeable analogue 8-bromo-cGMP to gastrointestinal muscles (e.g., Ward et al., 1992).

Evidence supporting a role for guanylyl cyclase in the actions of NO has been greatly strengthened with use of the guanylyl cyclase inhibitor 1H[1,2,4]oxadiazolo[4,3-a]quinoxalin-1-one (ODQ). ODQ abolishes the increase in tissue cGMP levels initiated by NO (Cellek et al., 1996; Bayguinov et al., 1999; Onoue and Katusic, 1998) and reduces or abolishes nitrergic relaxation in gastrointestinal muscles (Cellek et al., 1996; Fonseca et al., 1998; Young et al., 1996; Franck et al., 1997; Olgart et al., 1998; Bayguinov and Sanders, 1998). Likewise, ODQ abolishes hyperpolarization responses elicited by NO donors such as sodium nitroprusside (SNP) (Franck et al., 1997) or diethylenetriamine/NO adduct (DNO) (Goyal and He., 1998). Other compounds, such as LY83583 and methylene blue, are frequently used and assumed to be inhibitors of guanylyl cyclase, but these agents also generate superoxide, a known inactivator of NO (Barbier and Lefebvre, 1992; Marczin et al., 1992). Therefore the usefulness of LY83583 and methylene blue as pharmacological tools to dissect the role of guanylyl cyclase in NO-dependent effects is limited.

An important downstream mediator of cGMP is PKG (Francis and Corbin, 1994). PKG-I has been identified in gastrointestinal smooth muscles (Huber et al., 1998), and NO donors have been shown to activate PKG and relax gastrointestinal smooth muscles (Murthy and Makhlouf, 1995). Relaxation responses are inhibited by PKG blockers, such as KT5823 (Murthy and Makhlouf, 1995; Huber et al., 1998).

Some authors have suggested that a component of the NO-induced responses may be independent of cGMP generation and PKG activation

(Young et al., 1996; Goyal and He, 1998). A frequently cited mechanism to explain possible cGMP-independent responses in smooth muscles is direct activation of K^+ channels by NO (Bolotina et al., 1994; Koh et al., 1995; Goyal and He, 1998).

What Cellular Effectors Are Activated by Nitric Oxide- or cGMP-Dependent Mechanisms?

Ionic Conductances Affected by Nitric Oxide

Several studies have investigated the effects of NO and NO donors on ion channels in gastrointestinal muscles. Ca^{2+} channels, Cl^- channels, and a variety of K^+ channels may be targets for NO and cGMP-mediated pathways in generating enteric inhibitory responses.

Clapp and Gurney (1991) showed that SNP reduced inward currents due to L-type Ca^{2+} channels in rabbit pulmonary arterial cells. The effects were associated with an increased rate of inactivation of the Ca^{2+} current. Nitregic regulation has also been demonstrated in single-channel studies in basilar artery by showing that SNP causes a dose-dependent decrease in the open probability of L-type Ca^{2+} channels (10 nM–10 μM SNP) (Tewari and Simard, 1997). These effects were shown to be due to the cGMP/PKG pathway. The inhibitory effects of cGMP have also been studied in colonic muscles (Koh et al., 1995) and provide a means by which NO might directly decrease Ca^{2+} entry via this pathway.

A major response to application of NO or NO donors to visceral smooth muscles is hyperpolarization, and this could be coupled to inhibition of contraction by lowering membrane excitability and decreasing the net open probability of L-type Ca^{2+} channels. Most investigators agree that the hyperpolarization response is due to enhanced K^+ conductance, and several studies have attempted to characterize the conductances activated by NO and cGMP. For example, in guinea pig proximal colon NO, NO donors and electrical field stimulation of enteric inhibitory neurons induced hyperpolarizations that were associated with increased membrane conductance. This effect was proposed to be due to at least two K^+ conductances that were resistant to block by tetraethylammonium ions (Watson et al., 1996).

Several studies have shown that NO, via cGMP-dependent mechanisms, activates large conductance (BK) Ca^{2+}-activated K^+ channels (Thornbury et al., 1991; Koh et al., 1995; Lu et al., 1998). The first demonstration showed that authentic NO caused an increase in the open probability (NPo) of BK channels recorded from "on-cell" membrane patches in canine colonic myocytes (Thornbury et al., 1991). This group also showed that the effects of NO on BK in colonic myocytes required cellular second messengers, since NO did not directly activate BK in excised patches (Koh et al., 1995).

Nitric oxide donors were also shown to induce an outward current in myocytes isolated from the opossum esophagus (Murray et al., 1995). Inhibitors of BK channels or PKG blocked the effects of NO donors. In another study, authentic NO increased NPo of BK channels in "on-cell" recordings, and charybdotoxin, added to the pipette solution, blocked this effect (Lu et al., 1998). Membrane-permeable cGMP analogues and zaprinast, an inhibitor of type 5 (cGMP) phosphodiesterase, also increased NPo of BK channels, suggesting that the effects of NO could be mediated via cGMP. Nitric oxide had no effect on NPo of BK channels in excised patches, demonstrating the need for secondary cellular events in the regulation of these channels. Studies on BK channels from vascular smooth muscles have demonstrated that PKG activates BK channels (Robertson et al., 1993), and others have shown that PKG phosphorylation directly regulates BK, cloned from canine colonic smooth muscles, by phosphorylation of the α subunit at serine 1072 (see Figure 5.2 in Fukao et al., 1999). Other studies have suggested that NO can have direct effects on BK channels. For example, in a study of guinea pig colonic cells, NO was found to increase NPo of BK channels in cell attached and excised patches (Lang and Watson, 1998).

Activation of BK channels has been the main focus of studies of the effects of NO on K^+ conductances, but activation of these channels alone does not explain the physiological responses to enteric inhibitory neurotransmission. For example, neither tetraethylammonium (TEA) nor more specific inhibitors of BK channels block IJPs in gastrointestinal muscles. At least two additional K^+ conductances have been shown to be activated by NO. Koh and co-workers (1995) showed that NO and NO donors activated a 90 pS voltage-dependent conductance and a small conductance channel (approximately 2.5 pS) in on-cell patches of colonic myocytes. The small conductance channel was shown to be activated at negative potentials, suggesting that it could participate in hyperpolarization responses to NO. Membrane permeable cGMP analogues also activated these channels, and it was found that NO and NO donors could activate these channels in excised patches under conditions unfavorable to cGMP-dependent phosphorylation. BK channels were not affected under the same conditions. Thus it appears that some of the channels responsive to NO may be regulated by either cGMP-dependent or cGMP-independent mechanisms.

Some authors have also considered that the hyperpolarization response initiated by NO might be due to suppression of tonically activated Cl^- current. In a recent study it was reported that the NO donor DNO and 8-bromo-cGMP suppressed a Ca^{2+}-activated Cl^- current in esophageal smooth muscle cells (Zhang et al., 1998). This phenomenon was demonstrated in a rather indirect manner, and it is not clear that the Cl^- current suppressed by cGMP-dependent mechanisms is activated under normal

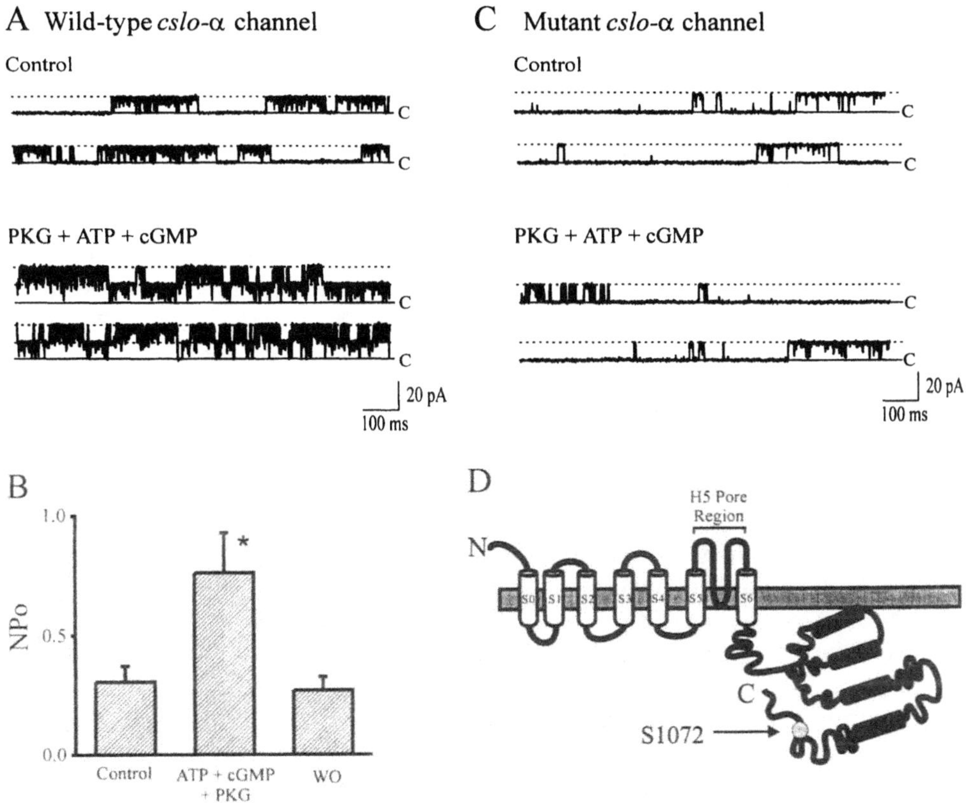

FIGURE 5.2. Cyclic GMP-dependent protein kinase (PKG-Iα) regulation of *cslo*-α channels. *cslo*-α channels were cloned from canine colon and expressed in HEK293 cells. (A) Representative recording of *cslo*-α channel activity obtained in an inside-out excised membrane patch at a holding potential of +40 mV. PKG increased *cslo*-α channel activity in the presence of ATP (1 mM) + cGMP (0.1 mM). Cytosolic Ca^{2+} concentration was clamped at 10^{-6} M. (B) Summary of the effect of PKG + ATP + cGMP ($n = 10$) on the open probability (NPo) of *cslo*-α channels. PKG led to a significant increase in *cslo*-α channel activity, which returned to control levels following washout of PKG. Shown are mean values ± SE (* indicates value significantly greater, $P < 0.05$, than control). (C) Representative recording of channel activity obtained from mutated *cslo*-α channels. In these channels serine 1072 within the consensus PKG phorphorylation site was replaced with alanine. In contrast to wild-type channels, application of PKG + ATP + cGMP to mutated *cslo*-α channels had no significant effect on channel open probability. (D) Schematic diagram of the *cslo*-α channel including seven transmembrane spanning groups and a large cytosolic C-terminal end. The high-affinity phosphorylation site for PKG is located near the C-terminus. Adapted from Fukao et al., 1999.

conditions (i.e., under resting Ca^{2+} levels in intact smooth muscle cells at physiological resting potentials).

Ca^{2+} Uptake and Efflux

The ultimate purpose of the electrophysiological effects described above appears to be regulation of cell Ca^{2+} concentration, and thus of contractile force. Besides effects on membrane conductances (which provide an integrated control of Ca^{2+} entry), Ca^{2+} removal or sequestration mechanisms also appear to be a target for the postjunctional effect of NO. Ca^{2+}-ATPases in both the plasma membrane (PMCA) and the sarcoplasmic reticulum (SERCA) play an important role in regulating intracellular Ca^{2+} concentration in smooth muscles (for review see Gonzalez et al., 1996). The evidence coupling NO and the cGMP/PKG pathway to changes in Ca^{2+}-ATPase activity is better established for SERCA than for PMCA. The NO/PKG pathway has been reported to lead to phosphorylation of the SERCA regulatory protein phospholamban, leading to an increase in Ca^{2+} sequestration and relaxation (see Lincoln et al., 1994; Karczewski et al., 1998; Petkov et al., 1998).

Calcium sequestered by the sarcoplasmic reticulum can be released in response to a variety of stimuli. Massive release of Ca^{2+} from the sarcoplasmic reticulum can lead to muscle contraction, but localized release of Ca^{2+} can affect ionic conductances in the plasma membrane and produce effects inhibitory to contraction (Nelson et al., 1995). Inositol trisphosphate (IP_3) receptors and ryanodine receptors control Ca^{2+} release. There is substantial evidence that NO can inhibit Ca^{2+} release from IP_3-sensitive stores (Felbel et al., 1988; Karaki et al., 1988; Cornwell and Lincoln, 1989). Protein kinase G appears to play an important role in regulating Ca^{2+} release (Clementi et al., 1995; Geiger et al., 1992; Komalavilas and Lincoln, 1996), and several different mechanisms have been proposed for regulation by PKG: inhibition of G proteins that activate phospholipase C (Hirata et al., 1990); inhibition of a downstream intermediate in the generation of IP_3 (Rapoport, 1986); and direct phosphorylation of the IP_3 receptor (Komalavilas and Lincoln, 1996).

Nitric oxide may also affect Ca^{2+} release from ryanodine receptors (Ca^{2+}-induced Ca^{2+} release or CICR channels). This mechanism may involve generation of cyclic adenosine diphosphate-ribose (cADPR), an endogenous modulator of CICR channels (Willmott et al., 1996; Galione, 1994; Clementi et al., 1996). Protein kinase G stimulates synthesis of cADPR from nicotinamide adenine dinucleotide by activating the adenosine diphosphate–ribosyl cyclase pathway. Nitric oxide may also directly activate ryanodine receptors, as in cardiac and skeletal muscles via *S*-nitrosylation (Stoyanovsky et al., 1997; Xu et al., 1998). Finally, there is evidence that NO may indirectly regulate Ca^{2+} release from CICR channels by increasing Ca^{2+} loading of the sarcoplasmic reticulum, thereby increasing the frequency of

"Ca sparks" (i.e., highly localized Ca^{2+} release; see Jaggar et al., 1998; Porter et al., 1998). The importance of each of these mechanisms has yet to be fully evaluated in gastrointestinal muscles.

Changes in Myofilament Sensitivity to Ca^{2+}

The mechanisms discussed above for NO-induced relaxation lead to changes in cytoplasmic Ca^{2+} concentration, a major regulator of contractile force in smooth muscle. It is also known that NO and cGMP-dependent mechanisms cause "desensitization" of the contractile apparatus to Ca^{2+} (Nishimura and van Breemen, 1989). Thus, a given concentration of cytoplasmic Ca^{2+} becomes less effective in coupling to force production. There is evidence that this mechanism contributes to the mechanical effects of NO in gastrointestinal muscles. Cyclic GMP reduces contractile force in Ca^{2+}-clamped, permeabilized gastric muscles (Ozaki et al., 1992), and NO released from enteric inhibitory neurons reduces the contractile force recorded at a given intracellular Ca^{2+} level (Bayguinov et al., 1998).

The mechanisms by which NO and cGMP decrease Ca^{2+} sensitivity are not fully understood, but they may involve inhibition of myosin light chain kinase (MLCK) activity as well as an increase in myosin light chain phosphatase (MLCP) activity (Somlyo and Somlyo, 1998; Murphy and Walker, 1998). It has also been suggested that PKG enhances MLCP activity via activation of telokin, a protein which is identical to the C-terminus of MLCK (Wu et al., 1998).

Relative Significance of Electrical Versus Biochemical Mechanisms in Eliciting Postjunctional Responses to Nitric Oxide

There have been a number of studies to address the question of which postjunctional mechanism mediates NO-dependent relaxation in enteric nerve stimulation. These studies have been frustrated by the lack of reagents that can specifically block one postjunctional pathway while leaving others intact. For example, there are no specific inhibitors of the K^+ channels that mediate NO-dependent hyperpolarization in gastrointestinal muscles. Therefore, it is difficult to block electrical responses to NO and test whether mechanical effects are inhibited to the same extent.

Some studies have compared the pharmacology of electrical and mechanical responses and concluded that electrical responses are not required for inhibition of contraction (see Bridgewater et al., 1995; Takeuchi et al., 1998). These studies have typically used single pulses or short trains of electrical field stimulation to elicit electrical responses and longer trains of stimuli to elicit mechanical responses. The results of these studies can be misleading, because multiple neurotransmitters mediate enteric inhibitory responses (see above). The fast IJPs observed in many gastrointestinal muscles following single stimuli are likely to be due to ATP. Therefore, single pulses or short trains may not be dependent upon NO. Nitric oxide-dependent responses, however, are quite apparent with multiple stimuli, and these effects can be reduced by inhibitors of NO synthesis (see Keef et al., 1993).

A recent study has addressed the role of membrane potential in NO-dependent mechanical responses in tonically active canine fundus (Bayguinov and Sanders, 1998). The nonselective K^+-channel blocker tetrapentylammonium (TPA) was used to block the hyperpolarization responses associated with nerve stimulation. In the presence of TPA, the electrical response to nerve stimulation was abolished while the inhibition of contraction persisted. These data suggest that in some gastrointestinal muscles a dissociation occurs between electrical and mechanical responses, and that other intracellular biochemical phenomena might be the dominant mechanism by which NO mediates relaxation. A different conclusion might be reached in phasic muscles in which periodic contractions are associated with spontaneous fluctuations in membrane potential. In these muscles the inhibition of contraction might be more closely associated with NO-induced electrical effects. The nonspecificity of TPA as a blocker of K^+ channels makes its use in studies of phasic muscles difficult to interpret.

Summary and Directions for Future Studies

The evidence is strong for NO as an inhibitory neurotransmitter in the gastrointestinal tract and other peripheral tissues. Nitric oxide appears to be synthesized on demand and released from neurons by diffusion. Postjunctional intracellular proteins appear to be the receptors for NO. All of these features make NO a unique transmitter substance. Many steps in NO-dependent neurotransmission are still in need of clarification. The precise prejunctional events that regulate NO synthesis, including specialized interactions of nNOS with the plasma membrane and specific Ca^{2+} channels, are yet to be clarified. Considerable work is needed to understand the extracellular environment that must be traversed by NO en route to postjunctional targets. At present we do not clearly understand what factors serve to hasten or delay the oxidation of NO. Many of the postjunctional mechanisms that mediate nitrergic effects are still unclear, and the specific effects of NO on ICC are an important new topic for research. At present, many of the effects of NO on gastrointestinal muscles appear similar to the mechanisms utilized by NO in the vasculature and in other smooth muscles. Thus, it is difficult to conceive of specific pharmacological agents that might be therapeutic in treating abnormal motility without widespread systemic consequences. Future studies may reveal mechanisms specific to gastrointestinal muscles, allowing nitrergic transmission to become a target for therapy in GI motility disorders.

Acknowledgments. The authors were supported by grants from NIDDK: DK 41315 to K.M.S. and K.D.K., DK40569 to K.M.S., and DK45376 to KDK.

References

Arnold, W.P., Mittal, C.K., Katsuki, S., and Murad, F. 1977. Nitric oxide activates guanylate cyclase and increases guanosine 3′:5′-cyclic monophosphate levels in various tissue preparations. *Proc. Natl. Acad. Sci. USA* 74:3203–3207.

Barbier, A.J., and Lefebvre, R.A. 1992. Effect of LY 83583 on relaxation induced by non-adrenergic non-cholinergic nerve stimulation and exogenous nitric oxide in the rat gastric fundus. *Eur. J. Pharmacol.* 219:331–334.

Barbier, A.J., and Lefebvre, R.A. 1993. Involvement of the L-arginine:nitric oxide pathway in nonadrenergic noncholinergic relaxation of the cat gastric fundus. *J. Pharmacol. Exp. Ther.* 266:172–178.

Barnette, M., Torphy, T.J., Grous, M., Fine, C., and Ormsbee, H.S. 1989. Cyclic GMP: a potential mediator of neurally- and drug-induced relaxation of opossum lower esophageal sphincter. *J. Pharmacol. Exp Ther.* 249:524–528.

Bayguinov, O., Keef, K.D., Hagen, B., and Sanders, K.M. 1999. Parallel pathways mediate inhibitory effects of vasoactive intestinal polypeptide and nitric oxide in canine funclus. *Br. J. Pharmacol.* 126:1543–1552.

Bayguinov, O., and Sanders, K.M. 1998. Dissociation between electrical and mechanical responses to nitrergic stimulation in the canine gastric fundus. *J. Physiol. (Lond).* 1:509 (Pt 2):437–438.

Belai, A., Schmidt, H.H., Hoyle, C.H., Hassall, C.J., Saffrey, M.J., Moss, J., Forstermann, U., Murad, F., and Burnstock, G. 1992. Colocalization of nitric oxide synthase and NADPH-diaphorase in the myenteric plexus of the rat gut. *Neurosci. Lett.* 143:60–64.

Bennett, M.R. 1997. Non-adrenergic non-cholinergic (NANC) transmission to smooth muscle: 35 years on. *Prog. Neurobiol.* 52:159–195.

Boeckxstaens, G.E., De Man, J.G., Pelckmans, P.A., Cromheeke, K.M., Herman, A.G., and Van Maercke, Y.M. 1993. Ca^{2+} dependency of the release of nitric oxide from non-adrenergic non-cholinergic nerves. *Br. J. Pharmacol.* 110:1329–1334.

Boeckxstaens, G.E., Pelckmans, P.A., Bult, H., De Man, J.G., Herman, A.G., and Van Maercke, Y.M. 1990. Nonadrenergic noncholinergic relaxation mediated by nitric oxide in the ileocolonic junction. *Eur. J. Pharmacol.* 190:239–246.

Boeckxstaens, G.E., Pelckmans, P.A., Ruytjens, I.F., Bult, H., De Man, J.G., Herman, A.G., and Van Maercke, Y.M. 1991. Bioassay of nitric oxide released upon stimulation of non-adrenergic non-cholinergic nerves in the canine ileocolonic junction. *Br. J. Pharmacol.* 103:1085–1091.

Bolotina, V.M., Najibi, S., Palacino, J.J., Pagano, P.J., and Cohen, R.A. 1994. Nitric oxide directly activates calcium-dependent potassium channels in vascular smooth muscle. *Nature* 368:850–853.

Bredt, D.S., Hwang, P.M., and Snyder, S.H. 1990. Localization of nitric oxide synthase indicating a neural role for nitric oxide. *Nature* 347:768–770.

Bredt, D.S., and Snyder, S.H. 1990. Isolation of nitric oxide synthetase, a calmodulin-requiring enzyme. *Proc. Natl. Acad. Sci. USA* 87:682–685.

Bridgewater, M., Cunnane, T.C., and Brading, A.F. 1995. Characteristic features of inhibitory junction potentials evoked by single stimuli in the guinea-pig isolated taenia caeci. *J. Physiol. (Lond).* 485:145–155.

Brookes, S.J. 1993. Neuronal nitric oxide in the gut. *J. Gastroenterol. Hepatol.* 8:590–603.

Bult, H., Boeckxstaens, G.E., Pelckmans, P.A., Jordaens, F.H., Van Maercke, Y.M., and Herman, A.G. 1990. Nitric oxide as an inhibitory nonadrenergic noncholinergic neurotransmitter. *Nature* 345:346–347.

Burns, A.J., Lomax, A.E.J., Torihashi, S., Sanders, K.M., and Ward, S.M. 1996. Interstitial cells of Cajal mediate inhibitory neurotransmission in the stomach. *Proc. Nat. Acad. Sci. USA* 93:12008–12013.

Burnstock, G. 1980. Purinergic nerves and receptors. *Prog. Biochem. Pharmacol.* 16:141–154.

Burnstock, G., and Kennedy, C. 1985. Is there a basis for distinguishing two types of P2-purinoceptor? *Gen. Pharmacol.* 16:433–440.

Busse, R., and Mulsch, A. 1990. Calcium-dependent nitric oxide synthesis in endothelial cytosol is mediated by calmodulin. *FEBS Lett.* 265:133–136.

Cellek, S., Kasakov, L., and Moncada, S. 1996. Inhibition of nitrergic relaxations by a selective inhibitor of the soluble guanylate cyclase. *Br. J. Pharmacol.* 118:137–140.

Chakder, S., and Rattan, S. 1993. Release of nitric oxide by activation of nonadrenergic noncholinergic neurons of internal anal sphincter. *Am. J. Physiol.* 264 (1 Pt 1):G7–G12.

Chakder, S., and Rattan, S. 1997. L-Arginine deficiency causes suppression of nonadrenergic noncholinergic nerve-mediated smooth muscle relaxation: role of L-citrulline recycling. *J. Pharmacol. Exp. Ther.* 282:378–384.

Clapp, L.H., and Gurney, A.M. 1991. Modulation of calcium movements by nitroprusside in isolated vascular smooth muscle cells. *Pflugers Arch.* 418:462–470.

Clementi, E. 1998. Role of nitric oxide and its intracellular signalling pathways in the control of Ca^{2+} homeostasis. *Biochem. Pharmacol.* 55:713–718.

Clementi, E., Riccio, M., Sciorati, C., Nistico, G., and Meldolesi, J. 1996. The type 2 ryanodine receptor of neurosecretory PC12 cells is activated by cyclic ADP-ribose. Role of the nitric oxide/cGMP pathway. *J. Biol. Chem.* 271:17739–17745.

Clementi, E., Vecchio, I., Sciorati, C., and Nistico, G. 1995. Nitric oxide modulation of agonist-evoked intracellular Ca^{2+} release in neurosecretory PC-12 cells: inhibition of phospholipase C activity via cyclic GMP-dependent protein kinase I. *Mol. Pharmacol.* 47:517–524.

Cornwell, T.L., and Lincoln, T.M. 1989. Regulation of intracellular Ca^{2+} levels in cultured vascular smooth muscle cells. Reduction of Ca^{2+} by atriopeptin and 8-bromo-cyclic GMP is mediated by cyclic GMP-dependent protein kinase. *J. Biol. Chem.* 264:1146–1155.

Costa, M., and Brookes, S.J. 1994. The enteric nervous system. *Am. J. Gastroenterol.* 89 (8 Suppl.):S129–S137.

Costa, M., Furness, J.B., Pompolo, S., Brookes, S.J., Bornstein, J.C., Bredt, D.S., and Snyder, S.H. 1992. Projections and chemical coding of neurons with immunoreactivity for nitric oxide synthase in the guinea-pig small intestine. *Neurosci. Lett.* 148:121–125.

Crist, J.R., He, X.D., and Goyal, R.K. 1992. Both ATP and the peptide VIP are inhibitory neurotransmitters in guinea-pig ileum circular muscle. *J. Physiol.* 447:119–131.

Dalziel, H.H., Thornbury, K.D., Ward, S.M., and Sanders, K.M. 1991. Involvement of nitric oxide synthetic pathway in inhibitory junction potentials in canine proximal colon. *Am. J. Physiol.* 260:G789–G792.

Daniel, E.E., and Posey-Daniel, V. 1984. Neuromuscular structures in opossum esophagus: role of interstitial cells of Cajal. *Am. J. Physiol.* 246:G305–G315.

Desai, K.M., Warner, T.D., Bishop, A.E., Polak, J.M., and Vane, J.R. 1994. Nitric oxide, and not vasoactive intestinal peptide, as the main neurotransmitter of vagally induced relaxation of the guinea-pig stomach. *Br. J. Pharmacol.* 113:1197–1202.

Felbel, J., Trockur, B., Ecker, T., Landgraf, W., and Hofmann, F. 1988. Regulation of cytosolic calcium by cAMP and cGMP in freshly isolated smooth muscle cells from bovine trachea. *J. Biol. Chem.* 263:16764–16771.

Fonseca, M., Uddin, N., and Gibson, A. 1998. No evidence for a significant non-nitrergic, hyperpolarising factor contribution to field stimulation-induced relaxation of the mouse anococcygeus. *Br. J. Pharmacol.* 124:524–528.

Francis, S.H., and Corbin, J.D. 1994. Structure and function of cyclic nucleotide-dependent protein kinases. *Annu. Rev. Physiol.* 56:237–272.

Franck, H., Sweeney, K.M., Sanders, K.M., and Shuttleworth, C.W. 1997. Effects of a novel guanylate cyclase inhibitor on nitric oxide-dependent inhibitory neurotransmission in canine proximal colon. *Br. J. Pharmacol.* 122:1223–1229.

Fukao, M., Kenyon, J., Mason, H., Britton, F., Horowitz, B., and Keef, K.D. 1999. Cyclic GMP-dependent protein kinase activates cloned BK_{Ca} channel expressed in mammalian cells by direct phosphorylation at serine 1072. *J. Biol. Chem.* 274:10927–10935.

Furness, J.B., Li, Z.S., Young, H.M., and Forstermann, U. 1994. Nitric oxide synthase in the enteric nervous system of the guinea-pig: a quantitative description. *Cell Tissue Res.* 277:139–149.

Furness, J.B., Morris, J.L., Gibbins, I.L., and Costa, M. 1989. Chemical coding of neurons and plurichemical transmission. *Annu. Rev. Pharmacol. Toxicol.* 29:289–306.

Furness, J.B., Young, H.M., Pompolo, S., Bornstein, J.C., Kunze, W.A., and McConalogue, K. 1995. Plurichemical transmission and chemical coding of neurons in the digestive tract. *Gastroenterology* 108:554–563.

Galione, A. 1994. Cyclic ADP-ribose, the ADP-ribosyl cyclase pathway and calcium signalling. *Mol. Cell Endocrinol.* 98:125–131.

Garthwaite, J., and Boulton, C.L. 1995. Nitric oxide signaling in the central nervous system. *Annu. Rev. Physiol.* 57:683–706.

Garthwaite, J., Charles, S.L., and Chess-Williams, R. 1988. Endothelium-derived relaxing factor release on activation of NMDA receptors suggests role as intercellular messenger in the brain. *Nature* 336:385–388.

Geiger, J., Nolte, C., Butt, E., Sage, S.O., and Walter, U. 1992. Role of cGMP and cGMP-dependent protein kinase in nitrovasodilator inhibition of agonist-evoked calcium elevation in human platelets. *Proc. Natl. Acad. Sci. USA* 89:1031–1035.

Gershon, M.D. 1997. Genes and lineages in the formation of the enteric nervous system. *Curr. Opin. Neurobiol.* 7:101–109.

Gillespie, J.S., Liu, X.R., and Martin, W. 1989. The effects of L-arginine and N^G-monomethyl-L-arginine on the response of the rat anococcygeus muscle to NANC nerve stimulation. *Br. J. Pharmacol.* 98:1080–1082.

Gonzalez, J.M., Jost, L.J., Rouse, D., and Suki, W.N. 1996. Plasma membrane and sarcoplasmic reticulum Ca-ATPase and smooth muscle. *Miner. Electrolyte Metab.* 22:345–348.

Goyal, R.K., and He, X.D. 1998. Evidence for NO redox form of nitric oxide as nitrergic inhibitory neurotransmitter in gut. *Am. J. Physiol* 275:G1185–G1192.

Grider, J.R., and Makhlouf, G.M. 1988. Vasoactive intestinal peptide. Transmitter of inhibitory motor neurons of the gut. *Ann. NY Acad. Sci.* 527:369–377.

Grider, J.R., Murthy, K.S., Jin, J.G., and Makhlouf, G.M. 1992. Stimulation of nitric oxide from muscle cells by VIP: prejunctional enhancement of VIP release. *Am. J. Physiol.* 262:G774–G778.

Gruetter, C.A., Barry, B.K., McNamara, D.B., Gruetter, D.Y., Kadowitz, P.J., and Ignarro, L. 1979. Relaxation of bovine coronary artery and activation of coronary arterial guanylate cyclase by nitric oxide, nitroprusside and a carcinogenic nitrosoamine. *J. Cyclic. Nucleotide Res.* 5:211–224.

He, X.D., and Goyal, R.K. 1993. Nitric oxide involvement in the peptide VIP-associated inhibitory junction potential in the guinea pig ileum. *J. Physiol. (Lond).* 461:485–499.

Hirata, M., Kohse, K.P., Chang, C.H., Ikebe, T., and Murad, F. 1990. Mechanism of cyclic GMP inhibition of inositol phosphate formation in rat aorta segments and cultured bovine aortic smooth muscle cells. *J. Biol. Chem.* 265:1268–1273.

Hirata, M., and Murad, F. 1994. Interrelationships of cyclic GMP, inositol phosphates, and calcium. *Adv. Pharmacol.* 26:195–216.

Huber, A., Trudrung, P., Storr, M., Franck, H., Schusdziarra, V., Ruth, P., and Allescher, H.D. 1998. Protein kinase G expression in the small intestine and functional importance for smooth muscle relaxation. *Am. J. Physiol.* 275:G629–G637.

Jaggar, J.H., Wellman, G.C., Heppner, T.J., Porter, V.A., Perez, G.J., Gollasch, M., Kleppisch, T., Rubart, M., Stevenson, A.S., Lederer, W.J., Knot, H.J., Bonev, A.D., and Nelson, M.T. 1998. Ca^{2+} channels, ryanodine receptors and Ca(2+)-activated K^+ channels: a functional unit for regulating arterial tone. *Acta. Physiol. Scand.* 164:577–587.

Jin, J.G., Murthy, K.S., Grider, J.R., and Makhlouf, G.M. 1996. Stoichiometry of neurally induced VIP release, NO formation, and relaxation in rabbit and rat gastric muscle. *Am. J. Physiol.* 271, G357–G369.

Karaki, H., Sato, K., Ozaki, H., and Murakami, K. 1988. Effects of sodium nitroprusside on cytosolic calcium level in vascular smooth muscle. *Eur. J. Pharmacol.* 156:259–266.

Karczewski, P., Hendrischke, T., Wolf, W.P., Morano, I., Bartel, S., and Schrader, J. 1998. Phosphorylation of phospholamban correlates with relaxation of coronary artery induced by nitric oxide, adenosine, and prostacyclin in the pig. *J. Cell Biochem.* 70:49–59.

Keef, K.D., Du, C., Ward, S.M., McGregor, D.B., and Sanders, K.M. 1993. Enteric inhibitory neural control of human colon: role of nitric oxide. *Gastroenterology* 105:1009–1016.

Keef, K.D., Shuttleworth, C.W., Xue, C., Bayguinov, O., Publicover, N.G., and Sanders, K.M. 1994. Relationship between nitric oxide and vasoactive intestinal polypeptide in enteric inhibitory neurotransmission. *Neuropharmacology* 33:1303–1314.

Koh, S.D., Campbell, J.D., Carl, A., and Sanders, K.M. 1995. Nitric oxide activates multiple potassium channels in canine colonic smooth muscle. *J. Physiol.* 489:735–743.

Komalavilas, P., and Lincoln, T.M. 1996. Phosphorylation of the inositol 1,4,5-trisphosphate receptor. Cyclic GMP-dependent protein kinase mediates cAMP and cGMP dependent phosphorylation in the intact rat aorta. *J. Biol. Chem.* 271:21933–21938.

Lang, R.J., and Watson, M.J. 1998. Effects of nitric oxide donors, *S*-nitroso-L-cysteine and sodium nitroprusside, on the whole-cell and single channel currents in single myocytes of the guinea-pig proximal colon. *Br. J. Pharmacol.* 123:505–517.

Lefebvre, R.A. 1993. Non-adrenergic non-cholinergic neurotransmission in the proximal stomach. *Gen. Pharmacol.* 24:257–286.

Lefebvre, R.A. 1995. Nitric oxide in the peripheral nervous system. *Ann. Med.* 27:379–388.

Lefebvre, R.A., Baert, E., and Barbier, A.J. 1992. Influence of N^{G}-nitro-L-arginine on non-adrenergic non-cholinergic relaxation in the guinea-pig gastric fundus. *Br. J. Pharmacol.* 106:173–179.

Lefebvre, R.A., Smits, G.J., and Timmermans, J.P. 1995. Study of NO and VIP as non-adrenergic non-cholinergic neurotransmitters in the pig gastric fundus. *Br. J. Pharmacol.* 116:2017–2026.

Li, C.G., and Rand, M.J. 1989. Evidence for a role of nitric oxide in the neurotransmitter system mediating relaxation of the rat anococcygeus muscle. *Clin. Exp. Pharmacol. Physiol.* 16:933–938.

Lincoln, T.M., Komalavilas, P., and Cornwell, T.L. 1994. Pleiotropic regulation of vascular smooth muscle tone by cyclic GMP-dependent protein kinase. *Hypertension* 23:1141–1147.

Lu, G., Mazet, B., Sarr, M.G., and Szurszewski, J.H. 1998. Effect of nitric oxide on calcium-activated potassium channels in colonic smooth muscle of rabbits. *Am. J. Physiol.* 274:G848–G856.

Lyster, D.J., Bywater, R.A., Taylor, G.S., and Watson, M.J. 1992. Effects of a nitric oxide synthase inhibitor on non-cholinergic junction potentials in the circular muscle of the guinea pig ileum. *J. Auton. Nerv. Syst.* 41:187–196.

Marczin, N., Ryan, U.S., and Catravas, J.D. 1992. Methylene blue inhibits nitrovasodilator- and endothelium-derived relaxing factor-induced cyclic GMP accumulation in cultured pulmonary arterial smooth muscle cells via generation of superoxide anion. *J. Pharmacol. Exp. Ther.* 263:170–179.

Mayer, B., Schmidt, K., Humbert, R., and Bohme, E. 1989. Biosynthesis of endothelium-derived relaxing factor: a cytosolic enzyme in porcine aortic endothelial cells Ca^{2+}-dependently converts L-arginine into an activator of soluble guanylyl cyclase. *Biochem. Biophys. Res. Commun.* 164:678–685.

McLaren, A., Li, C.G., and Rand, M.J. 1993. Mediators of nicotine-induced relaxations of the rat gastric fundus. *Clin. Exp. Pharmacol. Physiol.* 20:451–457.

Moncada, S., Palmer, R.M., and Higgs, E.A. 1991. Nitric oxide: physiology, pathophysiology, and pharmacology. *Pharmacol. Rev.* 43:109–142.

Murphy, R.A., and Walker, J.S. 1998. Inhibitory mechanisms for cross-bridge cycling: the nitric oxide-cGMP signal transduction pathway in smooth muscle relaxation. *Acta Physiol. Scand.* 164:373–380.

Murray, J., Bates, J.N., and Conklin, J.L. 1994. Nerve-mediated nitric oxide production by opossum lower esophageal sphincter. *Dig. Dis. Sci.* 39:1872–1876.

Murray, J.A., Du, C., Ledlow, A., Manternach, P.L., and Conklin, J.L. 1992. Guanylate cyclase inhibitors: effect on tone, relaxation, and cGMP content of lower esophageal sphincter. *Am. J. Physiol.* 263:G97–G101.

Murray, J.A., Shibata, E.F., Buresh, T.L., Picken, H., O'Meara, B.W., and Conklin, J.L. 1995. Nitric oxide modulates a calcium-activated potassium current in muscle cells from opossum esophagus. *Am. J. Physiol.* 269:G606–G612.

Murthy, K.S., Grider, J.R., Jin, J.G., and Makhlouf, G.M. 1995. Interplay of VIP and nitric oxide in the regulation of neuromuscular activity in the gut. *Arch. Int. Pharmacodyn. Ther.* 329:27–38.

Murthy, K.S., Jin, J.G., and Makhlouf, G.M. 1994. Inhibition of nitric oxide synthase activity in dispersed gastric muscle cells by protein kinase C. *Am. J. Physiol.* 266:G161–G165.

Murthy, K.S., and Makhlouf, G.M. 1995. Interaction of cA-kinase and cG-kinase in mediating relaxation of dispersed smooth muscle cells. *Am. J. Physiol.* 268:C171–C180.

Nelson, M.T., Cheng, H., Rubart, M., Santana, L.F., Bonev, A.D., Knot, H.J., and Lederer, W.J. 1995. Relaxation of arterial smooth muscle by calcium sparks. *Science* 270:633–637.

Niel, J.P., Bywater, R.A., and Taylor, G.S. 1983. Apamin-resistant post-stimulus hyperpolarization in the circular muscle of the guinea-pig ileum. *J. Auton. Nerv. Syst.* 9:565–569.

Nishimura, J., and van Breemen, C. 1989. Direct regulation of smooth muscle contractile elements by second messengers. *Biochem. Biophys. Res. Commun.* 163:929–935.

Olgart, C., Hallen, K., Wiklund, N.P., Iversen, H.H., and Gustafsson, L.E. 1998. Blockade of nitrergic neuroeffector transmission in guinea-pig colon by a selective inhibitor of soluble guanylyl cyclase. *Acta Physiol. Scand.* 162:89–95.

Onoue, H., and Katusic, Z.S. 1998. The effect of 1H-[1,2,4]oxadiazolo[4,3-a]quinoxalin-1-one (ODQ) and charybdotoxin (CTX) on relaxations of isolated cerebral arteries to nitric oxide. *Brain Res.* 785:107–113.

Ozaki, H., Blondfield, D.P., Hori, M., Publicover, N.G., Kato, I., and Sanders, K.M. 1992. Spontaneous release of nitric oxide inhibits electrical, Ca^{2+} and mechanical transients in canine gastric smooth muscle. *J. Physiol. (Lond).* 445:231–247.

Petkov, G.V., Spassov, G.D., and Boev, K.K. 1998. Role of sarcoplasmic reticulum in the myorelaxant activity of nitric oxide donors in guinea pig gastric fundus. *Eur. J. Pharmacol.* 354:59–66.

Porter, V.A., Bonev, A.D., Knot, H.J., Heppner, T.J., Stevenson, A.S., Kleppisch, T., Lederer, W.J., and Nelson, M.T. 1998. Frequency modulation of Ca^{2+} sparks is involved in regulation of arterial diameter by cyclic nucleotides. *Am. J. Physiol.* 274:C1346–C1355.

Publicover, N.G., Hammond, E.M., and Sanders, K.M. 1993. Amplification of nitric oxide signaling by interstitial cells isolated from canine colon. *Proc. Natl. Acad. Sci. USA* 90:2087–2091.

Ramagopal, M.V., and Leighton, H.J. 1989. Effect of N^G-monomethyl-L-arginine ion field stimulation induced decreases in cytosolic Ca^{2+} levels and relaxation in the rat anococcygeus muscle. *Eur. J. Pharmacol.* 19:147–169.

Rand, M.J., and Li, C.G. 1995. Nitric oxide as a neurotransmitter in peripheral nerves: nature of transmitter and mechanism of transmission. *Annu. Rev. Physiol.* 57:659–682.

Rapoport, R.M. 1986. Cyclic guanosine monophosphate inhibition of contraction may be mediated through inhibition of phosphatidylinositol hydrolysis in rat aorta. *Circ. Res.* 58:407–410.

Rattan, S., and Chakder, S. 1997. L-Citrulline recycling in opossum internal anal sphincter relaxation by nonadrenergic, noncholinergic nerve stimulation. *Gastroenterology* 112:1250–1259.

Robertson, B.E., Schubert, R., Hescheler, J., and Nelson, M.T. 1993. cGMP-dependent protein kinase activates Ca-activated K channels in cerebral artery smooth muscle cells. *Am. J. Physiol.* 265:C299–C303.

Sanders, K.M. 1996. A case for interstitial cells of Cajal as pacemakers and mediators of neurotransmission in the gastrointestinal tract. *Gastroenterology* 111:492–515.

Sanders, K.M., and Ward, S.M. 1992. Nitric oxide as a mediator of nonadrenergic noncholinergic neurotransmission. *Am. J. Physiol.* 262:G379–G392.

Shuttleworth, C.W., Burns, A.J., Ward, S.M., O'Brien, W.E., and Sanders, K.M. 1995a. Recycling of L-citrulline to sustain nitric oxide-dependent enteric neurotransmission. *Neuroscience* 68:1295–1304.

Shuttleworth, C.W., Conlon, S.B., and Sanders, K.M. 1997. Regulation of citrulline recycling in nitric oxide-dependent neurotransmission in the murine proximal colon. *Br. J. Pharmacol.* 120:707–713.

Shuttleworth, C.W., Weinert, J.S., Sanders, K.M., and Buxton, I.L. 1995b. Detection of nitric oxide release from canine enteric neurons. *J. Auton. Nerv. Syst.* 56:61–68.

Shuttleworth, C.W., Xue, C., Ward, S.M., de Vente, J., and Sanders, K.M. 1993. Immunohistochemical localization of 3′,5′-cyclic guanosine monophosphate in the canine proximal colon: responses to nitric oxide and electrical stimulation of enteric inhibitory neurons. *Neuroscience* 56:513–522.

Sneddon, P., and Graham, A. 1992. Role of nitric oxide in the autonomic innervation of smooth muscle. *J. Auton. Pharmacol.* 12:445–456.

Somlyo, A.P., and Somlyo, A.V. 1998. From pharmacomechanical coupling to G-proteins and myosin phosphatase. *Acta Physiol. Scand.* 164:437–448.

Stark, M.E., Bauer, A.J., and Szurszewski, J.H. 1991. Effect of nitric oxide on circular muscle of the canine small intestine. *J. Physiol.* 444:743–761.

Stark, M.E., Bauer, A.J., Sarr, M.G., and Szurszewski, J.H. 1993. Nitric oxide mediates inhibitory nerve input in human and canine jejunum. *Gastroenterology* 104:398–409.

Stark, M.E., and Szurszewski, J.H. 1992. Role of nitric oxide in gastrointestinal and hepatic function and disease. *Gastroenterology* 103:1928–1949.

Stoyanovsky, D., Murphy, T., Anno, P.R., Kim, Y.M., and Salama, G. 1997. Nitric oxide activates skeletal and cardiac ryanodine receptors. *Cell Calcium* 21:19–29.

Taha, Z., Kiechle, F., and Malinski, T. 1992. Oxidation of nitric oxide by oxygen in biological systems monitored by porphyrinic sensor. *Biochem. Biophys. Res. Commun.* 188:734–739.

Takeuchi, T., Niioka, S., Kishi, M., Ishii, T., Nishio, H., Hata, F., Takewaki, T., and Takatsuji, K. 1998. Nonadrenergic, noncholinergic relaxation mediated by nitric oxide with concomitant change in Ca^{2+} level in rectal circular muscle of rats. *Eur. J. Pharmacol.* 353:67–74.

Tewari, K., and Simard, J.M. 1997. Sodium nitroprusside and cGMP decrease Ca^{2+} channel availability in basilar artery smooth muscle cells. *Pflugers Arch.* 433:304–311.

Thornbury, K.D., Ward, S.M., Dalziel, H.H., Carl, A., Westfall, D.P., and Sanders, K.M. 1991. Nitric oxide and nitrosocysteine mimic nonadrenergic, noncholinergic hyperpolarization in canine proximal colon. *Am. J. Physiol.* 261:G553–G557.

Wang, X.Y., Sanders, K.M., and Ward, S.M. 1999. Intimate relationship between interstitial cells of Cajal and enteric nerves in the guinea-pig small intestine. *Cell Tissue Res.* 295:247–256.

Ward, S.M., Dalziel, H.H., Bradley, M.E., Buxton, I.L., Keef, K.D., Westfall, D.P., and Sanders, K.M. 1992a. Involvement of cyclic GMP in non-adrenergic, non-cholinergic inhibitory neurotransmission in dog proximal colon. *Br. J. Pharmacol.* 107:1075–1082.

Ward, S.M., Morris, G., Reese, L., Wang, X.Y., and Sanders, K.M. 1998. Interstitial cells of Cajal mediate enteric inhibitory neurotransmission in the lower esophageal and pyloric sphincters. *Gastroenterology* 115:314–329.

Ward, S.M., Xue, C., Shuttleworth, C.W., Bredt, D.S., Snyder, S.H., and Sanders, K.M. 1992b. NADPH diaphorase and nitric oxide synthase colocalization in enteric neurons of canine proximal colon. *Am. J. Physiol.* 263:G277–G284.

Watson, M.J., Bywater, R.A., Taylor, G.S., and Lang, R.J. 1996. Effects of nitric oxide (NO) and NO donors on the membrane conductance of circular smooth muscle cells of the guinea-pig proximal colon. *Br. J. Pharmacol.* 118:1605–1614.

Willmott, N., Sethi, J.K., Walseth, T.F., Lee, H.C., White, A.M., and Galione, A. 1996. Nitric oxide-induced mobilization of intracellular calcium via the cyclic ADP-ribose signaling pathway. *J. Biol. Chem.* 271:3699–3705.

Wood, J.D. 1994. Application of classification schemes to the enteric nervous system. *J. Auton. Nerv. Syst.* 48:17–29

Wu, X., Haystead, T.A., Nakamoto, R.K., Somlyo, A.V., and Somlyo, A.P. 1998. Acceleration of myosin light chain dephosphorylation and relaxation of smooth muscle by telokin. Synergism with cyclic nucleotide-activated kinase. *J. Biol. Chem.* 273:11362–11369.

Xu, L., Eu, J.P., Meissner, G., and Stamler, J.S. 1998. Activation of the cardiac calcium release channel (ryanodine receptor) by poly-*S*-nitrosylation. *Science* 279:234–237.

Xue, C., Pollock, J., Schmidt, H.H., Ward, S.M., and Sanders, K.M. 1994. Expression of nitric oxide synthase immunoreactivity by interstitial cells of the canine proximal colon. *J. Auton. Nerv. Syst.* 49:1–14.

Young, H.M., Ciampoli, D., Johnson, P.J., and Stebbing, M.J. 1996. Inhibitory transmission to the longitudinal muscle of the mouse caecum is mediated largely by nitric oxide acting via soluble guanylyl cyclase. *J. Auton. Nerv. Syst.* 61:103–108.

Young, H.M., Furness, J.B., Shuttleworth, C.W., Bredt, D.S., and Snyder, S.H. 1992. Co-localization of nitric oxide synthase immunoreactivity and NADPH diaphorase staining in neurons of the guinea-pig intestine. *Histochemistry* 97:375–378.

Young, H.M., McConalogue, K., Furness, J.B., and De Vente, J. 1993. Nitric oxide targets in the guinea-pig intestine identified by induction of cyclic GMP immunoreactivity. *Neuroscience* 55:583–596.

Zhang, Y., Vogalis, F., and Goyal, R.K. 1998. Nitric oxide suppresses a Ca(2+)-stimulated Cl^- current in smooth muscle cells of opossum esophagus. *Am. J. Physiol.* 274:G886–G890.

6
Nitric Oxide and Peripheral Neurotransmission in the Lung

MARIA G. BELVISI

Introduction

Nerves regulate several aspects of airway function (Barnes, 1992). However, for the purposes of this chapter, I will focus on the role of nitric oxide (NO)-containing nerves in the control of airway smooth muscle function. Neural control of airway smooth muscle is very complex, since in addition to cholinergic and adrenergic innervation there is a nonadrenergic, noncholinergic (NANC) innervation. The existence of a NANC nervous system in the gastrointestinal tract, which controls gut motility, sphincters, and secretions, had previously been established in vertebrates from fish to humans (Burnstock, 1972). The airways develop embryologically from the foregut, and so the existence of NANC nerves in the respiratory tract was not an unexpected finding.

The first conclusive evidence that pointed to the existence of a NANC relaxant response in airway smooth muscle came with the development of potent adrenergic receptor antagonists. Neural relaxation responses evoked by electrical field stimulation (EFS) in guinea pig trachea were not altered in the presence of the muscarinic antagonist atropine and were only partially inhibited by propranolol (Coburn and Tomita, 1973). This response had stimulus characteristics similar to those of NANC inhibitory nerves described in other tissues, such as the gut. Subsequently, NANC bronchodilator responses have been demonstrated in airways smooth muscle in vitro by EFS in the human, guinea pig, cat, ferret, sheep, horse, mouse, cow, and pig (Barnes et al., 1991). NANC relaxations can also be demonstrated in situ (Yip et al., 1981) or in vivo by electrical stimulation of the cervical vagus nerve (Diamond and O'Donnell, 1980; Irvin et al., 1980) and by reflex stimulation of the larynx (Ichinose et al., 1988; Lammers et al., 1988; Michoud et al., 1988). The relaxant response is abolished by tetrodotoxin and therefore is assumed to be neural in origin. In several species, both adrenergic and NANC pathways coexist, but in human airways, the NANC response is the only neural bronchodilator mechanism (Richardson, 1981). In contrast, the dog (Suzuki et al., 1976; Kannan and Daniel, 1980; Russell,

1980) and rat (Satchell, 1982) airways appear to lack a functional NANC relaxant response. However, the neural relaxant response is not always consistent throughout the airways in either the density of innervation or receptor population (Doidge and Satchell, 1982). In human airways the NANC relaxant response is greatly reduced in the peripheral compared with the central airways (Belvisi et al., 1995; Ward et al., 1995a).

On the whole, most experiments in the literature describing patterns of innervation in the airways have centred on developing in vitro systems of measuring smooth muscle relaxation. In this manner the effects of EFS, which stimulates all nerves in a preparation, on the development of isometric tension by the trachealis or bronchial smooth muscle have been determined in the presence and absence of various drugs. From these experiments it was elucidated that the smooth muscle of mammalian airways receives a dual contractile and relaxant innervation (Belvisi and Bai, 1994; Belvisi et al., 1995). In general, neural relaxation of airway smooth muscle is achieved via activation of adrenergic and NANC neural pathways (Ellis and Undem, 1994). However, the sympathetic innervation to airway smooth muscle is species-dependent and may be sparse or even absent (Mann, 1971). Moreover, in humans, sympathetic nerves innervate bronchial blood vessels, submucosal glands, and parasympathetic ganglia, and there are few, if any, nerve fibers supplying the airways smooth muscle (Richardson and Béland, 1976). Therefore, at least in human airways, the major neural bronchodilator pathway is the NANC system. This chapter will discuss the evidence that NO is the NANC neurotransmitter involved in neurally mediated relaxation of airways smooth muscle. Since in inflammatory airway diseases such as asthma changes in bronchial smooth muscle tone can occur very rapidly, it has been suggested that this could be due to a defect in the autonomic control of the airways smooth muscle (Barnes, 1986). This could manifest itself as an increase in the constrictor and a decrease in the dilator control of the airways. Therefore, if the NANC dilator innervation is dysfunctional in inflammatory conditions, its absence may lead to exaggerated bronchoconstriction (Belvisi et al., 1995).

Although identification of the mediators of this NANC response has been the subject of much research, the identity of the putative neurotransmitter or neurotransmitters has remained obscure until recently. Several candidates have been proposed to be mediators involved in the NANC response. γ-Amino-*n*-butyric acid (GABA), opiates, and the prostaglandins were thought to be unlikely candidates for the role of NANC transmitter. GABA failed to mimic the effects of nerve stimulation, and naloxone (opioid receptor antagonist) and indomethacin (an inhibitor of prostaglandin production) failed to reduce or abolish the inhibitory response in bovine trachea (Cameron et al., 1983). More promising candidates have included ATP and, more recently, vasoactive intestinal peptide (VIP) and NO. The reason these specific mediators were investigated to assess their involvement in the NANC relaxant response in the airways was because they have also been implicated in NANC neural relaxation responses of the gastrointestinal and genitourinary tract (Burnstock et al., 1970).

Role of Nitric Oxide in the Neural Bronchodilator Response

Nitric oxide formed from L-arginine by NO synthase (NOS) is released from a wide variety of cells (Nathan, 1992). Several isoforms of NOS have now been isolated, purified, cloned, and expressed (Forstermann et al., 1991). The isoform present in endothelial cells is a 135-kDa protein located in the membrane fraction (Pollock et al., 1991), whereas neuronal or brain NOS is a 155-kDa protein located in the soluble fraction (Bredt and Snyder, 1990; Mitchell et al., 1991; Schmidt et al., 1991). Bacterial lipopolysaccharide (LPS) or cytokines induce macrophages, vascular smooth muscle cells, endothelial cells, neutrophils and pulmonary epithelial cells (Hamid et al., 1993; Robbins et al., 1994), and other cells to express a different isoform of NOS (iNOS) (Forstermann et al., 1991; Nathan, 1992). Endogenously produced NO may play an integral role in many physiological and pathophysiological events in the lung. It seems to be involved in the neural NANC bronchodilator system in human airways, in vasodilator mechanisms, and in the regulation of airway and pulmonary blood flow and is known to be produced as a consequence of the inflammatory process (Barnes and Belvisi, 1993). All isoforms of NOS are inhibited by guanidino nitrogen-substituted L-arginine analogues such as N^G-monomethyl-L-arginine (L-NMMA), N^G-nitro-L-arginine (L-NNA), and N^G-nitro-L-arginine methylester (L-NAME). These compounds have been used as tools to demonstrate the role of NO in numerous physiological and pathophysiological events.

Nitric oxide synthase inhibitors have been shown to inhibit the NANC neural relaxation response evoked by EFS in guinea pig trachea in vitro by approximately 50% (Li and Rand, 1991), suggesting a role for NO in neurotransmission. Similar results have been observed in human, cat, pig and horse airways (Belvisi et al., 1995), although, in contrast to guinea pig airways, the inhibition evoked by NOS inhibitors was almost complete. Experimental evidence suggests that certain substances (hydroquinone, superoxide anions) reduce relaxations to exogenous NO but not to NANC nerve stimulation. Therefore, it is still in doubt as to whether it is NO itself that is released as the NANC transmitter or NO attached to a carrier molecule (e.g., released as a nitrosothiol).

Guinea Pig

There is a neural bronchodilator response in guinea pig airways. This neurogenic relaxation response, although partly noradrenergic, is also mediated by activation of NANC pathways (Coburn and Tomita, 1973; Coleman and Levy, 1974; Taylor et al., 1984). Epithelial removal has no effect on the NANC response in guinea pig airways. α-Chymotrypsin, a proteolytic enzyme that degrades VIP, inhibits the NANC neural response to EFS by approximately 35%, and incubation with VIP antisera markedly reduces the response. This suggests that VIP and related peptides such as peptide histidine isoleucine, released during EFS of guinea pig trachea, may partly mediate NANC relaxation responses (Ellis and Farmer, 1989a,b; Tucker et al., 1990; Li and Rand, 1991; Bai and Bramley, 1993). Therefore, it would appear that a substantial part of the NANC relaxant response in guinea pig trachea is mediated by a nonpeptide component.

The peptidase-resistant component of the NANC relaxation response to EFS, evoked in precontracted tissue, is attenuated in a concentration-dependent manner by L-NNA and L-NAME (Tucker et al., 1990; Li and Rand, 1991; Bai and Bramley, 1993). The inhibition observed was approximately 89%, but this was for relaxations elicited by low stimulation frequencies (4 Hz) (Tucker et al., 1990). However, in some reports, L-NAME completely abolished NANC relaxation responses at lower frequencies of stimulation (1 Hz) (Li and Rand, 1991). In addition, L-NAME was more potent than L-NMMA in reducing NANC relaxations. The reason for this potency difference is not clear, but it may be that L-NMMA is less effective, since it can also act as a substrate for NOS (Hecker et al., 1990), or it may be due to an effect other than inhibition of the enzyme. In fact, L-NMMA, but not L-NAME, has recently been shown to inhibit the endothelial cell L-arginine transporter (Bogle et al., 1993) and so may inhibit its own transport into the cell. The effect of these NOS inhibitors is stereoselective, since D-NNA, D-NAME, and D-NMMA are without effect (Li and Rand, 1991; Tucker et al., 1990). The inhibitory effects of L-NNA, L-NAME, and L-NMMA are partially reversed by L-arginine but not D-arginine (Tucker et al., 1990; Li and Rand, 1991; Bai and Bramley, 1993). There are several reasons why reversal by L-arginine is only partial. L-NAME and L-arginine may have different abilities to access intact cells. These enantiomer-specific effects are similar to those which have been observed in other tissues that exhibit NANC relaxant responses, such as the anococcygeus muscle (Li and Rand, 1989, 1990). Nitric oxide synthase inhibitors do not affect responses to sodium nitroprusside or isoprenaline; this is more evidence suggesting that a component of the NANC relaxation response in guinea pig trachea is mediated by NO or an NO-related compound.

Interestingly, there is some evidence in other organs, such as gastrointestinal tract, that VIP stimulates the release of NO from gastric muscle cells, so that NO acts as an indirect transmitter of relaxation (Grider et al., 1992). However, in the airways, L-NNA and L-NAME had no effect on relaxation responses to VIP (Tucker et al., 1990; Li and Rand, 1991; Bai and Bramley, 1993). Therefore, it is unlikely that NO is released as a secondary event by the release of VIP from airway nerves.

More evidence implicating NO in the neural control of airway tone comes from immunohistochemical studies describing the presence of the enzyme NOS in nerve fibers that project to the airways. In the guinea pig, the origin of NOS-containing nerves has been demonstrated, by NOS-immunoreactivity and NADPH-diaphorase staining, to be extrinsic ganglia (jugular, nodose, and stellate ganglia), with no positive staining in the intrinsic parasympathetic ganglia (Fischer et al., 1993).

The release of the NANC transmitter in guinea pig trachea is Ca^{2+}-dependent, since relaxant responses to NANC stimulation are reduced or abolished at low frequencies of stimulation by ω-conotoxin, which inhibits Ca^{2+} influx through neuronal N-type channels (De Luca et al., 1990; Li and Rand, 1991). With respect to the classical neurotransmitters, this could suggest that exocytotic release of transmitter is taking place. However, this may not be the case for NO, since constitutive NOS contained in neurons is a Ca^{2+}-dependent enzyme, and therefore the Ca^{2+} entry may be purely to activate the enzyme within the nerve terminals.

Cat

In the cat trachea, the NOS inhibitor L-NAME completely inhibited NANC responses as measured as changes in isometric force of contraction, evoked by EFS in tissues precontracted with 5-HT (Fisher et al., 1993). A 10-fold greater concentration of L-arginine, the substrate for NOS, reversed this inhibitory response. These results suggest that the NANC response evoked by EFS in cat trachea is mediated primarily by NO.

In contrast, other workers have demonstrated that NOS inhibitors failed to affect EFS-evoked NANC relaxation responses of cat intrapulmonary bronchi precontracted with 5-HT at concentrations that abolished acetylcholine-induced vascular relaxation in cat femoral artery and thoracic aorta (Diamond et al., 1992). In addition, NOS inhibitors had no effect on NANC relaxant responses evoked by vagal stimulation in mechanically ventilated cats in which airways tone had been elevated by 5-HT (Diamond et al., 1992). These results, in contrast to those of Fisher et al. (1993), do not appear to support a role for NO as a mediator of the NANC relaxant response in cat airways.

More recently, data have been presented that suggest that at least two neurotransmitters are involved in NANC neurotransmission (Jing et al., 1995). These workers demonstrated that EFS applied to the tracheal smooth muscle during contraction induced by 5-HT in the presence of atropine and guanethidine elicited a monophasic NANC relaxation. By contrast, NANC relaxation elicited in peripheral airways was biphasic, consisting of an initial fast component that was blocked by L-NAME, followed by a second slow component that was not affected by L-NAME (Takahashi et al., 1995). These results indicate that at least two neurotransmitters, possibly NO or NO-containing compounds and VIP, are involved in NANC neurotransmission and the distribution of the two components differs in the central and peripheral airways.

Pig

In pig tracheal smooth muscle that has been precontracted with carbachol and in which the isometric force of contraction is monitored, EFS evokes a frequency-dependent relaxation response that is NANC in origin (Kannan and Johnson, 1992a). This NANC response is completely inhibited by NOS inhibitors and reversed by L-arginine in a stereospecific manner (Kannan and Johnson, 1992b). In addition, in the presence of a NOS inhibitor, VIP, the nicotinic cholinoceptor agonist dimethylphenylpiperazinium bromide (DMPP) and isoprenaline relaxed carbachol-induced tone in pig trachea, implying that none of the aforementioned agents relax tracheal smooth muscle via a mechanism involving NO. These results seem to indicate that NO may be a transmitter involved in NANC neurotransmission in pig trachea. In fact, nerves immunoreactive for constitutive NOS have been localized in the bronchial wall of the pig adjacent to blood vessels, submucosa, and smooth muscle (Belvisi and Bai, 1994).

Rabbit

NANC relaxant responses, evoked by EFS, can also be demonstrated in rabbit smooth muscle but not in bronchial smooth muscle or in lung parenchymal strips (Doidge and Satchell, 1982; Fame et al., 1993).

Horse

In equine tracheal smooth muscle that has been pretreated with indomethacin, atropine, phentolamine, and propranolol, EFS evoked a frequency-dependent NANC relaxation response in vitro (Broadstone et al., 1991). This NANC relaxant innervation is mainly limited to the trachea and main bronchi. Interestingly, this response is absent in the third-generation airways of horses with recurrent obstructive disease (heaves). Recently, it has been demonstrated that the NANC relaxation response is completely abolished by inhibitors of NOS, suggesting that the NANC response is mediated by NO (Yu et al., 1994).

Ferret

Notric oxide synthase and VIP have been localized in a subpopulation of neurons within the plexus of the ferret trachea. The nerve cell bodies were located in specific ganglia and in the nerve fibers associated with tracheal smooth muscle and blood vessel walls (Dey et al., 1993). However, there is no functional evidence, as yet, for a NANC relaxant response.

Human

There is a prominent NANC response in human airways in vitro that is blocked in a concentration-dependent manner by the NOS inhibitor L-NAME (Belvisi et al., 1992a,b). This would seem to indicate that NO is the only demonstrable mediator involved in the NANC response in human tracheal smooth muscle. In these experiments L-NAME had no significant effect on relaxation response curves to sodium nitroprusside (SNP) in human tracheal and bronchial smooth muscle, demonstrating that L-NAME inhibits NOS and does not act via blockade of NO-dependent responses or by inhibition of any responses that are cyclic GMP-dependent (Bai and Bramley, 1993; Belvisi et al., 1992a). L-NAME was also without effect on relaxation responses to VIP and isoprenaline (Bai and Bramley, 1993), which is in agreement with the data described for guinea pig airways. D-NAME was ineffective at producing inhibition of the NANC response, and the inhibitory effect of L-NAME was partially reversed by L-arginine but not D-arginine (Belvisi et al., 1992a,b). These effects, which are enantiomer-specific, are similar to those described in guinea pig trachea (Li and Rand, 1991; Tucker et al., 1990).

NANC relaxant responses may also be evoked by EFS in human peripheral bronchioles (0.5–2 mm inner diameter) and central airways (5–12 mm inner diameter) (Ellis and Undem, 1992; Rand and Li, 1995). Ellis and Undem (1992) have suggested that the NANC innervation is quantitatively similar in central and peripheral airways. However, these authors did not compare NANC responses evoked by EFS in trachea and main bronchi. Other investigators have suggested that the NANC response diminishes as the size of the airway decreases (Palmer et al., 1986). Nitric oxide synthase inhibitors seem to inhibit NANC relaxant responses to EFS in human bronchial smooth muscle in vitro (Ellis and Undem, 1992; Bai and Bramley, 1993), and 3-morpholinosydnonimine (SIN-1), a NO donor, relaxes both central and peripheral airways (Ellis and Undem, 1992), suggesting that NANC responses may be mediated by NO. Ellis and Undem (1992) demonstrated that there was almost complete inhibition by L-NNA of the tetrodotoxin (TTX)-sensitive portion of the NANC relaxant response in human peripheral and central airways precontracted with histamine (3 μM). This study is in agreement with studies on NANC responses of human tracheal smooth muscle (Belvisi et al., 1992a). In contrast, Bai and Bramley (1993) found that L-NAME inhibited only approximately 50% of the neurally mediated airways smooth muscle relaxation in human bronchi (Bai and Bramley, 1993). This study seems to suggest that a large TTX-sensitive residual relaxation persists after NOS inhibition in human bronchi. However, in this later study, the tissues were precontracted with methacholine before NANC responses were elicited, and therefore atropine was not added to the bathing medium during the course of the experiment. The omission of atropine from the experiment could lead to the production of a certain amount of functional antagonism, which may have reduced the magnitude of the inhibitory effect. Alternatively, in these experiments, acetylcholine released from cholinergic nerve terminals could be acting at muscarinic cholinoceptors to release other neurotransmitters or mediators with the ability to relax human airways smooth muscle. Finally, differences between studies may simply reflect differences in tissue viability, the age group studied, the medical history of the patient, or the time from organ removal to the start of the experiment.

Nitric oxide activates soluble guanylyl cyclase after binding to its heme moiety to initiate a three-dimensional change in the shape of the enzyme, which increases its activity and consequently the production of cGMP. The rise in cGMP can initiate a whole series of events, including relaxation of smooth muscle (Rapoport and Murad, 1983), but the mechanism by which this occurs is unknown. However, it appears that neurally mediated NANC relaxations in human trachea are associated with a concomitant selective elevation of the level of cGMP, but not of cAMP, which is inhibited by L-NAME (Ward et al., 1995b). This confirms the hypothesis that the L-arginine-NO-cGMP pathway, and not the VIP-cAMP pathway, is responsible for mediating the NANC relaxant response in this tissue.

It is not certain where the NO is formed or the location of the NOS enzyme. However, the NO released on EFS does not appear to be localized in the epithelium, since its removal has no effect on the NANC response evoked by EFS, at least in guinea pig airways (Watson et al., 1993; Rhoden and Barnes, 1989). Recently, in human trachea obtained at autopsy, neuronal NOS immunoreactivity has been described in nerve fibers present in airway smooth muscle, around submucosal glands and blood vessels (Fischer and Hoffman, 1996), and in some cases NOS is colocalized with VIP. In addition, the density of nNOS immunoreactivity is reduced from proximal to distal airways, and these data correlate with the functional data demonstrating a reduced NANC relaxation response in peripheral compared with central airways (Ward et al., 1995a). Therefore, in view of the extensive array of studies describing the localization of nNOS in neurons within the airways of several species (Dey et al., 1993; Fischer et al., 1993; Hassall et al., 1993; Fischer and Hoffman, 1996) and its correlation with functional data, it is more likely that NO is released from nerves to evoke an NANC relaxant response than that another neurotransmitter substance induces the release of NO from another cell type, such as endothelial, epithelial, or airway smooth muscle cells.

Distribution of NANC Responses in the Human Respiratory Tract

In human airways in vitro, NANC responses evoked by EFS were progressively reduced from the main airways (trachea and main bronchi) through the peripheral airways (3–10mm) to the distal airways (<3mm) (Ward et al., 1995a) (Figure 6.1). This functional decrease was associated with a

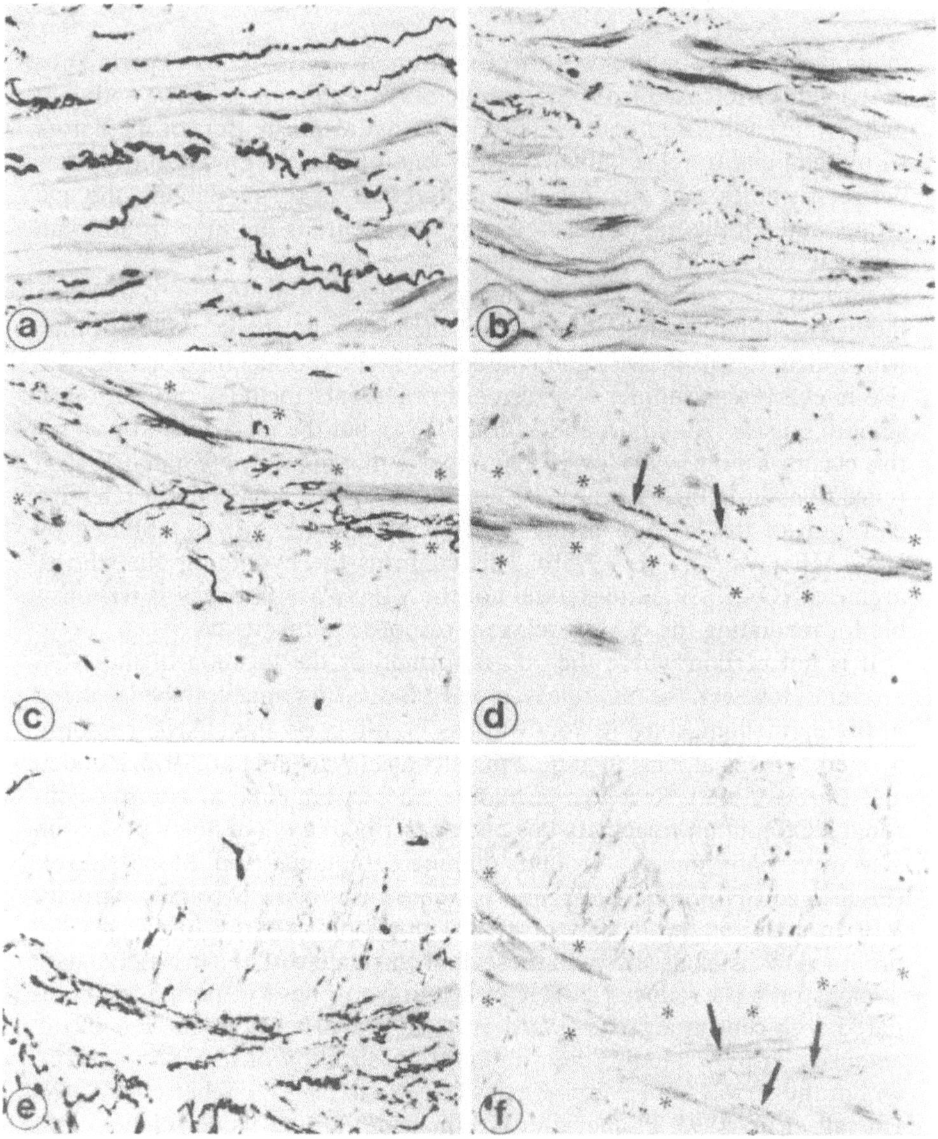

FIGURE 6.1. Nerve immunostaining at three levels of the human tracheobronchial tree. Nerves stained for protein gene product 9.5 (PGP) 9.5 (used as an estimate of total nerve density) (a,c,e) and NOS (b,d,f) are shown in equivalent areas of serial sections from main (a,b; trachea/main bronchi), proximal (c,d; 3–10 mm), and distal airways (e,f; <3 mm). Smooth muscle is recognizable by slight background staining, appearing in most of panels a and b and as a thiner band in panels c through f (edges defined by asterisks). A dense plexus of nerves is seen in the muscle at all airway levels, NOS staining appearing as fine varicose fibers (arrows) that become sparse in more distal airways (original magnification: ×240). From Ward et al. (1995a). *Am. J. Respir. Cell Mol. Biol.*, 13:175–184. Copyright 1995 by the American Lung Association.

decrease in the NOS-immunoreactive nerve density, suggesting that the NANC neural relaxations are reduced going down the tracheobronchial tree, apparently due to a decrease in the density of the "nitrergic" innervation (Ward et al., 1995a). In contrast, Ellis and Undem (1992) found no significant difference between NANC relaxations in human central (5–12 mm internal diameter) and peripheral (0.5–2 mm internal diameter) airways. However, responses in the smaller airways were not compared with those in the larger airways (trachea), where the differences may have been more profound.

The reduction in NANC responses down the human tracheobronchial tree observed by Ward et al. (1995a) in human airways is also consistent with a number of studies in other species. In feline airways both in vivo (Diamond and O'Donnell, 1980) and in vitro (Altiere and Diamond, 1984), the NANC response is reduced in the distal bronchi. Similar results were found for the NO-mediated NANC response in equine airways (Broadstone et al., 1991; Yu et al., 1993). In guinea pig airways NANC relaxant responses were obtained in trachea but not bronchial smooth muscle (Grundstrom et al., 1981). Undem et al. (1990), however, showed that when the noncholinergic contractions were inhibited by capsaicin desensitization and the tone was raised with histamine, NANC relaxations could be elicited by EFS in the mainstem bronchi. These data are supported by anatomical studies demonstrating the existence of NOS-positive nerves in the peripheral bronchi of the guinea pig (Fischer and Hoffman, 1996). However, the NANC relaxant response to EFS in the guinea pig trachea is still more prominent in the cervical than in the thoracic trachea (Ellis and Undem, 1990). Reduced NANC responses have also been demonstrated in rabbit, monkey (Doidge and Satchell, 1982), and bovine (Palmer et al., 1985) distal bronchi.

Therefore, the results obtained in several different mammalian species all seem to support the theory that NANC nerves exhibit their primary influence on airways located in the conducting airways rather than the gas-exchange regions of the lung. However, the functional significance of this pattern of innervation is unclear.

Involvement of Nitric Oxide in Modulation of Neurogenic Responses

Modulation of Cholinergic Neurotransmission

Nitric oxide synthase inhibition markedly potentiates cholinergic neural responses in guinea pig and human airways (Belvisi et al., 1991, 1993a; Brave et al., 1991; Ward et al., 1993) (Figure 6.2). This implies that endogenous NO, released by EFS, modulates cholinergic neurotransmission in guinea pig trachea. Several NOS inhibitors have been demonstrated

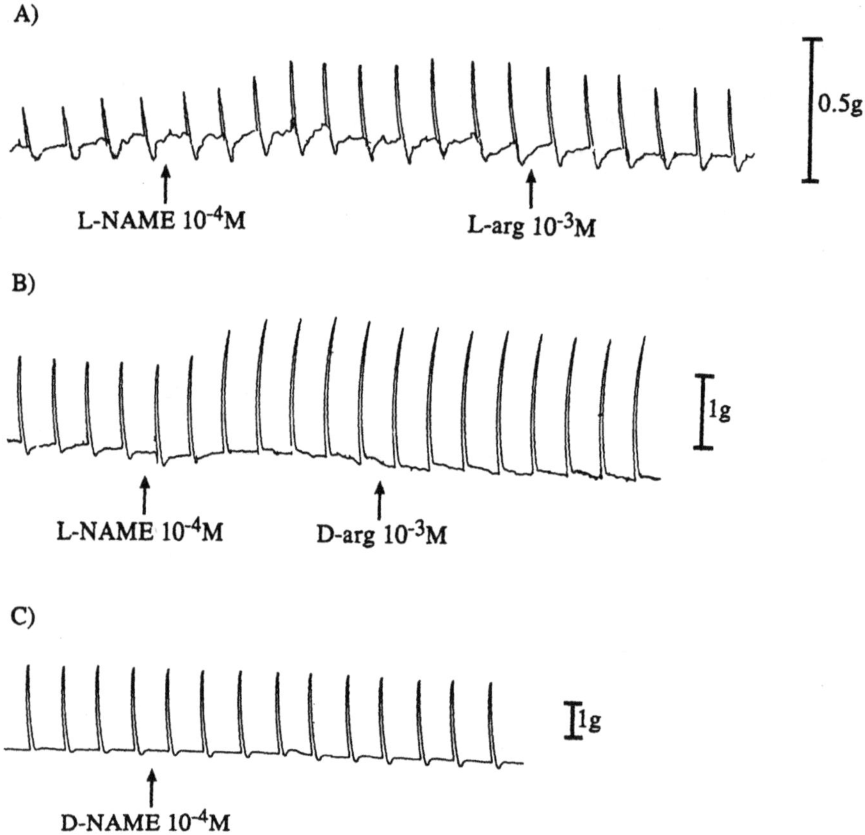

FIGURE 6.2. Trace showing the effects of D-/L-N^G-nitroarginine methyl ester (D-/L-NAME) on cholinergic contractile responses elicited to electrical field stimulation (EFS; 40 V, 0.5 ms, 4 Hz for 15 s in every 4 min) in human main stem bronchus. (A) and (B) The enhancement of the contractile response by L-NAME (10^{-4} M) and its reversal by (A) L-arginine (L-arg; 10^{-3} M), but not by (B) D-arginine (D-arg; 10^{-3} M). (C) D-NAME (10^{-4} M), the inactive isomer of L-NAME, has no effect on the contractile response. From Belvisi and Bai (1994). In *Airways Smooth Muscle: Structure, Innervation and Neurotransmission,* D. Raeburn and M.A. Giembycz eds. pp. 157–187. Copyright 1994 by Birkhäuser Verlag, Basel.

to enhance cholinergic responses to EFS in vitro in a concentration-dependent manner without affecting contractile responses to exogenously administered acetylcholine (Brave et al., 1991; Belvisi et al., 1993a; Ward et al., 1993). These effects were enantiomer-specific. The NOS inhibitors L-NAME and L-NMMA evoked maximal enhancement of cholinergic contractile responses in main airways, and this became smaller in segmental and subsegmental airways, suggesting that the NO-mediated NANC response is less prominent in lower airways. Epithelial removal did not influence the ability of NOS inhibitors to potentiate contractile responses to EFS (Brave et al., 1991), suggesting that the epithelium, in this case, was not acting as a diffusion barrier or as a site of metabolism, and even more importantly, it suggests that NO is not released from the epithelium by another transmitter.

The site of action of the endogenously released NO may be prejunctional to inhibit the release of acetylcholine and thereby reduce cholinergic responses. However, NOS inhibitors did not affect overflow of ^{3}H, evoked by EFS, from tissues previously loaded with [^{3}H]choline (Brave et al., 1991; Ward et al., 1993). These data suggest that EFS of guinea pig and human tracheal smooth muscle evokes the release of NO, which, by an inhibitory action on the smooth muscle, functionally antagonizes the actions of the parasympathetic neurotransmitter.

Modulation of Excitatory NANC Neurotransmission

Antidromic stimulation of C-fiber endings in rodents results in bronchoconstriction and an inflammatory response that is due to the release of neuropeptides from sensory nerve endings (Barnes et al., 1991). In guinea pig bronchi, EFS results in a rapid transient cholinergic response and a longer-lasting NANC contraction that is resistant to atropine, propranolol, prazocin, yohimbine, clemastine, and guanethidine (Grundstrom et al., 1981), but is blocked by neurokinin (NK) receptor antagonists (for NK_1 and NK_2 receptors), and is therefore thought to be due to the release of substance P (SP) and neurokinin A (NKA) (Maggi et al., 1991).

Interestingly, NOS inhibitors have no effect on NANC constrictor responses evoked by EFS in guinea pig bronchi (Belvisi et al., 1993a). This result may seem at odds with the observation that cholinergic responses are increased after NOS inhibition, probably by an effect involving functional antagonism of the cholinergic contractile response. One explanation for this apparent discrepancy between the modulatory effects of L-NAME on excitatory NANC and cholinergic neurotransmission could be related to the innervation of the tracheobronchial tree of the guinea pig. It has previously been demonstrated that guinea pig tracheal, but not bronchial, smooth muscle exhibits NANC relaxation responses to EFS (Grundstrom et al., 1981), so that L-NAME may have no effect on excitatory NANC neural responses in bronchi because NO may not be released at this level of the airways in the guinea pig.

Alternatively, L-NAME potentiated NANC neural bronchoconstrictor responses evoked by vagal stimulation in guinea pigs in vivo in a dose-related fashion (Lei et al., 1993). This suggests that endogenous NO, produced in association with stimulation of the vagus nerves, regulates the magnitude of the NANC bronchoconstriction. The specificity of the effect was confirmed by the ineffectiveness of D-NAME. However, bronchoconstrictor responses evoked by exogenous SP and NKA were not altered by L-NAME, indicating that NO is probably released in association with nerve activation rather than being released by tachykinins or as a result of the bronchoconstriction. It also suggests that NO is not regulating the excitatory NANC response by acting as a neurokinin receptor antagonist but is probably affecting NANC neurotransmitter release, or that functional antagonism is responsible for this modulatory effect as per the effects of NO on cholinergic neurotransmission. The two observations detailed above demonstrate that L-NAME potentiates NANC bronchoconstrictor responses evoked by vagal stimulation in vivo, measured as a change in overflow pressure, but has no effect on NANC constrictor responses evoked by EFS in main and hilar bronchi in vitro. This could indicate that the measured changes in overflow pressure reflect narrowing of airways other than extrapulmonary bronchi.

In addition, L-NAME attenuates the inhibitory effect of cigarette smoke on the excitatory NANC bronchoconstrictor response evoked by vagal stimulation in the guinea pig in vivo (Emms and Rogers, 1997). The authors concluded that endogenous NO and NO derived directly from cigarette smoke mediate inhibition of neurogenic bronchoconstriction.

Functional Significance of the NANC Response

The exact role of the NANC relaxant response in health and disease has not yet been defined. However, several theories have been put forward to explain the purpose of this phenomenon. The first and probably the most obvious explanation is that the NANC inhibitory system may play an important physiological role in the regulation of bronchomotor tone (Belvisi et al., 1995). Alternatively Coburn and Tomita (1973) hypothesized that it may be important in the control of the cough reflex. Finally, a more heretical explanation that has been put forward is that the NANC relaxant response is an innocuous response remaining from a primitive inhibitory system that has been conserved through the evolutionary process (Diamond and Altiere, 1989).

NANC Inhibitory Pathways in Disease

The NANC bronchodilator nerves are the only neural relaxant pathway in human airways. Therefore it is important to determine whether there is any defect in the ability of these nerves to function in diseased airways. In fact, it has been suggested that a defective function of the NANC nerves may contribute to bronchoconstriction and bronchial hyperresponsiveness in asthma (Lammers et al., 1992). On the basis of experiments performed in animals, it seemed as though this hypothesis could be true. NANC nerve stimulation potently inhibited antigen-induced bronchoconstriction and the increase in arterial plasma histamine in cats (Miura et al., 1990a), suggesting that the transmitter substances responsible for the NANC dilator response prevent the release of mediators such as histamine from activated sensitized mast cells (Undem et al., 1983). Furthermore, the same workers also demonstrated that the bronchodilator action of VIP and the neural relaxation response were reduced after allergen exposure, and that the protease inhibitor leupeptin abolished the allergen-induced NANC dysfunction in sensitized cats (Miura et al., 1990b). These results would seem to indicate that NANC relaxation is less effective in sensitized animals due to the degradation of the putative NANC neurotransmitter, such as VIP, by proteases released during the allergic response. VIP and related peptides are degraded by mast cell proteases such as tryptase and chymase (Tam and Caughey, 1990). This possible increase in mast cell proteases found in allergic conditions may contribute to bronchial hyperresponsiveness and to the decreased VIP immunoreactivity seen in nerves in asthmatic airways (Ollerenshaw et al., 1989), since mast cells are often found in close association with nerves (Kakuta et al., 1989). However, these observations may be more relevant in structures (e.g., human pulmonary vessels rather than airways) and species (guinea pig and cat airways) that receive a NANC innervation that is mediated by a neuropeptide that is susceptible to peptidases such as VIP. More recently it has also been demonstrated that airway allergic inflammation also affects NANC relaxant responses mediated by NO in tissues from antigen-exposed guinea pigs (Miura et al., 1997). This defect in the NANC relaxant response did not appear to be due to a decrease in the number of NOS-containing nerves, but rather to the scavenging of neural NO during the diffusion process from nerve endings to the effective sites of airway smooth muscle.

A reduction in VIP immunoreactivity has recently been reported in the airways of asthmatic patients with severe disease (Ollerenshaw et al., 1989). This loss of VIP may be due to the presence of human tryptase secreted from airway mast cells. However, more recently, preliminary data have emerged suggesting no difference in VIP immunoreactivity in nerves from biopsy samples from normal subjects and mild asthmatics (Howarth et al., 1995). If VIP is the neurotransmitter of NANC nerves in human airways, these data may suggest that there could be a decrease in the NANC dilator response in asthma according to the severity of the disease. However, as yet, there are no conclusive data implicating a role for VIP in NANC neurotransmission, at least in the nerves innervating the airway smooth muscle, in human airways.

However, in human airways in vitro, NANC responses do not appear to be impaired in the airways of patients with chronic airflow limitation (Taylor et al., 1985). Moreover, airways from mild asthmatic patients have been found to have a normal NANC response (Belvisi et al., 1993b). In addition, airways from patients who died during severe asthma attacks showed NANC inhibitory responses similar to those of control airways from nonasthmatic subjects (Bai, 1990). In agreement with the in vitro data, other investigators demonstrated that the degree of bronchodilator response observed in mild asthmatic patients was of similar duration and magnitude as that seen in normal subjects, suggesting that the NANC bronchodilator system was functioning in mild asthmatic subjects (Michoud et al., 1988; Lammers et al., 1992).

In contrast to asthmatic airways, NANC responses were significantly reduced in tissues from patients with cystic fibrosis compared with NANC responses in normal donor tissue (Belvisi et al., 1994). It is possible that "nitrergic" neurotransmission is impaired in inflammatory diseases of the airways, since production of superoxide anions by inflammatory cells, such as neutrophils and eosinophils, would lead to a rapid degradation of neurally released NO. This abnormality in the airway NANC innervation of patients with cystic fibrosis may lead to exaggerated bronchoconstrictor responses. Since the "nitrergic" innervation appears to be dysfunctional in some inflammatory diseases, it was tempting to suggest that NO functions as an endogenous braking mechanism in the airways and that its absence may therefore lead to exaggerated bronchoconstriction. We investigated the effect of NOS inhibition (i.e., effectively removing NANC relaxation responses) on cholinergic constrictor responses evoked by EFS in human donor tissue from trachea to peripheral airways. L-NAME produced a concentration-dependent enhancement of cholinergic neural constrictor responses to EFS, with no effect on cumulative concentration–response curves to acetylcholine in guinea pig and human airways (Brave et al., 1991; Belvisi et al., 1991, 1993a). In human airways, L-NAME evoked maximal enhancement of cholinergic contractile responses in main airways, and this became smaller in segmental and subsegmental airways, suggesting that the NO-mediated NANC response was less prominent in lower airways (Ward et al., 1993), and recently we have demonstrated this to be the case (Ward et al., 1995a). The mechanism of this modulation was determined by studying the effects of endogenously released NO on acetylcholine release evoked by EFS from strips of human tracheal smooth muscle that had been denuded of epithelium. Overflow of ^{3}H, evoked by EFS, in tissues previously loaded with [^{3}H]choline, which seems to be a good marker for measurement of neuronally evoked acetylcholine release, is not affected by NOS inhibitors (Ward et al., 1993). Therefore, it seems that endogenous NO does not modulate cholinergic contractile responses by prejunctional inhibition of acetylcholine release from the nerve terminal. In conclusion, it would appear that NO is probably modulating cholinergic neurotransmission postjunctionally by functional antagonism of acetylcholine at the level of the airway smooth muscle, which could, in theory, oppose cholinergic bronchoconstriction.

Conclusions

In this chapter I have illustrated the species differences in the neural control of the relaxation of airway smooth muscle. This serves to remind us of the problems that might be encountered when studying neural relaxation responses in animal airways and extrapolating the findings to the human condition.

In terms of the criteria for defining whether a substance is a neurotransmitter, it seems that NO differs radically from the classical neurotransmitters such as acetylcholine and noradrenaline. However, the criteria that are satisfied by NO for neurotransmitter status in the airways are as follows. The enzyme that is involved in the synthesis of NO from L-arginine has now been localized in neurones in the airways. Second, exogenously administered NO itself or, alternatively, nitrodilators have been shown to relax airway smooth muscle, and therefore NO is able to mimic the effects of NANC nerve stimulation. Furthermore, inhibition of NO formation with a NOS inhibitor results in the attenuation of the nerve-evoked relaxation of airway smooth muscle. However, this is where the similarity to classical neurotransmitter substances seems to end. The most difficult concept to reconcile, in terms of the classical ideas of neurotransmission, is the absence of a conventional stimulus-secretion coupling mechanism, since the release of NO does not appear to involve vesicular, quantal release of neurotransmitter

However, although this substance seems an unlikely candidate, in that it is a gas which is not stored in synaptic vesicles or released by exocytosis, and which does not act at typical cell membrane-associated receptors, NO may prove to have a more widespread and fundamental role than most classical neurotransmitters. The discovery of NO as a transmitter substance revolutionizes the classical pharmacological basis for neurotransmission and may lead to the identification of other equally unlikely candidates.

In conclusion, the NANC bronchodilator mechanism has been identified as the predominant system in the neural control of human airway smooth muscle relaxation. However, the precise physiological or pathophysiological role of this system remains to be defined. The identification of a disruption in this pathway in tissue from patients with airway inflammation is interesting, but the mechanism behind this dysfunction and the consequences of this are unknown and warrant further study.

References

Altiere, R.J., and Diamond, L. 1984. Relaxation of cat tracheobronchial and pulmonary arterial smooth muscle by vasoactive intestinal peptide: lack of influence of peptidase inhibitors. *Br. J. Pharmacol.* 82:321–328.

Bai, T.R. 1990. Abnormalities in airway smooth muscle in fatal asthma. *Am. Rev. Respir. Dis.* 141:552–557.

Bai, T.R., and Bramley, A.M. 1993. Effect of an inhibitor of nitric oxide synthase on neural relaxation of human bronchi. *Am. J. Physiol: Lung Cell Mol. Physiol.* 8:L425–L430.

Barnes, P.J. 1986. Neural control in health and disease. *Am. Rev. Respir. Dis.* 134:1289–1314.

Barnes, P.J. 1992. Modulation of neurotransmission in airways. *Physiol. Rev.* 72:699–729.

Barnes, P.J., Baraniuk, J.N., and Belvisi, M.G. 1991. Neuropeptides in the respiratory tract. *Am. Rev. Respir. Dis.* 144:1187–1198.

Barnes, P.J., and Belvisi, M.G. 1993. Nitric oxide and lung disease. *Thorax* 48:1034–1043.

Belvisi, M.G., and Bai, T.R. 1994. Inhibitory non-adrenergic non-cholinergic innervation of airways smooth muscle: role of nitric oxide. In: *Airways Smooth Muscle: Structure, Innervation and Neurotransmission* (eds. D. Raeburn and M.A. Giemybycz), pp. 157–187. Birkhauser Verlag, Basel.

Belvisi, M.G., Miura, M., Stretton, C.D., and Barnes, P.J. 1993a. Endogenous vasoactive intestinal peptide and nitric oxide modulate cholinergic neurotransmission in guinea-pig trachea. *Eur. J. Pharmacol.* 231:97–102.

Belvisi, M.G., Stretton, C.D., and Barnes, P.J. 1991. Nitric oxide as an endogenous modulator of cholinergic neurotransmission in guinea-pig airways. *Eur. J. Pharmacol.* 198:219–221.

Belvisi, M.G., Stretton, C.D., Miura, M., Verleden, G.M., Tadjkarimi, S., Yacoub, M.H., and Barnes, P.J. 1992a. Inhibitory NANC nerves in human tracheal smooth muscle: a quest for the neurotransmitter. *J. Appl. Physiol.* 73:2505–2510.

Belvisi, M.G., Stretton, C.D., Yacoub, M.H., and Barnes, P.J. 1992b. Nitric oxide is the endogenous neurotransmitter of bronchodilator nerves in humans. *Eur. J. Pharmacol.* 210:221–222.

Belvisi, M.G., Ward, J.K., Mitchell, J.A., and Barnes, P.J. 1995. Nitric oxide as a neurotransmitter in human airways, *Arch. Int. Pharmacodyn. Ther.* 329:97–110.

Belvisi, M.G., Ward, J.K., Springall, D.R., Tadjkarimi, S., Yacoub, M.H., Polak, J.M., and Barnes, P.J. 1994. Nitrergic innervation in the airways of patients with cystic fibrosis. *Am. J. Respir. Crit. Care Med.* 149:A675.

Belvisi, M.G., Ward, J.K., Tadjkarimi, S., Yacoub, M.H., and Barnes, P.J. 1993b. Inhibitory NANC nerves in human airways: differences in disease and after extrinsic denervation. *Am. Rev. Respir. Dis.* 147:A286.

Bogle, R.G., Moncada, S., Pearson, J.D., and Mann, G.E. 1993. Identification of inhibitors of nitric oxide synthase that do not interact with the endothelial cell L-arginine transporter. *Br. J. Pharmacol.* 105:768–770.

Brave, S.R., Hobbs, A.J., Gibson, A., and Tucker, J.F. 1991. The influence of L-N^G-nitro-arginine on field stimulation-induced contractions and acetylcholine release in guinea-pig isolated smooth muscle. *Biochem. Biophys. Res. Commun.* 179:1017–1022.

Bredt, D.S., and Snyder, S.H. 1990. Isolation of nitric oxide synthetase, a calmodulin-requiring enzyme. *Proc. Natl. Acad. Sci. USA* 87:682–685.

Broadstone, R.V., Leblanc, P.H., Derksen, F.J., and Robinson, N.E. 1991. In vitro responses of airway smooth muscle from horses with recurrent airway obstruction. *Pulmon. Pharmacol.* 4:191–202.

Burnstock, G. 1972. Purinergic nerves. *Pharmacol. Rev.* 24:247–324.

Burnstock, G., Campbell, G., Satchell, D.G., and Smythe, A. 1970. Evidence that adenosine triphosphate or a related nucleotide is the transmitter substance released by non-adrenergic inhibitory nerves in the gut. *Br. J. Pharmacol.* 40:668–675.

Cameron, A.R., Johnston, C.F., Kirkpatrick, C.T., and Kirkpatrick, M.C.A. 1983. The quest for the inhibitory neurotransmitter in bovine tracheal smooth muscle. *Q. J. Exp. Physiol.* 68:413–426.

Coburn, R.F., and Tomita, T. 1973. Evidence for nonadrenergic inhibitory nerves in the guinea-pig trachealis muscle. *Am. J. Physiol.* 224:1072–1080.

Coleman, R.A., and Levy, G.P. 1974. A non-adrenergic inhibitory nervous pathway in guinea-pig trachea. *Br. J. Pharmacol.* 52:167–174.

De Luca, A., Li, C.G., Rand, M.J., Reid, J.J., Thaina, P., and Wong-Dusting, H.K. 1990. Effects of ω-conotoxin GVIA on autonomic neuroeffector transmission in various tissues. *Br. J. Pharmacol.* 101:437–447.

Dey, R.D., Mayer, B., and Said, S.I. 1993. Colocalisation of vasoactive intestinal peptide and nitric oxide synthase in neurons of the ferret trachea. *Neuroscience* 54:839–843.

Diamond, L., and Altiere, R.J. 1989. Airway nonadrenergic noncholinergic inhibitory nervous system. In *Lung Biology in Health and Disease. The Airways: Neural Control in Health and Disease* (eds. M. Kaliner and P.J. Barnes). Marcel Dekker, New York, pp. 343–394.

Diamond, L., Lantta, J., Thompson, D., and Altiere, J. 1992. Nitric oxide synthase inhibitors fail to affect cat airway nonadrenergic noncholinergic inhibitory (NANCI) responses. *Am. Rev. Resp. Dis.* 145:A382.

Diamond, L., and O'Donnell, M. 1980. A non-adrenergic vagal inhibitory pathway to feline airway. *Science* 208:185–188.

Doidge, J.M., and Satchell, D.G. 1982. Adrenergic and nonadrenergic inhibitory nerves in mammalian airways. *J. Auton. Nerv. Syst.* 5:83–99.

Ellis, J.L., and Farmer, S.G. 1989a. Effect of peptidases on nonadrenergic, noncholinergic inhibitory responses of tracheal smooth muscle: a comparison with effects on VIP and PHI-induced relaxation. *Br. J. Pharmacol.* 96:521–526.

Ellis, J.L., and Farmer, S.G. 1989b. The effects of vasoactive intestinal peptide (VIP) antagonists, and VIP and peptide histidine isoleucine antisera on non-adrenergic, non-cholinergic relaxations of tracheal smooth muscle. *Br. J. Pharmacol.* 96: 513–520.

Ellis, J.L., and Undem, B.J. 1990. Non-adrenergic, non-cholinergic contractions in the electrically field stimulated guinea-pig trachea. *Br. J. Pharmacol.* 101:875–880.

Ellis, J.L., and Undem, B.J. 1992. Inhibition by L-N^G-nitro-L-arginine of nonadrenergic noncholinergic relaxations of human isolated central and peripheral airways. *Am. Rev. Respir. Dis.* 146:1543–1547.

Ellis, J.L., and Undem, B.J. 1994. Pharmacology of non-adrenergic, non-cholinergic nerves in airway smooth muscle. *Pulmon. Pharmacol.* 7:205–223.

Emms, J.C., and Rogers, D.F. 1997. Cigarette smoke-induced inhibition of neurogenic bronchoconstriction in guinea-pigs in vivo: involvement of exogenous and endogenous nitric oxide. *Br. J. Pharmacol.* 122:779–785.

Fame, T.M., Loader, J.E., Graves, J., Colasurdo, G.N., and Larsen, G.L. 1993. Decrease in the airways' non-adrenergic, non-cholinergic inhibitory system in allergen sensitised rabbit. 1991. *Am. Rev. Respir. Dis.* 147:A285.

Fischer, A., and Hoffman, B. 1996. Nitric oxide synthase in neurons and nerve fibres of lower airways and vagal sensory ganglia of man. *Am. J. Respir. Crit. Care Med.* 154:209–216.

Fischer, A., Mundel, P., Mayer, B., Preissler, U., Philippin, B., and Kummer, W. 1993. Nitric oxide synthase in guinea-pig lower airway innervation. *Neurosci. Lett.* 149:157–160.

Fisher, J.T., Anderson, J.W., and Waldron, M.A. 1993. Nonadrenergic noncholinergic neurotransmitter of feline trachealis: VIP or nitric oxide. *J. Appl. Physiol.* 74:31–39.

Forstermann, U., Schmidt, H.H.H.W., Pollock, J.S., Sheng, H., Mitchell, J.A., Warner, T.D., Nakane, M., and Murad, F. 1991. Isoforms of nitric oxide synthase: characterisation and purification from different cell types. *Biochem. Pharmacol.* 42:1849–1857.

Grider, J.J., Murthy, K.S., Jin, J.G., and Makhlouf, G.M. 1992. Stimulation of nitric oxide from muscle cells by VIP: pre-junctional enhancement of VIP release. *Am. J. Physiol.* 262:G774–G778.

Grundstrom, N., Andersson, R.G.G., and Wikberg, J.E.S. 1981. Pharmacological characterisation of the autonomous innervation of guinea-pig tracheobronchial smooth muscle. *Acta. Pharmacol. Scand.* 49:150–157.

Hamid, Q., Springall, D.R., Riveros-Moreno, V., Chanez, P., Howarth, P., Redington, A., Bousquet, J., Godard, P., Holgate, S., and Polats, J.M. 1993. Induction of nitric oxide synthase in asthma. *Lancet* 342:1510–1513.

Hassall, C.J.S., Saffrey, M.J., and Burnstock, G. 1993. NADPH-diaphorase activity by guinea-pig paratracheal neurones. *NeuroReport* 4:49–52.

Hecker, M., Mitchell, J.A., Harris, H.J., Katsura, M., Thiemermann, C., and Vane, J.R. 1990. Endothelial cells metabolize N^G-monomethyl-L-arginine to L-citrulline and subsequently to L-arginine. Biochem. *Biophys. Res. Commun.* 167:1037–1043.

Howarth, P., Springall, D.R., Redington, A.E., Djukanovic, R., Holgate, S.T., and Polak, J.M. 1995. Neuropeptide-containing nerves in endobronchial biopsies from asthmatic and nonasthmatic subjects. *Am. J. Respir. Cell Mol. Biol.* 13:288–296.

Ichinose, M., Inoue, H., Miura, M., and Takishima, T. 1988. Nonadrenergic bronchodilation in normal subjects. *Am. Rev. Respir. Dis.* 138:31–34.

Irvin, C.G., Boileau, R., Tremblay, J., Martin, R.R., and Macklem, P.T. 1980. Bronchodilatation: noncholinergic, nonadrenergic mediation demonstrated in vivo in the cat. *Science* 207:791–792.

Jing, L., Inoue, R., Tashiro, K., Takahashi, S., and Ito, Y. 1995. Role of nitric oxide in inhibition and modulation of excitatory neuroeffector transmission in cat airway. *J. Physiol.* 481:225–237.

Kakuta, Y., Stead, R.H., Perdue, M.H., Marshall, J.S., and Bienenstock, J. 1989. Micro-anatomical relationship of mast cells and nerves in rat lung and trachea. *Am. Rev. Respir. Dis.* 139:A118.

Kannan, M.S., and Daniel, E.E. 1980. Structural and functional study of control of canine tracheal smooth muscle. *Am. J. Physiol.* 238:C27–C33.

Kannan, M.S., and Johnson, D.E. 1992a. Functional innervation of pig tracheal smooth muscle: neural and non-neural mechanisms of relaxation. *J. Pharmacol. Exp. Ther.* 260:1180–1184.

Kannan, M.S., and Johnson, D.E. 1992b. Nitric oxide mediates the neural nonadrenergic, non-cholinergic relaxation of pig tracheal smooth muscle. *Am. J. Physiol.* 262:L511–L514.

Lammers, J.-W., Minette, M., McCusker, M., Chung, K.F., and Barnes, P.J. 1988. Non-adrenergic bronchodilator mechanisms in normal human subjects in vivo. *J. Appl. Physiol.* 64:1817–1822.

Lammers, J.-W.J., Barnes, P.J., and Chung, K.F. 1992. Non-adrenergic, non-cholinergic airway inhibitory nerves. *Eur. Respir. J.* 5:239–246.

Lei, Y.-H., Barnes, P.J., and Rogers, D.F. 1993. Regulation of NANC neural bronchoconstriction in vivo in the guinea-pig: involvement of nitric oxide, vasoactive intestinal peptide and soluble guanylyl cyclase. *Br. J. Pharmacol.* 108:228–235.

Li, C.G., and Rand, M.J. 1989. Evidence for a role of nitric oxide in the NANC-mediated relaxations in rat anococcygeus muscle. *Clin. Exp. Physiol. Pharmacol.* 16:933–938.

Li, C.G., and Rand, M.J. 1990. Nitric oxide and vasoactive intestinal polypeptide mediate non-adrenergic, non-cholinergic inhibitory transmission to smooth muscle of the rat gastric fundus. *Eur. J. Pharmacol.* 191:303–309.

Li, C.G., and Rand, M.J. 1991. Evidence that part of the NANC relaxant response of guinea-pig trachea to electrical field stimulation is mediated by nitric oxide. *Br. J. Pharmacol.* 102:91–94.

Maggi, C.A., Patacchini, R., Rovero, P., and Santicioli, P. 1991. Tachykinin receptors and non-cholinergic bronchoconstriction in the guinea-pig isolated bronchi. *Am. Rev. Respir. Dis.* 144:363–367.

Mann, S.P. 1971. The innervation of mammalian bronchial smooth muscle, the localisation of catecholamines and cholinesterases. *Histochem. J.* 3:319–331.

Michoud, M.-C., Amyot, R., Jeanneret-Grosjean, A., and Couture, J. 1988. Reflex decrease of histamine-induced bronchoconstriction after laryngeal stimulation in humans. *Am. Rev. Respir. Dis.* 136:616–622.

Mitchell, J.A., Sheng, H., Forstermann, U., and Murad, F. 1991. Characterisation of NO synthase in non-adrenergic non-cholinergic nerve containing tissue from the rat anoccocygeus muscle. *Br. J. Pharmacol.* 104:289–291.

Miura, M., Inoue, H., Ichinose, M., Kimura, K., Katsumata, U., and Takishima, T. 1990a. Effect of nonadrenergic, noncholinergic inhibitory nerve stimulation on the allergic reaction in cat airways. *Am. Rev. Respir. Dis.* 141:29–32.

Miura, M., Kimura, K., Takahashi, T., Inoue, H., and Takishima, T. 1990b. Possible mechanisms of the antigen-induced dysfunction of nonadrenergic, noncholinergic inhibitory nervous system. *Am. Rev. Respir. Dis.* 141:A387.

Miura, M., Yamauchi, H., Ichinose, M., Ohuchi, Y., Kageyama, N., Tomaki, M., Endoh, N., and Shirato, K. 1997. Impairment of neural nitric oxide-mediated relaxation after antigen exposure in guinea-pig airways in vitro. *Am. J. Respir. Crit. Care Med.* 156:217–222.

Nathan, C.F. 1992. Nitric oxide as a secretory product of mammalian cells. *FASEB J.* 6:3051–3064.

Ollerenshaw, S., Jarvis, D., Woolcock, A., Sullivan, C., and Scheibner, T. 1989. Absence of immunoreactive vasoactive intestinal polypeptide in tissue from the lungs of patients with asthma. *N. Engl. J. Med.* 320:1244–1248.

Palmer, J.B.D., Cuss, F.M.C., and Barnes, P.J. 1986. VIP and PHM and their role in non-adrenergic inhibitory responses in isolated human airways. *J. Appl. Physiol.* 61:1322–1328.

Palmer, J.B.D., Sampson, A.P., and Barnes, P.J. 1985. Cholinergic and non-adrenergic inhibitory responses in bovine airways: distribution and functional association. *Am. Rev. Respir. Dis.* 128:A283.

Pollock, J.S., Forstermann, U., Mitchell, J.A., Warner, T.D., Schmidt, H.H., Nakane, M., and Murad, F. 1991. Purification and characterisation of particulate endothelium-derived relaxant factor synthase from cultured and native bovine aortic endothelial cells. *Proc. Natl. Acad. Sci. USA* 88:10480–10484.

Rand, M.J., and Li, C.G. 1995. Nitric oxide as a neurotransmitter in peripheral nerves: nature of transmitter and mechanism of transmission. *Annu. Rev. Physiol.* 57:659–682.

Rapoport, R.M., and Murad, F. 1983. Agonist induced endothelium-dependent relaxation in rat thoracic aorta may be mediated through cGMP. *Circ. Res.* 52: 352–357.

Rhoden, K.J., and Barnes, P.J. 1989. Epithelial modulation of NANC and VIP-induced responses: role of neutral endopeptidase. *Eur. J. Pharmacol.* 171:247–250.

Richardson, J.B. 1981. Nonadrenergic inhibitory innervation of the lung. *Lung* 159:315–322.

Richardson, J.B., and Béland, J. 1976. Nonadrenergic inhibitory nervous system in human airways. *J. Appl. Physiol.* 41:764–771.

Robbins, R.A., Barnes, P.J., Springall, D.R., Warren, J.B., Kwon, O.J., Buttery, L.D., Wilson, A.J., Geller, D.A., and Polak, J.M. 1994. Expression of inducible nitric oxide synthase in human bronchial epithelial cells. *Biochem. Biophys. Res. Commun.* 203:209–218.

Russell, J. 1980. Noradrenergic inhibitory innervation of canine airways. *J. Appl. Physiol.* 48:16–22.

Satchell, D. 1982. Non-adrenergic non-cholinergic nerves in mammalian airways: their function and the role of purines. *Comp. Biochem. Physiol.* 72:189–196.

Schmidt, H.H.H.W., Pollock, J.S., Nakane, M., Gorsky, L.D., Forstermann, U., and Murad, F. 1991. Purification of a soluble isoform of guanylyl cyclase-activating-factor synthase. *Proc. Natl. Acad. Sci. USA* 88:365–369.

Suzuki, H., Morita, K., and Kuriyama, H. 1976. Innervation and properties of the smooth muscle of the dog trachea. *Jpn. J. Physiol.* 26:303–320.

Takahashi, N., Tanaka, H., Abdullah, N., Jing, L., Inoue, R., and Ito, Y. 1995. Regional difference in the distribution of L-NAME-sensitive and -insensitive NANC relaxations in cat airway. *J. Physiol.* 488:709–720.

Tam, E.K., and Caughey, G.H. 1990. Degradation of airway neuropeptides by human lung typtase. *Am. J. Respir. Cell Mol. Biol.* 3:27–32.

Taylor, J.F., Paré, P.D., and Schellenburg, R.R. 1984. Cholinergic and non-adrenergic mechanisms in human and guinea-pig airways. *J. Appl. Physiol.* 56:958–965.

Taylor, S.M., Paré, P., Armour, C.L., Hogg, J.C., and Schellenburg, R.R. 1985. Airway reactivity in chronic obstructive pulmonary disease. Failure of in vivo methacholine responsiveness to correlate with cholinergic, adrenergic, or non-adrenergic responses in vitro. *Am. Rev. Respir. Dis.* 132:30–35.

Tucker, J.F., Brave, S.R., Charalambous, L., Hobbs, A., and Gibson, A.J. 1990. L-N^G-Nitro arginine inhibits nonadrenergic, noncholinergic relaxations of guinea-pig tracheal smooth muscle. *Br. J. Pharmacol.* 100:663–664.

Undem, B.J., Dick, E.C., and Buckner, C.K. 1983. Inhibition by vasoactive intestinal peptide of antigen-induced histamine release from guinea-pig minced lung. *Eur. J. Pharmacol.* 88:247–250.

Undem, B.J., Myers, A.C., Barthlow, H., and Weinrelch D. 1990. Vagal innervation of guinea-pig bronchus. *J. Appl. Physiol.* 1336–1346.

Ward, J.K., Barnes, P.J., Springall, D.R., Abelli, L., Tadjkarimi, S., Yacoub, M.H., Polak, J.M. and Belvisi M.G. 1995a. Distribution of human i-NANC bronchodilator and nitric oxide-immunoreactive nerves. *Am. J. Respir. Cell. Mol. Biol.* 13: 175–184.

Ward, J.K., Barnes, P.J., Tadjkarimi, S., Yacoub, M.H., and Belvisi, M.G. 1995b. Neural relaxation in human tracheal smooth muscle (HTSM) is mediated by an increase in cGMP: further evidence for the role of nitric oxide. *J. Physiol.* 483:525–536.

Ward, J.K., Belvisi, M.G., Fox, A.J., Miura, M., Tadjkarimi, S., Yacoub, M.H., and Barnes, P.J. 1993. Modulation of cholinergic bronchoconstrictor responses by endogenous nitric oxide and vasoactive intestinal peptide in human airways in vitro. *J. Clin. Invest.* 92:736–742.

Watson, N.J., Maclagan, J., and Barnes, P.J. 1993. Vagal control of guinea-pig tracheal smooth muscle: lack of involvement of VIP or nitric oxide. *J. Appl. Physiol.* 74: 1964–1971.

Yip, P., Palambini, B., and Coburn, R.F. 1981. Inhibitory innervation of the guinea-pig trachealis muscle. *J. Appl. Physiol.* 50:373–382.

Yu, M., Robinson, N.E., and Wang, Z. 1993. Regional distribution of nitroxidergic and adrenergic nerves in equine airway smooth muscle. *Am. Rev. Respir. Dis.* 147:A286.

Yu, M., Wang, Z., Robinson, N.E., and Leblanc, P.H. 1994. Inhibitory nerve distribution and mediation of NANC relaxation by nitric oxide in horse airways. *J. Appl. Physiol.* 76:339–344.

7
Nitrergic Neurogenic Control of Resistance Blood Vessels

Tony J.-F. Lee and Raj Mishra

Introduction

Until two decades ago, chemical transmission from autonomic innervation to vascular effector tissues was thought to be mediated mainly by norepinephrine and acetylcholine. However, it was noted that despite pharmacological blockade of adrenergic receptors, electrical stimulation of sympathetic nerves still evoked residual vasoconstriction in several vascular beds in experimental animals (Folkow and Uvnas, 1948; Lundberg and Tatemoto, 1982). Direct evidence from examining neuroeffector transmission in isolated vascular preparations using in vitro tissue bath techniques has indicated that contraction and relaxation of isolated blood vessels upon field electrical stimulation of the intramural nerves persist in the presence of adrenoceptor and muscarinic receptor antagonists (Lee and Bevan, 1974; Lee et al., 1975, 1976, 1978, 1980; Von Kugelgen and Starke, 1985; Burnstock and Kennedy, 1986; Morris and Murphy, 1988). These findings, particularly those from in vitro studies using isolated preparations, have led to the evolution of new paradigms for neuroeffector communication in which norepinephrine and acetylcholine are not the respective primary vasoconstrictor and dilator transmitters, for example, in cerebral vascular beds, and a second transmitter substance from each type of nerve is predominant in mediating vasoconstriction and dilation (Lee and Bevan, 1974; Lee et al., 1975; Lee, 1986). Accruing evidence has now clearly indicated that several substances other than norepinephrine and acetylcholine are involved in autonomic neurotransmission in various vascular beds (Burnstock, 1972; Su and Lee, 1976; Lee, 1994). Foremost among this has been the discovery of nitric oxide (NO) as a cellular messenger.

The discovery that NO functions as a signaling molecule has widened the concept of cellular communication. Nitric oxide is a pluripotent molecule, ubiquitously produced in response to both physiological and pathological stimuli. Recent literature is inundated with information about the molecular characteristics, enzymatic synthesis, and physiological and pathophysiological roles of NO. The most significant finding of the physiological role of

NO is the recognition that this simple gas is a mediator of relaxation of peripheral and cerebral blood vessels (Furchgott and Zawadzki, 1980; Lee, 1980; Ignarro et al., 1987; Moncada et al., 1991; Snyder, 1992; Toda and Okamura, 1990a,b; Lee and Sarwinski, 1991).

Following its original discovery as a mediator in endothelial cells and macrophages, NO was recognized as a prominent neuronal messenger. Its involvement in neuroeffector transmission was first suggested in the cerebral arteries of the pig (Lee et al., 1989) and the anococcygeus muscles of the rat (Gillespie et al., 1989; Li and Rand, 1989; Ramgopal and Leighton, 1989) and mouse (Gibson et al., 1990). Subsequently, its wide distribution in the autonomic nervous system has been well documented (Sanders and Ward, 1992; Sneddon and Graham, 1992): the anococcygeus and retractor penis muscles (Martin and Gillespie, 1991), the cerebral vasculature of different species (Bredt et al., 1990; Toda and Okamura, 1990a,b; Lee and Sarwinski, 1991; Kimura et al., 1997; Yu et al., 1997a,b, 1998; Ishine et al., 1999), the respiratory tract (Belvisi et al., 1993), esophageal muscle (Tottrup, 1993), the gastrointestinal tract (Stark and Szurszewski, 1992; Lefebvre, 1993), the ileocolonic junction (Boeckstaens et al., 1991), the urinary tract (Andersson and Persson, 1993), and the genital system (Gillespie et al., 1990; Andersson and Stief, 1997; Sjostrand et al., 1998). Nitric oxide synthesized in CNS neurons also has been shown to act as a neuromediator subserving many physiological functions, including memory formation and pain modulation (Garthwaite and Boulton, 1995; Snyder, 1992; Moncada et al., 1991). The characterization of the biological activities of NO further indicates that the classical paradigm of neurotransmission no longer holds, since signaling by NO disregards the spatial constraints on neurotransmitter activity normally imposed by membranes, transporters, and inactivating enzymes (Garthwaite and Boulton, 1995). The evolution of the concept of the role of "NO-producing nerves" in mediating neurogenic relaxation of blood vessels as well as visceral smooth muscle has therefore culminated in the recognition of these "nitrergic nerves" (Rand and Li, 1995).

Resistance vessels are generally those vessels smaller than 400 μm in diameter. These vessels are important in the regulation of the microcirculation to organs. The functional state of these vessels is critical in determining organ blood flow as well as total peripheral resistance, which is crucial in maintaining blood pressure and flow. Although vasodilation in these vessels is induced by NO tonically released from endothelial cells in response to vasoactive substances and shear stress, relaxation of these vessels has also been shown to be dependent on NO release from perivascular nitrergic nerves. Recent literature suggests that the bulk of this evidence is observed initially in major arteries at the base of the brain, which are by definition considered resistance vessels. The following discussion will examine and evaluate the data substantiating the evidence for the presence of a neural control of resistance blood vessels in cerebral and peripheral circulation by nitrergic nerves.

Cerebral Blood Vessels

The resistance of large arteries in the cerebral circulation appears to be greater than that in other vascular beds in several species, and large cerebral arteries are important determinants of local microvascular pressure contributing significantly to the total cerebrovascular resistance (Faraci and Heistad, 1990). The cerebral circulation is regulated by neural pathways "extrinsic" and "intrinsic" to the CNS (Edvinsson et al., 1993; Iadecola et al., 1994). Neural pathways extrinsic to the CNS originate in cranial autonomic ganglia, give rise to perivascular plexuses, and innervate cerebral arteries down to a diameter of 50 μm. The cerebral veins are generally sparsely innervated compared with the cerebral arteries (Edvinsson et al., 1983; Itakura, 1983; Ishine et al., 1999). Both vasoconstrictor nerves of sympathetic origin and dilator nerves of nonsympathetic origin are present in cerebral blood vessels (Lee, 1994). Under normal resting conditions, the nonsympathetic vasodilator innervation appears to be predominant (Lee et al., 1975; Edvinsson et al., 1993).

The presence of a vasodilator innervation to the cerebral circulation was first described in 1930 (Chorobski and Penfield, 1932). The authors suggested that acetylcholine was the transmitter mediating cholinergic nerve-elicited vasodilation. This concept was accepted until the early 1970s. Lee et al. (1975), using isolated cerebral arterial ring preparations, for the first time demonstrated that acetylcholine was not the dilator transmitter in the cat cerebral arteries. This was based on the finding that neurogenic vasodilation in the cat cerebral arteries was not blocked by atropine (Lee et al., 1975). This hypothesis was generally well received when it was further shown that acetylcholine induced constriction of cerebral arteries without endothelial cells, whereas transmural nerve stimulation (TNS) still elicited exclusive relaxation in the same preparation (Lee, 1980) (Figure 7.1). This result provided convincing evidence that the endogenous transmitter mediating vasodilation was not acetylcholine. Similar findings have been reported in different species (Lee, 1994). This initial finding prompted new thinking in the search for this dilator substance. It was then proposed that the dilator transmitter was VIP (vasoactive intestinal peptide) and CGRP (calcitonin gene-related peptide), since fibers immunoreactive to both peptides were shown to be present in cerebral arteries and veins, and both induced endothelium-independent vasodilation (Lee et al., 1984b; Saito et al., 1989; Lee, 1994).

The role of neuropeptides as the primary vasodilator transmitters in cerebral arteries, however, was demonstrated to be unlikely. Cerebral neurogenic vasodilation elicited by TNS was accompanied by cGMP synthesis in vascular smooth muscle, whereas vasodilation induced by VIP and CGRP was accompanied by an increase in cAMP synthesis (Lee, 1987; Lee et al., 1989; Gonzalez et al., 1997). When arterial rings became tachyphylactic to repeated applications of these peptides, TNS-elicited relaxation persisted

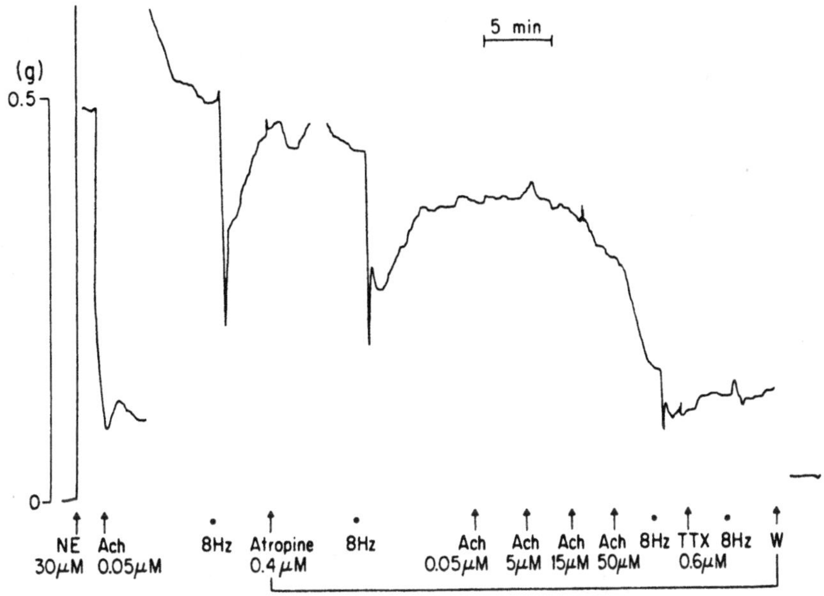

FIGURE 7.1. Effect of atropine on the dilator responses of cerebral (basilar) artery ring. Atropine did not affect the response to TNS but blocked the response to acetylcholine (Ach) up to 5 μM. NE, norepinephrine, TTX, tetrodotoxin. From Lee et al., 1975.

(Lee, 1987). Furthermore, the time course of relaxation induced by TNS and these two peptides was significantly different (Lee, 1987).

Subsequent studies demonstrated that hemolysates (the membrane-free erythrocyte extracts) as well as hemoglobin (the main constituent of hemolysates) abolished TNS-elicited relaxation without affecting VIP- or CGRP-induced relaxation (Lee et al., 1984a; Linnik and Lee, 1989). These original findings ruled out the possibility that these two peptides were the primary transmitters in mediating cerebral neurogenic vasodilation. It was proposed that the yet-to-be-determined vasodilator transmitters were trapped by hemoglobin extracellularly, since hemoglobin does not cross the cell membrane easily (Lee et al., 1984a). This conclusion was later supported by the finding that hemoglobin binds endothelium-derived relaxing factor (EDRF) or NO (Martin et al., 1985; Asada et al., 1997). These effects of hemolysates and hemoglobin have been suggested to play a key role in the pathogenesis of cerebral vasospasm following subarachnoid hemorrhage (Lee et al., 1988).

The mode of action of TNS-induced vasodilation and that of EDRF/NO on vascular smooth muscle relaxation in the cerebral arteries have been shown to be similar. Both were shown to act by stimulating cGMP synthesis, and vasodilator responses and cGMP synthesis to both were nearly completely blocked by methylene blue, hemoglobin, and NO synthesis inhibitors and were reversed by L-arginine and L-citrulline (Linnik and Lee, 1989; Chen and Lee, 1993; Ignarro et al., 1987; Lee et al., 1996). In addition, the time courses of nerve-stimulation-induced relaxation in isolated arteries and that induced by exogenously applied NO were identical (Chen and Lee, 1993). These results led to the suggestion that TNS-induced cerebral vasodilation was mediated by a substance similar to EDRF or NO (Lee et al., 1989).

Results from immunocytochemical studies have indicated that large cerebral arteries and pial arterioles are innervated by nitric oxide synthase

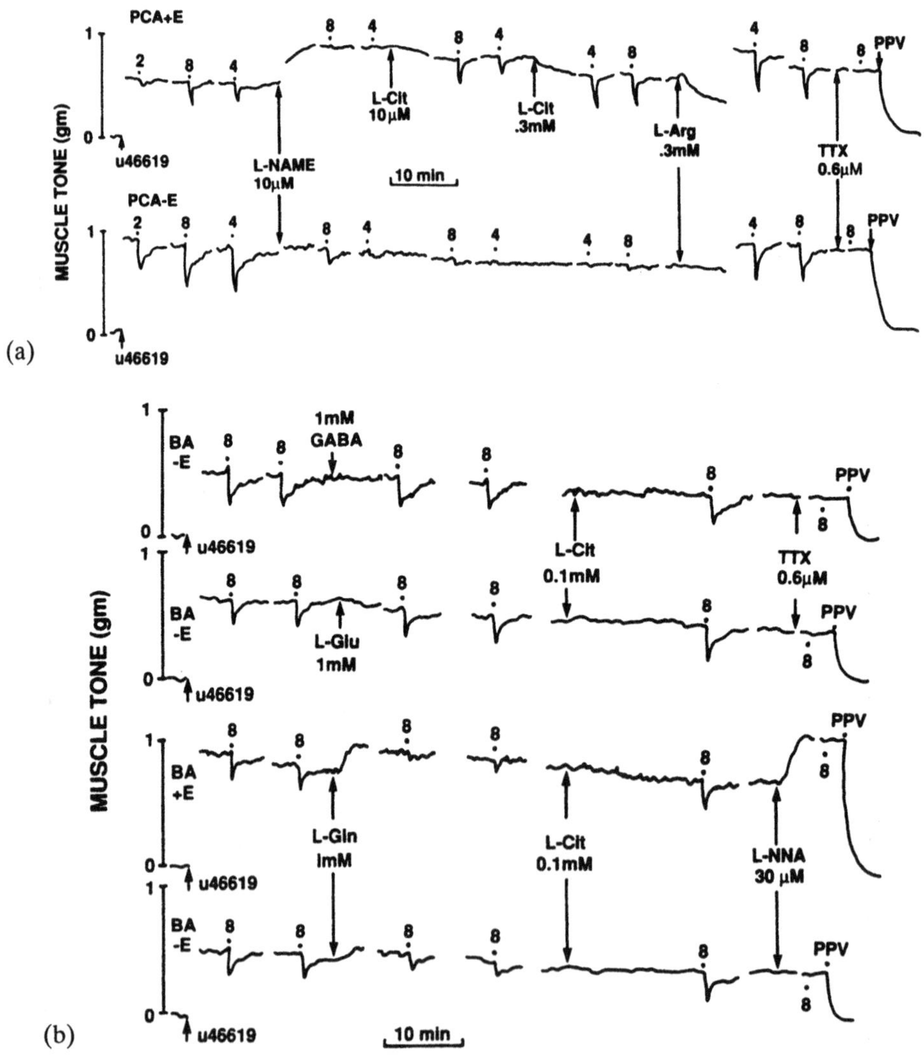

(NOS)-containing nerve fibers in many species, including humans (Bredt et al., 1990; Faraci and Brian, 1994; Nozaki et al., 1993; Estrada et al., 1993; Ceccatelli et al., 1992; Poeggel et al., 1992; Suzuki et al., 1993; Toda et al., 1993a,b; Tomimoto et al., 1993; Chen and Lee, 1993; Saito and Goto, 1994; Kimura et al., 1997;Yu et al., 1997a,b, 1998; Gorelova at al., 1996; Ando, 1998). These cerebral perivascular NOS-I nerves have been shown to originate in the sphenopalatine ganglia, with some contribution from the otic ganglia and the intrinsic miniganglia found on the large arteries at the base of the brain (Nozaki et al., 1993; Suzuki and Hardebo, 1993; Iadecola et al., 1994; Okuno et al., 1994; Kimura et al., 1997). Transmural nerve stimulation of these intramural nerves elicited relaxation that was almost abolished by inhibitors of NO synthesis (Toda and Okamura, 1990a,b; Lee and Sarwinski, 1991) and L-citrulline uptake (Lee et al., 1996) (Figure 7.2). The inhibition was reversed by L-arginine and L-citrulline, but not by their D-enantiomers. These results suggest that NO or a NO-releasing substance mediates a major component (>90%) of the cerebral neurogenic vasodilation. Similar results have been found in cerebral arteries of different species (Lee, 1994).

Despite compelling anatomical and pharmacological evidence for NO as the transmitter for vasodilation in the cerebral vascular bed, the neuronal source of NO in mediating cerebral neurogenic vasodilation has been questioned. One report suggested that certain peptide transmitters, such as VIP and CGRP, which contain two molecules of L-arginine, might release NO at the extraneuronal or synaptic regions to induce relaxation in sheep cere-

---

FIGURE 7.2(a) In the presence of active muscle tone induced by U-46619 (0.1–1 μM), the posterior communicating artery (PCA) with intact endothelial cells (+E; upper tracing) or without endothelial cells (–E, lower tracing) relaxed frequency-dependently upon TNS at 2, 4 and 8 Hz. L-NAME, which increased the basal tone in the artery with endothelial cells, but did not affect that in the artery without endothelial cells, abolished the TNS-induced relaxation. The inhibition was reversed by L-citrulline (L-Cit) and L-arginine (L-Arg). Both L-Cit and L-Arg decreased the basal tone in the artery with endothelial cells, but did not affect that in the artery without endothelial cells. The TNS-induced relaxation was abolished by tetrodotoxin (TTX). Papaverine (PPV, 300 μM) induced maximum relaxation. From Lee et al., 1996.

(b) Effect of L-glutamine, L-glutamate (L-Glu), and GABA on neurogenic vasodilations in porcine basilar arteries (BA) with (+E) or without (–E) endothelial cells. In the presence of active muscle tone induced by U-46619 (0.1–1 μM), TNS at 8 Hz induced relaxations in arteries both with and without endothelial cellls. In both preparations, relaxations were blocked by L-Gln, which increased basal tone in both preparations. The inhibition by L-Gln was reversed by L-citrulline (L-Cit). On the other hand, TNS-induced relaxation was not blocked by L-Glu or GABA. The residual relaxation was abolished by TTX and L-NNA. Papaverine (PPV, 300 μM) induced maximum relaxation.

bral arteries (Gaw et al., 1991). Another report suggested that an unidentified substance released from the perivascular nerves in the adventitia might diffuse through the media to release NO from the endothelial cells and, therefore, cause vasodilation in bovine cerebral arteries (Gonzalez and Estrada, 1991). Both possibilities are unlikely explanations for neurogenic vasodilation in cerebral arteries. Denuding endothelial cells has not been shown to block TNS-induced cerebral vasodilation in several species, including the cat, dog, and pig (Lee, 1980; Lee et al., 1982; Toda et al., 1997), suggesting that neurogenic vasodilation is independent of the endothelium. Furthermore, hemoglobin and N^G-nitro-L-arginine (L-NNA) at concentrations that abolished TNS-induced relaxation of the cerebral arteries in the cat and pig, did not affect relaxation induced by exogenous VIP and CGRP in the same preparations (Lee et al., 1993), indicating that endogenous VIP and CGRP do not release NO at the extraneuronal regions in the cerebral arteries of the cat and pig (Lee, 1994). Although species variation may be a factor in the differences between these findings, dense NO synthase (NOS)-I fibers are found to innervate bovine and sheep cerebral arteries (Gonzalez et al., 1997; Matthew et al., 1997), as in all other species examined (Kimura et al., 1997; Nozaki et al., 1993; Yu et al., 1998).

Furthermore, the results from examinination of cerebral perivascular neuronal uptake of L-citrulline by cerebral arteries have led to the conclusion that cerebral perivascular nerves can recycle L-citrulline, the byproduct of NO synthesis, to L-arginine for synthesis of more NO (Chen and Lee, 1995a; Lee et al., 1996; Yu et al., 1997a,b). This is further supported by the findings that all the enzymes necessary for recycling L-citrulline to L-arginine (argininosuccinate synthase and argininosuccinate lyase) and for synthesizing NO (NOS) are axoplasmic enzymes that are costored in the same neurons (Yu et al., 1997a,b) (Figure 7.3). This L-citrulline–L-arginine cycle does not take place in vascular smooth muscle of normal subjects (Chen and Lee, 1995a). Therefore, blocking of TNS-elicited neurogenic vasodilation by glutamine, possibly due to its inhibition of L-citrulline uptake and redistribution, should occur at the presynaptic neurons. These results provide strong evidence that NO is synthesized in and released from perivascular nerves to induce cerebral vasodilation (Chen and Lee, 1995a; Yu et al., 1997a,b; Ishine et al., 1999).

As has already been described, one major source of cerebral parasympathetic vasodilator innervation is the sphenopalatine ganglion. Stimulation of this ganglion and its pathway has been shown to increase cortical blood flow in anesthetized rats and cats (Seylaz et al., 1988; Suzuki et al., 1990; Goadsby, 1990). The sphenopalatine ganglion has also been shown to be the origin of cerebral vascular perivascular cholinergic, VIPergic, and nitric oxidergic nerves (Edvinsson et al., 1993; Kimura et al., 1997). Since VIP and NOS, and choline acetyltransferase (ChAT) and NOS, but not VIP and NOS, are colocalized in the same nerve fibers, nitrergic fibers originating in the sphenopalatine ganglion have been identified to include cholinergic-

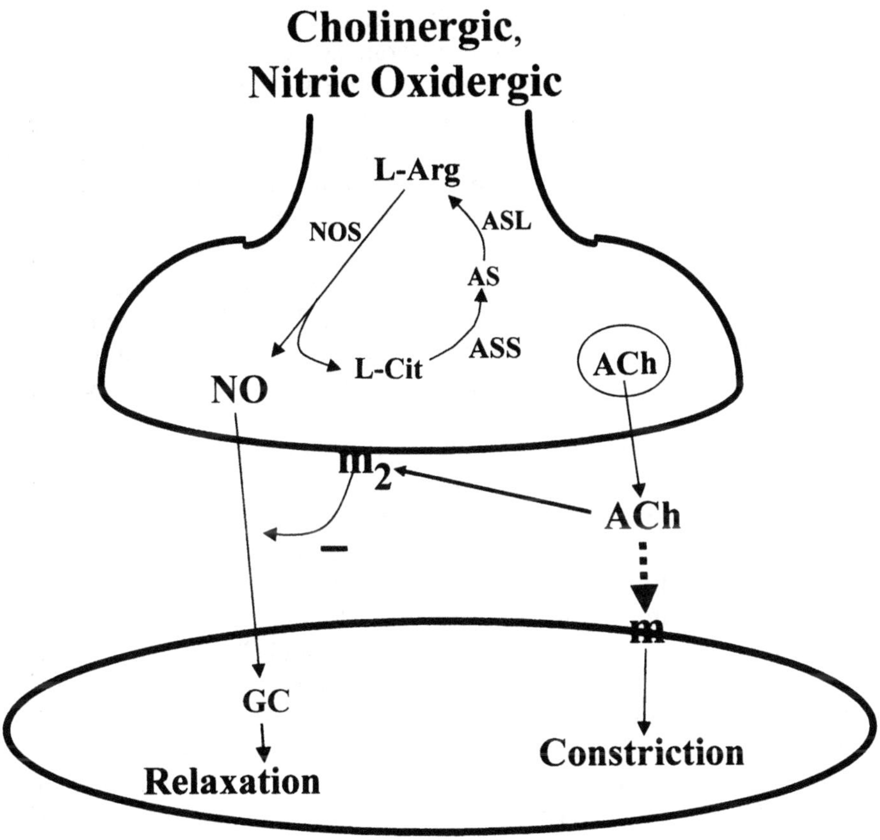

FIGURE 7.3. Diagram showing a cholinergic-nitric oxidergic nerve terminal and its relationship to a smooth muscle cell in cerebral arteries. Nitric oxide, which is not stored in vesicles, is colocalized and coreleased with vesicle-bound acetylcholine (ACh) in cholinergic-nitric oxidergic nerves. Nitric oxide is synthesized from L-arginine (L-Arg) catalyzed by NOS. L-Citrulline (L-Cit), the by-product of NO synthesis, is converted to argininosuccinate (AS) in the presence of ASS (argininosuccinate synthase), and is further converted to L-Arg catalyzed by ASL (argininosuccinate lyase). This L-Cit-L-Arg cycle argues for the neuronal source of NO. Acetylcholine released from cholinergic-nitric oxidergic nerves acts on presynaptic m_2 (muscarinic m_2 receptors), resulting in inhibition (–) of NO release accompanied by decreased relaxation. Although the direct effect of exogenous ACh on smooth muscle is constriction, endogenous ACh does not directly affect smooth muscle tone (as indicated by dotted lines). GC, guanylate cyclase. From Lee et al., 1996.

nitric oxidergic and VIPergic-nitric oxidergic nerves (Yu et al., 1998). Increased cortical blood flow upon electrical stimulation of the sphenopalatine ganglion was shown to be attenuated by inhibitors of NOS, such as N^G-nitro-L-arginine methylester (L-NAME). The inhibition was reversed by

L-arginine administration (Morita-Tsuzuki et al., 1993; Goadsby et al., 1996). Although the distribution of nitrergic nerves of the sphenopalatine ganglion to pial vessels has been well described, the distribution of these nerves to the cortical circulation has not been defined. For this reason, infusion of nicotine into the vertebral arteries was demonstrated in the dog (Toda et al., 1993a). The result indicated that nicotine infusion dilated the basilar arteries. The dilation was also blocked by nitro-L-arginine. This dilation was expected to be due to release of NO from perivascular nerves, since nicotine has been shown to release neuronal NO from isolated cerebral arteries (Toda and Okamura, 1990a; Zhang et al., 1998). These findings from in vitro and in vivo studies indicate that nitrergic innervation in cerebral resistance blood vessels plays a role in regulating cerebral vascular tone and circulation.

Because of its physiochemical properties, NO is not stored in vesicles and is not released by exocytotic mechanisms as are classical neurotransmitters such as acetylcholine (Lundberg, 1996). Pharmacological studies in isolated arterial preparation have shown that NO mediates the major component (>90%) of the neurogenic vasodilator response (Lee, 1994), whereas endogenous acetylcholine has no direct effect on the smooth muscle in cerebral arteries (Lee, 1982). Also, colocalization of NOS and ChAT in the parasympathetic ganglia and perivascular nerves in the cerebral blood vessels of several species (Kimura et al., 1997; Suzuki and Hardebo, 1993) has led to the proposal that endogenous acetylcholine plays a modulatory role (Lee, 1986) in inhibiting release of its cotransmitter NO (Ayajiki et al., 1994; Liu and Lee, 1999). Acetylcholine acts on presynaptic m_2 receptors on the nitrergic nerves to inhibit N-type Ca^{2+}channel-mediated NO release (Liu and Lee, 1999) (Figure 7.3). Modulation of release of neuronal NO by other transmitters such as norepinephrine and VIP also has been reported (Zhang et al., 1998; Gonzalez et al., 1997). Both norepinephrine released from sympathetic nerves and VIP coreleased with NO facilitate release of the transmitter NO and induce vasodilation. A complete understanding of the complexities of the neurotransmitter interactions in modulating cerebrovascular tone, therefore, will provide information to further clarify the significance of the nitrergic innervation in regulating the cerebral vasculature in health and disease.

Peripheral Blood Vessels

Mesenteric Circulation

Mesenteric vascular tone is regulated by sympathetic vasoconstrictor nerves and nonadrenergic, noncholinergic (NANC) vasodilator nerves. Although norepinephrine is the primary transmitter for vasoconstriction, neurogenic vasodilation is mediated by multiple transmitters similar to those found in the cerebral circulation. The primary transmitters for mesenteric vasodilation have been shown to vary among species. For example, CGRP in the rat (Takenaga et al., 1995; Peng et al., 1995), CGRP and ATP in the rabbit (Kakuyama et al., 1998), ATP but not CGRP in the golden hamster (Hill et al., 1996), CGRP and NO in the guinea pig (Gyoda et al., 1995), and NO in the dog (Yoshida et al., 1993), monkey (Tsuchiya et al., 1993; Yoshida et al., 1994) and guinea pig (Zheng et al., 1997) have been shown to mediate neurogenic vasodilation in the mesentery.

The nitrergic nerves in mesenteric arteries, however, have been demonstrated in most species examined (Yoshida et al., 1994; Gyoda et al., 1995; Zheng et al., 1997). Nitric oxide released from the vasodilator nerves and norepinephrine from sympathetic nerves appear to play reciprocal roles in regulating mesenteric vascular tone. Tsuchiya et al. (1993), Nase and Boegehold (1996), and Peng et al. (1995) have demonstrated that mesenteric vasoconstriction induced by stimulation of sympathetic nerves is balanced by release of neuronal NO from nitrergic nerves. Yamamoto et al. (1997), however, reported an interesting finding that the release of norepinephrine, which caused vasoconstriction of the mesenteric vascular bed was in part dependent on NO. When production of NO was inhibited by L-NNA, norepinephrine overflow upon electrical field stimulation (EFS) in the mesenteric vascular bed was diminished by 50% at 37°C but remained unchanged at 24°C. Norepinephrine release elicited by EFS was significantly less (about 50%) at 24 °C than at 37°C. These findings were consistent with the observation that synthesis of NO in mesenteric vasculature was almost diminished at 24°C (Yamamoto et al., 1997). The perfusion pressure response of the mesenteric vascular bed upon EFS, however, was higher at 24°C than at 37°C. The greater increase in perfusion pressure at 24°C than at 37°C was suggested to be due to the reduction of NO (a potent vasodilator) at 24°C, thus unmasking the vasoconstricting effect of norepinephrine. These results indicated that norepinephrine release upon EFS of perivascular nerves in the mesenteric vasculature was partially NO-dependent.

Evidence for a modulatory role of endogenous NO on the release of peptide transmitters in the intestinal microvasculature also has been reported (Peng et al., 1995). In the rat perfused mesenteric vascular beds, when adrenergic nerves were blocked by guanethidine, EFS produced a frequency-dependent vasodilator response, which was due to release of CGRP (Peng et al., 1995). Pretreatment with ginsenosides (3–30 μg/ml), which are saponins from *Panax ginseng*, significantly inhibited the vasodilator response. The effect of ginsenosides was abolished by inhibition of NO synthesis with L-NAME and of cGMP synthesis with methylene blue. Ginsenosides, however, did not affect relaxation induced by exogenous CGRP. These results suggest that in the rat perfused mesenteric vascular bed, ginsenosides inhibit the vasodilator response by acting on presynaptic nerve terminals to decrease the release of transmitter CGRP, which is mediated by NO.

Using double-labeling immunohistochemistry, Zheng et al. (1997) demonstrated that neuronal NOS immunoreactivity was found in 12% of cells of guinea pig thoracic and lumbar dorsal root ganglia. In 95% of these cells, there was colocalization of NOS and substance P (SP). Furthermore, neuronal NOS immunoreactivity was localized in nerve fibers surrounding guinea pig mesenteric artery, and 25% of these were double stained with SP immunoreactivity. In endothelium-denuded inferior mesenteric arterial preparations, repetitive EFS produced a frequency-dependent vasodilation that was significantly inhibited by the NOS inhibitor L-NNA (Zheng et al., 1997). These results suggest that NO released from primary sensory nerves also causes vasodilation.

The role of neural reflexes in the regulation of gastrointestinal vascular function has been well established, especially the mucosal hyperemia following a meal (Vanner and Surprenant, 1996). Stimulation of vagal nerves is known to cause noncholinergic (atropine-resistant) dilation of gastric arterioles (Morishita and Guth, 1986). A clear delineation of the reflex pathways involved in the control of mucosal blood flow has not been presented (Vanner and Surprenant, 1996). It has recently been demonstrated that neurogenic vasodilation of submucosal arterioles is mediated by release of acetylcholine and/or neuropeptides such as SP and CGRP from extrinsic sensory nerves (Vanner and Surprenant, 1996). It is not known, however, whether neuronal NO is involved in this reflex hyperemia

Cardiopulmonary Circulation

The control of the cardiovascular system is known to involve humoral and neuronal mechanisms comprising several nonclassical transmitters, including NO (Ahluwalia and Cellek, 1997). There is accumulating evidence for modulation of sympathetic neurotransmission by endogenously produced NO in the vascular tree of the heart. Results from morphological studies on the distribution of endothelial NOS (eNOS) and neuronal NOS (nNOS) indicated that eNOS expression in the rat heart depended on blood vessel diameter (Addicks et al., 1994). The arteries with diameters greater than 100 μm consistently showed intense staining, whereas smaller arteries and veins exhibited much weaker staining. Neuronal NOS was found to localize in autonomic nerve fibers along the entire vascular tree. Ultrastructural NADPH diaphorase staining revealed the presence of adjacent localization of NOS-positive and -negative axons, suggesting the possibility of interaxonal modulation of adjacent non-nitrergic autonomic nerve fibers by NO released from nitrergic nerves, and vice versa (Addicks et al., 1994). Additionally, it was noted that intracoronary application of NO markedly decreased coronary perfusion pressure, accompanied by an elevation in intraaxonal catecholamine levels in adrenergic nerve fibers (Addicks et al., 1994), further suggesting the modulatory role of NO in the sympathetic neurotransmission.

It is well established that the control of the coronary vascular resistance is mediated by a number of reflexes that function through the action of the sympathetic and parasympathetic nervous systems. It has been reported that activation of a number of reflexes in conscious dogs, especially the Bezold-Jarisch reflex and the carotid chemoreflex, results in cholinergic coronary vasodilation that is blocked by L-NNA (Zhao et al., 1996a). It was found that whereas pacing-induced heart failure abolished cholinergic dilation following activation of the Bezold-Jarisch or carotid chemoreflex, exercise training potentiated the Bezold-Jarisch reflex-induced coronary vasodilation. Since the Bezold-Jarisch reflex may be a significant compensatory phenomenon during acute myocardial infarction and the carotid chemoreflex is important for ameliorating systemic hypoxemia, the neurally released NO in reflex cholinergic coronary vasodilation may be crucial in the compensatory vascular homeostatic responses evoked by these reflexes (Zhao et al., 1996a). Substantiating this view is the report by the same group that the loss of NO production in the coronary circulation after the development of cardiomyopathy involves a specific defect in the neural regulation of coronary blood flow (Zhao et al., 1996b).

The presence of sympathetic adrenergic and parasympathetic cholinergic innervations in pulmonary vasculature and their role in pulmonary circulation have been well described (Su and Bevan, 1976; Knight et al., 1981; Amenta et al., 1983; Donald et al., 1990). The presence of nitrergic nerves in the pulmonary vasculature was first described in the guinea pig (Haberberger et al., 1997). The authors reported that the pulmonary vasculature was innervated by autonomic noradrenergic, nitrergic, and cholinergic axons. The cholinergic axons are present in the arteries down to 100 μm in diameter, and they innervate up to two-thirds of the small arteries (50–100 μm). The innervation of the nitrergic axons, on the other hand, is restricted to the large (>700 μm) extrapulmonary arteries. Upon EFS and in the presence of adrenergic and cholinergic receptor blockade, the isolated pulmonary arteries relaxed exclusively, and the relaxation was largely (about 74% of the dilator response) mediated by NO (Liu et al., 1992). Endogenously released NO has been shown to inhibit adrenergic (Cederqvist et al., 1991) and cholinergic (Belvisi et al., 1993) neurotransmission. The exact mechanism of this inhibition and its consequences remain to be determined. In the pulmonary vasculature of the semiarboreal snake *Elaphe obsoleta*, although extensive functional innervation of the pulmonary vasculature by the adrenergic and cholinergic nerves was demonstrated, no evidence for NANC transmission to the vasculature was found (Donald et al., 1990). Variation in nitrergic innervation to pulmonary vasculature among other species may exist.

Penile Blood Vessels

The functional state of the penis, i.e., detumescence and flaccidity versus tumescence and erection, is determined by the balance between contracting and relaxing factors controlling the smooth muscle of the corpus cavernosum (Andersson and Stief, 1997). Nerve-released NO has been shown to be the most important factor for relaxation of penile vessels and the corpus cavernosum (Giuliano et al., 1995; Andersson and Stief, 1997; Simonsen et al., 1997), although other transmitter substances have also been shown to be involved (Toda et al., 1996; Ehmke et al., 1995). Nerve fibers containing NOS immunoreactivity have been demonstrated in the corpus cavernosum, artery, and vein (Burnett et al., 1993; Hayashida et al., 1996). It also has been suggested that human and rabbit genitourinary organs have a powerful nitrergic innervation that does not merely modulate, but actually controls, the sympathetic responses (Cellek and Moncada, 1997, 1998). Electrical field stimulation causes NO-mediated relaxation of the corpus cavernosum accompanied by an increase in cGMP synthesis (Ignarro et al., 1990; Chen and Lee, 1995b). This NO-cGMP pathway has been the basis for the development of drugs that promote penile erection. Sildenafil, a selective inhibitor of phosphodiesterase (PDE) type 5, promotes penile erection by increasing intracellular cGMP in response to sexual stimulation and potentiating relaxation of the corpus cavernosum (Moreland et al., 1998; Chuang et al., 1998). Additional evidence supporting the role of NO in penile erection is the observation of relaxation of the corpus cavernosum induced by ginsenosides, saponin extracts from *panax ginseng,* which have been an essential constituent in traditional Chinese aphrodisiacs. Chen and Lee (1995b) demonstrated that ginsenosides released NO from the perivascular nerves and endothelial cells and enhanced NO release from endothelial cells induced by other vasoactive substances in isolated corpus cavernosum of the rabbit. These neurogenic and endothelial effects of ginsenosides in inducing relaxation of the corpus cavernosum have been suggested to account for the aphrodisiac effect of *Panax ginseng* (Chen and Lee, 1995b).

Renal Circulation

The efferent renal innervation is composed of postganglionic sympathetic fibers to the renal arterioles, juxtaglomerular apparatus, and renal tubules. Increased efferent sympathetic nerve activity results in increased renal vascular resistance, renin release, and sodium retention (Katholi, 1983). Enhanced activity of renal sympathetic nerves to both afferent and efferent arterioles has been implicated in the pathogenesis of hypertension in experimental animals and humans (Katholi, 1985; Katholi and Woods, 1987). Marked hypertension in rats following chronic inhibition of NO synthesis by L-NAME was first attributed to the acute elevation of sympathetic nerve activity (Matsuoka et al., 1994). This assumption, however, was not supported by subsequent studies demonstrating that hypertension following NOS inhibition was independent of activation of the renal sympathetic nervous system (Granger et al., 1996; Reinhart et al., 1997). In fact, NO has been demonstrated to be an inhibitory modulator of renal noradrenergic neurotransmission at the prejunctional level in the rabbit and dog (Egi et al., 1994, 1995; Maekawa et al., 1996; Vials et al., 1997). Thus, marked hypertension following inhibition of NOS by L-NAME most likely is due to inhibition of the synthesis of endothelium-derived NO (Vials et al., 1997). This raised the question of whether the renal vasculature receives authentic "nitrergic nerves" that may modulate the sympathetic activity. Both NOS-I and NADPH diaphorase-reactive fibers were found to innervate renal arteries of the rabbit (Vials et al., 1997), dog, and monkey (Okamura et al., 1995). The authors suggested that depression of the vasodilator nerve function by NOS inhibition potentiated the contraction caused by adrenergic nerve excitation. However, whether this nitrergic innervation of the renal arteries extends to the level of the arterioles has not been clarified. Neuronal NOS (bNOS), however, has been demonstrated in the macula densa of the rat, mouse, guinea pig, rabbit, pig, and human (Bachmann et al., 1995; Ichihara et al., 1998). The intimate spatial relation between NOS-I and renin-containing preglomerular afferent arterioles suggests a potential effect of macula densa-derived NO on the juxtaglomerular granular cells (Bachmann et al., 1995). A subpopulation of renal nerves containing NADPH diaphorase and NOS immunoreactivity was found in the perivascular connective tissue and near pelvic epithelium (Bachmann et al., 1995). These results indicate that NO released from nitrergic nerves can play a crucial role in local glomerular signaling events. Thus, nitrergic nerves could be important in the physiological renal mechanisms regulating blood pressure and in the pathogenesis of hypertension.

Conclusions

In addition to the resistance arteries in vital organs in which nitrergic innervation appears to play an important role in controlling regional vascular tone and blood flow, other vascular beds have been shown to receive nitrergic nerves. These include the human vagina (Hoyle et al., 1996), canine uterine arteries (Okamura et al., 1995), dog retinal arteries and arterioles (Toda et al., 1994), monkey central retinal arteries (Toda et al., 1996), guinea pig choroidal arterioles (Hashitani et al., 1998), bovine ciliary arteries (Wiencke et al., 1994), porcine ciliary arteries (Toda et al., 1997), rat skin microvasculature (Merhi et al., 1998), and the arteriovenous anastomoses of the tongue, ear, eye, and glomus organ of the finger in different species, including the rat, rabbit, dog, and human (Funk et al., 1994). The demonstration of active recycling of L-citrulline to form L-arginine for synthesis of more NO in the axoplasm, for example in the cerebral circulation, provides strong evidence for the hypothesis that NO of neuronal origin acts as the transmitter. Transmitter NO and EDRF/NO stimulate soluble guanylyl cyclase and synthesis of cGMP in smooth muscle and elicit vasodilation. Stimulation of nitrergic innervation or its ganglion causes vasodilation such as that found in the cerebral circulation. The demonstration of cotransmission of NO and several classical transmitters (adrenergic, cholinergic, and peptidergic) in cerebral and peripheral vascular beds suggests the intimate relationship among these transmitters. Since NO is a potent vasodilator, interaction between NO and other transmitters may play an important role in the integrative control of resistance arterial tone in health and disease.

Acknowledgments. This work was supported by NIH grants (HL27763 and HL47574), AHA/IHA (9807871), and SIU-CRC/EAM.

References

Addicks, K., Bloch, W., and Feelisch, M. 1994. Nitric oxide modulates sympathetic neurotransmission at the prejunctional level. *Microsc. Res. Tech.* 29:161–168.

Ahluwalia, A., and Cellek, S. 1997. Regulation of the cardiovascular system by non-adrenergic, non-cholinergic nerves. *Curr. Opin. Nephrol. Hypertens.* 6:74–79.

Amenta, F., Cavalloti, C., Ferrante, F., and Tonelli, F. 1983. Cholinergic innervation of the human pulmonary circulation. *Acta. Anat. (Basel)* 117:58–64.

Andersson, K.E., and Persson, K. 1993. The L-arginine/nitric oxide pathway and NANC relaxation of the lower urinary tract. *Gen. Pharmacol.* 24:833–839.

Andersson, K.E., and Stief, C.G. 1997. Neurotransmission and the contraction and relaxation of penile erectile tissues. *World J. Urol.* 15:14–20.

Ando, K. 1998. Nitrergic innervation of the cerebral arterial tree in the bent-winged bat. *J. Comp. Neurol.* 390:366–376.

Asada, Y., Yu, J.G., and Lee, T.J.-F. 1997. Myo-endothelial junctions in endothelium-dependent vasodilation. *Cell Vision* 4:308–317.

Ayajiki, K., Okamura, T., and Toda, N. 1994. Neurogenic relaxations caused by nicotine in isolated cat middle cerebral arteries. *J. Pharmacol. Exp. Ther.* 270:795–801.

Bachmann, S., Bosse, H.M., and Mundel, P. 1995. Topography of nitric oxide synthesis by localising constitutive NO synthases in mammalian kidney. *Am. J. Physiol.* 268:F885–F898.

Belvisi, M.G., Miura, M., Stretton, D., and Barnes, P.J. 1993. Endogenous vasoactive intestinal polypeptide and nitric oxide modulate cholinergic neurotransmission in guinea-pig trachea. *Eur. J. Pharmacol.* 231:97–102.

Boeckstaens, G.E., Pelckmans, P.A., Rampart, M., Verbeuren, T.J., Herman, A.G., and Van Maercke, Y.M. 1991. NANC mechanisms in the ileocolonic junction. *Arch. Int. Pharmacodyn. Ther.* 303:270–281.

Bredt, D.S., Hwang, P.M., and Snyder, S.H. 1990. Localisation of nitric oxide synthase indicating a neural role for nitric oxide. *Nature* 347:768–770.

Burnett, A.L., Tillman, S.L., Chang, T.S., Epstein, J.L., Lowenstein, C.J., Bredt, D.S., Snyder, S.H., and Walsh, P.C. 1993. Immunohistochemical localization of nitric oxide synthase in the autonomic innervation of the human penis. *J. Urol.* 50:73–76.

Burnstock, G. 1972. Purinergic nerves. *Pharmacol. Rev.* 24:509–560.

Burnstock, G., and Kennedy, C. 1986. A dual function for adenosine 5′triphosphate in the regulation of vascular tone: excitatory cotransmitter with noradrenaline from perivascular nerves and locally released inhibitory intravascular agent. *Circ. Res.* 58:319–330.

Ceccatelli, S., Lundberg, J.M., Fahrenkrug, J., Bredt, D.S., Snyder, S.H., and Hokflet, T. 1992. Evidence for involvement of nitric oxide in the regulation of hypothalamic portal blood flow. *Neuroscience* 51:769–772.

Cederqvist, B., Wiklund, N.P., Persson, M.G., and Gustafsson, L.E. 1991. Modulation of neuroeffector transmission in the guinea-pig pulmonary artery by endogenous nitric oxide. *Neurosci. Lett.* 127:67–69.

Cellek, S., and Moncada, S. 1997. Nitrergic control of peripheral sympathetic responses in the human corpus cavernosum: a comparison with other species. *Proc. Natl. Acad. Sci. USA* 94:8226–8231.

Cellek, S., and Moncada, S. 1998. Nitrergic neurotransmission mediates the non-adrenergic non-cholinergic responses in the clitoral corpus cavernosum of the rabbit. *Br. J. Pharmacol.* 125:1627–1629.

Chen, F.Y., and Lee, T.J.-F. 1993. Role of nitric oxide in neurogenic vasodilation of porcine cerebral artery. *J. Pharmacol. Exp. Ther.* 265:339–345.

Chen, F.Y., and Lee, T.J.-F. 1995a. Arginine synthesis from citrulline in perivascular nerves of cerebral artery. *J. Pharmacol. Exp. Ther.* 273:895–901.

Chen, X., and Lee, T.J.-F. 1995b. Ginsenosides-induced nitric oxide-mediated relaxation of the rabbit corpus cavernosum. *Br. J. Pharmacol.* 115:15–18.

Chorobski, J., and Penfield, W. 1932. Cerebral vasodilator nerves and their pathway from the medulla oblongata. *Arch. Neurol. Psychiat.* 28:1257–1289.

Chuang, A.T., Strauss, J.D., Murphy, R.A., and Steers, W.D. 1998. Sildenafil, a type-5 CGMP phosphodiesterase inhibitor, specifically amplifies endogenous cGMP-dependent relaxation in rabbit corpus cavernosum smooth muscle in vitro. *J. Urol.* 160:257–261.

Donald, J.A., O'Shea, J.E., and Lillywhite, H.B. 1990. Neural regulation of the pulmonary vasculature in a semi-arboreal snake, *Elaphe obsoleta*. *J. Comp. Physiol. (B)* 159:677–685.

Edvinsson, L., Hogestatt, E.D., Udmann, R., and Auer, L.M. 1983. Cerebral veins: fluorescence histochemistry, electron microscopy, and in vitro reactivity. *J. Cereb. Blood Flow Metab.* 3:226–330.

Edvinsson, L., Mackenzie, E.T., and McCulloch, J. 1993. *Cerebral Blood Flow and Metabolism.* Raven Press, New York.

Egi, Y., Matsumura, Y., Miura, A., Murata, S., and Morimoto, S. 1995. Interaction between nitric oxide and angiotensin II on antidiuresis and norepinephrine overflow induced by stimulation of renal nerves in anesthetized dogs. *J. Cardiovasc. Pharmacol.* 25:187–193.

Egi, Y., Matsumura, Y., Murata, S., Umekawa, T., Hisaki, K., Takaoka, M., and Morimoto, S. 1994. The effects of N^G-nitro-L-arginine, a nitric oxide synthase inhibitor, on norepinephrine overflow and antidiuresis induced by stimulation of the renal nerves in anesthetized dogs. *J. Pharmacol. Exp. Ther.* 269:529–535.

Ehmke, H., Junemann, K.P., Mayer, B., and Kummer, W. 1995. Nitric oxide synthase and vasoactive intestinal polypeptide colocalisation in neurons innervating the human penile circulation. *Int. J. Impot. Res.* 7:147–156.

Estrada, C., Mengual, E., and Gonzalez, C. 1993. Local NADPH-diaphorase neurons innervate pial arteries and lie close or project to intracranial blood vessels: a possible role for nitric oxide in the regulation of cerebral blood flow. *J. Cereb. Blood. Flow. Metab.* 13:978–984.

Faraci, F.M., and Brian, J.E. 1994. Nitric oxide and the cerebral circulation. *Stroke* 25:692–703.

Faraci, F.M., and Heistad, D.D. 1990. Regulation of large cerebral arteries and cerebral microvascular pressures. *Circ. Res.* 66:8–17.

Folkow, B., and Uvnas, B. 1948. The chemical transmission of vasoconstrictor impulses to the hind limbs and the splanchnic region of the cat. *Acta Physiol. Scand.* 15:365–388.

Funk, R.H., Mayer, B., and Worl, J. 1994. Nitrergic innervation and nitrergic cells in arteriovenous anastomoses. *Cell Tissue Res.* 277:477–484.

Furchgott, R.F., and Zawadzki, J.V. 1980. The obligatory role of endothelial cells in the relaxation of arterial smooth muscle by acetylcholine. *Nature* 288:373–376.

Garthwaite, J., and Boulton, C.L. 1995. Nitric oxide signaling in the central nervous system. *Annu. Rev. Physiol.* 57:683–706.

Gaw, A.J., Aberdeen, J., Humphrey, P.P.A., Wadsworth, R.M., and Burnstock, G. 1991. Relaxation of sheep cerebral arteries by vasoactive intestinal polypeptide and neurogenic stimulation: inhibition by L-N^G-monomethyl arginine in endothelium denuded vessels. *Br. J. Pharmacol.* 102:567–572.

Gibson, A., Mirzazadehs, Hobbs, A.J., and Moore, P.K. 1990. L-N^G monomethylarginine and L-N^G nitroarginine inhibit nonadrenergic, noncholinergic relaxation of the mouse anococcygeus. *Br. J. Pharmacol.* 99:602–606.

Gillespie, J.S., Liu, X.R., and Martin, W. 1989. The effects of L-arginine and N^G-monomethyl L-arginine on the response of the rat anococcygeus muscle to NANC nerve stimulation. *Br. J. Pharmacol.* 98:1080–1082.

Gillespie, J.S., Liu, X.R., and Martin, W. 1990. The neurotransmitter of the NANC nerves to smooth muscle of the genital system. In: *Nitric Oxide from L-Arginine. A Bioregulatory System* (ed. S. Moncada), pp. 147–164. Elsevier, Amsterdam.

Giuliano, F.A., Rampin, O., Benoit, G., and Jardin, A. 1995. Neural control of penile erection. *Urol. Clin. North Am.* 22:747–766.

Goadsby, P.J. 1990. Sphenopalatine ganglion stimulation increases regional cerebral blood flow independent of glucose utilization in the cat. *Brain Res.* 506:145–148.

Goadsby, P.J., Uddman, R., Edvinsson, L. 1996. Cerebral vasodilation in the cat involves nitric oxide from parasympathetic nerves. *Brain Res.* 707:110–118.

Gonzalez, C., Barroso, C., Martin, G., Gulbenkian, S., and Estrada, C. 1997. Neuronal nitric oxide synthase activation by vasoactive intestinal peptide in bovine cerebral arteries. *J. Cereb. Blood Flow Metab.* 17:977–984.

Gonzalez, C., and Estrada, C. 1991. Nitric oxide mediates the neurogenic vasodilation of bovine cerebral arteries. *J. Cereb. Blood Flow Metab.* 11:366–370.

Gorelova, E., Loesch, A., Bodin, P., Chadwick, L., Hamlyn, P.J., and Burnstock, G. 1996. Localisation of immunoreactive factor VIII, NOS, substance P, endothelin and 5-hydroxy-tryptamine in human postmortem middle cerebral artery. *J. Anat.* 188:97–107.

Granger, J., Novak, J., Schnackenberg, C., Williams, S., and Reinhart, G.A. 1996. Role of renal nerves in mediating the hypertensive effects of nitric oxide synthesis inhibition. *Hypertension* 27:613–618.

Gyoda, Y., Tsukada, Y., Saito, A., and Goto, K. 1995. Role of nitric oxide and neuropeptides in neurogenic vasodilatation of the guinea pig mesenteric artery. *Eur. J. Pharmacol.* 279:83–92.

Haberberger, R., Schemann, M., Sann, H., and Kummer, W. 1997. Innervation pattern of guinea pig pulmonary vasculature depends on vascular diameter. *J. Appl. Physiol.* 82:426–434.

Hashitani, H., Windle, A., and Suzuki, H. 1998. Neuroeffector transmission in arterioles of the guinea pig choroid. *J. Physiol. (Lond.)* 510:209–223.

Hayashida, H., Okamura, T., Tomoyoshi, T., and Toda, N. 1996. Neurogenic nitric oxide mediates relaxation of canine corpus cavernosum. *J. Urol.* 155:1122–1127.

Hill, B., Ralevic, V., Crowe, R., and Burnstock, G. 1996. Innervation and nitric oxide modulation of mesenteric arteries of the golden hamster. *Eur. J. Pharmacol.* 317:275–283.

Hoyle, C.H., Stones, R.W., Robson, T., Whitley, K., and Burnstock, G. 1996. Innervation of the vasculature and microvasculature of the human vagina by NOS and neuropeptide-containing nerves. *J. Anat.* 188:633–644.

Iadecola, C., Pellegrino, D.A., Moskowitz, M.A., and Lassen, N.A. 1994. Nitric oxide synthase inhibition and cerebrovascular regulation. *J. Cereb. Blood Flow Metab.* 14:175–192.

Ichihara, A., Inscho, E.W., Imig, J.D., and Navar, L.G. 1998. Neuronal nitric oxide synthase modulates rat renal microvascular function. *Am. J. Physiol.* 274:F516–F524.

Ignarro, L.J., Bush, P.A., Buga, G.M., Woods, K.S., Fukuto, J.M., and Rajefer, J. 1990. Nitric oxide and cGMP formation upon electrical field stimulation cause relaxation of corpus cavernosum smooth muscle. *Biochem. Biophys. Res. Commun.* 170:843–850.

Ignarro, L.J., Byrns, R.E., Buga, G.M., and Wood, K.S. 1987. Endothelium-derived relaxing factor from pulmonary artery and vein possesses pharmacologic and chemical properties identical to those of nitric oxide radical. *Circ. Res.* 60:82–92.

Ishine, T., Asada, Y., Yu, J.G., and Lee, T.J.F. 1999. Nitric oxide is predominant mediator for neurogenic vasodilation in the porcine pial veins. *J. Pharmacol. Exp. Ther.* 289(1):398–404.

Itakura, T. 1983. Aminergic and cholinergic innervations of the spinal cord blood vessels of cats. *J. Neurosurg.* 58:900–905.

Kakuyama, M., Vallance, P., and Ahluwalia, A. 1998. Endothelium-dependent sensory NANC vasodilation: involvement of ATP, CGRP and a possible NO store. *Br. J. Pharmacol.* 123:310–316.

Katholi, R.E. 1983. Renal nerves in the pathogenesis of hypertension in experimental animals and humans. *Am. J. Physiol.* 245:F1–F14.

Katholi, R.E. 1985. Renal nerves and hypertension: an update. *Fed. Proc.* 44: 2846–2850.

Katholi, R.E., and Woods, W.T. 1987. Afferent renal nerves and hypertension. *Clin. Exp. Hypertens.* A.9 Suppl 1:211–226.

Kimura, T., Yu, J.G., Edvisson, L., and Lee, T.J.-F. 1997. Cholinergic, nitric oxidergic innervation in cerebral arteries of the cat. *Brain Res.* 773:117–124.

Knight, D.S., Ellison, J.P., Hibbs, R.G., Hyman, A.L., and Kadowitz, P.J. 1981. A light and electron microscopic study of the innervation of pulmonary arteries in the cat. *Anat. Rec.* 201:513–521.

Lee, T.J.-F. 1980. Direct evidence against acetylcholine as the dilator transmitter in the cat cerebral artery. *Eur. J. Pharmacol.* 68:393–394.

Lee, T.J.-F. 1982. Cholinergic mechanisms in the large cat cerebral arteries. *Circ. Res.* 50:870–879.

Lee, T.J.-F. 1986. Sympathetic and nonsympathetic transmitter mechanisms in cerebral vasodilation and constriction. *Proceedings of the Erin K. Fernstrom Sympo-*

sium on Neural Regulation of Brain Circulation (eds. C. Owman et al.), pp. 285–296. Elsevier Biomedical Press.

Lee, T.J.-F. 1987. Evidence for and against VIP as a transmitter for vasodilation in cerebral blood vessels. In: *Peptidergic Mechanisms in the Cerebral Circulation* (eds. L. Edvinsson and J. McCulloch), pp. 65–74. New York: Ellis Horwood.

Lee, T.J.-F. 1994. Putative transmitters in cerebral neurogenic vasodilation. In: *The Human Brain Circulation: Functional Changes in Disease* (eds. J.A. Bevan and R.D. Bevan), pp. 73–91, Totowa, New Jersey: Humana Press.

Lee, T.J.-F., and Bevan, J.A. 1974. The effect of nerve stimulation and pharmacological agents on the contractile activity of the isolated basilar artery of the rabbit. *Proc. West. Pharmacol. Soc.* 17:129–131.

Lee, T.J.-F., Chen, F.Y., and Sarwinski, S. 1993. Role of NO, VIP and CGRP in cerebral neurogenic vasodilation. *J. Cereb. Blood Flow Metab.* 13 (Suppl 1):S264.

Lee, T.J.-F., Chiueh, C.C., and Adams, M. 1980. Synaptic transmission of vasoconstrictor nerves in the rabbit basilar artery. *Eur. J. Pharmacol.* 61:55–70.

Lee, T.J.-F., Fang, Y.X., and Nickils, G.A. 1989. Cyclic nucleotides and cerebral neurogenic vasodilation. In: *Neurotransmission and Cerebrovascular Function* (eds. J. Seylaz and E.T. Mackenzie), pp. 277–280. Amsterdam: Elsevier.

Lee, T.J.-F., Hume, W.R., Su, C., and Bevan, J.A. 1978. Neurogenic vasodilation of cat cerebral arteries. *Circ. Res.* 42:535–542.

Lee, T.J.-F., Kinkhead, L., and Sarwinski, S. 1982. Norepinephrine and acetylcholine transmitter mechanism in pig cerebral blood vessels. *J. Cerebral Blood Flow Metab.* 2:439–450.

Lee, T.J.-F., Linnik, M.D., and Miao, F.J.-P. 1988. Erythrocyte extracts and cerebral vascular function. In: *Vascular Neuroeffector Mechanisms* (eds. J.A. Bevan, H. Majewski, R.A. Maxwell, and D.F. Story), pp. 299–309. Oxford: IRL Press.

Lee, T.J.-F., McIlhany, M.P., and Sarwinski, S. 1984a. Erythrocyte extracts enhance neurogenic vasoconstriction of cerebral arteries in vitro. *J. Cerebral Blood Flow Metab.* 4:474–476.

Lee, T.J.-F., Saito, A., and Beresein, I. 1984b. Vasoactive intestinal polypeptide-like substance: the potential cerebral vasodilator transmitter. *Science* 224:898–901.

Lee, T.J.-F., and Sarwinski, S.J. 1991. Nitric oxidergic neurogenic vasodilation in the porcine basilar artery. *Blood Vessels* 28:402–412.

Lee, T.J.-F., Sarwinski, S., Ishine, T., Lai, C., and Chen, F.Y. 1996. Inhibition of cerebral neurogenic vasodilation by L-glutamine and nitric oxide synthase inhibitors. *J. Pharmacol. Exp. Ther.* 267:353–358.

Lee, T.J.-F., Su, C., and Bevan, J.A. 1975. Nonsympathetic dilator innervation of cat cerebral arteries. *Experientia* 31:1424–1425.

Lee, T.J.-F., Su, C., and Bevan, J.A. 1976. Neurogenic sympathetic vasoconstriction of the rabbit basilar artery. *Circ. Res.* 39:120–126.

Lefebvre, R.A. 1993. NANC neurotransmission in the proximal stomach. *Gen. Pharmacol.* 24:257–266.

Li, C.G., and Rand, M.J. 1989. Evidence for a role of nitric oxide in the neurotransmitter system mediating relaxation of the rat anococcygeus muscle. *Clin. Exp. Pharmacol. Physiol.* 16:933–938.

Linnik, M.D., and Lee, T.J.-F. 1989. Effects of hemoglobin on neurogenic responses and cholinergic parameters in porcine cerebral arteries. *Cerebral Blood Flow Metab.* 9:219–225.

Liu, J., and Lee, T.J.-F. 1999. Mechanism of prejunctional muscarinic receptor-mediated inhibition of neurogenic vasodilation in cerebral arteries. *Am. J. Physiol.* 276:H194–H204.

Liu, S.F., Crawley, D.E., Rohde, J.A., Evans, T.W., and Barnes, P.J. 1992. Role of nitric oxide and guanosine 3′,5′-cyclic monophosphate in mediating nonadrenergic, noncholinergic relaxation in guinea-pig pulmonary arteries. *Br. J. Pharmacol.* 107:861–866.

Lundberg, J.M. 1996. Pharmacology of cotransmission in the autonomic nervous system: integrative aspects on amines, neuropeptides, adenosine triphosphate, amino acids and nitric oxide. *Pharmacol. Rev.* 48:113–178.

Lundberg, J.M., and Tatemoto, K. 1982. Pancreatic polypeptide family (APP, BPP, NPY and PYY) in relation to sympathetic vasoconstriction resistant to adrenoceptor blockade. *Acta. Physiol. Scand.* 116:393–402.

Maekawa, H., Matsumura, Y., Matsuo, G., and Morimoto, S. 1996. Effect of sodium nitroprusside on norepinephrine overflow and antidiuresis induced by stimulation of renal nerves in anesthetised dogs. *J. Cardiovasc. Pharmacol.* 27:211–217.

Martin, W., and Gillespie, J.S. 1991. L-Arginine derived nitric oxide: the basis of inhibitory transmission in the anococcygeus and retractor penis muscle. In: *Novel Peripheral Neurotransmitters* (ed. C. Bell), pp. 65–79. Pergamon, New York.

Martin, W., Villani, G.M., Jothianandan, D., and Furchgott, R.F. 1985. Selective blockade of endothelium-dependent and glyceryl trinitrate-induced relaxation by hemoglobin and by methylene blue in the rabbit aorta. *J. Pharmacol. Exp. Ther.* 232:708–716.

Matsuoka, H., Nishida, H., Nomura, G., Van Vliet, B.N., and Toshima, H. 1994. Hypertension induced by nitric oxide synthesis inhibition is renal nerve dependent. *Hypertension* 23:971–975.

Matthew, J.D., Wadsworth, R.M., and McPhaden, A.R. 1997. Inhibition of vasodilator neurotransmission in the sheep middle cerebral artery by VIP antiserum. *J. Auton. Pharmacol.* 17:13–19.

Merhi, M., Dusting, G.J., and Khalil, Z. 1998. CGRP and nitric oxide of neuronal origin and their involvement in neurogenic vasodilation in rat skin microvasculature. *Br. J. Pharmacol.* 123:863–868.

Moncada, S., Palmer, R.M., and Higgs, E.A. 1991. Nitric oxide: physiology, pathophysiology, and pharmacology. *Pharmacol. Rev.* 43:109–142.

Moreland, R.B., Goldstein, I., and Traish, A. 1998. Sildenafil, a novel inhibitor of phosphodiesterase type 5 in human corpus cavernosum smooth muscle cells. *Life Sci.* 62:309–318.

Morishita, T., and Guth, P.H. 1986. Vagal nerve stimulation causes noncholinergic dilatation of gastric arterioles. *Am. J. Physiol.* 250:G660–G664.

Morita-Tsuzuki, Y., Hardebo, J.E., and Bouskela, E. 1993. Inhibition of nitric oxide synthase attenuates the cerebral blood flow response to stimulation of postganglionic parasympathetic nerves in the rat. *J. Cereb Blood Flow Metab.* 13(6):993–997.

Morris, J.L., and Murphy, R. 1988. Evidence that neuropeptide Y released from noradrenergic axons causes prolonged contraction of the guinea pig uterine artery. *J. Auton. Nerv. Syst.* 24:241–249.

Nase, G.P., and Boegehold, M.A. 1996. Nitric oxide modulates arteriolar responses to increased sympathetic nerve activity. *Am J. Physiol.* 271:H860–H869.

Nozaki, K., Moskowitz, M.A., Maynard, K.I., Koketsu, N., Dawson, T.M., Bredt, D.S., and Snyder, S.H. 1993. Possible origins and distribution of immunoreactive nitric oxide synthase-containing nerve fibers in cerebral arteries. *J. Cereb. Blood. Flow. Metab.* 13:70–79.

Okamura, T., Yoshida, K., and Toda, N. 1995. Nitricoxidergic innervation in the dog and monkey renal arteries. *Hypertension* 25:1090–1095.

Okuno, T., Itakura, T.L., Lee, T.J.-F., Masami, U., Shimizu, M., and Komai, N. 1994. Cerebral pial arterial innervation with specific reference to GABAergic innervations. *J. Auton. Nerv. Syst.* 49:S105–S110.

Peng, C.F., Li, Y.J., Li, Y.J., and Deng, H.W. 1995. Effects of ginsenosides on vasodilator nerve actions in the rat perfused mesentery are mediated by nitric oxide. *J. Pharm. Pharmacol.* 47:614–617.

Poeggel, G., Muller, M., Seidel, I., Rechardt, L., and Bernstein, H.G. 1992. Histochemistry of guanylate cyclase, phosphodiesterase, and NADPH-diaphorase (nitric oxide synthase) in rat brain vasculature. *J. Cardiovasc. Phrmacol.* 20 (Suppl 12):S76–S79.

Ramgopal, M.V., and Leighton, H.J. 1989. Effect of N^G monomethyl-L-arginine on field stimulation induced decreases in cytosolic Ca^{2+} levels and relaxation in the rat anococcygeus muscle. *Eur. J. Pharmacol.* 174:297–299.

Rand, M.J., and Li, C.G. 1995. Nitric oxide as a neurotransmitter in peripheral nerves: nature of transmitter and mechanism of transmission. *Annu. Rev. Physiol.* 57:659–682.

Reinhart, G.A., Lohmeier, T.E., and Mizelle, H.L. 1997. Temporal influence of the renal nerves on excretory function during chronic inhibition of nitric oxide synthesis. *Hypertension* 29:199–204.

Saito, A., and Goto, K. 1994. Vasodilator innervation of small cerebral arteries of guinea pigs. *J. Auton. Nerv. Syst.* 49 (Suppl.):S59–S62.

Saito, A., Masaki, T., Uchiyama, Y., Lee, T.J.-F., and Goto, K. 1989. Calcitonin gene-related peptide (CGRP) and vasodilator nerves in large cerebral arteries of cats. *J. Pharmacol. Exp. Ther.* 248:455–462.

Sanders, K.M., and Ward, S.M. 1992. Nitric oxide as a mediator of NANC neurotransmission. *Am J. Physiol.* 262:G379–G392.

Seylaz, J., Hara, H., Pinard, E., Moravitch, S., MacKenzie, E.T., and Edvinsson, L. 1988. Effect of stimulation of the sphenopalatine ganglion on cortical blood flow in the rat. *J. Cereb. Blood Flow Metab.* 8:875–878.

Simonsen, U., Prieto, D., Hernandez, M., Saenz de Tejada, I., and Garcia-Sacristan, A. 1997. Prejunctional $alpha_2$-adrenoceptors inhibit nitrergic neurotransmission in horse penile resistance arteries. *J. Urol.* 157:2356–2360.

Sjostrand, N.O., Ehren, I., Eldh, J., and Wiklund, N.P. 1998. NADPH-diaphorase in glandular cells and nerves and its relation to acetylcholinesterase-positive nerves in the male reproductive tract of man and guinea-pig. *Urol. Res.* 26:181–188.

Sneddon, P., and Graham, A. 1992. Role of nitric oxide in the autonomic innervation of smooth muscle. *J. Auton. Pharmacol.* 12:445–456.

Snyder, S.H. 1992. Nitric oxide: first in a new class of neurotransmitters. *Science* 257:494–496.

Stark, M.E., and Szurszewski, J.H. 1992. Role of nitric oxide in gastrointestinal and hepatic function and disease. *Gastroenterology* 103:1928–1949.

Su, C., and Bevan, J.A. 1976. Pharmacology of pulmonary blood vessels. *Pharmacol. Ther. [B]* 2:275–288.

Su, C., and Lee, T.J.-F. 1976. Regional variation of adrenergic and nonadrenergic nerves in blood vessels. *Second International Symposium on Vascular Neuroeffector Mechanism*, University of Odense, Denmark, pp. 35–42. Karger, Basel.

Suzuki, N., Fukuuchi, Y., Koto, A., Naganuma, Y., Isozumi, K., Matsuoko, S., Gotoh, J., and Shimuzu, T. 1993. Cerebrovascular NADPH diaphorase-containing nerve fibers in the rat. *Neurosci. Lett.* 151:1–3.

Suzuki, N., and Hardebo, J.E. 1993. The cerebrovascular parasympathetic innervation. *Cerebrovasc. Brain Metab. Rev.* 5:33–46.

Suzuki, N., Hardebo, J.E., Kahrstrom, J., and Owman, C. 1990. Selective electrical stimulation of postganglionic cerebral vascular parasympathetic nerve fibers originating from the sphenopalatine ganglion enhances the cortical blood flow in the rat. *J. Cereb. Blood Flow Metab.* 10:383–391.

Takenaga, M., Kawasaki, H., Wada, A., and Eto, T. 1995. Calcitonin gene-related peptide mediates acetylcholine-induced endothelium-independent vasodilation in mesenteric resistance blood vessels of the rat. *Circ. Res.* 76:935–941.

Toda, M., Okamura, T., Azuma, I., and Toda, N. 1997. Modulation of neurogenic acetylcholine of nitroxidergic function in porcine ciliary arteries. *Invest. Ophthalmol Vis. Sci.* 38:2261–2269.

Toda, N., Ayajiki, K., and Okamura, T. 1993a. Neural mechanism underlying basilar arterial constriction by intracisternal L-NNA in anesthetized dogs. *Am. J. Physiol.* 265:H103–H107.

Toda, N., Ayajiki, K., Yoshida, K., Kimura, H., and Okamura, T. 1993b. Impairment by damage of the pterygopalatine ganglion of nitroxidergic vasodilator nerve function in canine cerebral and retinal arteries. *Circ. Res.* 72:206–213.

Toda, N., Kitamura, Y., and Okamura, T. 1994. Role of nitroxidergic nerve in dog retinal arterioles in vivo and arteries in vitro. *Am. J. Physiol.* 266:H1985–H1992.

Toda, N., and Okamura, T. 1990a. Possible role of nitric oxide in transmitting information from vasodilator nerve to cerebroarterial muscle. *Biochem. Biophys. Res. Commun.* 170:308–313.

Toda, N., and Okamura, T. 1990b. Mechanism underlying the response to vasodilator nerve stimulation in isolated dog and monkey cerebral arteries. *Am. J. Physiol.* 259:H1511–H1517.

Toda, N., Toda, M., Ayajiki, K., and Okamura, T. 1996. Monkey central retinal artery is innervated by nitroxidergic vasodilator nerves. *Invest. Ophthalmol Vis. Sci.* 37: 2177–2184.

Toda, N., Toda, M., Ayajika, K., and Okamura, T. 1998. Cholinergic nerve function in monkey ciliary arteries innervated by nitroxidergic nerve. *Am. J. Physiol.* 274:H1582–H1589.

Tomimoto, H., Akiguchi, I., Wakita, H., Nakamura, S., and Kimura, J. 1993. Distribution of NADPH diaphorase in the cerebral blood vessels of rats: a histochemical study. *Neurosci. Lett.* 156:105–108.

Tottrup, A. 1993. The role of nitric oxide in oesophageal motor function. *Dis. Esophagus* 6:2–10.

Tsuchiya, K., Urabe, M., Yamamoto, R., Asada, Y., and Lee, T.J.-F. 1993. Effects of L-N^{w}-nitro-arginine and capsaicin on neurogenic vasomotor responses in isolated mesenteric arteries of the monkey. *J. Pharmac. Pharmacol.* 46:155–157.

Vanner, S., and Surprenant, A. 1996. Neural reflexes controlling intestinal microcirculation. *Am. J. Physiol.* 271:G223–G230.

Vials, A.J., Crowe, R., and Burnstock, G. 1997. A neuromodulatory role for neuronal nitric oxide in the rabbit renal artery. *Br. J. Pharmacol.* 121:213–220.

Von Kugelgen, I., and Starke, K. 1985. Noradrenaline and adenosine triphosphate as cofactors of neurogenic vasoconstriction in rabbit mesenteric artery. *J. Physiol. (Lond.)* 367:435–455.

Wiencke, A.K., Nilsson, H., Nielsen, P.J., and Nyborg, N.C. 1994. Nonadrenergic noncholinergic vasodilation in bovine ciliary artery involves CGRP and neurogenic nitric oxide. *Invest. Ophthalmol Vis. Sci.* 35:3268–3277.

Yamamoto, R., Wada, A., Asada, Y., Yanagita, T., Yuhi, T., Niina, H., Sumiyoshi, A., Kobayashi, H., and Lee, T.J.-F. 1997. Nitric oxide-dependent and -independent norepinephrine release in rat mesenteric arteries. *Am. J. Physiol.* 273:H207–H210.

Yoshida, K., Okamura, T., Kimura, H., Bredt, D.S., Snyder, S.H., and Toda, N. 1993. Nitric oxide synthase-immunoreactive nerve fibers in dog cerebral and peripheral arteries. *Brain Res.* 629:67–72.

Yoshida, K., Okamura, T., and Toda, N. 1994. Histological and functional studies on the nitroxidergic nerve innervating monkey cerebral, mesenteric and temporal arteries. *Jpn. J. Pharmacol.* 65:351–359.

Yu, J.G., Ishine, T., Kimura, T., O'Brien, W.E., and Lee, T.J.-F. 1997a. L-Citrulline conversion to L-arginine in sphenopalatine ganglia and cerebral perivascular nerves in the pig. *Am. J. Physiol.* 273 (*Heart Circ. Physiol.* 42):H2192–H2199.

Yu, J.G., Kimura, T., Chang, X.-F., and Lee, T.J.-F. 1998. Segregation of VIPergic-nitric oxidergic and cholinergic-nitric oxidergic innervation in porcine middle cerebral arteries. *Brain Res.* 801:78–87.

Yu, J.G., O'Brien, W.E., and Lee, T.J.-F. 1997b. Morphologic evidence for L-citrulline conversion to L-arginine via the argininosuccinate pathway in porcine cerebral perivascular nerves. *J. Cereb. Blood Flow Metab.* 17:884–893.

Zhang, W., Edvinsson, L., and Lee, T.J.-F. 1998. Mechanism of nicotine-induced relaxation in the porcine basilar artery. *J. Pharmacol. Exp. Ther.* 284:790–797.

Zhao, G., Hintze, T.H., and Kaley, G. 1996a. Neural regulation of coronary vascular resistance: role of nitric oxide in reflex cholinergic coronary vasodilation in normal and pathophysiologic states. *EXS* 76:1–19.

Zhao, G., Shen, W., Zhang, X., Smith, C.J., and Hintze, T.H. 1996b. Loss of nitric oxide production in the coronary circulation after the development of dilated cardiomyopathy: a specific defect in the neural regulation of coronary blood flow. *Clin. Exp. Pharmacol. Physiol.* 23:715–721.

Zheng, Z., Shimamura, K., Anthony, T.L., Travagli, R.A., and Kreulen, D.L. 1997. Nitric oxide is a sensory nerve neurotransmitter in mesenteric artery of guinea pig. *J. Auton. Nerv. Syst.* 67:137–144.

8
Is There Neural Control of Peripheral Resistance Arteries by Nitrergic Nerves?

Hiromu Kawasaki and Makoto Takenaga

Resistance blood vessels play an important role in maintenance of blood pressure and regulation of tissue blood flow (Mulvany and Aalkjaer, 1990). Resistance vascular tone is mainly controlled by sympathetic adrenergic nerves through the release of the neurotransmitter norepinephrine (Mulvany and Aalkjaer, 1990). Indeed, electrical field stimulation (EFS) of perivascular nerves in blood vessels isolated from various species evokes vasoconstriction, which is blocked by α-adrenoceptor antagonists (prazosin), adrenergic neuron blockers (guanethidine), and the neurotoxin tetrodotoxin (TTX). However, EFS of precontracted blood vessels isolated from various species causes vasodilation (Bevan and Bryden, 1987), which can be abolished by TTX but not by β-adrenoceptor antagonists or muscarinic cholinoceptor antagonists (atropine). These results suggest nonadrenergic, noncholinergic (NANC) vasodilator innervation. However, these studies have been performed on relatively large arteries and/or conduit arteries in vitro, and evidence of NANC innervation and NANC neurotransmitter in small arteries and/or resistance arteries is scarce.

Recently, nitric oxide (NO), an endothelium-derived relaxing factor (EDRF), has been implicated as a possible vasodilator transmitter for NANC nerves in cerebral arteries isolated from several species (Toda and Okamura, 1990a,b, 1992; Lee and Sarwinski, 1991). A previous study by us, however, showed that rat mesenteric resistance blood vessels are innervated by NANC nerves in which calcitonin gene-related peptide (CGRP), a sensory neuropeptide and potent vasodilator, acts as a vasodilator transmitter (Kawasaki et al., 1988).

The present chapter focuses on innervation by nitrergic nerves and then physiological role in neural control of peripheral resistance blood vessels. Innervation by CGRP-containing vasodilator nerves and their role in neural control of the tone of resistance blood vessels from reciprocal interaction with sympathetic nerves are also discussed.

Nitrergic Innervation in Various Blood Vessels

As shown in Table 8.1, many reports suggest the presence of nitrergic innervation in peripheral blood vessels isolated from various species. Original evidence that there is nitrergic innervation in the blood vessel has been provided by the study of Toda and Okamura (1990a) on the dog cerebral artery. In the precontracted cerebral artery, EFS induces TTX-sensitive relaxation, which is NANC in nature and endothelium-independent (Toda and Okamura, 1990b; Lee and Sarwinski, 1991; Toda and Okamura, 1992). This EFS-induced relaxation is inhibited by NO synthase (NOS) inhibitors such as N^G-nitro-L-arginine (L-NA), N^G-nitro-L-arginine methylester (L-NAME), and N^G-monomethyl-L-arginine (L-NMMA), and the inhibition can be reversed by addition of L-arginine but not an enantioisomer of D-arginine. Also, the guanylate cyclase inhibitor methylene blue and the NO scavenger oxyhemoglobin inhibit EFS-induced relaxation. Furthermore, NOS-immunoreactive fibers are observed in the cerebral arteries from several species (Yoshida et al., 1993). These results strongly suggest that the cerebral artery is innervated by nitrergic nerves in which NO acts as neurotransmitter. It is likely that the nitrergic nerves in the cerebral artery play a physiological role in the regulation of cerebral circulation.

Nitrergic innervation in peripheral arteries remains controversial, and there are large species and tissue differences. As shown in Table 8.1, although the sizes of the blood vessels used in these studies were relatively large, EFS of precontracted ciliary, retinal, and ophthalmic arteries isolated from monkeys (Toda et al., 1996, 1998), cows (Wiencke et al., 1994), pigs (Su et al., 1994; Toda et al., 1997), and dogs (Toda et al., 1994, 1995) induces TTX-sensitive and endothelium-independent vasorelaxation. EFS-induced relaxation is inhibited by NOS inhibitors and reversed by addition of L-arginine but not by D-arginine. These studies suggest that nitrergic nerves are present in ophthalmic arteries. However, analysis of bovine ciliary arteries (Wiencke et al., 1994) indicates that the EFS-induced relaxation is largely reduced by NOS inhibitors. However, this neurogenic response is also sensitive to capsaicin, a toxin for CGRP-containing nerves, and to CGRP(8–37), a CGRP receptor antagonist, suggesting that CGRP is also involved in neurogenic relaxation. In dog lingual arteries without endothelium, Kobayashi et al. (1995) provided evidence that the EFS-induced TTX-sensitive NANC relaxation was not inhibited by NOS inhibitors such as L-NAME and L-NMMA, but is blocked by CGRP(8–37) and by capsaicin. This study indicates that there is little nitrergic innervation and that CGRP-containing nerves mainly mediate EFS-induced relaxation. On the contrary, Okamura et al. (1998) found that EFS-induced NANC relaxation in the dog lingual artery was inhibited by NOS inhibitor (L-NA) and the inhibition was restored by L-arginine, and further that the CGRP receptor antagonist did not affect the neurogenic response. This study suggests innervation by nitrergic nerves and no involvement of CGRP or other peptides in this artery. The reason for these large discrepancies in the results of these studies remains unresolved. In studies on pulmonary arteries, EFS-induced NANC relaxation was shown to be partially inhibited by NOS inhibitors and reversed by L-arginine, suggesting partial involvement of nitrergic nerves in the pulmonary artery (Scott et al., 1996).

CGRP-Containing Innervation in Mesenteric Blood Vessels

The NANC nerves in mesenteric arteries are different from those in other peripheral arteries (Table 8.1). Neurogenic NANC relaxation in bovine mesenteric arteries (Ahlner et al., 1991) was inhibited by methylene blue and potentiated by zaprinast, a selective inhibitor of cGMP degradation. A NOS inhibitor was ineffective, but pyrogallol, a generator of superoxide anions, was a potent inhibitor of relaxation induced by EFS. The results obtained in this study suggest that cGMP produced by NO-like substances mediates NANC neurogenic relaxation. However, in studies on mesenteric arteries of rat and rabbit (Kawasaki et al., 1988; Li et al., 1993; Amerini et al., 1993; Ahluwalia et al., 1996; Kakuyama et al., 1998), EFS induced TTX-sensitive NANC relaxation; this response was not inhibited by NOS inhibitors and was sensitive to capsaicin and CGRP(8–37). These data provide evidence that the mesenteric artery is mostly innervated with capsaicin-sensitive sensory nerves in which CGRP acts as a neurotransmitter. Gyoda et al. (1995) found that EFS of the guinea pig mesenteric artery induced TTX-sensitive NANC relaxation, which is more sensitive to NOS inhibitors and partially to capsaicin or CGRP(8–37). Zheng et al. (1997) showed that NOS inhibitors partially inhibited EFS-induced NANC relaxation of the guinea pig mesenteric artery, which is sensitive to capsaicin and TTX. These data indicate that NO is released from primary sensory nerves in the guinea pig mesenteric arteries. Therefore, there might be regional and segmental differences in the innervation of mesenteric arteries.

In the isolated, perfused rat mesenteric vascular bed, which includes small resistance arteries, EFS caused TTX-sensitive NANC vasodilation in the presence of methoxamine and guanethidine (Kawasaki et al., 1988). This NANC response was abolished by CGRP(8–37), CGRP receptor desensitization, and pretreatment with capsaicin (Kawasaki et al., 1988; Han et al., 1990). In addition, EFS in the mesenteric vascular bed evoked an increase in CGRP release, which was sensitive to TTX and capsaicin and was abolished by removal of calcium from the medium (Fujimori et al., 1987). Furthermore, there was dense innervation with fibers containing CGRP-like immunoreactivity, and exposure of the mesenteric artery to capsaicin abolished the intensity of these fibers (Kawasaki et al., 1988). These data strongly suggest that rat mesenteric resistance arteries have major NANC innervation with CGRP-containing nerves.

TABLE 8.1. Vascular responses induced by electrical field stimulation (EFS) and effect of nitric oxide synthase (NOS) inhibitor in different arteries.

Artery	Species	EFS-induced	TTX sensitivity
Ciliary	Monkey	Relaxation	Yes
Ciliary	Cow	Relaxation	Yes
Ciliary	Pig	Relaxation	Yes
Central retinal	Monkey	Relaxation	Yes
Ophthalmic	Dog	Relaxation	Yes
Retinal	Dog	Relaxation	Yes
Lingual	Monkey	Relaxation	Yes
Lingual	Dog	Relaxation	Yes
Lingual	Dog	Relaxation	Yes
Pulmonary	Human	Relaxation	Yes
Pulmonary	Guinea pig	Relaxation	Yes
Mesenteric	Guinea pig	Relaxation	
Mesenteric	Guinea pig	Relaxation	Yes
Mesenteric	Guinea pig	Relaxation	Yes
Mesenteric	Rat	Contraction	Yes
Mesenteric	Rat	Relaxation	Yes
Mesenteric	Rat	Relaxation	Yes
Mesenteric	Rat	Relaxation	
Mesenteric	Rat	Relaxation	Yes
Mesenteric	Rabbit	Relaxation	Yes
Mesenteric	Cow	Relaxation	Yes
Mesenteric	Cow	Relaxation	Yes
Mesenteric	Monkey	Relaxation	Yes
Mesenteric	Monkey	Relaxation	Yes
Mesenteric	Monkey	Contraction	Yes
Renal	Rabbit	Contraction	
Renal	Monkey	Relaxation	
Renal	Dog	Relaxation	

TTX, tetrodotoxin; CGRP, calcitonin gene-related peptide; L-NA, N^G-nitro-L-arginine; L-NAME, N^G-nitro-L-arginine methylester; L-NMMA, N^G-monomethyl-L-arginine.

NOS inhibitor	Transmitter	Reference
Inhibition (L-NA)	NO	Toda et al., 1998
Inhibition (L-NA)	NO (largely), CGRP(partially)	Wiencke et al., 1994
Partial inhibition (L-NAME)	NO, unknown	Su et al., 1994
Inhibition (L-NA)	NO	Toda et al., 1996
Inhibition (L-NA)	NO	Toda 1995
Inhibition (L-NA)	NO	Toda et al., 1994
Inhibition (L-NA)	NO	Toda et al., 1997
No effect (L-NMMA, L-NAME)	CGRP	Kobayashi et al., 1995
Inhibition (L-NA)	NO	Okamura et al., 1998
Inhibition (L-NAME)	NO (partially)	Scott et al., 1996
Inhibition (L-NMMA, L-NAME)	NO	Liu et al., 1992
No effect (L-NA)	CGRP	Matsuda et al., 1995
Inhibition (L-NA)	NO, CGRP	Gyoda et al., 1995
Partial inhibition (L-NA)	NO	Zheng et al., 1997
Potentiation (L-NAME, L-NMMA)	NO	Marin and Baffagon, 1998
No effect (L-NMMA)	CGRP, NO-endothelium	Ahluwalia and Vallance, 1996
No effect (L-NAME, L-NMMA)	CGRP	Amerini et al., 1993
Potentiation (L-NAME)	CGRP	Li 1993
No effect (L-NA)	CGRP	Kawasaki et al., 1994
No effect (L-NA, L-NAME)	CGRP, ATP	Kakuyama et al., 1998
No effect (L-NA, L-NAME)	Unknown	Ahlner et al., 1991
Inhibition (L-NAME, L-NMMA)	NO, peptide	Leckstrom et al., 1993
Inhibition (L-NA)	NO	Okamura 1996
Inhibition (L-NA)	NO	Toda and Nakamusa, 1992
Potentiation (L-NA)	NO, peptide (partial)	Tsuchiya et al., 1994
Potentiation (L-NAME)	NO	Vials et al., 1997
Inhibition (L-NA)	NO	Toda et al., 1995
Inhibition (L-NA)	NO	Okamura et al., 1995

As shown in Figure 8.1, the rat perfused mesenteric vascular bed was precontracted with methoxamine in the presence of guanethidine. In this preparation, EFS induced neurogenic vasodilation mediated by CGRP-containing nerves, as described before. Removal of endothelial cells by perfusion with sodium deoxycholate abolished acetylcholine-induced vasodilation but did not inhibit neurogenic vasodilation or CGRP-induced vasodilation. On the contrary, vasodilation induced by EFS and exogenous CGRP injection in the endothelium-removed preparation was significantly potentiated (Figures 8.1 and 8.2). These results indicate that both CGRP-containing nerve-mediated and CGRP-induced vasodilation are independent on the endothelium, and suggest that the endothelium may play an inhibitory role in CGRP-containing nerve function. Furthermore, in preparations without endothelium, perfusion with L-NA had no effect on the neurogenic vasodilation (Figures 8.1C and 8.2). These data indicate that nitrergic nerves are probably not present in rat mesenteric resistance vessels.

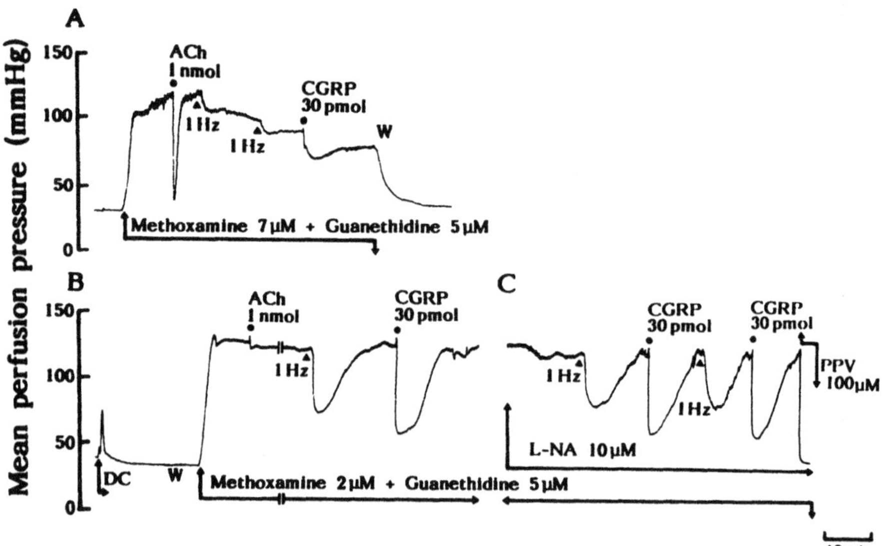

FIGURE 8.1. Typical records showing the effects of endothelium removal and N^G-nitro-L-arginine (L-NA) on vasodilation induced by acetylcholine (ACh), periarterial nerve stimulation (PNS), and CGRP injection in the rat perfused mesenteric vascular bed with active tone produced by methoxamine in the presence of guanethidine. Closed circles indicate bolus injection of acetylcholine (1 nmol) and CGRP (30 pmol), and triangles indicate PNS (1 Hz). Endothelium was removed by perfusion with sodium deoxycholate (DC), 1.8 mg/ml in saline, for 30 seconds. (A) Control responses. (B) Responses after endothelium removal. (C) Effect of L-NA. W, washing with Krebs solution; PPV, papaverine perfusion.

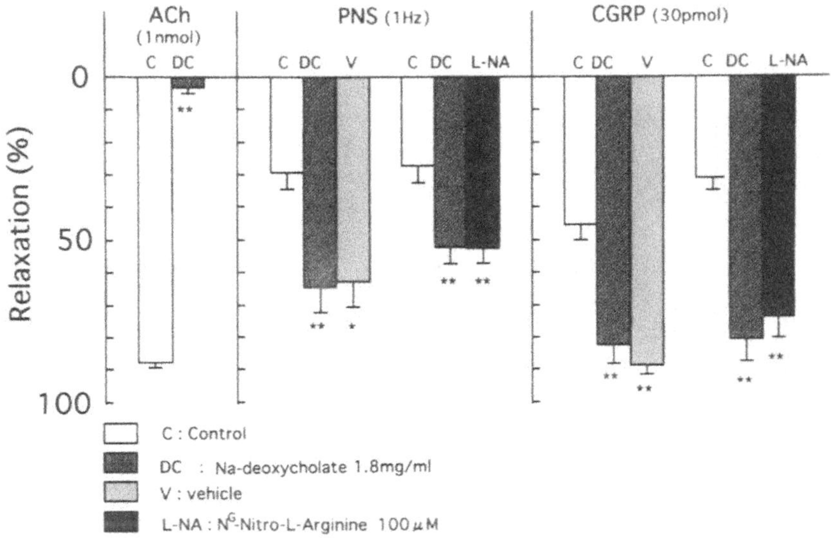

FIGURE 8.2. Effect of N^G-nitro-L-arginine (L-NA) on the neurogenic vasodilation induced by periarterial nerve stimulation (PNS) in the rat perfused mesenteric vascular bed with active tone produced by methoxamine in the presence of guanethidine. ACh, acetylcholine. $*P < 0.05$, $**P < 0.01$ vs. control.

Toda and Okamura (1992) reported that L-NA potentiated the vasoconstrictor response to adrenergic nerve stimulation in dog and monkey mesenteric arteries without endothelium. Because L-NA did not modify norepinephrine-induced vasoconstriction in these arteries, they suggested that EFS activated nitrergic nerves to release NO and then caused vasodilation and counteracted the adrenergic nerve function. Similar results in the denuded mesenteric artery of the monkey have been reported by Tsuchia et al. (1994), who found that L-NA potentiated EFS-induced adrenergic vasoconstriction and that the potentiating effect of L-NA was greater than that of capsaicin. Therefore, they suggested that nerve-derived NO played a larger role than capsaicin-sensitive vasodilator transmitters in neurogenic regulation of monkey mesenteric arterial tone.

As shown in Figure 8.3, in the rat perfused mesenteric vascular bed with resting tone, endothelium removal markedly potentiated both vasoconstriction mediated by adrenergic nerve stimulation and vasoconstriction mediated by exogenously injected norepinephrine. Perfusion with L-NA slightly, but not significantly, caused further potentiation of the responses to EFS and norepinephrine (Figures 8.3C and 8.4). The augmented response by L-NA was slightly reversed by additional perfusion with L-NA and L-arginine, especially the response to high-frequency stimulation

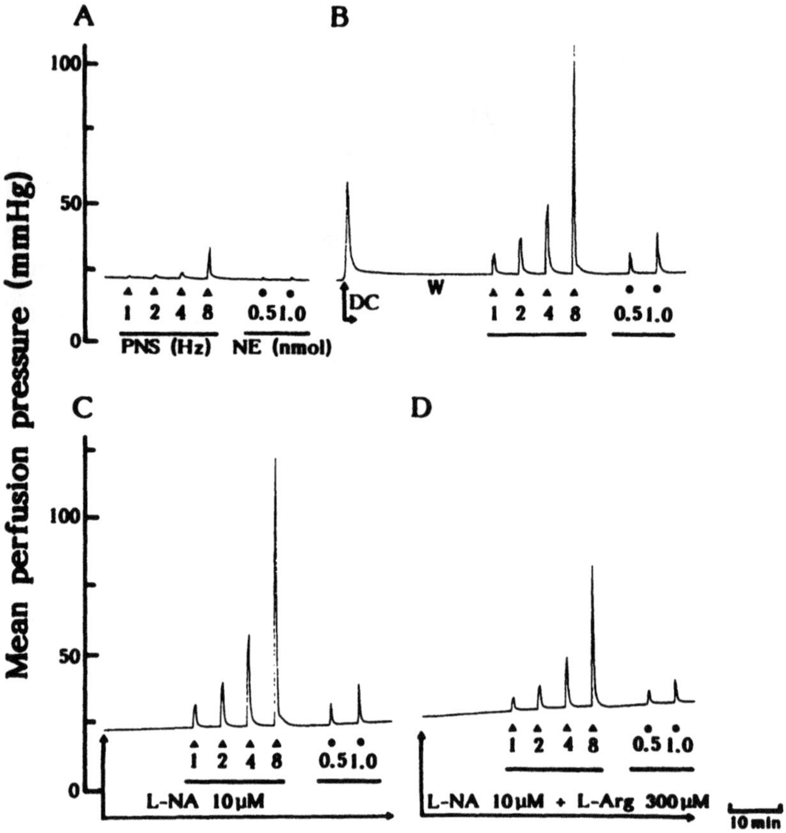

FIGURE 8.3. Typical records showing the effects of N^G-nitro-L-arginine (L-NA) and L-NA plus L-arginine on vasoconstrictor responses to periarterial nerve stimulation (PNS) and norepinephrine (NE) injection in the rat perfused mesenteric vascular bed without endothelium. Closed triangles and circles indicate PNS and bolus injection of NE, respectively. (A) Control responses. (B) Responses after sodium deoxycholate (DC) perfusion. (C) Effect of L-NA. (D) Effect of L-NA plus L-arginine. W, washing with normal Krebs solution.

(8 Hz), without change in norepinephrine-induced vasoconstriction. Therefore, these data cannot exclude the possibility of nitrergic nerves in rat mesenteric resistance vessels. However, L-NA has been shown to suppress the neurogenic release of norepinephrine in rat perfused mesenteric vascular beds with attached intestine, and L-arginine has been shown to attenuate the inhibition of norepinephrine overflow by L-NA (Yamamoto et al., 1993), suggesting that NO increases the release of norepinephrine in rat mesenteric vasculature. Although this study did not identify the source of NO (neuron or endothelium), it is possible that the inhibitory effect of L-NA on the augmented vasoconstrictor response to periarterial nerve stim-

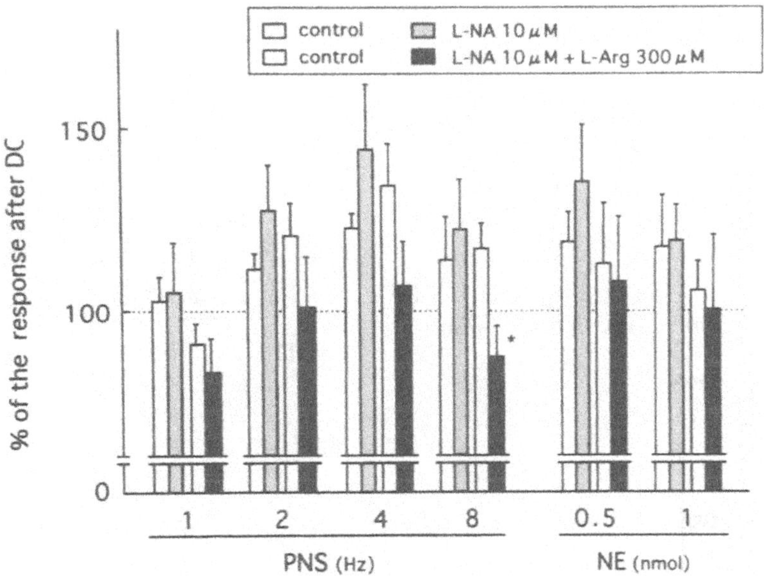

FIGURE 8.4. Effects of N^G-nitro-L-arginine (L-NA) and L-NA plus L-arginine on vasoconstrictor responses to periarterial nerve stimulation (PNS) and norepinephrine (NE) injection in the rat perfused mesenteric vascular bed without endothelium. The response after endothelium removal by sodium deoxycholate (DC) was taken as 100%. $*P < 0.05$ vs. control.

ulation in mesenteric vascular beds is due to decrease in norepinephrine release by L-NA. Therefore, it seems likely that nitrergic nerves may be sparse or absent and play a minor role in the control of vascular tone. As shown in Figure 8.5, capsaicin treatment in the mesenteric vascular bed without endothelium caused more marked potentiation of vasoconstriction induced by adrenergic nerve stimulation than L-NA perfusion did. This further potentiation was reversed by additional treatment with exogenous CGRP. Thus, these results indicate that CGRP-containing nerves play a major role in mesenteric resistance blood vessels in the rat and there might be large species differences.

Recently, surprising data from studies on the porcine basilar artery were reported by Lee and co-workers (Zhang et al., 1998). They found that nicotine-induced but not EFS-induced vasorelaxation, which is sensitive to NOS inhibitor and is mediated by neuronally released NO, was inhibited by guanethidine and by the adrenergic neurotoxin 6-hydroxydopamine. These results suggest that nicotine-induced NO-mediated vasorelaxation is dependent on intact adrenergic innervation, and that nicotine acts on nicotinic cholinoceptor receptors on presynaptic adrenergic nerve terminals to

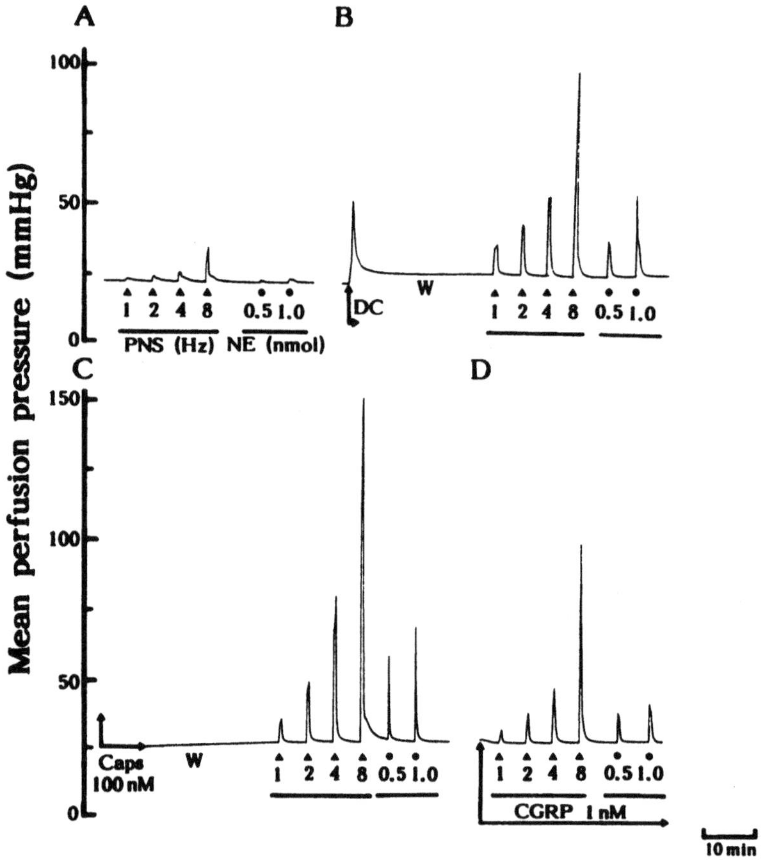

FIGURE 8.5. Typical records showing the effects of capsaicin (Caps) treatment and exogenous CGRP on vasoconstrictor responses to periarterial nerve stimulation (PNS) and norepinephrine (NE) injection in the rat perfused mesenteric vascular bed with resting tone. Closed triangles and circles indicate PNS and bolus injection of NE, respectively.

release norepinephrine, which then stimulates release of NO from nitrergic nerves. Thus, this study might explain why, in rat mesenteric arteries without endothelium, L-NA inhibits vasoconstriction induced by intact adrenergic stimulation but does not suppress vasorelaxation induced by NANC nerve stimulation in the presence of guanethidine. Norepinephrine released by EFS in the intact adrenergic nerves may indirectly activate nitrergic nerves to release NO, which then inhibits adrenergic vasoconstriction. In a preliminary study on rat mesenteric vascular beds, we found that perfusion with nicotine elicited a concentration-dependent vasodilation, which was blocked by hexamethonium, a nicotinic cholinoceptor blocker, and by guanethidine, as reported by Zhang et al. (1998). This nicotine-induced vasodilation was abolished by capsaicin, but NOS inhibitor (L-NA) or endothelium removal was without effect. Therefore, this study supports the view that nitrergic nerves may be sparse or absent and play a minor role in the control of mesenteric vascular tone. It is possible that in rat mesenteric arteries nicotine indirectly stimulates CGRP-containing nerves via adrenergic nerves to cause vasodilation.

Physiological Role of Nitrergic Nerves and CGRP-Containing Nerves in Regulation of Resistance Vessel Tone

In an in vivo study, Okamura et al. (1996) investigated whether nitrergic nerves participated in systemic blood pressure regulation. They found that intravenous injection of L-NA but not D-NA in the anesthetized Japanese monkey elevated the mean blood pressure. The magnitude of the pressor response by L-NA after treatment with pentolinium (an autonomic ganglion blocker) was less than that after treatment with phentolamine (an α-adrenoceptor antagonist), suggesting that high blood pressure produced by a NOS inhibitor is associated with elimination of nitrergic nerve function together with impairment of basal release of NO from the endothelium. The authors proposed that NO is continuously liberated from nitrergic nerves by tonic discharges from the central nervous system. However, they did not rule out the possibility that autonomic nerves can promote NO release, as reported by Habler et al. (1997), who demonstrated in hindlimb microvascular beds of the rat in vivo that sympathetic nerve impulses led to NO release from the endothelium, which in turn attenuated neurogenic vasoconstriction.

The CGRP-containing nerve is sensitive to capsaicin and is thought to be the primary sensory afferent nerve. It, therefore, raises the question how the sensory nerve is able to participate in the regulation of resistance vessel tone, like the adrenergic efferent nerve. This question was resolved by the in vivo study of Taguchi et al. (1992) on the pithed rat, in which blood pressure was artificially elevated by methoxamine in the presence of hexamethonium, an autonomic ganglion blocker. The study showed that electrical stimulation of the lower thoracic spinal cord (Th9–12) but not the upper thoracic cord (Th1–4) caused a frequency-dependent fall in blood pressure without changing the heart rate. This depressor response is sensitive to TTX and capsaicin, and it is abolished by CGRP(8–37) but not by muscarinic cholinoceptor antagonists, β-adrenoceptor antagonists, or histamine H1 and H2 receptor antagonists. These results indicate that the depressor response to spinal cord stimulation is NANC in nature and is mediated by endogenous CGRP, which is released by CGRP-containing nerves. Direct stimulation of the spinal cord and of the peripheral part of the cut dorsal roots or sensory nerves has been shown to result in local vasodilation (skin and skeletal muscle) produced by antidromic conduction of sensory afferent nerves but not to changes in systemic blood pressure (Hilton and Marshall, 1980; Linderoth et al., 1991). The regional vasodilation induced by antidromic stimulation is sensitive to atropine or histamine receptor antagonists. Therefore, the CGRP-containing nerve-mediated depressor response to limited (Th8–12) spinal cord stimulation is unlikely to be due to antidromic stimulation. There may be outflow of CGRP-containing nerves from the spinal cord to the blood vessels. This study suggests that CGRP-containing nerves participate in the neuronal control of the tone of resistance blood vessels.

In rat mesenteric resistance vessels, Kawasaki et al. (1990a) provided evidence that the adrenergic neurotransmitter norepinephrine exerts an inhibitory influence upon CGRP release from CGRP-containing nerves by activating presynaptic α_2-adrenoceptors. In addition, neuropeptide Y, a vasoconstrictor neuropeptide colocalized with norepinephrine in peripheral adrenergic nerves, inhibited neurogenic vasodilation and CGRP release in these vessels (Kawasaki et al., 1991). On the other hand, inhibition of neurotransmission of CGRP-containing nerves by capsaisin and CGRP(8–37) potentiated adrenergic nerve-mediated vasoconstriction without changing neuronal release of norepinephrine, suggesting that CGRP-containing nerves, by releasing CGRP from the nerve, postsynaptically inhibit noradrenergic nerve-mediated vasoconstriction in these vessels (Kawasaki et al., 1990b; Takenaga and Kawasaki, 1999). Therefore, it has been proposed that adrenergic and CGRP-containing nerves regulate the tone of mesenteric resistance by reciprocal interaction (Kawasaki et al., 1990b). Nuki et al. (1994) provide additional evidence that exogenously applied CGRP causes inhibition of CGRP-containing nerve-mediated vasodilation, which is antagonized by CGRP(8–37), and low doses of CGRP(8–37) enhance CGRP-containing nerve-mediated vasorelaxation. This study indicates that CGRP-containing nerves supplying the rat mesenteric resistance vessel are endowed with presynaptic CGRP receptors that regulate the release of CGRP from the nerve by a negative feedback mechanism. This study supports a physiological role for CGRP as a neurotransmitter in the regulation of peripheral resistance blood vessel tone.

Conclusions

Evidence has accumulated that peripheral blood vessels isolated from various species are supplied with nitrergic nerves in which NO acts as a vasodilator neurotransmitter. However, the physiological role of nitrergic nerves in peripheral vascular function still remains unclear. In mesenteric resistance arteries, CGRP-containing nerves are the main NANC vasodilator nerves and play a role in the regulation of mesenteric resistance blood vessel tone. It is still not clear from the available evidence whether there is neural control of the resistance arteries by nitrergic nerves. Further investigations are needed to provide evidence that nitrergic nerves actively participate in the neural control of the tone of peripheral resistance blood vessels.

References

Ahlner, J., Ljusegren, M.E., Grundstrom, N., and Axelsson, K.L. 1991. Role of nitric oxide and cyclic GMP as mediators of endothelium-independent neurogenic relaxation in bovine mesenteric artery. *Circ. Res.* 68:756–762.

Ahluwalia, A., and Vallance, P. 1996. Interaction between sympathetic and sensory nerves in rat small arteries: involvement of nitric oxide. *Am. J. Physiol.* 271:H969–H976.

Amerini, S., Mantelli, L., and Ledda, F. 1993. Nitric oxide is not involved in the effects induced by non-adrenergic non-cholinergic stimulation and calcitonin gene-related peptide in the rat mesenteric vascular bed. *Neuropeptides* 25:51–55.

Bevan, J.A., and Bryden. J.E. 1987. Nonadrenergic neural vasodilator mechanisms. *Circ. Res.* 60:309–326.

Fujimori, A., Saito, A., Kimura, S., Watanabe, T., Uchiyama, Y., Kawasaki, H., and Goto, H. 1987. Neurogenic vasodilation and release of CGRP from perivascular nerves in the rat mesenteric artery. *Biochem. Biophys. Res. Commun.* 164:394–399.

Gyoda, Y., Tsukada, Y., Saito, A., and Goto, K. 1995. Role of nitric oxide and neuropeptides in neurogenic vasodilation of the guinea pig mesenteric artery. *Eur. J. Pharmacol.* 279:83–92.

Habler, H.J., Wasner, G., and Janig, W. 1997. Attenuation of neurogenic vasoconstriction by nitric oxide in hindlimb microvascular beds of the rat in vivo. *Hypertension* 30:957–961.

Han, S.P., Haes, L., and Westfall, T.C. 1990. Inhibition of periarterial nerve stimulation-induced vasodilation of the mesenteric arterial bed by CGRP8-37 and CGRP receptor desensitization. *Biochem. Biophys. Res. Commun.* 168:786–791.

Hilton, S.M., and Marshall, J.M. 1980. Dorsal root vasodilation in cat skeletal muscle. *J. Physiol.* 299:277–288.

Kakuyama, M., Vallance, P., and Ahluwalia, A. 1998. Endothelium-dependent sensory NANC vasodilation: involvement of ATP, CGRP and a possible NO store. *Br. J. Pharmacol.* 123:310–316.

Kawasaki, H., Nuki, C., Saito, A., and Takasaki, K. 1990a. Adrenergic modulation of CGRP-containing nerve-mediated vasodilation in the rat mesenteric resistance vessel. *Brain Res.* 506:287–290.

Kawasaki, H., Nuki, C., Saito, A., and Takasaki, K. 1990b. Role of CGRP-containing nerves in the vascular adrenergic neurotransmission. *J. Pharmacol. Exp. Ther.* 252:403–409.

Kawasaki, H., Nuki, C., Saito, A., and Takasaki, K. 1991. NPY modulates neurotransmission of CGRP-containing vasodilator nerves in rat mesenteric arteries. *Am. J. Physiol.* 261:H683–H690.

Kawasaki, H., Takasaki, K., Saito, A., and Goto, K. 1988. Calcitonin gene-related peptide acts as a novel vasodilator neurotransmitter in mesenteric resistance vessels of the rat. *Nature* 335:164–167.

Kawasaki, H., Takenaga, M., Nuki, C., and Wada, A. 1994. Neuronal control of resistance blood vessel tone by calcitonin gene-related peptide- and nitric oxide-nerves. In: *Nitric Oxide: Roles in Neuronal Communication and Neurotoxicity* (eds. H. Takagi, N. Toda, and R.D. Hawkins), pp. 89–99. Japan Scientific Societies Press and CRC Press, Tokyo.

Kobayashi, D., Todoki, K., Ozono, S., and Okabe, E. 1995. Calcitonin gene-related peptide mediated neurogenic vasorelaxation in the isolated canine lingual artery. *Jpn. J. Pharmacol.* 67:329–339.

Leckstrom, A., Ahlner, J., Grundstrom, N., and Axles, K.L. 1993. Involvement of nitric oxide and peptides in the inhibitory non-adrenergic, non-cholinergic (NANC) responses in bovine mesenteric artery. *Pharmacol. Toxicol.* 72:194–198.

Lee, T.J.-F., and Sarwinski, S. 1991. Nitric oxidergic neurogenic vasodilation in the porcine basilar artery. *Blood Vessels* 28:407–412.

Li, Y.J., Yu, X.J., and Deng, H.W. 1993. Nitric oxide modulates responses to sensory nerve activation of the perfused rat mesentery. *Eur. J. Pharmacol.* 239:127–132.

Linderoth, B., Fedoresak, T., and Meyerson, B.A. 1991. Peripheral vasodilation after spinal cord stimulation: animal studies of putative effector mechanisms. *Neurosurgery* 28:187–195.

Liu, S.F., Crawley, D.E., Rohde, J.A., Evans, T.W., and Barnes, P.J. 1992. Role of nitric oxide and guanosine 3′, 5′-cyclic monophosphate in mediating nonadrenergic, noncholinergic relaxation in guinea-pig pulmonary arteries. *Br. J. Pharmacol.* 107:861–866.

Marin, J., and Balfagon, G. 1998. Effect of clenbuterol on non-endothelial nitric oxide release in rat mesenteric arteries and the involvement of beta-adrenoceptors. *Br. J. Pharmacol.* 124:473–478.

Matsuda, H., Ohmori, F., Uematsu, M., Nagata, S., and Miyatake, K. 1995. Different modes of sensory neuropeptides and nitric oxide involvement in relaxation of guinea-pig vessels. *J. Auton. Nerv. Syst.* 55:115–122.

Mulvany, M.J., and Aalkjaer, C. 1990. Structure and function of small arteries. *Physiol. Rev.* 70:921–961.

Nuki, C., Kawasaki, H., Takasaki, K., and Wada, A. 1994. Pharmacological characterization of presynaptic calcitonin gene-related peptide (CGRP) receptors on CGRP-containing vasodilator nerves in rat mesenteric resistance vessels. *J. Pharmacol. Exp. Ther.* 268:59–64.

Okamura, T., Ayajiki, K., and Toda, N. 1996. Neural mechanism of pressor action of nitric oxide synthase inhibitor in anesthetized monkeys. *Hypertension* 28:341–346.

Okamura, T., Ayajiki, K., Uchiyama, M., Kagami, K., and Toda, N. 1998. Mechanisms underlying constrictor and dilator responses to perivascular nerve stimulation in canine lingual arteries. *Eur. J. Pharmacol.* 354:43–50.

Okamura, T., Ayajiki, K., and Toda, N. 1996. Neural mechanism of pressor action of nitric oxide synthase inhibitor in anesthetized monkeys. *Hypertension* 28:341–346.

Okamura, T., Yoshida, K., and Toda, N. 1995. Nitroxidergic innervation in dog and monkey renal arteries. *Hypertension* 25:1090–1095.

Scott, J.A., Craig, I., and McCormack, D.K. 1996. Nonadrenergic noncholinergic relaxation of human pulmonary arteries is partially mediated by nitric oxide. *Am. J. Respir. Crit. Care Med.* 154:629–632.

Su, E.N., Alder, V.A., Yu, D.Y., and Cringle, S.J. 1994. Adrenergic and nitrergic neurotransmitters are released by the autonomic system of the pig long posterior ciliary artery. *Curr. Eye Res.* 13:907–917.

Taguchi, T., Kawasaki, H., Imamura, T., and Takasaki, K. 1992. Endogenous calcitonin gene-related peptide mediated nonadrenergic noncholinergic depressor response to spinal cord stimulation in the pithed rat. *Circ. Res.* 71:357–364.

Takenaga, M., and Kawasaki, H. 1999. Endogenous calcitonin gene-related peptide suppresses vasoconstriction mediated by adrenergic nerves in rat mesenteric resistance blood vessels. *Eur. J. Pharmacol.* 367:239–245.

Toda, N., Ayajiki, K., Uchiyama, M., and Okamura, T. 1997. Nitric oxide-mediated neurogenic vasodilation in isolated monkey lingual arteries. *Am. J. Physiol.* 272:H1582–H1588.

Toda, N., Kitamura, Y., and Okamura, T. 1994. Role of nitroxidergic nerve in dog retinal arterioles in vivo and arteries in vitro. *Am. J. Physiol.* 266:H1985–H1992.

Toda, N., Kitamura, Y., and Okamura, T. 1995. Functional role of nerve-derived nitric oxide in isolated dog ophthalmic arteries. Invest. *Ophthalmol. Vis. Sci.* 36:563–570.

Toda, N., and Okamura, T. 1990a. Modulation by L-N^G-monomethyl arginine (L-NMMA) of the response to nerve stimulation in isolated dog mesenteric and cerebral arteries. *Jpn. J. Pharmacol.* 52:170–173.

Toda, N., and Okamura, T. 1990b. Mechanism underlying the response to vasodilator nerve stimulation in isolated dog and monkey cerebral arteries. *Am. J. Physiol.* 259:H1511–H1517.

Toda, N., and Okamura, T. 1992. Mechanism of neurally induced monkey mesenteric artery relaxation and contraction. *Hypertension* 19:161–166.

Toda, M., Okamura, T., Azuma, I., and Toda, N. 1997. Modulation by neurogenic acetylcholine of nitroxidergic nerve function in porcine ciliary arteries. *Invest. Ophthalmol. Vis. Sci.* 38:2261–2269.

Toda, N., Toda, M., Ayajiki, K., and Okamura, T. 1996. Monkey central retinal artery is innervated by nitroxidergic vasodilator nerves. *Invest. Ophthalmol. Vis. Sci.* 37:2177–2184.

Toda, N., Toda, M., Ayajiki, K., and Okamura, T. 1998. Cholinergic nerve function in monkey ciliary arteries innervated by nitroxidergic nerve. *Am. J. Physiol.* 274:H1582–H1589.

Tsuchia, K., Urabe, M., Yamamoto, R., Asada, Y., and Lee, T.J. 1994. Effects of N^{ω}-nitro-L-arginine and capsaicin on neurogenic vasomotor responses in isolated mesenteric arteries of the monkey. *J. Pharm. Pharmacol.* 46:155–157.

Vials, A..J., Crowe, R., and Burnstock, G. 1997. A neuromodulatory role for neuronal nitric oxide in the rabbit renal artery. *Br. J. Pharmacol.* 121:213–220.

Wiencke, A.K., Nilsson, H., Nielsen, P.J., and Nylborg, N.C. 1994. Nonadrenergic noncholinergic vasodilation in bovine ciliary artery involves CGRP and neurogenic nitric oxide. *Invest. Ophthalmol. Vis. Sci.* 35:3268–3277.

Yamamoto, R., Wada, A., Asada, Y., Niina, H., and Sumiyoshi, A. 1993. N^{ω}-nitro-L-arginine, an inhibitor of nitric oxide synthase, decreases noradrenaline outflow in rat isolated perfused mesenteric vasculature. *Naunyn-Schmiedebergs Arch. Pharmacol.* 347:238–240.

Yoshida, K., Okamura, T., Kimura, H., Bredt, D.S., Snyder, S.H., and Toda, N. 1993. *Brain Res.* 629:67–72.

Zhang, W., Edvinsson, L., and Lee, T.J-F. 1998. Mechanism of nicotine-induced relaxation in the porcine basilar artery. *J. Pharmacol. Exp. Ther.* 284:790–797.

Zheng, Z., Shimamura, K., Anthony, T.L., Travagli, R.A., and Kreulen, D.L. 1997. Nitric oxide is sensory nerve neurotransmitter in the mesenteric artery of guinea pig. *J. Auton. Nerv. Syst.* 67:137–144.

9
Nitric Oxide, Perivascular Neural Transmission, and Migraine

Jes Olesen, Lars Lykke Thomsen, and Inger Jansen-Olesen

Neurovascular Transmission and Migraine

Migraine attacks are known to involve alterations or even dysfunction in the regulation of tone in intra- and/or extra-cranial blood vessels (Iversen et al., 1990; Friberg et al., 1991; Olesen 1991; Thomsen et al., 1995). The nature of these alterations is not clearly established, and a number of candidate neurotransmitters have been discussed (Edvinsson et al. 1988). Some of these candidates have their origin in sympathetic, parasympathetic, or sensory neurons; endothelial cells; mast cells; or on blood-borne cells. An increasing number of perivascular neurotransmitters have been detected, and their receptor mechanisms and effects have been elucidated during the last decades (Jansen-Olesen and Edvinsson, 1999). In addition, endothelium-derived relaxing factors (EDRFs) have been demonstrated; today these include nitric oxide (NO) prostaglandin I_2 (PGI_2) and endothelium-derived hyperpolarizing factor (EDHF). These agents are released following the stimulation of receptors located on endothelial cells or following shear-stress phenomena causing relaxation. Thus, a number of mechanisms exist for regulation of tone in cranial blood vessels. Neurogenic inflammation has been suggested as a mechanism of migraine (Moskowitz, 1993). It involves the liberation of several neuropeptides, primarily substance P (SP) and calcitonin-gene-related peptide (CGRP) (Buzzi and Moskowitz 1990). Unfortunately, neurogenic inflammation has never been shown to occur around cranial blood vessels during migraine attacks and the concentration of SP in external and internal jugular venous blood remains normal during migraine attacks (Goadsby et al., 1990; Friberg et al., 1994). Furthermore, the endothelin-receptor antagonist Bosentan and the SP-antagonist RPR 100983–201 both block neurogenic inflammation but have no effect against migraine attacks (May et al., 1996; Diener Poster communication at the 6th International Headache Research seminar, Copenhagen 1995). Also neuropeptide Y (NPY) and vasoactive interstitial polypeptide (VIP), which are important neuropeptides found in sympathetic and parasympathetic nerve fibers around the intra- and extra-cranial blood vessels, remain normal in blood from the internal and external jugular veins during migraine attacks (Goadsby et al., 1990; Friberg et al., 1994). The only peptide known to be released during migraine attacks is CGRP. The concentration of this peptide is increased in the external but not in the internal jugular venous blood (Goadsby et al., 1990; Friberg et al., 1994). However, neither when infused intravenously nor when injected into the superficial temporal muscle does CGRP cause pain (Shekar et al., 1991; Pedersen-Bjerregård et al., 1991). It does not even potentiate pain induced by other substances (Pedersen-Bjerregård et al., 1991.) The spectacular therapeutic effect of 5-hydroxytryptamine (5-HT) receptor agonists in migraine might indicate that 5-HT is involved. However, 5-HT injected intravenously or locally into the superficial temporal muscle does not cause pain (Jensen et al., 1990, Roberts 1992), and it only slightly aggravates bradykinin-induced pain in the temporal muscle (Jensen et al., 1990). It may be concluded that none of the above-mentioned peptides or monoamines are likely to cause the nociception responsible for migraine pain.

Based on previous studies to be discussed later in this chapter we have suggested that nitric oxide (NO) is a likely candidate molecule in triggering migraine attacks. NO may induce migraine secondary to endothelial receptors, secondary to activation in non-adrenergic non-cholinergic (NANC) nerves or—less likely—as a result of an action on the central nervous system.

The relaxant effects of ACh, SP, and neurokinin A (NKA) are known to involve the release from vascular endothelial cells of a short-lived relaxant factor (EDRF) (Furchgott, 1984). It was suggested that EDRF may be NO because endothelium-independent vasodilators such as sodium nitroprusside and nitroglycerine (NTG) are known to generate free NO. The free NO rapidly diffuses to the smooth muscle cells where it stimulates guanylyl cyclase to produce cyclic guanosine monophosphate (cGMP). Vasodilatation induced by the production of NO can be antagonised by N^c-monomethyl-L-arginine, an inhibitor of NO synthase that converts L-arginine to NO and L-citrulline (Palmer et al. 1987). NOS-ir perivascular nerves have been shown in human cerebral vessels (Tomimoto et al., 1994) and temporal arteries (Gulbenkian, unpublished). We could not show any perivascular NOS-ir in the middle meningeal artery, but found ample NOS-ir in the endothelium (Edvinsson et al. 1998). In a still unpublished study we have shown that mRNA for neuronal nitric oxide synthase (NOS) is present in the human middle meningeal artery. In preliminary studies of the superficial temporal arteries we observed that the vasodilator response to ACh, SP, and NKA was lost after removal of the endothelium and after treatment with a nitric oxide synthase inhibitor. The response was restored by L-arginine (Jansen and Edvinsson, unpublished).

NANC nerve fibers contain NOS and liberate NO when activated. The action potential opens voltage-gated calcium channels in the terminals and the increased intracellular calcium turns on nNOS to produce NO. NANC nerves all are found in the parasympathetic supply to cranial blood vessels in most species including primates and activity in these nerves could cause migraine pain via an increase in perivascular NO. This, in turn, could directly activate sensory trigeminal nerve terminals to liberate CGRP, causing dilatation of cranial blood vessels (Toda, 1997).

Glyceryl Trinitrat and Histamine in Experimental Headache

Glyceryl trinitrate (GTN) and histamine are substances that reliably and dose-dependently produce headache in normal volunteers and migraine sufferers (Krabbe and Olesen, 1980; Iversen, Olesen, et al., 1989; Olesen et al., 1993). GTN itself has no known action in the human body but acts via liberation of nitric oxide (NO) and is thus generally regarded as a NO donor (Ignarro et al., 1981; Grutter et al., 1981; Feelisch and Noack, 1987). GTN is the most suitable substance for experimental studies of NO-induced headache since it is well tolerated and diffuses freely across membranes owing to its lipid solubility. It may thus deliver NO to several tissues including those protected by the blood—brain barrier. The following observations support that GTN induces headache by liberating NO: GTN-induced headache in normal controls is very short lived and is therefore unlikely to be caused by metabolites other than NO since these have a longer half-life (Iversen, Olesen, et al., 1989). The long-acting nitrate 5-isosorbide-mononitrate (5-ISMN) induces a dose-dependent headache and arterial dilatation, but its metabolites apart from NO are different from those of GTN (glyceryl trinitrate) (Iversen et al., 1992). N-Acetylcysteine, which augments GTN effects in the heart by increasing the formation of NO or by enhancing the effect of NO itself, also augments the headache response to GTN and prolongs GTN-induced arterial dilatation of the superficial temporal artery but not of the radial arteries (Iversen 1992).

The hemodynamic responses during GTN-induced headache in healthy subjects and migraine without aura are very similar with unchanged rCBF and dilatation of both intra- and extra-cranial large arteries (Iversen, Holm, et al., 1989; Dahl et al., 1989). Interestingly, administration of sumatriptan in combination with nitroglycerin reduces GTN-induced headaches in healthy subjects and prevents large artery dilatation (Iversen and Olesen, 1996; Schmetterer et al., 1996). Histamine also seems to induce headache via NO, and in human cerebral blood vessels histamine stimulates an endothelial H_1 receptor, which activates NOS (Toda, 1990; Ottosson et al., 1991; Jansen-Olesen et al., 1997). Thus, histamine stimulates the endogenous formation of NO, whereas GTN delivers NO directly.

Glyceryl Trinitrate and Histamine in Migraine Patients

Previous studies have suggested that migraine patients more often experience a migraine-like headache in association with nitroglycerine administration than do non-migraineurs (Sicuteri et al., 1987). Most recently it has been confirmed in controlled double blind trials that migraines with a time-delay of several hours (peak intensity 5.5 hours after nitroglycerine infusion) actually develop a genuine migraine attack after nitroglycerine infusion (Olesen et al., 1993; Thomsen et al., 1994). This induced migraine headache is preceded by an immediate headache response during the infusion, resembling but not fulfilling diagnostic criteria for migraine without aura (IHS criteria). The immediate headache response, which also is seen in nonmigrainers, is more severe in migrainers (Olesen et al., 1993). Thus, migrainers are hypersensitive to nitroglycerine-induced headache and most likely, therefore, to nitric oxide (Figure 9.1). An increased headache response could, however, reflect a greater general sensitivity to pain, or it

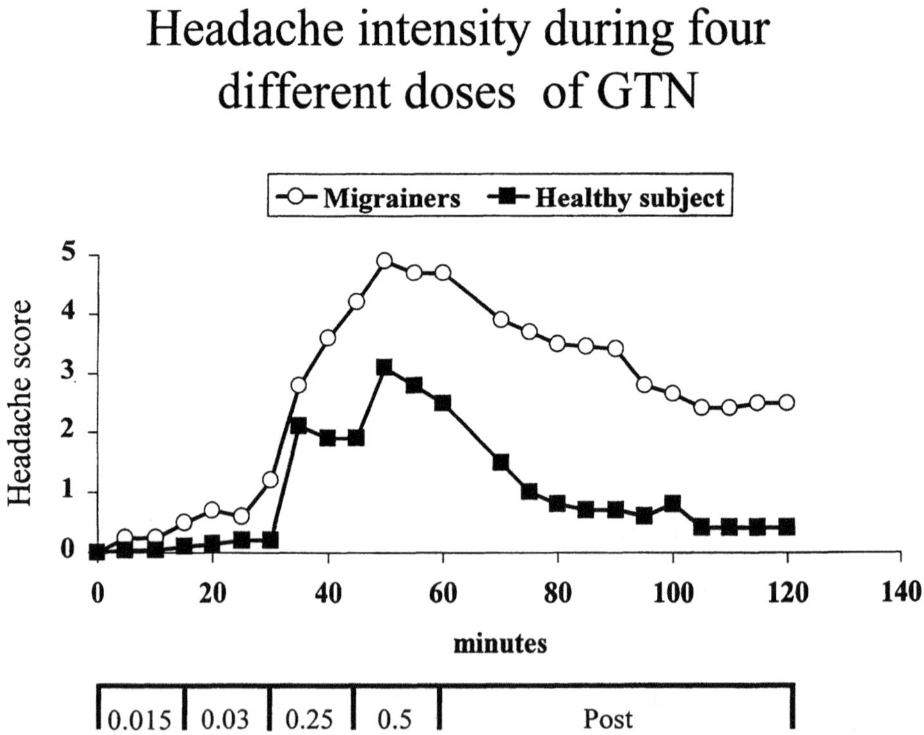

FIGURE 9.1. Headache intensity over time during four different doses of nitroglycerine. Comparison between responses in migraners (n = 17) and healthy subjects (n = 17). During doses above 0.015 µg/kg/min migraine patients experienced significantly more headache than controls ($p < 0.05$) (Kruskal Wallis test).

could be due to increased physiological sensitivity to nitric oxide. It is well known that nitroglycerine dilates the middle cerebral artery via NO without affecting cerebral blood flow (Dahl et al., 1989; Iversen Holm, et al., 1989). Applying the ultrasound technique transcranial doppler (Thomsen and Iversen, 1993), which provides an indirect measure of large intracranial artery diameters in situations of unchanged blood flow (Busija et al., 1981; Dahl et al., 1989; Iversen Holm, et al., 1989; Friberg et al., 1991), we examined whether the increased sensitivity to the nitric oxide donor was reflected not only in increased headache in migrainers but also in increased dilatation of the middle cerebral artery. Indeed migrainers were found to be more sensitive in this aspect as well. During a 3-hour observation period the time profile of the nitroglycerine-induced middle cerebral artery dilatation corresponded with the headache response (Thomsen et al., 1993; Thomsen et al.; 1994). After sublingual administration of isosorbide dinitrate, which also acts as a NO donor, middle cerebral artery responses has been followed even longer during nitrate-induced delayed migraine headache in sufferers of migraine without aura. Compared to baseline, middle cerebral artery blood velocities were till reduced at this time (3–5 hours after isosorbide dinitrate administration) but only on the customary side of the head pain of a spontaneous migraine attack (Bellantonio et al., 1997).

Moreover, migrainers have been found to be hypersensitive to histamine in controlled trials (Krabbe and Olesen, 1980; De-Marinis et al., 1990). The headache induced by histamine was almost completely blocked by the histamine H_1-blocker mepyramine, whereas the H_2-blocker cimetidine only had a small effect (Krabbe and Olesen, 1980). As mentioned above, activation of endothelial H_1-receptors induce the formation of nitric oxide (Toda, 1990; Jansen-Olesen 1997). Thus, the increased sensitivity to histamine in migrainers may also be explained by hypersensitivity to activation of the NO pathway. However, the accompanying symptoms characteristic of migraine headache (nausea, vomiting, photophobia, and phonophobia) were not reported in the short study period of the above mentioned studies. In a more recent double blind 12-hour study migrainers suffering from migraine without aura were randomized to pretreatment with either mepyramine or placebo before histamine infusion (Lassen et al., 1995). Seven out of 10 patients, who received histamine in combination with placebo, experienced a migraine attack fulfilling the IHS criteria for migraine without aura (Figure 9.2). Interestingly, the time profile of the histamine-induced headache in migrainers was exactly the same as after GTN (Figure 9.3). These similarities suggest common mechanisms. GTN may induce liberation of histamine from mast cells and basophils. However, whereas mepyramine blocked both the immediate and the delayed histamine response (Lassen et al., 1995), it did not prevent GTN-induced immediate headache (Iversen and Olesen 1994) or the delayed GTN-induced migraine (Lassen et al., 1996). Thus, a likely final common mechanism of

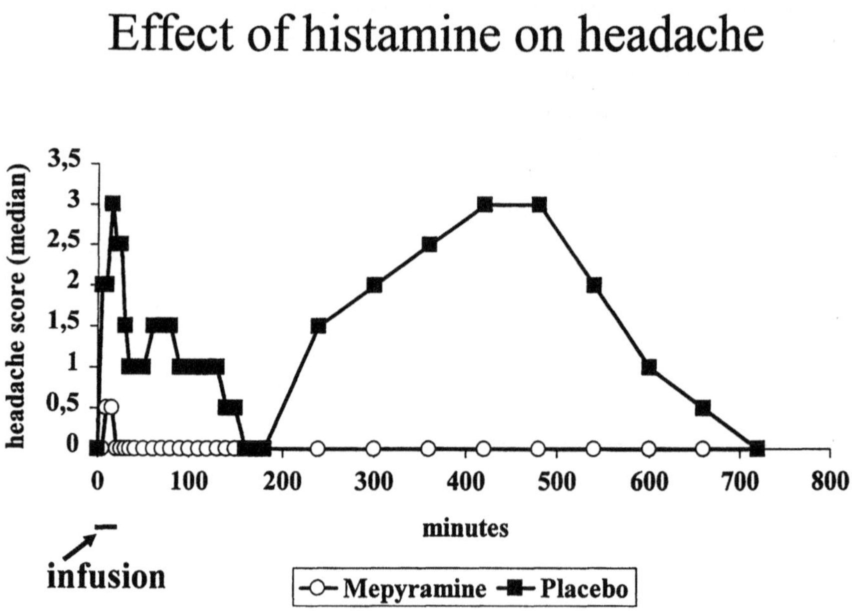

FIGURE 9.2. Time course of histamine-induced headache in migrainers when patients were pretreated with either placebo or mepyramine.

histamine- and GTN-induced immediate headache and delayed migraine is not explained by histamine release after GTN infusion. Instead activation of the NO-cGMP pathway and/or of NO-mediated toxicity provide possible common molecular mechanisms for GTN- (exogenous NO) and histamine- (endogenous NO) induced immediate headache and arterial responses and for the triggering of GTN- and histamine-induced delayed headache responses (Olesen et al., 1994; Thomsen et al., 1994). More direct evidence for the causal relationship between histamine-induced headache and H_1 receptor mediated activation of the NO pathway would be provided if inhibition of nitric oxide synthase (NOS) proved to have any effect on histamine induced headaches. A recent study, however, did not show any effect of the competitive NOS inhibitor N-monomethyl-L-arginin (L-NMMA) on histamine-induced headaches in healthy subjects. One explanation for this may be that the administered doses of L-NMMA were too low to outbalance the administered histamine doses.

Other Headache Triggerers

Beside GTN and histamine other substances that have been shown to reliably cause headache, including reserpine, meta-chlorphenylpiperazine (mCPP), fenfluramine, and, less convincingly, prostacyclin and hypoxia, may

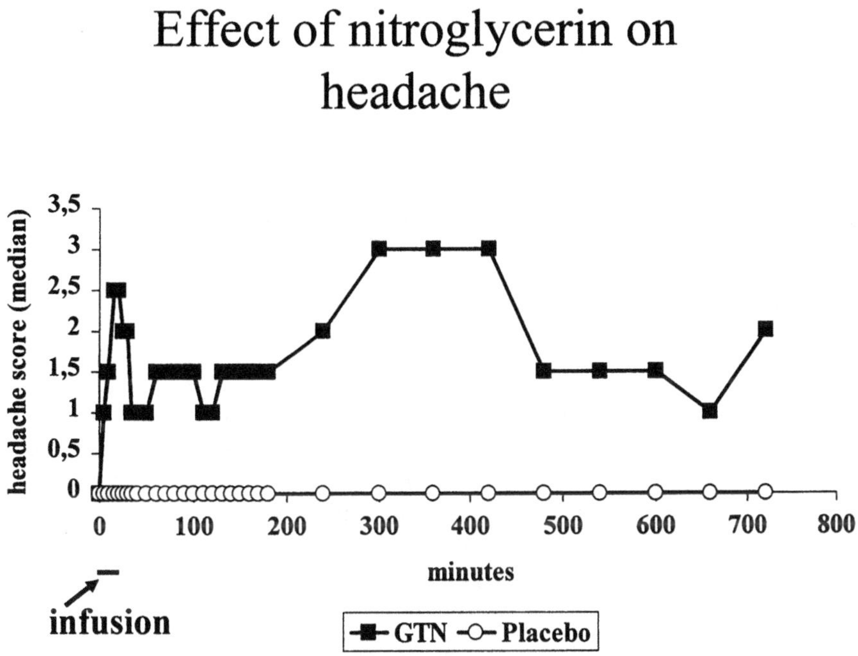

FIGURE 9.3. Time course of headache after a 20 min, glyceryl trinitrate infusion in migrainers as compared to headache after infusion of placebo.

cause headache via NO (Olesen et al., 1995). The best known example of hypoxic headache is high-altitude headache. However, no formal study of the effects of hypoxia in migraine sufferers is available. Recently Arregui et al. (1991) showed that persons living at high altitude had a huge increase in migraine prevalence. Hypoxia increases longevity of NO, whereas pure oxygen acts as a NO scavenger reducing the lifetime and thereby the effect of NO (Rengasamy and Johns, 1991). Hypoxia-induced migraine may thus be due to increased spontaneous NO concentration.

Prostacyclin has been shown in one study to cause headache in migrainers (Peatfield et al., 1981). Prostacyclin is a vasodilatator and may act directly on smooth muscle receptors but, in cerebral vessels protected by the blood-brain-barrier, is more likely to act via endothelial receptors by liberating NO (Lücher and Vanhoutte, 1990).

In migrainers reserpine has been shown to cause headache with some migrainous features (Lance, 1991). Reserpine depletes not only platelets but also presynaptic nerve terminals of their content of monoamines. Substances released include 5-HT. Another liberator of 5-HT is fenfluramine which induces migrainous headache in migrainers (Glover et al., 1996). The 5-HT_{2B} (former called 5-HT_{1c}) receptor has recently been suggested as playing a crucial role in the initiation of spontaneous migraine attacks (Fozard and Kalkman, 1994). 5-HT caused an EDRF response in a number of vessels from different species, and this effect was mediated via the 5-

HT_{2B} receptor. The vascular response to 5-HT_{2B} activation, at least in the pig, is primarily a consequence of the release of NO (Glusa and Richter, 1993). mCPP is a direct agonist at the 5-HT_{2B} receptor and, therefore, is likely to cause vascular headache via NO synthesis (Brewerton et al., 1988; Fozard and Kalkman 1994). We concluded that NO is a likely common denominator for headaches induced by GTN, histamine, reserpine, fenfluramine mCPP, prostacyclin, and hypoxia (Olesen et al., 1995).

Possible Mechanisms in Nitric Oxide-Induced Migraine Headache

The mechanisms by which NO may trigger migraine headache are at present speculative. In theory several neurotransmitters in brain tissue, periarterial cerebral nerves, and in the blood stimulate the formation of NO in brain neurons and arterial endothelium and possibly also interact with NOS-containing nerve terminals (Lücher and Vanhoutte, 1990; Moncada et al., 1991; Garthwaite et al., 1988). Thus, fluctuations in neurotransmitter concentration both in brain and blood may trigger spontaneous migraine headache in migraine patients due to their supersensitivity to nitric oxide. Formation of nitric oxide may also be elicited by pathological reactions such as (a) spreading depression of Leao (Goadsby et al., 1992); (b) activation of the trigeminovascular system with liberation of substance P, for example: (c) fever and inflammation via interleukines and histamine and so forth (Olesen et al., 1994). At present it is not known in further detail how activation of the NO pathway may cause migraine headache. Dilatation of large intra and extra-cranial arteries may be involved because: (a) arterial dilatation is induced by nitric oxide liberated from endothelium and probably perivascular nerve endings (Moncada et al., 1991); (b) cephalic vasodilatation has been reported during spontaneous migraine headache (Iversen et al., 1990; Friberg et al., 1991; Thomsen et al., 1995); (c) mechanical dilatation of intracranial arteries causes referred pain in the areas where most patients feel their pain during migraine attacks (Nichols et al., 1990); and (d) agents such as ergotamine, DHE, and sumatriptan, which constrict arteries, are effective in the treatment of the acute migraine attack. On the other hand, the moderate mechanical arterial dilatation reported during migraine attacks may not be enough to cause severe pain. Another possibility is central pain modulating the effects of nitric oxide. Direct activation of perivascular sensory nervefibers and/or initiation of perivascular neurogenic inflammation (Moskowitz, 1993) by nitric oxide may be other possibilities, and also the direct noxious and cytotoxic effect of nitric oxide should be considered. It is striking that nitric oxide causes migraine with a delay of up to several hours (Thomsen et al., 1994), a time course that mimics the often slowly progressing development of migraine pain during spontaneous migraine attacks. NO is an unstable free radical with a very short half-life. Other mediators or mechanisms therefore seem to be involved in the rather slow cascade of events set up by activation of the nitric oxide pathway and eventually leading to a migraine attack.

Traditionally the mechanism of action of acute migraine therapy (except generally acting analgesics) has been ascribed to constriction of pathologically dilated cranial arteries. Thus, i.v. administration of the effective antimigraine drug ergotamine was shown long ago to reduce temporal artery pulsations in parallel with a decrease in headache intensity. It has recently been suggested that one important action of ergotamine may be the inhibition of antidromic release of trigeminal neuropeptides as part of a neurogenic inflammation (for review, see Moskowitz 1993). Ergotamine interacts with both 5-HT-, dopamine and noradrenaline receptors. Thus, elucidation of a specific receptor involvement requires more specific pharmacological tools. So far the most specific and highly effective acute migraine treatments available are the 5-$HT_{1B/1D}$ receptor agonists. Several mechanisms of action of the 5-$HT_{1B/1D}$ receptor agonist, sumatriptan, have been proposed. A direct pain modulating effect in the central nervous system is unlikely since sumatriptan is soluble in water and crosses the blood—brain barrier very slowly. One possibility is that constriction of dilated, large intracranial arteries provides the causative mechanism (Friberg et al., 1991). However, as with ergotamine another possible mechanism of action is blockade of the release of sensory neuropeptides as part of a neurogenic inflammation (for review, see Moskowitz 1993). Since these neuropeptides also induce vasodilatation, the observed vasoconstrictive effect may be a secondary phenomenon. At the molecular level it is possible that sumatriptan intereferes with the synthesis of nitric oxide.

Regarding drugs with established prophylactic effect in migraine their mechanism of action has for long been an enigma. These drugs include beta-adrenergic blocking drugs without partial agonist activity (i.e., propranolol, metroprolol, atenol, nadolol, and timolol), antiserotonergic drugs (i.e., methysergide, pizotifen) and calcium antagonists (i.e., flunarizine and verapamil). Many observations suggest that all of these drugs interact with the NO-triggered cascade of reactions. Thus, calcium antagonists block the effect of activity in non-adrenergic non-cholinergic perivascular nerves by inhibiting voltage gated t-channels and therefore the activation of neuronal NOS (Toda 1997). Methysergide and pizotifen are 5-HT antagonists that do not discriminate between the 5-HT_1 and the 5-HT_2 receptors. It has recently been suggested that their effect is via 5-HT_{2B} receptor antagonism. 5-HT_{2B} receptor stimulation liberates NO. Thus, amine antagonists may well exert their action by reducing NO production. Propranolol blocks isoprenaline-induced relaxation of rat thoracic aorta in an endothelium-dependent fashion. The response is also blocked by the NOS inhibitor L-NOARG. The prophylactic effect of beta-adrenergic blockers in migraine may thus result from blockade of beta-adrenoceptor-induced NO production.

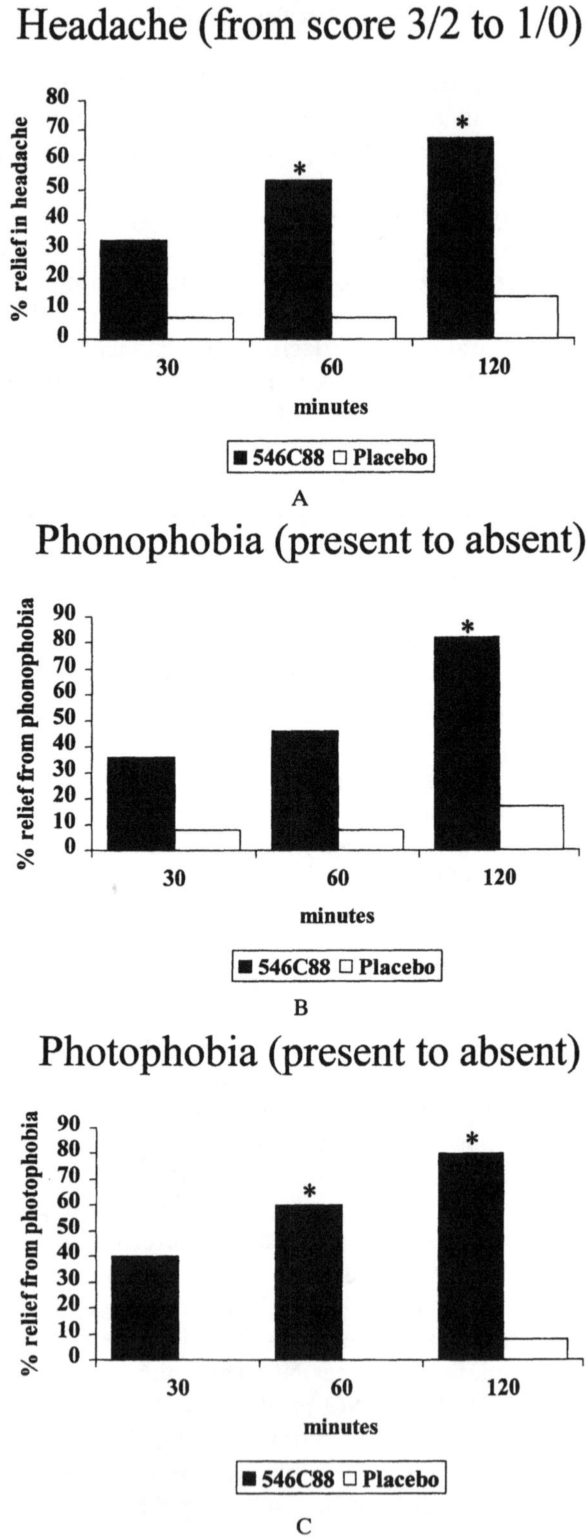
Headache (from score 3/2 to 1/0)
% relief in headache
80
70
60
50
40
30
20
10
0
*
*
30
60
120
minutes
546C88 Placebo
A
Phonophobia (present to absent)
% relief from phonophobia
90
80
70
60
50
40
30
20
10
0
*
30
60
120
minutes
546C88 Placebo
B
Photophobia (present to absent)
% relief from photophobia
90
80
70
60
50
40
30
20
10
0
*
*
30
60
120
minutes
546C88 Placebo
C

FIGURE 9.4.

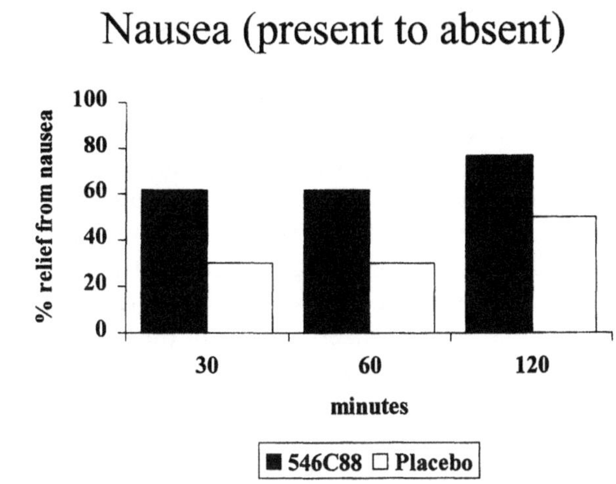

D

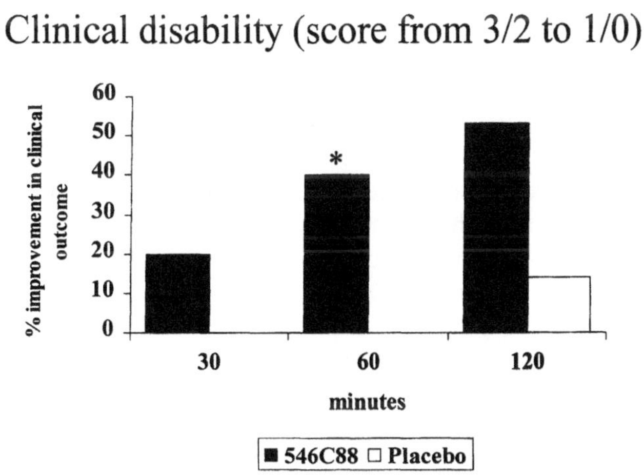

E

FIGURE 9.4. Percent relief in A. headache, B. phonophobia, C. photophobia, D. nausea, and E. improvement in clinical outcome in 546C88 and placebo-treated patients at 30, 60, and 120 min after of infusion of 546C88/placebo.

Propranolol also antagonizes the 5-HT_{2B} receptor on the endothelium. This is another mechanism whereby it may reduce endothelial NO production. In contrast to propranolol, pindolol, which is ineffective in migraine, lacks affinity to the 5-HT_{2B} receptor (for review, see Olesen et al., 1995).

The central role of NO in migraine pain does not only provide possible explanations as to the mechanisms of action of existing antimigraine drugs but is also likely to offer future therapeutic possibilities. Thus, drugs which directly counteract the NO-activated cascade (NOS inhibitors, nitric oxide scavengers, guanylate cyclase inhibitors, etc.) may be effective in migraine.

NOS Inhibitors in the Treatment of the Acute Migraine Attack

In a recent study the effect of the nonspecific NOS inhibitor L-N[c] methylarginine hydrochloride (546C88) was studied on spontaneous migraine attacks in a double blind study (Lassen et al., 1997). Fifteen patients received 546C88 (6 mg/kg), and 3 received a placebo (5% dextrose) given intravenously over 15 minutes for a single migraine attack. Furthermore, 11 placebo-treated patients from previous double blind trials with almost identical design were added to the placebo group in the statistical evaluation. Headache severity and clinical outcome was assessed by the patient on a 4-point verbal scale from 0 (no headache / able to work and function normally) to 3 (severe headache / bed rest required). Nausea, vomiting, phonophobia, and photophobia were scored as present or absent.

The number of patients who experienced a relief in headache 30, 60, and 120 minutes after 546C88 infusion is shown in Figure 9.4A. A significant relief in the 546C88 group was found 60 min after infusion. Two hours after infusion of 546C88, 10 of 15 patients experienced headache relief compared to 2 of 14 placebo-treated patients. Symptoms such as phonophobia and photophobia were also significantly improved (Figures 9.4B and C). However, as a high degree of the placebo-treated patients (5 of 10) experienced relief of nausea, there was no significant difference as compared to the 546C88-treated group, in which 10 of 13 experienced relief of nausea (Lassen et al., 1997) (Figure 9.4D). There was a significant improvement in clinical disability of the 546C88-treated group 60 min after treatment as compared to the placebo group, but improvement of clinical disability in the placebo group 120 min after the infusion caused only a nonsignificant difference between the two groups at this time (Figure 9.4E).

The mean blood pressure (MAP) and heart rate (HR) changed significantly over time in the 546C88-treated patients. The maximum increase in MAP was 17% and occurred 15 min after start of the infusion. The maximum decrease in HR was 21% and occurred 20 min after start of the infusion.

Owing to the nonspecific nature of this compound, systemic side effects are, however, likely to limit its clinical usefulness. More specific compounds are therefore warranted.

Selective NOS Inhibitors in the Future Treatment of Migraine

Drugs that selectively inhibit the different NOS enzymes are already under development. The kind of selectivity we should look for in order to achieve optimal inhibition of NO production in migraine patients can, however, be discussed.

Endothelial NOS continuously produces NO in the vascular endothelium, and thus exerts a continuous dilatory effect. Using an inhibitor selective for eNOS will, as noted in the experiment using the nonselective NOS inhibitor 546C88 (Lassen et al., 1997), cause an increase in mean arterial blood pressure and a decrease in heart rate. Unless a selective eNOS inhibitor were to be effective in a low dose, the increase in blood pressure would be prohibitive for its use in migraine.

Neuronal NOS may mediate vasodilatation when it releases NO from perivascular nerves, and in the central nervous system it is important for pain perception. nNOS selective compounds seem to have no effect on blood pressure and may, therefore, be given in doses that result in a more complete inhibition of nNOS than can be accomplished with an eNOS-selective compound.

The last possibility is the development of inhibitors selective for iNOS. When induced by endotoxins and cytokines, it causes the release of large quantities of NO from macrophages, astrocytes, and microglia. iNOS activation may also be involved in the prolonged production of NO after cortical spreading depression (CSD) (Read et al., 1997). This is of particular interest since CSD probably is the underlying mechanism of the migraine aura (Lauritzene, 1996). Inhibitors selective for iNOS could also be of interest if the migraine attack is caused by neurogenic inflammation or by another inflammatory component.

It remains to be shown whether selective inhibitors are as effective as nonselective inhibitors. If so, which of these types of NOS inhibitors will then be most valuable in the future treatment of migraine attacks? When these compounds enter clinical trials, they will no doubt shed further light on our hypothesis and make possible a further elucidation of the mechanisms involved in the complex pathophysiology of migraine.

References

Arregui, A., Carera, J., Leon-Velarde, F., Paredes, S., Viscarra, D., and Arbaiza, D. 1991. High prevalence of migraine in high-altitude population. *Neurology* 41:1668–1669.

Bellantonio, P., Micieli, G., Buzzi, M.G., Marcheselli, S., Castellano, A.E., Rossi, F., and Nappi, G. 1997. Hemodynamic correlates of early and delayed responses to sublingual administration of isosorbide dinitrate in migraine patients: A transcranial Doppler study. *Cephalalgia* 17:183–187.

Brewerton, T.D., Murphy, D.L., Mueller, E.A., and Jimerson, D.C. 1988. Induction of migraine like headaches by the serotonin agonist m-chlorophenylpiperazine. *Clin. Pharmacol. Ther.* 43:605–609.

Busija, D.W., Heistad, D.D., and Marcus, M.L. 1981. Continuous measurements of cerebral blood flow in anesthetized cats and dogs. *Am. J. of Physiol.* 241:H228–H234.

Buzzi, G., and Moskowitz, M.A. 1990. The antimigraine drug sumatriptan (GR43175) selectively blocks neurogenic plasma extravasation from blood vessels in dura mater. *Br. J. Phamacol.* 99:202–206.

Dahl, A., Russell, D., Nyberg-Hansen, R., and Rootwell, K. 1989. Effect of nitroglycerin on cerebral circulation measured by transcranial Doppler and SPECT. *Stroke* 20:1733–1736.

De-Marinis, M., Feliciani, M., Janiri, L., Cerbo, R., and Agnoli, A. 1990. Increased reactivity to a met-enkephalin analog in the control of autonmic responses in migraine patients. *Clin. Neuropharmacol.* 13:507–521.

Edvinsson, L., Gulbenkian, S., Barroso, C.P., Jansen-Olesen, I., and Polak, J.M. 1998. Innervation of the human middle meningeal artery: Immunohistochemistry, ultrastructure, and role of endothelium for vasomotility. *Peptides* 19:1213–1225.

Edvinsson, L., MacKenzie, E.T., McCulloch, J., and Uddman, R. 1988. Nerve supply and receptor mechanisms in intra- and extracerebral blood vessels. In (eds J. Olesen, and L. Edvinsson) *Basic Mechanisms of Headache*, pp. 129–144 Amsterdam: Elsevier.

Feelisch, M., and Noack, E.A. 1987. Correlation between nitric oxide formation during degradation of organic nitrates and activation of guanylate cyclase. *Eur. J. Pharmacol.* 139:19–30.

Fozard, J.R., and Kalkman, H.O. 1994. 5-Hydroxytryptamine (5-HT) and the initiation of migraine: New perspectives. *Naunyn-Schmiedeberg's Arch. Pharmacol.* 350:225–229.

Friberg, L., Olesen, J., Iversen, H.K., and Sperling, B. 1991. Migraine pain associated with middle cerebral artery dilatation: Reversal by sumatriptan. *Lancet* 338: 13–17.

Friberg, L., Olesen, J., Skyhøj Olsen, T., Karle, A., Ekman, R., and Fahrenkrug, J. 1994. Absence of vasoactive peptide release from brain to cerebral circulation during onset of migraine with aura. *Cephalalgia* 14:47–54.

Furchgott, R.F. 1984. The requirement for endothelial cells in the relaxation of arteries by acetylcholine and some other vasodilators. *Trends Pharmacol. Sci.* 2:173–176.

Garthwaite, J., Charles, S.L., and Chess-Williams, R. 1988. Endothelium-derived relaxing factor release on activation of NMDA receptors suggests a role as intercellular messenger in the brain. *Nature* 336:385–388.

Glover, V., Ahmed, F., Hussein, N., Jarman, J., and Peatfield, R. 1996. Central 5-hydroxytrptamine supersensitivity in migraine. In (eds. M. Sandler, M. Ferrari, and S. Harnett) *Migraine: Pharmacology and Genetics*, pp. 117–126. London: Chapman & Hall.

Glusa, E., and Richter, M. 1993. Endothelium-dependent relaxation of porcine pulmonary arteries via 5-Ht_{lc}-like receptors. *Naunyn-Schmiedeberg's Arch. Pharmacol.* 347:471–477.

Goadsby, P.J., Edvinsson, L., and Ekman, R. 1990. Vasoactive peptide release in the extracerebral circulation of humans during migraine headache. *Ann. Neurol.* 28:183–187.

Goadsby, P.J., Kaube, H., and Hoskin, K.L. 1992. Nitric oxide synthesis couples cerebral blood flow and metabolism. *Brain Res.* 595:167–170.

Gruetter, C.A., Kadowitz, P.J., and Ignarro, L.J. 1981. Methylene blue inhibits coronary arterial relaxation and guanylate cyclase activation by nitroglycerin, sodium nitrate and amyl nitrate. *Canadian J. Physiol. Pharmacol.* 59:150–156.

Ignarro, L.J., Lipton, H., Edwards, J.C., Baricos, W.H., Hyman, A.L., Kadowitz, P.J., and Gruetter, C.A. 1981. Mechanisms of vascular smooth muscle relaxation by organic nitrates, nitrites, nitroprusside and nitric oxide: evidence for the involvement of S-nitrosothiols as active intermediates. *J. Pharmacol. Exp. Ther.* 218:739–749.

Iversen, H.K. 1992. N-acetylcysteine enhances nitroglycerin-induced headache and cranial arterial responses. *Clin. Pharmacol. Ther.* 52:125–133.

Iversen, H.K., and Olesen, J. 1994. Nitroglycerin-induced headache is not dependent on histamine release: Support for a direct nociceptive action of nitric oxide. *Cephalalgia* 14:437–442.

Iversen, H.K., and Olesen, J. 1996. Headache induced by a nitric oxide donor (nitroglycerin) responds to sumatriptan. A human model for development of migraine drugs. *Cephalalgia* 16:412–418.

Iversen, H.K., Holm, S., and Friberg, L. 1989. Intracranial hemodynamics during intravenous nitroglycerin infusion. *Cephalalgia* 9 (Suppl 10):84–85.

Iversen, H.K., Nielsen, T.H., Garre, K., Tfelt-Hansen, P., and Olesen, J. 1992. Dose-dependent headache response and dilatation of limb and extracranial arteries after three doses of 5-isosorbide-mononitrate. *Eur. J. Clin. Pharmacol.* 42:31–35.

Iversen, H.K., Nielsen, T.H., Olesen, J., and Tfelt-Hansen, P. 1990. Arterial responses during migraine headache. *Lancet* 336:837–839.

Iversen, H.K., Olesen, J., and Tfelt-Hansen, P. 1989. Intravenous nitroglycerin as an experimental model of vascular headache. *Basic Characteristics. Pain* 38:17–24.

Jansen-Olesen, I., and Edvinsson, L. 2000. Cephalic Neurovascular Transmitters and Receptor. In (eds. J. Olesen, P. Tfelt-Hansen, and K.M.A. Welch) *The Headaches (2nd edition)*, pp. 87–98. New York: Raven Press, Ltd.

Jansen-Olesen, I., Ottosson, A., Cantera, L., Strunk, S., Lassen, L.H., Olesen, J., Mortensen, A., Engel, U., and Edvinsson, L. 1997. Role of endothelium and nitric oxide in histamine-induced responses in human cranial arteries and detection of mRNA encoding H1 and H2 receptors by RT-PCR. *Br. J. Pharmacol.* 121:41–48.

Jensen, K., Tuxen C., Pedersen-Bjerregård, U., Jansen, I., Edvinsson, L., and Olesen, J. 1990. Pain and tenderness in human temporal muscle induced by bradykinin and 5-hydroxytryptamine. *Peptides* 11:1127–1132.

Krabbe, A.Æ., and Olesen, J. 1980. Headache provocation by continuous intravenous infusion of histamine. Clinical results and receptor mechanisms. *Pain* 8:253–259.

Lance, J.W. 1991. 5-hydroxytrptamine and its role in migraine. *Eur. Neurology* 31:279–281.

Lassen, L.H., Ashina, M., Christiansen, I., Ulrich, V., and Olesen, J. 1997. Nitric oxide synthase inhibition in migraine. *Lancet* 349:401–402.

Lassen, L.H., Thomsen, L.L., Kruuse, C., Iversen, H.K., and Olesen, J. 1996. H_1 receptor blockade does not prevent nitroglycerin induced migraine: Support for the NO-hypothesis of migraine. *Eur. J. Pharmacol.* 49:335–340.

Lassen, L.H., Thomsen, L.L., and Olesen, J. 1995. Histamine induces migraine via the H1-receptor: Support for the NO-hypothesis of migraine. *Neuroreport* 6:1475–1479.

Lauritzen, M. 1996. Pathophysiology of the migraine aura. *Science & Medicine* 3:31–41.

Lüscher, T.F., and Vanhoutte, P.M. 1990. The endothelium: Modulator of cardiovascular functions. Ann Abor, Boston: CRC Press.

May, A., Gijsman, H.J., Wallnofer, A., Jones, R., Diener, H.C., and Ferrari, M.D. 1996. Endothelin antagonist bosentan blocks neurogenic inflammation, but is not effective in aborting migraine attacks. *Pain* 67:375–378.

Moncada, S., Palmer, R.M.J., and Higgs, E.A. 1991. Nitric oxide: Physiology, pathophysiology and pharmacology. *Pharmacological Reviews* 43:109–142.

Moskowitz, M.A. 1993. Neurogenic inflammation in the pathophysiology and treatment of migraine. *Neurology* 43 (Suppl 3):S16–S20.

Nichols, F.T. III., Mawad, M., Mohr, J.P., Stein, B., Hilal, S., and Michelsen, J. 1990. Focal headache during balloon inflation in the internal carotid and middle cerebral arteries. *Stroke* 21:555–559.

Olesen, J. 1991. Cerebral and extracranial circulatory disturbances in migraine: Pathophysiology implications. *Cerebrovasc. Brain Metab. Rev.* 3:1–28.

Olesen, J., Iversen, H.K., and Thomsen, L.L. 1993. Nitric oxide supersensitivity. A possible molecular mechanism of migraine pain. *Neuroreport* 4:1027–1030.

Olesen, J., Thomsen, L.L., and Iversen, H.K. 1994. Nitric oxide is a key molecule in migraine and other vascular headaches. *TIPS* 15:149–153.

Olesen, J., Thomsen, L.L., Lassen, L.H., and Jansen-Olesen, I. 1995. The nitric oxide hypothesis of migraine and other vascular headaches. *Cephalalgia* 15:94–100.

Ottosson, A.L.P., Jansen, I., Langemark, M., Olesen, J., and Edvinsson, L. 1991. Histamine receptors in the isolated human middle meningeal artery: A comparison with cerebral and temporal arteries. *Cephalalgia* 11:183–188.

Palmer, R.M.J., Ferrige, A.G., and Moncada, S. 1987. Nitric oxide release accounts for the biological activity of endothelium-derived relaxing factor. *Nature* 327:524–526.

Peatfield, R.C., Gawel, M.J., and Rose, F.C. 1981. The effect of infused prostacyclin in migraine and cluster headache. *Headache* 21:190–195.

Pedersen-Bjerregård, U., Nielsen, L.B., Jensen, K., Edvinsson, L., Jansen, I., and Olesen, J. 1991. Calcitonin gene-related peptide, neurokinin A and substance P: Effects on nociception and neurogenic inflammation in human skin and temporal muscle. *Peptides* 12:333–337.

Read, S.J., Smith, M.I., Hunter, A.J., and Parsons, A.A. 1997. The dynamics of nitric oxide release measured in real time following repeated waves of cortical spreading depression in the anaesthetised cat. *Neurosci. lett.* 232:127–130.

Rengasamy, A., and Johns, R.A. 1991. Characterization of endothelium-derived relaxing factor/nitric oxide synthase from bovine cerebellum and mechanism of modulation by high and low oxygen tensions. *J. Pharmacol. Exper. Ther.* 259:310–316.

Roberts, M.H.T. 1992. 5-Hydroxytryptamine in nociception and antinociception. In (eds. J. Olesen, and P.R. Saxena) *5-HT Mechanisms of Primary Headaches*, pp. 69–76. New York: Raven Press.

Schmetterer, L., Wolzt, M., Krejcy, K., Graselli, U., Findl, O., Eichler, H.G., and Singer, E.A. 1996. Cerebral and ocular hemodynamic effects of sumatriptan in the nitroglycerine headache model. *Clin. Pharmcol. Ther.* 60:199–205.

Shekar, Y.C., Anand, I.S., Sarma, R., Ferrari, R., Wahi, P.L., and Poole-Wilson, P.A. 1991. Effect of prolonged infusion of human alpha calcitonin gene-related peptide on hemodynamics, renal bloodflow and hormone levels in congestive heart failure. *Am J. Cardiol.* 67:732–736.

Sicuteri, F., Del Bene, E., Poggioni, M., and Bonazzi, A. 1987. Unmasking latent dysnociception in healthy subjects. *Headache* 27:180–185.

Thomsen, L.L., and Iversen, H.K. 1993. Experimental and biological variation of three dimensional transcranial doppler measurements. *J. Appl. Physiol.* 75:2805–2810.

Thomsen, L.L., Iversen, H.K., Brinck, T.A., and Olesen, J. 1993. Arterial supersensitivity to nitric oxide (nitroglycerin) in migraine sufferers. *Cephalalgia* 13:395–399.

Thomsen, L.L., Kruuse, C., Iversen, H.K., and Olesen, J. 1994. A nitric oxide donor (nitroglycerin) triggers genuine migraine attacks. *Eur. J. Neurol.* 1:73–80.

Toda, N. 1990. Mechanism underlying responses to histamine of isolated monkey and human cerebral arteries. *Am J. Physiol.* 258:H311–H317.

Toda, N. 1997. Nitroxidergic nerve in cranial arteries. In (eds. J. Olesen and L. Edvinsson) *Headache Pathogenesis: Monoamines, Neuropeptides, Purines, and Nitric Oxide*, pp. 241–245. Philadelphia: Lippincott-Raven Publishers.

Tomimoto, H., Nishimura, M., Suenaga, T., Nakamura, S., Akiguchi, I., Wakita, H., Kimura, J., and Mayer, B. 1994. Distribution of nitric oxide synthase in the human cerebral blood vessels and brain tissues. *J. Cereb. Blood Flow Metab.* 14:930–938.

10 Presynaptic Modulation of Peripheral Nitrergic Neurotransmission

R.A. LEFEBVRE

Introduction

Through the work of authors such as K. Starke and S. Langer, the concept of presynaptic modulation of the release of classic neurotransmitters such as noradrenaline, acetylcholine, and dopamine has developed (Langer, 1981; Starke, 1981; Starke et al., 1989). In both the central and peripheral nervous system, the release of these neurotransmitters can, indeed, be regulated via presynaptic receptors on the axon terminals. Modulation can be exerted by the neuron's own neurotransmitter via presynaptic autoreceptors or by other neurotransmitters or locally produced substances via presynaptic heteroreceptors. In general, stimulation of presynaptic receptors will lead to inhibition of transmitter release; via presynaptic autoreceptors the transmitter is thus able to exert negative feedback on the axon terminals from which it is released. Presynaptic facilitation has also been described, such as via presynaptic β-receptors on peripheral noradrenergic nerve endings. Classic neurotransmitters are stored in vesicles in the axon terminals, and when the terminals are depolarized by an action potential, an influx of calcium through voltage-gated calcium channels will occur, which is the trigger for the process of exocytosis of the vesicles and the release of neurotransmitter into the synaptic cleft. The calcium channels and the release mechanism must be localized close to each other and might be part of a multimolecular entity (Stanley, 1997). The principal mechanism by which presynaptic receptors inhibit neurotransmitter release is through inhibition, often via a G protein, of the voltage-gated calcium channels, leading to a decrease in electrosecretory coupling (Starke et al., 1989; Stjarne, 1989). The inverse relation between the intensity of nerve activation and the effectiveness of presynaptic receptors to inhibit transmitter release might be due to the increasing intracellular calcium concentration upon high stimulation intensity (Duckles and Budai, 1990).

Nitrergic neurotransmission differs from classic neurotransmission in several aspects (Gibson and Lilley, 1997). Two aspects are important with regard to the possibility of presynaptic modulation of nitrergic nerves: the nitrergic neurotransmitter is not stored in the nerve terminals but is synthesized when needed, and nitric oxide (NO) does not interact with membrana receptors but diffuses out of the neuron and into the effector cell, where the principal mechanism of action is through activation of soluble guanylate cyclase. This implies that, when negative feedback on the nitrergic nerves occurs, NO will not do so via presynaptic autoreceptors, and that the mechanism of inhibition of NO release by NO itself or by other substances cannot be through inhibition of exocytosis.

The evidence for negative feedback and heteroregulation of peripheral nitrergic neurotransmission is summarized below. The term presynaptic is used only to indicate that the effect described is due to an interaction not at the level of the muscle cells but at the level of the neurons, and it does not imply knowledge of specialized structures between these neurons and the muscle cells. In many publications where it is referred to, the term prejunctional instead of presynaptic has been used. It should also be realized that in peripheral nonadrenergic, noncholinergic (NANC) neurotransmission, NO is often coacting with other NANC neurotransmitters, such as vasoactive intestinal peptide (VIP) or ATP. The interaction with these other NANC neurotransmitters is discussed in chapter 3 by Cocks and Selemidis in this volume.

Negative Feedback of Nitrergic Neurotransmission

In the central as well as in the peripheral nervous system, neuronal NO is thought to be synthesized by one of the two constitutive isoforms of NO synthase (NOS), neuronal NOS (nNOS) or NOS I (Förstermann and Kleinert, 1995), although the endothelial constitutive isoform (eNOS or NOS III) has also been found in neurons of the rat brain (Dinerman et al., 1994). In the peripheral nervous system, NO is thought to act as a classical anterograde transmitter, diffusing from a presynaptic nerve ending to the postsynaptic effector cell, mainly smooth muscle cells in the gastrointestinal, respiratory, and urogenital tracts (Boeckxstaens and Pelckmans, 1997). In the brain, NO acts as a retrograde transmitter, being released from a postsynaptic neuron and diffusing to the presynaptic neuron, where it could be involved in long-term potentiation (Hölscher, 1997), but anterograde signaling is also possible (Park et al., 1998).

Experiments with crude preparations of rat and bovine cerebellar NOS, where NOS activity was assessed by measurement of the conversion of L-[^{3}H]arginine to L-[^{3}H]citrulline, suggested that NO is able to inhibit nNOS activity. The rate of citrulline formation with the rat cerebellar preparation was nonlinear during the time of incubation, and this was not due to substrate depletion, since increasing the L-arginine concentration did not prevent nonlinearity (Rogers and Ignarro, 1992). Since superoxide dismutase, which reduces the interaction of NO with superoxide anions, increased the nonlinearity of the reaction, whereas oxyhemoglobin, which binds and inactivates NO, made the rate of citrulline formation linear, it was suggested that NO itself, produced during the progression of the reaction, inhibits NOS activity, the inhibition appearing reversible. Exogenous authentic NO and two NO donors, *S*-nitroso-*N*-acetylpenicillamine (SNAP) and *N*-methyl-*N*′-nitro-*N*-nitrosoguanidine, also inhibited NOS activity, whereas L-citrulline was without effect (Rogers and Ignarro, 1992). Similar results and conclusions were soon reported for bovine cerebellar NOS (Rengasamy and Johns, 1993). The group of Ignarro (Griscavage et al., 1994) continued work with a more purified preparation of rat cerebellar NOS. Nitric oxide synthase activity was reduced by NO, and this effect was mimicked by the other heme ligands carbon monoxide and cyanide ion. The inhibitory effect of NO was enhanced by oxidizing agents, which also had an inhibitory effect per se, but it was prevented by the reducing agent and NOS cofactor tetrahydrobiopterin. It was, therefore, suggested that NO inhibits NOS activity by formation of a nitrosyl complex with the heme iron in the enzyme. This reduces the binding of oxygen to the heme iron, which is required for incorporation into one of the nitrogen atoms in the guanidine function of L-arginine leading to the formation of L-citrulline and NO. Nitric oxide seems to bind preferentially with the heme iron in the oxidized ferric state (Fe^{III}), and the effect of tetrahydrobiopterin might then be explained by reduction of the heme iron to the ferrous state (Fe^{II}) (Ignarro, 1995). From these experiments, it was implied that in NANC transmission, NO might regulate its own synthesis. In a theoretical model on concentration profiles of NO, the concentration of NO rises linearly with time when multiple evenly distributed NO point sources are active in a given tissue, assuming that there is no decline in source strength and no decay. Negative feedback of NO on NOS activity would thus provide an adequate mechanism to make sure that NO concentrations do not build up to toxic concentrations (Wood and Garthwaite, 1994). Still, all the above-described experiments were performed using cell-free enzyme preparations, and surprising results were obtained when NO donors were tested on intact neuroblastoma N1E115 cells, which express endogenous NOS (Hu and El-Fakahany, 1995). The NO donors sodium nitroprusside, 3-morpholinosydnonimine (SIN-1), and SNAP reduced NOS activity in cytosolic preparations of the neuroblastoma cells, but in intact cells sodium nitroprusside and SIN-1 enhanced NOS activity when incubated for more than 90 minutes, whereas SNAP had no effect. Neither the inhibitory action in the cytosolic preparations nor the stimulatory one in the intact cells corresponded with the potency of the NO donors to increase cGMP formation in intact cells. The stimulation of NOS activity in intact cells was dependent on the presence of extracellular calcium. Since the stimulatory effect was not present with all NO donors, it was suggested that this might be related to differences in the oxidation state of the species of NO released by the NO donors (Hu and El-Fakahany, 1996).

Few attempts have been made to investigate the possibility of NO-mediated inhibition of nitrergic neurotransmission in intact peripheral tissues. In longitudinal muscle strips of the rat gastric fundus, incubation for 30 minutes with SIN-1 or nitroglycerin reduced short-lasting NANC relaxations induced by electrical field stimulation with 10-second trains, and the degree of inhibition was comparable to that obtained with the NOS inhibitor N^G–nitro-L-arginine (L-NA) (De Man et al., 1995). Since the relaxant responses to exogenous NO were not influenced, a postsynaptic site of action seems excluded, and it was suggested that NO released from the NO donors exerts a presynaptic inhibitory effect on nitrergic neurotransmission. A nonspecific neurotoxic effect seemed less likely, since the effect of the NO donor was almost completely reversed after 90 minutes of rinsing and peptidergic neurotransmission (of VIP) was not influenced. The amplitude of sustained relaxations, obtained with 3-minute trains of stimulation, was indeed not influenced; it has been shown before that these sustained relaxations are largely due to VIP release, except for the initial part. It was thus suggested that the effect of the NO donors was due to a persistent inhibitory effect of NO on NOS activity. Similar results were recently obtained in circular muscle strips of the rat gastric body (Hosoda et al., 1998). Incubation of the tissues with SIN-1, sodium nitroprusside, or SNAP for 15 minutes reduced in a concentration-dependent manner the NANC relaxations induced by 30-second trains of electrical field stimulation. The effect was also observed with the membrane-permeable analogue 8-bromo-cGMP. The possibility that these effects were due to inhibition of NO synthesis at the presynaptic level was confirmed directly by studying the effect of SIN-1 on NO formation, by measuring L-[^{3}H]-citrulline formation evoked by electrical field stimulation after preloading the tissues with L-[^{3}H]-arginine. Incubation with SIN-1 concentration dependently reduced the electrically induced formation of L-[^{3}H]-citrulline. Again, the effect was also obtained with 8-bromo-cGMP, and that of SIN-1 was partly prevented by pretreatment with the guanylate cyclase inhibitor 1H-[1,2,4]oxadiazolo[4,3-a]quinoxalin-1-one (ODQ). These results demonstrate that exogenous NO is able to inhibit NOS activity in the rat gastric myenteric plexus. The effect seems at least partly due to activation of guanylate cyclase and generation of cGMP, which differs from the direct binding of NO with NOS as proposed on the basis of the results with enzyme preparations (see above). In the experiments of Hosoda et al. (1998), the sustained relaxation induced by a 3-minute train of stimulation was also reduced by SIN-1, whereas the effect of exogenous VIP was not influenced, so that it was suggested that NO also exerts a presynaptic inhibitory effect on VIPergic neurotransmission, which contrasts with the results of De Man et al. (1995). In circular muscle strips of the pig gastric fundus, no evidence for inhibition of nNOS activity by NO could be obtained (Lefebvre and Vandekerckhove, 1998). In this tissue, both short-lasting and sustained relaxations induced by electrical field stimulation are largely due to NO release. Incubation of the tissues with nitroglycerin for 30 minutes reduced both types of electrically induced relaxations, but it also reduced relaxations induced by exogenous NO (short-lasting by bolus administration of NO and sustained by infusion of NO into the organ bath). Relaxations to exogenous VIP and to the potassium channel opener lemakalim were not influenced by nitroglycerin. These results suggest that nitroglycerin induces postsynaptic tolerance to the relaxant effect of the nitrates, a phenomenon well known in vascular smooth muscle (Axelsson and Ahlner, 1987). In the study on the pig gastric fundus, it was also investigated whether long-term contact with endogenous NO might lead to inhibition of nNOS activity, but long-term electrical stimulation for 30 minutes did not influence the relaxant responses to short trains of stimulation, NO, or VIP (Lefebvre and Vandekerckhove, 1998).

In biochemical assays, NO also inhibits the other constitutive isoform eNOS. The rate of conversion of L-arginine to L-citrulline in endothelial cell homogenates was nonlinear during the incubation period, and similar results with superoxide dismutase, oxyhemoglobin, exogenous NO, and a NO donor as described above for nNOS were obtained (Buga et al., 1993). Ravichandran et al. (1995) confirmed a direct and reversible inhibition of eNOS using a partly purified enzyme preparation from cultured endothelial cells. In view of the rather limited evidence for NO-mediated regulation of peripheral nNOS in intact tissues, it is interesting to note that this inhibition of eNOS by NO has also been observed in vitro and in vivo in the cardiovascular system by several groups. Nitric oxide production from endothelial cells attached to microcarrier beads, packed in a chromatography column and superfusing a detector tissue, was induced by increasing the perfusion rate or by perfusion with bradykinin; this effect was markedly inhibited when the endothelial cells were pretreated for 15 minutes with the NO donor SNAP. The same result was obtained in precontracted rings of bovine intrapulmonary artery, where the endothelium-dependent and NO-mediated relaxation by acetylcholine, bradykinin, and electrical field stimulation was reduced by pretreatment with SNAP. The endothelium-dependent relaxation of arterial rings and the generation of NO from intact endothelial cells were thus inhibited by a NO donor, but without changing the sensitivity of the vascular smooth muscle cells to the relaxant effect of NO (Buga et al., 1993). Ma et al. (1996) measured the increase in vascular tone in rat aortic rings and the hypertensive response in anesthetized rats induced by the NOS inhibitor N^G–nitro-L-arginine methylester (L-NAME) as an indication of basal NO release in vitro and in vivo. When the rings were preincubated or the rats were infused with SNAP, these effects of L-NAME were blunted without interfering with the NO-independent vasoconstrictor responses to phenylephrine, unless vasoactive concentrations of SNAP were used. This demonstrates that exogenous NO inhibits basal endothelial NO release in vitro and in vivo. Further in vivo evidence was obtained in the vascular bed of the rabbit hindquarters, where an infusion of SNAP reversibly reduced the endothelium-dependent vasodilatory response to acetylcholine and bradykinin, without influencing the response to SNAP itself or to 8-bromo-cGMP. The pressor response to L-NAME, both in the hindquarters and in the systemic vascular bed, was markedly attenuated during the infusion with SNAP (Cohen et al., 1996). From these studies with exogenous NO donors, it was suggested that endogenous NO might also act as a negative feedback modulator of eNOS, although no absolute evidence is available for this. The mechanism might be important for reduction of endogenous endothelial NO production by administration of exogenous NO and NO donors. It has been suggested, for example, that the rebound vasoconstriction upon sudden discontinuation of NO inhalation therapy for pulmonary hypertension is related to inhibition of eNOS by NO (Oka et al., 1996), but the mechanism seems to differ at least partly from that proposed by Ignarro (1995), i.e., direct binding of NO to the heme iron in the enzyme, since NO would interact with superoxide anions to produce peroxynitrite, which would inactivate the enzyme (Sheehy et al., 1998). In contrast, Frank et al. (1998) did not find evidence for a down-regulation of eNOS by prolonged inhaled NO.

For completeness, it can be mentioned that the activity of the inducible isoform of NOS (iNOS) can also be inhibited by NO (Griscavage et al., 1993; Assreuy et al., 1993; Kiff et al., 1994), but that NO also inhibits iNOS gene expression by interference with the NF-κB interaction with the iNOS gene (Park et al., 1997; Taylor et al., 1997).

Presynaptic Heteroregulation of Nitrergic Neurotransmission

The activity of nNOS is dependent upon the intracellular calcium concentration, since an increase in intracellular calcium is required to bind calmodulin to the enzyme (Bredt and Snyder, 1994). It is usually accepted that an action potential invading the peripheral nitrergic nerve ending will lead to calcium entry through voltage-gated calcium channels and activation of the enzyme (Sneddon and Graham, 1992). In many tissues, ω-conotoxin GVIA reduces electrically induced nitrergic relaxation (Boeckxstaens et al., 1993a; Kasakov et al., 1995; Toda et al., 1995; Werkström et al., 1995) or nitrergic neurotransmitter release (Boeckxstaens et al., 1993a), suggesting that presynaptic calcium influx through N-type voltage-gated calcium channels is involved. The incomplete blockade by ω-conotoxin GVIA might be related to the additional involvement of non-N-type calcium channels, such as Q-type channels (Frew and Lundy, 1995; Zygmunt et al., 1995). In some tissues, ω-conotoxin GVIA does not affect the electrically induced nitrergic responses (Zagorodnyuk and Maggi, 1994; Matthew and Wadsworth, 1997). The possibility of calcium-independent activation of nNOS has been proposed (Kostka et al., 1993; Zygmunt et al., 1995).

Corresponding to presynaptic inhibition of classic neurotransmitter release, which is due to inhibition of presynaptic calcium influx (see Introduction), it might theoretically be expected that NOS activity and thus NO release are inhibited by substances that decrease calcium influx at presynaptic nerve endings, on condition that the relevant receptors are expressed on the nitrergic neurons. Synaptic contacts between NOS-immunoreactive nerve fibers and non-NOS-immunoreactive neurons have, indeed, been shown in the intestine (Llewellyn-Smith et al., 1992). Phosphorylation of the enzyme is another mechanism that can lead to a decrease in enzymatic activity and NO release (Hu and El-Fakahany, 1996).

Cholinergic Modulation

In systems with NANC neurotransmission, especially the gastrointestinal tract, acetylcholine is the principal excitatory neurotransmitter and NO (sometimes in cooperation with VIP or ATP) is the principal inhibitory neurotransmitter (Sanders and Ward, 1992). Functional antagonism between both neurotransmitters at the muscular level has been described (Ward et al., 1993; Milenov and Kalfin, 1996; Leclere and Lefebvre, 1998), but it is also conceivable that both neurotransmitters inhibit each other's release at the presynaptic level. There is evidence that NO can inhibit acetylcholine release. When electrically induced acetylcholine release was assessed by measurement of the outflow of [^{3}H]acetylcholine after preincubation of longitudinal muscle–myenteric plexus preparations of the guinea pig ileum with [^{3}H]choline, the depolarization-evoked release of acetylcholine was reduced by NO donors (Hebeiss and Kilbinger, 1996) but enhanced by NOS inhibitors, suggesting that endogenous NO also decreases electrically induced acetylcholine release (Kilbinger and Wolf, 1994). The inhibitory effect of NO seems mediated through activation of guanylate cyclase (Hebeiss and Kilbinger, 1998). In other tissues, such as the canine ileum, NO has also been shown to modulate cholinergic responses by presynaptic inhibition of acetylcholine release (Hryhorenko et al., 1994). The question is whether the opposite is also possible: can acetylcholine inhibit NO release from nitrergic neurons?

In 1989 Li and Rand reported that the muscarinic receptor agonist McN-A-343 inhibited electrically induced NANC relaxations in the rat anococcygeus muscle, without influencing the tone of the tissues or relaxations induced by sodium nitroprusside. The inhibitory effect of McN-A-343 on NANC relaxations was decreased by the muscarinic antagonists atropine, pirenzepine, and gallamine, which had no effect per se. The electrically induced NANC relaxations in the rat anococcygeus muscle were soon proven to be completely nitrergic in nature (Gillespie et al., 1989; Hobbs and Gibson, 1990), so that these findings suggest that activation of presynaptic muscarinic receptors by an exogenous agonist inhibits NO release. The observation that atropine did not influence the electrically induced responses per se demonstrates that the presynaptic muscarinic receptors are not activated by endogenous acetylcholine; the motor innervation of the rat anococcygeus is, indeed, noradrenergic and not cholinergic (Gillespie, 1980). Nitrergic perivascular nerves have been shown in cerebral and ocular arteries from different species, and also in these tissues, indirect evidence for presynaptic inhibition of NO release via muscarinic receptors was obtained. In endothelium-denuded helical strips of bovine basilar artery, contracted with prostaglandin $F_{2\alpha}$, transmural electrical stimulation induced relaxations that were nearly abolished by L-NA. Acetylcholine did not influence the tone of the strips but concentration dependently attenuated the electrically induced relaxations, without influencing those induced by exogenous NO; the effects of acetylcholine were antagonized by atropine but not by hexamethonium. This suggests the impairment of NO release by activation of presynaptic muscarinic receptors. The stimulation-induced relaxation was also reduced by physostigmine but potentiated by atropine, and both agents were without influence on relaxations induced by exogenously applied NO. These results seem to indicate that endogenous acetylcholine, released upon electrical stimulation, also inhibits NO release, and that the effect is enhanced when the degradation of acetylcholine is reduced with physostigmine and decreased when muscarinic receptors are blocked with atropine (Ayajiki et al., 1993). Potentiation by atropine and inhibition by physostigmine of the relaxations induced by nitrergic nerve stimulation were not found in the canine cerebral artery (Toda et al., 1995) but were found in monkey cerebral and ciliary artery and porcine ciliary artery (Toda et al., 1997a,b, 1998). In the latter studies, the presence of perivascular nitrergic and cholinergic nerves was confirmed by histochemical examination for NADPH diaphorase or NOS and acetylcholinesterase, respectively. From a study of the influence of the M_1, M_2, and M_3 receptor antagonists pirenzepine, AF-DX116, and 4-DAMP on the effect of acetylcholine in the monkey cerebral artery, the muscarinic receptors involved in presynaptic inhibition of nitrergic nerves were considered to be predominantly of the M_2 type (Toda et al., 1997b). This corresponds to the subtype frequently involved in presynaptic inhibition of sympathetic and cholinergic outflow (Barnes, 1990; Eglen and Whiting, 1990). Except for the monkey ciliary artery (Toda et al., 1998), exogenous acetylcholine did not contract the tissues, and no cholinergic postsynaptic component was observed during electrical stimulation; neuronally released acetylcholine would thus mainly serve to regulate NO release at a presynaptic level. Colocalization of choline acetyltransferase and NOS has been shown in cerebral arteries (Yu et al., 1998), so that the acetylcholine influencing NO release might originate in the same nerve (Toda et al., 1997b).

All the above-cited experiments were performed by measuring NO release indirectly via the muscular response. In contrast, when NO release from the guinea pig ileum was evaluated by measurement of nitrite via chemiluminescence analysis, atropine was found to inhibit spontaneous

NO/nitrite release. Carbachol reversed the inhibitory effect of atropine but had no influence per se on NO/nitrite release. The possibility of near-maximal muscarinic stimulation of NO release under resting conditions was suggested (Wiklund et al., 1993). By use of the same method in the distal part of the guinea pig colon, NO/nitrite release induced by electrical field stimulation was found to be reduced by atropine but enhanced by physostigmine. Exogenous acetylcholine evoked a concentration-dependent increase in NO/nitrite overflow. It was suggested that part of the nerve-induced NO formation in the guinea pig colon depends on activation of muscarinic receptors belonging to the M_1 class, as assessed from the study of a set of muscarinic receptor antagonists (Iversen et al., 1997). In the rabbit anococcygeus muscle, nitrergic relaxations by muscarinic receptor activation were also observed. The investigation of a large set of muscarinic agonists and antagonists led to the conclusion that M_4 receptors were located on nitrergic neurons (Gross et al., 1997). The presence of facilitatory presynaptic muscarinic receptors on nitrergic neurons was considered as a possible explanation for the fact that the inhibitory effect of L-NA and oxyhemoglobin on nerve-induced membrane hyperpolarization in the guinea pig taenia coli was no longer observed in the presence of atropine (Ward et al., 1996). Thus, evidence in favor of both muscarinic inhibition and stimulation of peripheral NO release has been provided in different tissues and species. As far as they have been assessed, the receptors involved in muscarinic inhibition belong to the M_2, those involved in muscarinic stimulation to the M_1/M_4 class. A possible way of combining these regulatory mechanisms in the same tissue might be the presence of ganglionic stimulatory M_1 (or M_4) receptors on the somatodendritic region of nitrergic neurons and presynaptic inhibitory M_2 receptors on nitrergic nerve endings, as has been proposed for the parasympathetic nerve pathways in the airways (Barnes, 1990). On the other hand, different subtypes of presynaptic muscarinic receptors with an opposite effect seem possible, as has been proposed in the guinea pig ileum with regard to acetylcholine release (Soejima et al., 1993).

The influence of nicotinic receptor activation on nitrergic neurotransmission seems less complex, in that it is consistently stimulatory. When tested in tissues with nitrergic neurotransmission, stimulation of nicotinic receptors with nicotine or 1,1-dimethyl-4-phenylpiperazinium (DMPP) induced neurogenic nitrergic relaxations (Kojima et al., 1993; McLaren et al., 1993; Kortezova et al., 1994; Mizhorkova et al., 1994; Tanobe et al., 1995; Börjesson et al., 1997) and NO release (Boeckxstaens et al., 1991b; Chakder and Rattan, 1993; Shuttleworth et al., 1995). Ganglionic nicotinic receptors are involved in vagally induced nitrergic relaxations (Lefebvre et al., 1992; Takahashi and Owyang, 1995). Recently, it was proposed that the vagal nerve is important via nicotinic synapses for the regulation of nNOS expression in the myenteric plexus. Truncal vagotomy and administration of hexamethonium reduced the number of NOS-immunoreactive cells and the density of NOS-immunoreactive bands and NOS mRNA bands in the rat gastric myenteric plexus, whereas the opposite was observed upon nicotinic receptor stimulation in cultured gastric myenteric ganglia (Nakamura et al., 1998). Tetrodotoxin-resistant relaxations with high concentrations of nicotinic receptor agonists have been reported, implying that neuronal action potential propagation is not required, and the possibility of nicotinic receptors on terminal varicosities is considered (Shuttleworth et al., 1995; Börjesson et al., 1997). In the canine ileocolonic junction, NO release induced by DMPP was not reduced by the N-type calcium channel blocker ω-conotoxin GVIA, whereas it was upon electrical stimulation (Boeckxstaens et al., 1993a). The same applied to nitrergic relaxation in canine cerebral arteries (Toda et al., 1995). This might imply that electrical and chemical stimulation of nitrergic neurons opens other calcium channels. But in the rat anococcygeus, ω-conotoxin GVIA clearly reduced nicotine-induced nitrergic relaxation (Rand and Li, 1992).

Noradrenergic Modulation

In the gut, the majority of postganglionic sympathetic fibers synapse with intrinsic enteric neurons and not with smooth muscle cells (Wood, 1987). The inhibitory effect of the sympathetic nervous system on nonsphincteric gastrointestinal motility can largely be ascribed to inhibition of acetylcholine output by noradrenaline via presynaptic α_2-receptors on the postganglionic cholinergic neurons (McIntyre and Thompson, 1992). In some tissues, stimulation of α_2-receptors apparently also leads to inhibition of NO release. In the rat gastric fundus, the α_2-adrenoceptor agonist 5-bromo-6-[-2-imidazolin-2-ylamino]-quinoxaline (UK-14,304) inhibited electrically induced nitrergic relaxations without influencing those induced by administration of exogenous NO. The inhibitory effect of UK-14,304 was antagonized by the α_2-adrenoceptor antagonist rauwolscine but not by the α_1-adrenoceptor antagonist prazosin, and the inhibition of NO release via presynaptic α_2-adrenoceptors was suggested (Lefebvre and Smits, 1992). In a region where sympathetic nerve stimulation has a stimulatory effect (McIntyre and Thompson, 1992), the canine ileocolonic junction, similar results were obtained with the α_2-adrenoceptor agonists clonidine and UK-14,304, the effect of UK-14,304 being antagonized by the α_2-adrenoceptor antagonist yohimbine (Boeckxstaens et al., 1993b). The inhibition of electrically induced NO release by activation of α_2-adrenoceptors in the canine ileocolonic junction was further corroborated by showing the reduction of actual NO release, as measured by bioassay, under the influence of UK-14,304. In the presence of the nonselective K^+-channel blocker 4-aminopyridine or the Ca^{2+}-activated K^+-channel blocker charybdotoxin, which enhanced NO release per se, UK-14,304 no longer inhibited NO release. The inhibition of NO release via presynaptic α_2-adrenoceptors might thus result from an increase in K^+ conductance (De Man et al., 1994). In the three above-cited studies, the inhibitory effects were inversely related to the stimulus frequency, as is usual for presynaptic inhibitory modulation (Duckles and Budai, 1990). The studies were performed in the presence of guanethidine to block noradrenergic neurotransmission, so that a possible influence of endogenous noradrenaline could not be assessed. In the guinea pig proximal colon, α_2-adrenoceptor agonists inhibited electrically induced NANC relaxations by activation of presynaptic α_2-receptors (Kojima et al., 1988). Since it was later shown that NO is involved in this NANC inhibitory transmission (Iversen et al., 1994), α_2-receptor-mediated inhibition of NO release seems also possible at this level. In the opossum internal anal sphincter, the α_2-receptor agonist clonidine concentration-dependently inhibited the relaxation induced by rectal balloon distension, but not that induced by electrical stimulation of the sacral nerve or local intramural stimulation (Yamato and Rattan, 1990), whereas the responses by the three methods were all shown to be nitrergic (Rattan et al., 1992). This suggests that in this tissue there is no α_2-receptor-mediated inhibition of NO release at the neuromuscular level. Also in canine cerebral arteries, clonidine did not inhibit electrically induced nitrergic relaxations (Toda et al., 1995). It did in the rat anococcygeus muscle, but it also increased the tone of the tissues, and the effect was mimicked by the α_1-receptor agonist methoxamine; the effect was ascribed to the increase in tone and not to inhibition of transmitter release (Li and Rand, 1989). Thus, presynaptic α_2-mediated inhibition of NO release seems not to be a general phenomenon. In the porcine basilar artery, it was recently proposed that noradrenaline released from noradrenergic nerve terminals is even able to stimulate NO release from neighboring nitrergic nerves (Zhang et al., 1998).

There are also indications that NO can modulate noradrenergic neurotransmission. Presynaptic inhibition of peripheral sympathetic vasoconstriction by neuronally and endothelially produced NO might contribute to the nitrergic control of vessel tone (Addicks et al., 1994; Vials et al., 1997). In the rat vas deferens, in contrast, NO was shown to have a presynaptic facilitatory effect on sympathetic neurotransmission, probably by enhancing neurotransmitter release (Postorino et al., 1998). In the rat, rabbit, and mouse anococcygeus muscle and in the rabbit and human corpus cavernosum, nitrergic transmission is thought to control noradrenergic transmission by postsynaptic functional antagonism (Brave et al., 1993; Kasakov et al., 1994; Cellek and Moncada, 1997).

Nonadrenergic, Noncholinergic Modulation

In some tissues, agents other than muscarinic or α_2-receptor agonists have been reported to modulate nitrergic neurotransmission, but the information is still more fragmentary. For example, in canine lower esophageal sphincter, electrically induced NANC relaxations, which are now considered to be mainly nitrergic, were inhibited by opioid peptides via presynaptic opioid receptors (Barnette et al., 1990). Inhibition of NANC or nitrergic relaxations by opioid agonists was also reported in the guinea pig ileum and the human colon (Angel et al., 1993; Ivancheva and Radomirov, 1996). Prostaglandin E_2 was shown to modulate NANC relaxation in the rabbit stomach (Baccari et al., 1996), and P_1-purinoceptor agonists inhibited NANC inhibitory junction potentials in the guinea pig caecum via presynaptic P_1-purinoceptors (Hoyle et al., 1988); these inhibitory junction potentials are at least partly nitrergic (Watson et al., 1996).

On the other hand, NO is the mediator of neurogenic relaxations by agents such as 5-HT, γ-aminobutyric acid (GABA), and ATP (see, e.g., Boeckxstaens et al., 1991a; Bogers et al., 1991; Briejer et al., 1992; Ševčik et al., 1998). In addition to nicotinic receptors, nitrergic neurons might thus be entailed with stimulatory 5-HT, GABA, and purinergic receptors.

Conclusions

The presynaptic regulation of peripheral nitrergic neurotransmission is not as well studied and well known as that of noradrenergic or cholinergic transmission. Some evidence for negative feedback by NO itself and for noradrenergic and cholinergic modulation of NO release has been pub-

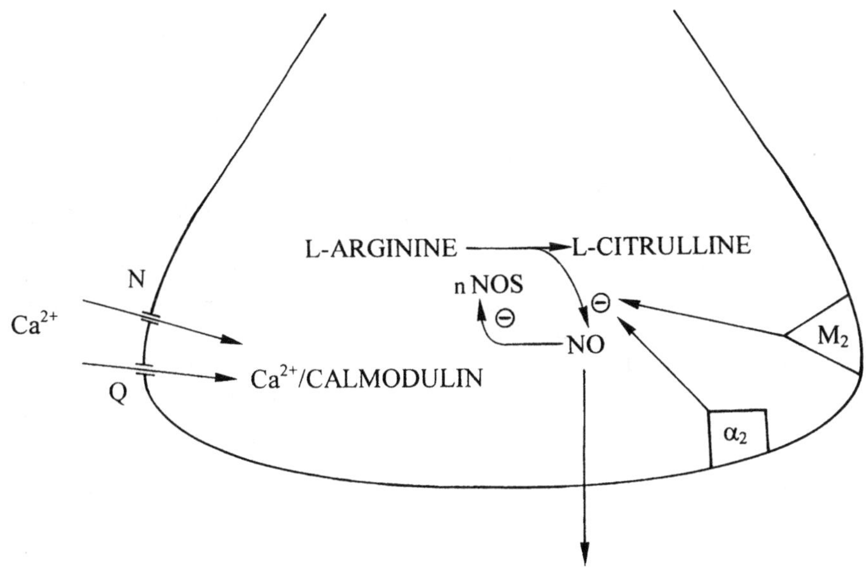

FIGURE 10.1. Theoretical scheme of the presynaptic regulation of a nitrergic nerve terminal. Data obtained in different tissues have been schematized together. Stimulatory nicotinic and M_1 or M_4 receptors might be localized at the somatodendritic region.

lished (Figure 10.1). Although it seems clear that NO can inhibit nNOS activity when tested with enzyme preparations, the number of studies assessing this phenomenon in intact tissues is very limited, and the studies have not yielded consistent results. In the publications on noradrenergic and cholinergic modulation, nNOS activity and NO release are usually assessed indirectly via measurement of muscular relaxation. The results do not allow exact localization of a presynaptic site of action at the level of the axon terminals or at the somatodendritic region, and the mechanism of presynaptic modulation is nearly totally unstudied. Systematic studies, preferably with inclusion of direct measurement of NO release, seem required to fully evaluate negative feedback and crosstalk with neighboring cholinergic and noradrenergic neurons in nitrergic neurons.

References

Addicks, K., Bloch, W., and Feelisch, M. 1994. Nitric oxide modulates sympathetic neurotransmission at the prejunctional level. *Microsc. Res. Techn.* 29:161–168.

Angel, F., Chamouard, P., Klein, A., and Martin, E. 1993. Opioid agonists modulate excitatory and inhibitory neurotransmission in human colon. *J. Gastroint. Mot.* 5:289–297.

Assreuy, J., Cunha, F.Q., Liew, F.Y., and Moncada, S. 1993. Feedback inhibition of nitric oxide synthase activity by nitric oxide. *Br. J. Pharmacol.* 108:833–837.

Axelsson, K.L., and Ahlner, J. 1987. Nitrate tolerance from a biochemical point of view. *Drugs* 33 (Suppl. 4):63–68.

Ayajiki, K., Okamura, T., and Toda, N. 1993. Nitric oxide mediates, and acetylcholine modulates, neurally induced relaxation of bovine cerebral arteries. *Neurosci.* 54:819–825.

Baccari, M.C., Calamai, F., and Staderini, G. 1996. Prostaglandin E_2 modulates neurally induced nonadrenergic noncholinergic gastric relaxations in the rabbit in vivo. *Gastroenterology* 110:129–138.

Barnes, P.J. 1990. Muscarinic receptors in airways: recent developments. *J. Appl. Physiol.* 68:1777–1785.

Barnette, M.S., Grous, M., Manning, C.D., Callahan, J.F., and Barone, F.C. 1990. Inhibition of neuronally induced relaxation of canine lower esophageal sphincter by opioid peptides. *Eur. J. Pharmacol.* 182:363–368.

Boeckxstaens, G.E., De Man, J.G., Pelckmans, P.A., Cromheeke, K.M., Herman, A.G., and Van Maercke, Y.M. 1993a. Ca^{2+} dependency of the release of nitric oxide from non-adrenergic non-cholinergic nerves. *Br. J. Pharmacol.* 110:1329–1334.

Boeckxstaens, G.E., De Man, J.G., Pelckmans, P.A., Herman, A.G., and Van Maercke, Y.M. 1993b. α_2-Adrenoceptor mediated modulation of the nitrergic innervation of the canine isolated ileocolonic junction. *Br. J. Pharmacol.* 109:1079–1084.

Boeckxstaens, G.E., and Pelckmans, P.A. 1997. Nitric oxide and the non-adrenergic non-cholinergic neurotransmission. *Comp. Biochem. Physiol.* 118A:925–937.

Boeckxstaens, G.E., Pelckmans, P.A., Bult, H., De Man, J.G., Herman, A.G., and Van Maercke, Y.M. 1991a. Evidence for nitric oxide as mediator of non-adrenergic, non-cholinergic relaxations induced by ATP and GABA in the canine gut. *Br. J. Pharmacol.* 102:434–438.

Boeckxstaens, G.E., Pelckmans, P.A., Ruytjens, I.F., Bult, H., De Man, J.G., Herman, A.G., and Van Maercke, Y.M. 1991b. Bioassay of nitric oxide released upon stimulation of non-adrenergic non-cholinergic nerves in the canine ileocolonic junction. *Br. J. Pharmacol.* 103:1085–1091.

Bogers, J.J., Pelckmans, P.A., Boeckxstaens, G.E., De Man, J.G., Herman, A.G., and Van Maercke, Y.M. 1991. The role of nitric oxide in serotonin-induced relaxations in the canine terminal ileum and ileocolonic junction. *Naunyn-Schmiedeberg's Arch. Pharmacol.* 344:716–719.

Börjesson, L., Nordgren, S., and Delbro, D.S. 1997. DMPP causes relaxation of rat distal colon by a purinergic and a nitrergic mechanism. *Eur. J. Pharmacol.* 334:223–231.

Brave, S.R., Bhat, S., Hobbs, A.J., Tucker, J.F., and Gibson, A. 1993. The influence of L-N^G-nitro-arginine on sympathetic nerve induced contraction and noradrenaline release in the rat isolated anococcygeus muscle. *J. Auton. Pharmacol.* 13:219–225.

Bredt, D.S., and Snyder, S.H. 1994. Nitric oxide: a physiologic messenger molecule. *Annu. Rev. Biochem.* 63:175–195.

Briejer, M.R., Akkermans, L.M.A., Meulemans, A.L., Lefebvre, R.A., and Schuurkes, J.A.J. 1992. Nitric oxide is involved in 5-HT-induced relaxations of the guinea-pig colon ascendens in vitro. *Br. J. Pharmacol.* 107:756–761.

Buga, G.M., Griscavage, J.M., Rogers, N.E., and Ignarro, L.J. 1993. Negative feedback regulation of endothelial cell function by nitric oxide. *Circ. Res.* 73:808–812.

Cellek, S., and Moncada, S. 1997. Nitrergic control of peripheral sympathetic responses in the human corpus cavernosum: a comparison with other species. *Proc. Natl. Acad. Sci. USA* 94:8226–8231.

Chakder, S., and Rattan, S. 1993. Release of nitric oxide by activation of nonadrenergic noncholinergic neurons of internal anal sphincter. *Am. J. Physiol.* 264:G7–G12.

Cohen, G.A., Hobbs, A.J., Fitch, R.M., Zinner, M.J., Chaudhuri, G., and Ignarro, L.J. 1996. Nitric oxide regulates endothelium-dependent vasodilator responses in rabbit hindquarters vascular bed in vivo. *Am. J. Physiol.* 271:H133–H139.

De Man, J.G., Boeckxstaens, G.E., De Winter, B.Y., Moreels, T.G., Herman, A.G., and Pelckmans, P.A. 1995. Inhibition of non-adrenergic non-cholinergic relaxations by nitric oxide donors. *Eur. J. Pharmacol.* 285:269–274.

De Man, J.G., Boeckxstaens, G.E., Herman, A.G., and Pelckmans, P.A. 1994. Effect of potassium channel blockade and α_2-adrenoceptor activation on the release of nitric oxide from non-adrenergic non-cholinergic nerves. *Br. J. Pharmacol.* 112:341–345.

Dinerman, J.L., Dawson, T.M., Schell, M.J., Snowman, A., and Snyder, S.H. 1994. Endothelial nitric oxide synthase localized to hippocampal pyramidal cells: implications for synaptic plasticity. *Proc. Natl. Acad. Sci. USA* 91:4214–4218.

Duckles, S.P., and Budai, D. 1990. Stimulation intensity as critical determinant of presynaptic receptor effectiveness. *Trends Pharmacol. Sci.* 11:440–443.

Eglen, R.M., and Whiting, R.L. 1990. Heterogeneity of vascular muscarinic receptors. *J. Auton. Pharmacol.* 19:233–245.

Förstermann, U., and Kleinert, H. 1995. Nitric oxide synthase: expression and expressional control of the three isoforms. *Naunyn-Schmiedeberg's Arch. Pharmacol.* 352:351–364.

Frank, D.U., Horstman, D.J., Morris, G.N., Johns, R.A., and Rich, G.F. 1998. Regulation of the endogenous NO pathway by prolonged inhaled NO in rats. *J. Appl. Physiol.* 85:1070–1078.

Frew, R., and Lundy, P.M. 1995. A role for Q type Ca^{2+} channels in neurotransmission in the rat urinary bladder. *Br. J. Pharmacol.* 116:1595–1598.

Gibson, A., and Lilley, E. 1997. Superoxide anions, free-radical scavengers, and nitrergic neurotransmission. *Gen. Pharmacol.* 28:489–493.

Gillespie, J.S. 1980. The physiology and pharmacology of the anococcygeus muscle. *Trends Pharmacol. Sci.* 2:453–457.

Gillespie, J.S., Liu, X., and Martin, W. 1989. The effects of L-arginine and N^G-monomethyl L-arginine on the response of the rat anococcygeus muscle to NANC nerve stimulation. *Br. J. Pharmacol.* 98:1080–1082.

Griscavage, J.M., Fukuto, J.M., Komori, Y., and Ignarro, L.J. 1994. Nitric oxide inhibits neuronal nitric oxide synthase by interacting with the heme prosthetic group. *J. Biol. Chem.* 269:21644–21649.

Griscavage, J.M., Rogers, N.E., Sherman, M.P., and Ignarro, L.J. 1993. Inducible nitric oxide synthase from a rat alveolar macrophage cell line is inhibited by nitric oxide. *J. Immunol.* 151:6329–6337.

Gross, J., Mutschler, E., and Lambrecht, G. 1997. Evidence for muscarinic M_4 receptors mediating nonadrenergic noncholinergic relaxations in rabbit anococcygeus muscle. *Naunyn-Schmiedeberg's Arch. Pharmacol.* 356:505–516.

Hebeiss, K., and Kilbinger, H. 1996. Differential effects of nitric oxide donors on basal and electrically evoked release of acetylcholine from guinea-pig myenteric neurones. *Br. J. Pharmacol.* 118:2073–2078.

Hebeiss, K., and Kilbinger, H. 1998. Nitric oxide-sensitive guanylate cyclase inhibits acetylcholine release and excitatory motor transmission in the guinea-pig ileum. *Neuroscience* 82:623–629.

Hobbs, A.J., and Gibson, A. 1990. L-N^G-nitro-arginine and its methyl ester are potent inhibitors of non-adrenergic, non-cholinergic transmission in the rat anococcygeus. *Br. J. Pharmacol.* 100:749–752.

Hölscher, C. 1997. Nitric oxide, the enigmatic neuronal messenger: its role in synaptic plasticity. *Trends Neurosci.* 20:298–303.

Hosoda, K., Takahashi, T., Fujino, M.A., and Owyang, C. 1998. Inhibitory effects of nitric oxide donors on nitric oxide synthesis in rat gastric myenteric plexus. *J. Pharmacol. Exp. Ther.* 286:1222–1230.

Hoyle, C.H.V., Vladimirova, I.A., and Burnstock, G. 1988. Pre- and postjunctional actions of purine and xanthine compounds in the guinea-pig caecum circular muscle. *Br. J. Pharmacol.* 95:653–663.

Hryhorenko, L.M., Woskowska, Z., and Fox-Threlkeld, J.-A.E.L. 1994. Nitric oxide (NO) inhibits release of acetylcholine from nerves of isolated circular muscle of the canine ileum: relationship to motility and release of nitric oxide. *J. Pharmacol. Exp. Ther.* 271:918–926.

Hu, J., and El-Fakahany, E.E. 1995. Anomalous increase in nitric oxide synthase activity by certain nitric oxide-generating compounds in intact neuronal cells. *J. Neurochem.* 65:117–124.

Hu, J., and El-Fakahany, E.E. 1996. Intricate regulation of nitric oxide synthesis in neurons. *Cell Signal.* 8:185–189.

Ignarro, L.J. 1995. Negative modulation of nitric oxide synthase by nitric oxide and nitroso compounds. *Adv. Pharmacol.* 34:215–234.

Ivancheva, C., and Radomirov, R. 1996. Met-enkephalin-dependent nitrergically mediated relaxation in the guinea pig ileum. *Meth. Find. Exp. Clin. Pharmacol.* 18:521–525.

Iversen, H.H., Wiklund, N.P., and Gustafsson, L.E. 1994. Nitric oxide-like activity in guinea pig colon as determined by effector responses, bioassay and chemiluminescence analysis. *Acta Physiol. Scand.* 152:315–322.

Iversen, H.H., Wiklund, N.P., Olgart, C., and Gustafsson, L.E. 1997. Nerve stimulation-induced nitric oxide release as a consequence of muscarinic M_1 receptor activation. *Eur. J. Pharmacol.* 331:213–219.

Kasakov, L., Belai, A., Vlaskovska, M., and Burnstock, G. 1994. Noradrenergic-nitrergic interactions in the rat anococcygeus muscle: evidence for postjunctional modulation by nitric oxide. *Br. J. Pharmacol.* 112:403–410.

Kasakov, L., Cellek, S., and Moncada, S. 1995. Characterization of nitrergic neurotransmission during short- and long-term electrical stimulation of the rabbit anococcygeus muscle. *Br. J. Pharmacol.* 115:1149–1154.

Kiff, R.J., Moss, D.W., and Moncada, S. 1994. Effect of nitric oxide gas on the generation of nitric oxide by isolated blood vessels: implications for inhalation therapy. *Br. J. Pharmacol.* 113:496–498.

Kilbinger, H., and Wolf, D. 1994. Increase by NO synthase inhibitors of acetylcholine release from guinea-pig myenteric plexus. *Naunyn-Schmiedeberg's Arch. Pharmacol.* 349:543–545.

Kojima, S., Ishizaki, R., and Shimo, Y. 1993. Investigation of nicotine-induced relaxation of circular smooth muscle of the guinea-pig gastric fundus. *Eur. J. Pharmacol.* 241:171–175.

Kojima, S., Sakato, M., and Shimo, Y. 1988. An α_2-adrenoceptor-mediated inhibition of non-adrenergic non-cholinergic inhibitory responses of the isolated proximal colon of the guinea-pig. *Asia Pacific J. Pharmacol.* 3:69–75.

Kortezova, N., Velkan, V., Mizhorkova, Z., Bredy-Dobreva, G., Vizi, E.S., and Papasova, M. 1994. Participation of nitric oxide in the nicotine-induced relaxation of the cat lower esophageal sphincter. *J. Auton. Nerv. Syst.* 50:73–78.

Kostka, P., Jang, E., Watson, E.G., Stewart, J.L., and Daniel, E.E. 1993. Nitric oxide synthase in the autonomic nervous system of canine ileum. *J. Pharmacol. Exp. Ther.* 264:234–239.

Langer, S.Z. 1981. Presynaptic regulation of the release of catecholamines. *Pharmacol. Rev.* 32:337–362.

Leclere, P., and Lefebvre, R.A. 1998. Investigation of the interaction between cholinergic and nitrergic neurotransmission in the pig gastric fundus. *Br. J. Pharmacol.* 125:1779–1787.

Lefebvre, R.A., and Smits, G.J.M. 1992. Modulation of non-adrenergic non-cholinergic inhibitory neurotransmission in rat gastric fundus by the α_2-adrenoceptor agonist, UK-14 304. *Br. J. Pharmacol.* 107:256–261.

Lefebvre, R.A., Hasrat, J., and Gobert, A. 1992. Influence of N^G-nitro-L-arginine methyl ester on vagally induced gastric relaxation in the anaesthetized rat. *Br. J. Pharmacol.* 105:315–320.

Lefebvre, R.A., and Vandekerckhove, K. 1998. Effect of nitroglycerin and long-term electrical stimulation on nitrergic relaxation in the pig gastric fundus. *Br. J. Pharmacol.* 123:143–149.

Li, C.G., and Rand, M.J. 1989. Prejunctional inhibition of non-adrenergic non-cholinergic transmission in the rat anococcygeus muscle. *Eur. J. Pharmacol.* 168:107–110.

Llewellyn-Smith, I.J., Song, Z.M., Costa, M., Bredt, D.S., and Snyder, S.H. 1992. Ultrastructural localization of nitric oxide synthase immunoreactivity in guinea-pig enteric neurons. *Brain Res.* 577:337–342.

Ma, X.-L., Lopez, B.L., Christopher, T.A., Birenbaum, D.S., and Vinten-Johansen, J. 1996. Exogenous NO inhibits basal NO release from vascular endothelium in vitro and in vivo. *Am. J. Physiol.* 271:H2045–H2051.

Matthew, J.D., and Wadsworth, R. 1997. The role of nitric oxide in inhibitory neurotransmission in the middle cerebral artery of the sheep. *Gen. Pharmacol.* 28:393–397.

McIntyre, A.S., and Thompson, D.G. 1992. Review article: adrenergic control of motor and secretory function in the gastrointestinal tract. *Aliment. Pharmacol. Ther.* 6:125–142.

McLaren, A., Li, C.G., and Rand, M.J. 1993. Mediators of nicotine-induced relaxations of the rat gastric fundus. *Clin. Exp. Pharmacol. Physiol.* 20:451–457.

Milenov, K., and Kalfin, R. 1996. Cholinergic-nitrergic interactions in the guinea-pig gastric fundus. *Neuropeptides* 30:365–371.

Mizhorkova, Z., Kortezova, N., Bredy-Dobreva, G., and Papasova, M. 1994. Role of nitric oxide in mediating non-adrenergic non-cholinergic relaxation of the cat ileocecal sphincter. *Eur. J. Pharmacol.* 265:77–82.

Nakamura, K., Takahashi, T., Taniuchi, M., Hsu, C.X., and Owyang, C. 1998. Nicotinic receptor mediates nitric oxide synthase expression in the rat gastric myenteric plexus. *J. Clin. Invest.* 101:1479–1489.

Oka, M., Ohnishi, M., Takahashi, H., Soma, S., Hasunuma, K., Sato, K., and Kira, S. 1996. Altered vasoreactivity in lungs isolated from rats exposed to nitric oxide gas. *Am. J. Physiol.* 271:L419–L424.

Park, J.-H., Straub, V.A., and O'Shea, M. 1998. Anterograde signaling by nitric oxide: characterization and in vitro reconstitution of an identified nitrergic synapse. *J. Neurosci.* 18:5463–5476.

Park, S.K., Lin, H.L., and Murphy, S. 1997. Nitric oxide regulates nitric oxide synthase-2 gene expression by inhibiting NF-κB binding to DNA. *Biochem. J.* 322:609–613.

Postorino, A., Vetri, T., Leggio, L., Serio, R., and Bonvissuto, F. 1998. Nitric oxide as neuromodulator of sympathetic transmission in rat vas deferens. *J. Auton. Pharmacol.* 18:21–29.

Rand, M.J., and Li, C.G. 1992. Activation of noradrenergic and nitrergic mechanisms in the rat anococcygeus muscle by nicotine. *Clin. Exp. Pharmacol. Physiol.* 19:103–111.

Rattan, S., Sarkar, A., and Chakder, S. 1992. Nitric oxide pathway in rectoanal inhibitory reflex of opossum internal anal sphincter. *Gastroenterology* 103:43–50.

Ravichandran, L.V., Johns, R.A., and Rengasamy, A. 1995. Direct and reversible inhibition of endothelial nitric oxide synthase by nitric oxide. *Am. J. Physiol.* 268:H2216–H2223.

Rengasamy, A., and Johns, R.A. 1993. Regulation of nitric oxide synthase by nitric oxide. *Mol. Pharmacol.* 44:124–128.

Rogers, N.E., and Ignarro, L.J. 1992. Constitutive nitric oxide synthase from cerebellum is reversibly inhibited by nitric oxide formed from L-arginine. *Biochem. Biophys. Res. Commun.* 189:242–249.

Sanders, K.M., and Ward, S.M. 1992. Nitric oxide as a mediator of nonadrenergic noncholinergic neurotransmission. *Am. J. Physiol.* 262:G379–H392.

Ševčik, J., Růžička, V., Slánský, J., Šmejkal, J., and Mašek, K. 1998. Is nitric oxide involved in 5-HT_3 receptor-mediated neurogenic relaxation of guinea pig proximal colon? *Jpn. J. Pharmacol.* 77:265–270.

Sheehy, A.M., Burson, M.A., and Black, S.M. 1998. Nitric oxide exposure inhibits endothelial NOS activity but not gene expression: a role for superoxide. *Am. J. Physiol.* 274:L833–L841.

Shuttleworth, C.W.R., Weinert, J.S., Sanders, K.M., and Buxton, I.L.O. 1995. Detection of nitric oxide release from canine enteric neurons. *J. Auton. Nerv. Syst.* 56:61–68.

Sneddon, P., and Graham, A. 1992. Role of nitric oxide in the autonomic innervation of smooth muscle. *J. Auton. Pharmacol.* 12:445–456.

Soejima, O., Katsuragi, T., and Furukawa, T. 1993. Opposite modulation by muscarinic M_1 and M_3 receptors of acetylcholine release from guinea pig ileum as measured directly. *Eur. J. Pharmacol.* 249:1–6.

Stanley, E.F. 1997. The calcium channel and the organization of the presynaptic transmitter release face. *Trends Neurosci.* 20:404–409.

Starke, K. 1981. Presynaptic receptors. *Annu. Rev. Pharmacol. Toxicol.* 21:7–30.

Starke, K., Göthert, M., and Kilbinger, H. 1989. Modulation of neurotransmitter release by presynaptic autoreceptors. *Physiol. Rev.* 69:864–989.

Stjarne, L. 1989. Basic mechanisms and local modulation of nerve impulse-induced secretion of neurotransmitters from individual sympathetic nerve varicosities. *Rev. Physiol. Biochem. Pharmacol.* 112:4–137.

Takahashi, T., and Owyang, C. 1995. Vagal control of nitric oxide and vasoactive intestinal polypeptide release in the regulation of gastric relaxation in rat. *J. Physiol.* 484:481–492.

Tanobe, Y., Okamura, T., Fujimara, M., and Toda, N. 1995. Functional role and histological demonstration of nitric oxide-mediated inhibitory nerves in dog sphincter of Oddi. *Neurogastroenterology* 7:219–227.

Taylor, B.S., Kim, Y.-M., Wang, Q., Shapiro, R.A., Billiar, T.R., and Geller, D.A. 1997. Nitric oxide down-regulates hepatocyte-inducible nitric oxide synthase gene expression. *Arch. Surg.* 132:1177–1183.

Toda, M., Okamura, T., Azuma, I., and Toda, N. 1997a. Modulation by neurogenic acetylcholine of nitroxidergic nerve function in porcine ciliary arteries. *Invest. Ophthalmol. Vis. Sci.* 38:2261–2269.

Toda, N., Ayajiki, K., and Okamura, T. 1997b. Inhibition of nitroxidergic nerve function by neurogenic acetylcholine in monkey cerebral arteries. *J. Physiol.* 498:453–461.

Toda, N., Toda, M., Ayajiki, K., and Okamura, T. 1998. Cholinergic nerve function in monkey ciliary arteries innervated by nitroxidergic nerves. *Am. J. Physiol.* 274:H1582–H1589.

Toda, N., Uchiyama, M., and Okamura, T. 1995. Prejunctional modulation of nitroxidergic nerve function in canine cerebral arteries. *Brain Res.* 700:213–218.

Vials, A.J., Crowe, R., and Burnstock, G. 1997. A neuromodulatory role for neuronal nitric oxide in the rabbit renal artery. *Br. J. Pharmacol.* 121:213–220.

Ward, J.K., Belvisi, M.G., Fox, A.J., Miura, M., Tadjkarimi, S., Yacoub, M.H., and Barnes, P.J. 1993. Modulation of cholinergic neural bronchoconstriction by endogenous nitric oxide and vasoactive intestinal peptide in human airways in vitro. *J. Clin. Invest.* 92:736–742.

Ward, S.M., Dalziel, H.H., Khoyi, M.A., Westfall, A.S., Sanders, K.M., and Westfall, D.P. 1996. Hyperpolarization and inhibition of contraction mediated by nitric oxide released from enteric inhibitory neurones in guinea-pig taenia coli. *Br. J. Pharmacol.* 118:49–56.

Watson, M.J., Bywater, R.A.R., Taylor, G.S., and Lang, R.J. 1996. Effects of nitric oxide (NO) and NO donors on the membrane conductance of circular smooth muscle cells of the guinea-pig proximal colon. *Br. J. Pharmacol.* 118:1605–1614.

Werkström, V., Persson, K., Ny, L., Bridgewater, M., Brading, A.F., and Andersson, K.E. 1995. Factors involved in the relaxation of female pig urethra evoked by electrical field stimulation. *Br. J. Pharmacol.* 116:1599–1604.

Wiklund, C.U., Wiklund, N.P., and Gustafsson, L.E. 1993. Modulation of neuroeffector transmission by endogenous nitric oxide: a role for acetylcholine receptor-activated nitric oxide formation, as indicated by measurements of nitric oxide/nitrite release. *Eur. J. Pharmacol.* 240:235–242.

Wood, J.D. 1987. Physiology of the enteric nervous system. In: *Physiology of the Gastrointestinal Tract*, 2nd ed. (ed. L.R. Johnson), pp. 67–109. Raven Press, New York.

Wood, J., and Garthwaite, J. 1994. Models of the diffusional spread of nitric oxide: implications for neural nitric oxide signalling and its pharmacological properties. *Neuropharmacology* 33:1235–1244.

Yamato, S., and Rattan, S. 1990. Role of alpha adrenoceptors in opossum internal anal sphincter. *J. Clin. Invest.* 86:424–429.

Yu, J.-G., Kimura, T., Chang, X.-F., and Lee, T.J.F. 1998. Segregation of VIPergic-nitric oxidergic and cholinergic-nitric oxidergic innervation in porcine middle cerebral arteries. *Brain Res.* 801:78–87.

Zagorodnyuk, V., and Maggi, C.A. 1994. Electrophysiological evidence for different release mechanisms of ATP and NO as inhibitory NANC transmitters in guinea-pig colon. *Br. J. Pharmacol.* 112:1077–1082.

Zhang, W., Edvinsson, L., and Lee, T.J. 1998. Mechanism of nicotine-induced relaxation in the porcine basilar artery. *J. Pharmacol. Exp. Ther.* 284:790–797.

Zygmunt, P.K.E., Zygmunt, P.M., Högestätt, E.D., and Andersson, K.-E. 1995. NANC neurotransmission in lamina propria of the rabbit urethra: regulation by different subsets of calcium channels. *Br. J. Pharmacol.* 115:1020–1026.

11 Neural Nitric Oxide Synthase and Neutrophil Function

STAN S. GREENBERG, XINFANG ZHAO, JIE OUYANG, AND THOMAS D. GILES

Introduction

Nitric oxide synthases (NOS, EC 1.14.13.39) are members of a family of cytochrome P450-like reductases linked to an NADPH oxidase enzyme. Nitric oxide synthase converts L-arginine to nitric oxide (NO) and L-citrulline utilizing the electrons generated by the reduction of oxygen to superoxide, hydrogen peroxide, and ultimately water (Wang and Marsden, 1995, Xie and Nathan, 1994; Forstermann and Kleinert, 1995). At least three isozymes of NOS exist, each of which is derived from a different gene (Forstermann and Kleinert, 1995). Endothelial NOS (ecNOS) and neural NOS (nNOS) are constitutive, calcium-activated isozymes found in endothelial and endocardial cells, airway epithelial cells, adrenal cortical and medullary cells, neurons, skeletal muscle, and many soft tissues (Wang and Marsden, 1995; Xie and Nathan, 1994; Forstermann and Kleinert, 1995). Inducible NOS (iNOS), primarily regulated by transcription, is elicited in inflammatory and immunocompetent cells; cardiac, striated, and smooth muscles; pancreatic beta cells; and a wide variety of tissues by bacterial endotoxins, cytokines, and autacoids (Wang and Marsden, 1995; Xie and Nathan, 1994; Forstermann and Kleinert, 1995; Greenberg et al., 1996, 1997; Nichols et al., 1988; Kwon et al., 1996).

The regulation and function of NOS isozymes have been recently reviewed (Wang and Marsden, 1995; Xie and Nathan, 1994; Forstermann and Kleinert, 1995). Nitric oxide plays an important role in regulating airway function and in the pathophysiology of respiratory tract inflammation and infections (Gross and Wolin, 1995; Hogg and Duerscuk, 1995; Lu et al., 1998; Rand and Li, 1995; Sheridan et al., 1996, 1998). However, the NOS derived from each of the three isozymes of NOS appear to have distinct functions within the lung.

Inducible Nitric Oxide Synthase and the Lung

Nitric oxide derived from endotoxin and cytokine-inducible NOS II is believed to play a role in the bactericidal activity of polymorphonuclear neutrophils (PMNs), macrophages, and lymphocytes within the lung (Gross and Wolin, 1995; Lu et al., 1998; Kubes et al., 1991; Kuo et al., 1997; McCall et al., 1989). The mechanism of cytotoxicity and bacterial killing is independent of cellular cyclic GMP and may result from at least five possible actions of NO. First, NOS II-derived NO can interact with superoxide to generate cytotoxic peroxynitrite and hydroxyl radical-like mediators (Wang and Marsden, 1995; Xie and Nathan, 1994; Forstermann and Kleinert, 1995). Nitric oxide may also inhibit Na^+/H^+ exchange by bacterial H^+-ATPase, thereby impairing bacterial cytoplasmic pH regulation (Swallow et al., 1991). In addition NO may reduce enzymatic Fe^{3+} to inactive Fe^{2+} in bacterial enzymes, thereby inhibiting bacterial enzyme activity (Wang and Marsden, 1995; Xie and Nathan, 1994; Forstermann and Kleinert, 1995). The ability of NO to act as an antioxidant within bacteria may suppress the formation of oxidant-activated transcription factors involved in cellular growth, cell replication, or toxin formation (Wang and Marsden, 1995; Xie and Nathan, 1994; Forstermann and Kleinert, 1995; Moffat et al., 1996). Finally, peroxynitrite or NO may nitrosate DNA, resulting in mutations, cellular damage, or cell death (Wang and Marsden, 1995; Xie and Nathan, 1994; Forstermann and Kleinert, 1995).

Endothelial Nitric Oxide Synthase and the Lung

Constitutive endothelial nitric oxide synthase (NOS III) is located primarily in the endothelial cells of the pulmonary and bronchial circulation of the lung (Wang and Marsden, 1995; Xie and Nathan, 1994; Forstermann and Kleinert, 1995; Gross and Wolin, 1995; Rand and Li, 1995; Geiger et al., 1997) and in rat alveolar macrophages (Miles et al., 1998). NOS III-derived NO is produced in the endothelium and macrophages by the oxidation of the amino acid L-arginine (Gross and Wolin, 1995; Rand and Li, 1995). Nitric oxide derived from NOS III appears to be important as a modulator of intrapulmonary blood flow distribution and, indirectly, plasma exudation. Nitric oxide diffuses from the abluminal surface of the endothelium to the smooth muscle where it activates the enzyme guanylate cyclase, with the resultant formation of the potent vasodilator cyclic GMP (Wang and Marsden, 1995; Xie and Nathan, 1994; Forstermann and Kleinert, 1995; Gross and Wolin, 1995; Rand and Li, 1995). NOS III-derived NO also modulates the thrombogenicity of the endothelium and the ability of the formed elements of the blood to adhere to the endothelium. Nitric oxide released at the luminal surface of the endothelium can diffuse into the blood where it can elevate cGMP in platelets or inactivate NADPH or cytochrome *c* oxidase in circulating PMNs, monocytes, and platelets (Gross and Wolin, 1995; Murad et al., 1993; Murad, 1994; Rubanyi et al., 1991). These effects of NO can diminish platelet and phagocyte cell excitability and inhibit the expression of adhesion molecules (Wang and Marsden, 1995; Xie and Nathan, 1994; Forstermann and Kleinert, 1995; McCall et al., 1989; De Caterina et al., 1995; Peng et al., 1995). This in turn, would attenuate the adhesion of the platelets and phagocytes to the endothelium, thereby preventing the formation of thrombi and the transport of monocytes and macrophages through the endothelial membrane to the subintimal tissue (Hoshiga et al., 1995; Schwartz et al., 1996). In addition, NO derived from NOS III can act on endothelial cells in an autocrine and paracrine manner to up-regulate the inhibitory transcription factor kappa Bα (IκBα) and inhibit the nuclear translocation of nuclear factor kappa B (NFκB). This is believed to inhibit the endothelial expression of intracellular cell adhesion molecule (ICAM), E-selectin, and vascular cell adhesion molecule-1 (VCAM-1) involved in binding of leukocytes to the endothelium (Wang and Marsden, 1995; Biffl et al., 1996; Spiecker et al., 1997). Adherence of PMNs to the pulmonary vascular endothelial membrane is the initial step in the recruitment of PMNs to the site of the infection or injury (Hogg and Duerscuk, 1995; McCall et al., 1989). The function of NOS III-derived NO in alveolar macrophages, if any, is currently unknown (Miles et al., 1998).

Neural Nitric Oxide Synthase and the Lung

Nitric oxide derived from nNOS (NOS I) is the neurotransmitter of bronchodilator nerves in human airways. Nitric oxide derived from nNOS located within these neurons appears to counteract the bronchial constriction due to cholinergic neural mechanisms and circulating and locally produced bronchial constrictor substances (McIlveen et al., 1997; Schütte et al., 1998). However, constitutive NOS I is not localized only to neurons but, as has been described in this volume, is present in skeletal muscle (Capanni et al., 1998; El-Dwairi et al., 1998) and, as we will review here, is present in the PMNs of at least three animal species: rats, mice, and humans (Chen and Mehta, 1996; Greenberg et al., 1998, 1999; Zhao et al., 1999). Freshly isolated oyster glycogen-elicited peritoneal PMNs and circulating PMNs perfused with Krebs-Henseleit buffer or added to a muscle chamber elicit a vasodilator substance that relaxes rings of endothelium-denuded blood vessels. A similar substance released from rat PMNs and vascular smooth muscle cells was also found to inhibit platelet aggregation (Kadota et al., 1991; Lee et al., 1990; Clancy et al., 1992; Liu et al., 1994; Chen et al., 1994; Wood et al., 1990). These substances behaved as if they were endothelium-derived relaxing factor, now believed to be L-arginine-derived NO or an NO-like compound (Wang and Marsden, 1995; Xie and Nathan, 1994; Forstermann and Kleinert, 1995). Bacterial exotoxins and endotoxin as well as cytokines can induce iNOS in immunocompetent and vascular smooth muscle cells. Moreover, similar to the PMN-derived factor, they can relax as well as inhibit relaxation of vascular smooth muscle (Parker et al., 1994; Wylam et al., 1990). Inducible NOS mRNA and protein were found in rat and human PMNs and vascular smooth muscle cells. Subsequent biochemical and molecular biologic analyses of rat and human PMNs and rat vascular smooth muscle demonstrated the existence of a 130-kDa protein that was immunologically indistinct from iNOS protein (McCall et al., 1991). Therefore, it has generally been assumed that the presence of the PMN-derived relaxing factor probably resulted from oyster glycogen- or airborne bacteria-induced activation of iNOS in the PMNs (McCall et al., 1991; Yui et al., 1991). However, the results of several studies challenged this assumption. Chen and Mehta (1996), using the reverse transcription polymerase chain reaction (RT-PCR), demonstrated the existence of constitutive nNOS mRNA in human PMNs. Although the PMNs generated L-citrulline from L-arginine, the constitutive NOS inhibitors N^G-nitro-L-arginine methylester (L-NAME) and N^G-nitro-L-arginine (L-NA) inhibited arginine uptake into the PMNs but failed to inhibit the formation of L-citrulline. It was concluded that human PMNs contain nNOS mRNA, but that the protein behaves differently to the inhibitors than eNOS, which was found to be constitutive in the platelet. Constitutive NOS I mRNA was also found in human and rat PMNs by using a competitor equalized RT-PCR (Greenberg et al., 1998, 1999a,b; Zhao et al., 1999). Although PMNs constitutively express NOS I and produce concentrations of NO capable of producing vasorelaxation, the function of nNOS in PMNs has remained undefined.

Properties of Neural Nitric Oxide Synthase Within Polymorphonuclear Neutrophils

Neural NOS mRNA was found in PMNs obtained from rats and humans. The mRNA from each species was specific for that species when compared with authentic human and rat nNOS mRNA. Neural NOS protein was present in rat PMNs but was below detectable limits in human PMNs (Figure 11.1). Circulating PMNs obtained from control rats spontaneously produced RNI nitrate and nitrite in excess of that found in normal physiologic buffer solution. Spontaneous production of RNI was increased ($P < 0.05$) when measured in the incubates of PMN obtained from lipopolysaccharide (LPS) treated rats.

Incubation of PMNs obtained from human volunteers and monocytes and macrophages obtained from control rats failed to demonstrate any increase in incubate reactive nitrogen intermediates (RNI) over that of the buffer (Figure 11.2). Incubation of PMNs spontaneously producing nitrite and nitrate anions with Liminoethyllysine (L-NIL), a selective inhibitor of inducible iNOS (Moore et al., 1994), or with L-NA (150 μM) failed to inhibit

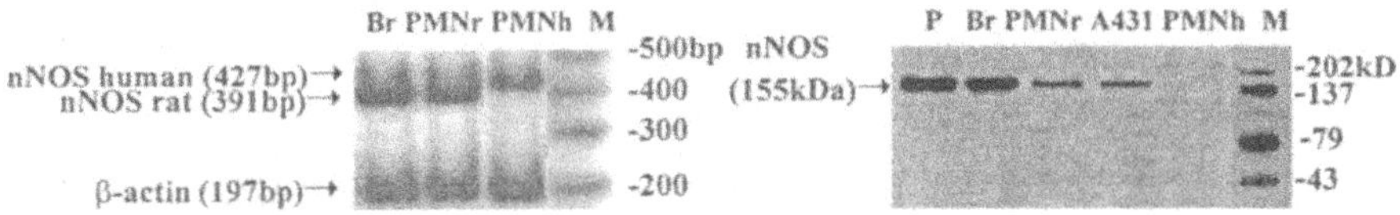

FIGURE 11.1. Top panel: representative PCR gel of the mRNA of NOS isozymes and β-actin mRNA (left) and Western blot (right) of nNOS protein in control rat brain (Br), rat PMN (PMNr), and human A431 cell lysates (A431) and its absence in circulating human PMNs (PMNh). From Greenberg et al. (1998), Copyright 1998 by ••. Reprinted with permission.

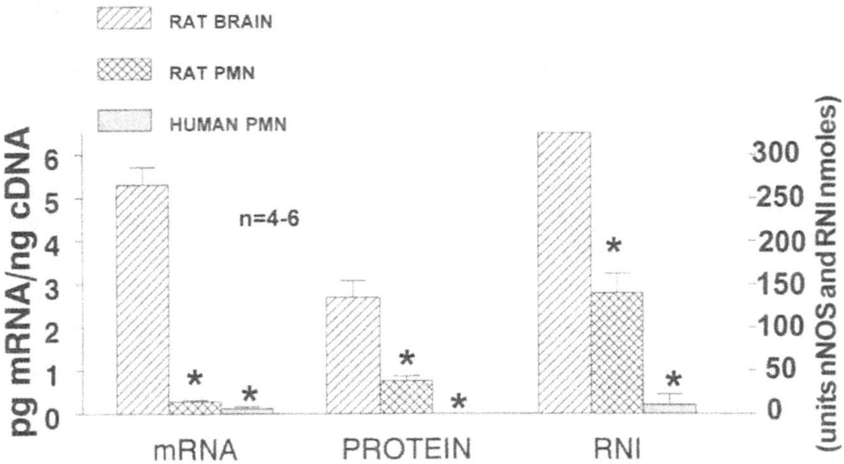

FIGURE 11.2. Mean and SD of the mRNA of nNOS isozyme (gel values/ng of β-actin), nNOS protein (gel values/50 μg of protein), and spontaneous nitrate and nitrite production (nmol/10^7 PMN/h) when obtained from rat brain and PMNs and circulating PMNs obtained from human volunteers. An asterisk denotes that the responses differ from brain ($P < 0.05$). Modified from Greenberg et al. (1998, 1999a,b).

the spontaneous production of RNI in the incubates of PMNs obtained from control rats, whereas RNI production by PMNs or monocytes and macrophages obtained by bronchoalveolar lavage from the lung of LPS-treated rats was attenuated following incubation with L-NIL or L-NA (Figure 11.3). In contrast, 7-nitroindazole (7-NI), an inhibitor of nNOS (Moore and Bland-Ward, 1996), attenuated spontaneous production of RNI in PMNs obtained from control rats as well as in PMNs obtained by bronchoalveolar lavage from LPS-treated rats and preincubated with L-NIL to inhibit iNOS-derived NO (Figures 11.3 and 11.4). Finally, basal production of nitrate and nitrite by the nNOS of PMNs, similar to that in brain (Chandler et al., 1997; Fitzgerald et al., 1995), was augmented following in vivo

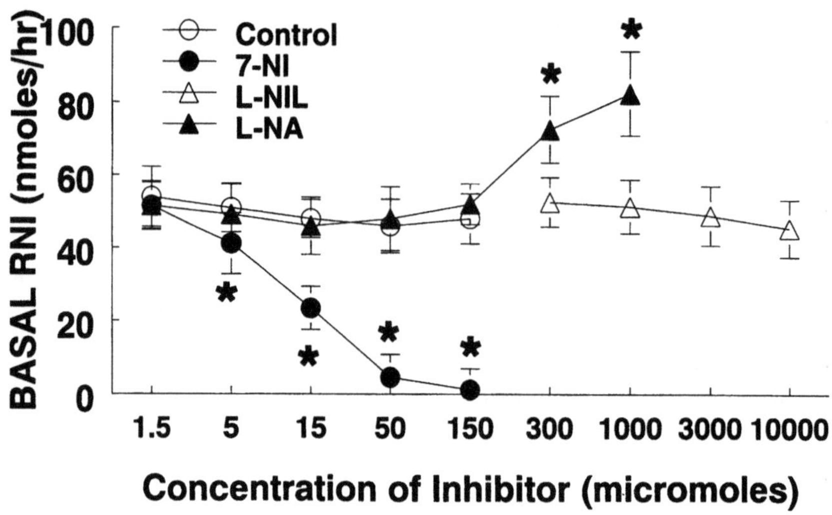

FIGURE 11.3. Effect of 7-nitroindazole (7-NI) L-iminoethyllysine (L-NIL), and L-NA on spontaneous RNI nitrate and nitrite formation in ex vivo incubates of rat PMNs. The ordinate shows the changes in spontaneous RNI production by the PMNs (nmol/h). The abscissa shows the concentrations of inhibitor tested. Vertical lines are SDs from 5 or 6 rats per group. An asterisk denotes that the responses in the presence of inhibitor differ from those in the absence of inhibitor ($P < 0.05$).

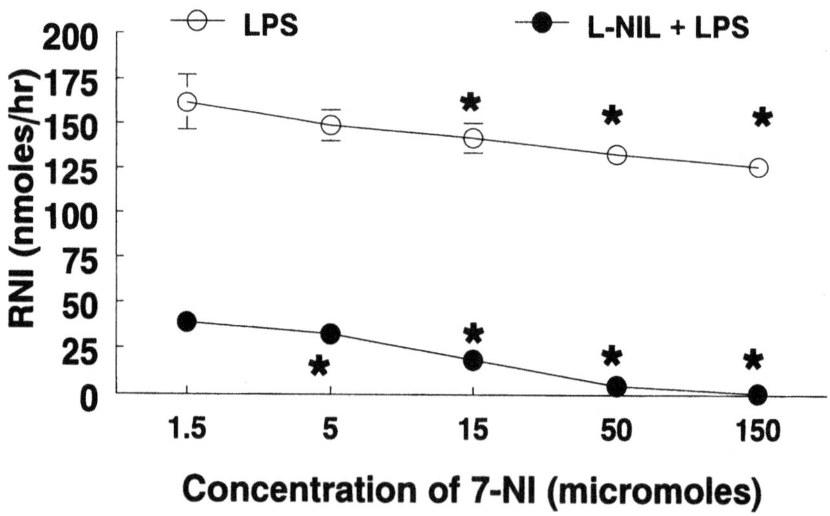

FIGURE 11.4. Effect of 7-NI on spontaneous and LPS-stimulated RNI formation in ex vivo incubates of rat recruited PMNs. The ordinate shows the changes in RNI concentration in recruited PMNs obtained by lung lavage from LPS-treated rats in which RNI consisted of nNOS-generated and iNOS-generated RNI (LPS-treated), and nNOS-generated RNI in which iNOS activity was maximally inhibited with 10 mM of L-NIL (L-NIL + LPS). The abscissa snows the concentrations of 7-NI tested. Vertical lines are SDs from 5 or 6 rats per group. An asterisk denotes that the responses in the presence of 7-NI differ from those in the absence of 7-NI ($P < 0.05$).

administration of ethanol (2–5.4 g/kg, i.p.) to rats and in vitro incubation of PMNs with ethanol (Figure 11.5). Ethanol did not affect nNOS transcription or translation, but rather the enzymatic activity of this enzyme (Greenberg et al., 1998, 1999a; Zhao et al., 1999). This is in contrast to the documented inhibitory effect of in vivo administration of ethanol on LPS- and cytokine-inducible NOS transcription (Greenberg et al., 1997; Xie et al., 1995).

The differential effects of these inhibitors on nNOS in the rat PMN could provide us with a tool to study the role of the nNOS-derived NO in PMN function, provided we were reasonably certain that L-NA inhibited nNOS in the lung and not in the PMN. To evaluate this possibility, we anesthetized a cohort of six rats per group, removed their lungs, and, after washing out the blood, homogenized the lungs and incubated the homogenate with buffer containing 150 μM of L-arginine and 10 μCi of [^{14}C]arginine. We then measured the formation of [^{14}C]citrulline (Bredt and Schmidt, 1996) in the presence and absence of L-NA, L-NAME, 7-NI, or L-NIL for 30 minutes. Since the lavaged lungs of normal rats contain few intramural PMNs, the majority of nNOS activity must be located within the lung matrix and nerves innervating the lung. The experiment was then repeated with the addition of 1×10^7 intact PMNs to the lung homogenate. Thus, the increase in nNOS activity must reflect the NOS within the PMNs.

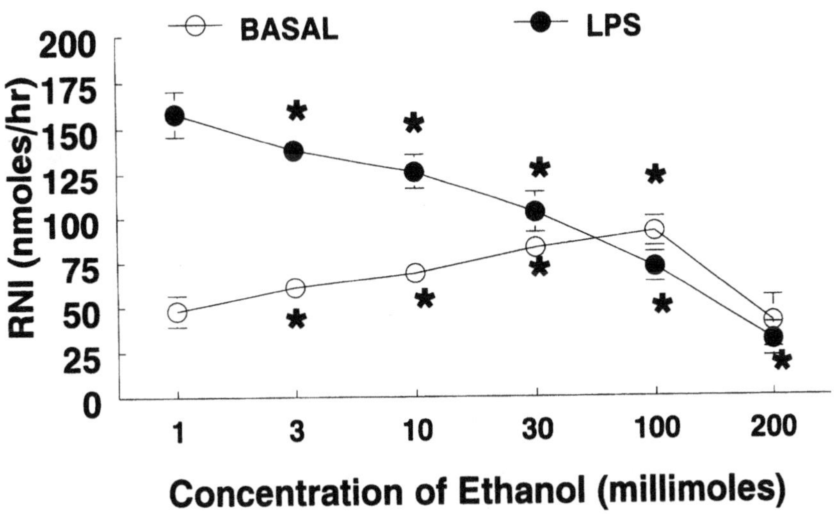

FIGURE 11.5. Effect of ethanol on spontaneous and LPS-stimulated RNI formation in ex vivo incubates of rat PMNs. The ordinate shows the changes in RNI concentration in recruited PMNs obtained by lung lavage from LPS-treated rats in which RNI consisted of nNOS-derived NO and iNOS-generated RNI (LPS-treated) and circulating PMNs in which the sole source of the generated NO was nNOS (basal). The abscissa shows the concentrations of ethanol tested. Vertical lines are the SDs from 5 or 6 rats per group. An asterisk denotes that the responses in the presence of ethanol differ from those in the absence of ethanol ($P < 0.05$).

Extracts of lung and incubates of PMNs both converted [^{14}C]arginine to [^{14}C]citrulline. In lung extracts the formation of citrulline was maximally inhibited by 83 ± 12% with L-NA, 76 ± 14% with L-NAME, 11 ± 6% with L-NIL, and 44 ± 9% with 7-NI. However, in incubates of PMNs the spontaneous production of [^{14}C]citrulline from [^{14}C]arginine was maximally inhibited by 7 ± 9% with L-NA, 5 ± 12% with L-NAME, 4 ± 7% with L-NIL, and 78 ± 7% with 7-NI. Addition of PMNs to aliquots of lung extract increased the total conversion of [^{14}C]arginine to [^{14}C]citrulline within the eluent from the mixture. 7-NI-mediated inhibition of [^{14}C]citrulline formation was augmented in mixtures of lung and PMNs, whereas the absolute magnitude of inhibition of [^{14}C]citrulline formation was unchanged for L-NA, L-NAME, and L-NIL (Figure 11.6). Thus, the data strongly suggest that

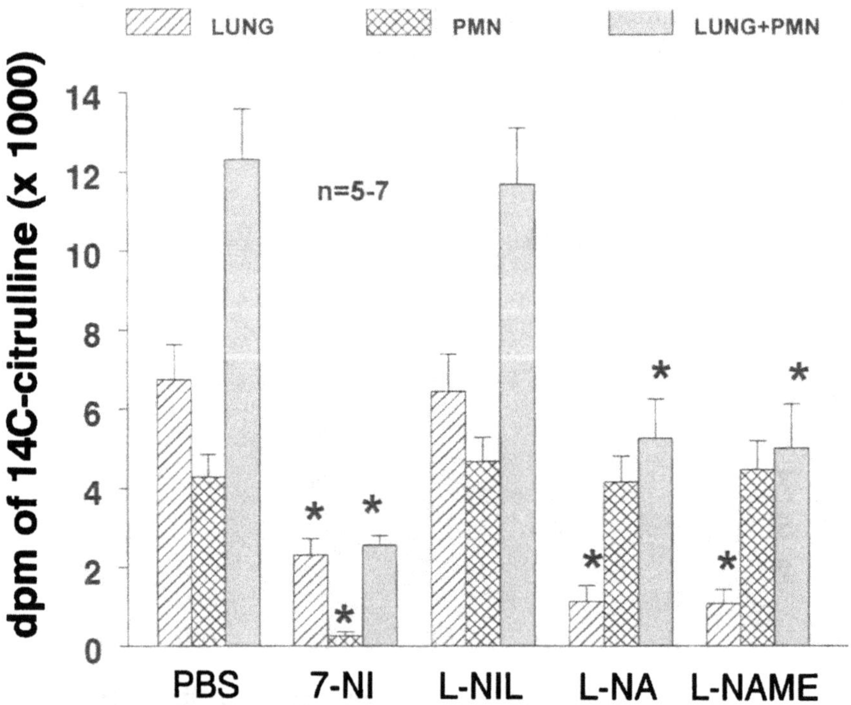

FIGURE 11.6. Effect of inhibitors on the formation of [^{14}C]citrulline from [^{14}]C-arginine in lung homogenates or lung homogenates incubated with circulating PMNs. The ordinate shows the content of [^{14}C]citrulline. The abscissa shows the drug tested: 7-NI (150 μM), L-NIL (2 mM), L-NAME (150 μM), or L-NA (50 μM). PBS is phosphate-buffered saline. Vertical lines are the SDs from 5 or 6 rats per group. An asterisk denotes that the responses in the presence and absence of inhibitor differ ($P < 0.05$).

L-NA and L-NAME inhibit nNOS in lung but not in PMNs, whereas 7-NI inhibits neural NOS in PMNs and lung.

Potential Roles of Neural Nitric Oxide Synthase in Neutrophils

Within the central nervous system, nNOS-derived NO serves not only as a chemical mediator of cell communication (Maines, 1996) but also as a microbicidal agent involved in the control of growth and progression of several viral and fungal infections and as a modulator of cellular ischemia and infarction (Zaharchuk et al., 1997; James, 1995; Barna et al., 1996; Dietzschold, 1995). Within the lung, nNOS-derived NO within the PMNs may serve an antiadhesion role in place of eNOS-derived NO, which is present in limited amount in the lung arterial microvasculature (Geiger et al., 1997). Thus, within the PMNs, nNOS-derived NO could act to modulate recruitment of the PMNs to the site of injury or infection, phagocytosis of bacteria or foreign particulate matter, or microbial killing. The latter could be achieved by NO-mediated changes in enzymatic activity of the PMNs or by NO-mediated inactivation of the activity of endogenous NADPH oxidases and superoxide generation (Clancy et al., 1992; Kaplan et al., 1996).

Neural NOS and Lung Recruitment of PMN

In the systemic circulation, antigen-stimulated PMN adhesion and recruitment to the site of inflammation or infection occurs on the postcapillary venous endothelium, which contains significant amounts of constitutive eNOS (Kubes et al., 1991; Hickey and Kubes, 1997). In contrast, the recruitment of circulating PMNs into the site of infection within the lung parenchyma and alveolar space begins with adhesion of the PMNs to the microvascular capillaries of the arterial microcirculation of the lung and not on the NO-producing postcapillary venous endothelium characteristic of the systemic circulation (Hogg and Duerscuk, 1995). The arterial microcirculation of the lung produces significantly less NO from constitutive eNOS relative to the endothelium of the larger conduit vessels or the systemic vessels (Wang and Marsden, 1995; Xie and Nathan, 1994; Forstermann and Kleinert, 1995; Geiger et al., 1997). In the systemic circulation, eNOS-derived NO is believed to limit PMN adhesion to the endothelium by increasing the synthesis of the inhibitory transcription factor IκBα. This, in turn, is believed to inhibit the synthesis of the adhesion molecules on the endothelium and the circulating PMNs and thus limit PMN recruitment to the site of injury (Kubes et al., 1991; Hickey and Kubes, 1997). In support of this hypothesis were the findings that stimulation of endogenous NO prevented the formation of the adhesion molecule ICAM-1 in isolated vascular endothelial cells, but that administration of NO could not reverse the formation of these adhesion molecules once their synthesis was initiated (De Caterina et al., 1995; Peng et al., 1995; Biffl et al., 1996; Spiecker et al., 1997). In addition, administration of the NO precursor L-arginine to rats prevented accumulation of PMNs in the lungs and preserved pulmonary endothelial vasodilation (Sheridan et al., 1996, 1998). It was concluded that L-arginine up-regulated NOS III activity in the pulmonary vascular endothelium. Moreover, it was suggested that NOS III-derived NO inhibited PMN adhesion to the vascular endothelium and restored endothelium-derived relaxation (Sheridan et al., 1996, 1998). The alternative possibility that L-arginine provided a substrate for nNOS-derived NO produced within the PMN was not considered. However, in vitro and in vivo studies demonstrated that nNOS activity within the PMN was susceptible to inhibition by 7-NI but was refractory to inhibition by the nonselective NOS inhibitors L-NA and its methylester L-NAME (Chen and Mehta, 1996; Greenberg et al., 1998, 1999). Moreover, systemic administration of L-NA and L-NAME paradoxically decreased PMN adhesion to the endothelium at the site of inflammation while increasing PMN adhesion to the vascular endothelium in remote organs in a rat model of peritonitis (Fukatsu et al., 1997). This latter effect results from inhibitor-induced sequestration of PMNs in the postcapillary venules of the peripheral circulation of the infected rats.

The differential effects of the inhibitors on PMN adhesion and the documented localized lung recruitment of PMNs following intratracheal injections of endotoxin to rats provided us with a model to test the hypothesis that nNOS within the lung, and more specifically within the PMN, may modulate PMN recruitment to the site of infection within the lung. Intratracheal administration of LPS stimulates local recruitment of PMNs into the alveolar space of the rat lung. If nNOS within the PMN plays a specific role in PMN adhesion to the endothelium and subsequent diapedesis and recruitment within the lung, then it would be expected that pretreatment of rats with L-NA, L-NAME, 7-NI, L-NIL, or ethanol should exert distinct effects on LPS-induced lung recruitment of PMNs. Intraperitoneal administration of phosphate-buffered saline (PBS), the inducible NOS inhibitor L-NIL (10 mg/kg), ethanol (4.5 g/kg), or the nNOS I inhibitor 7-NI (25 mg/kg) did not affect the basal circulating PMN content of plasma obtained from peripheral blood or the bronchoalveolar lavage (BAL) fluid content of alveolar macrophages and recruited PMNs obtained from the lung of normal rats. In contrast, treatment of normal rats with L-NA (25 mg/kg), which can inhibit constitutive NOS III (Gross and Wolin, 1995), decreased the number of circulating PMNs in peripheral blood and the alveolar macrophages recoverable in the BAL fluid. However, pretreatment of rats with L-NA did not increase the basal PMN content of the lung (Table 11.1). Although administration of a nNOS inhibitor did not affect the lung

TABLE 11.1. Effect of treatments on lung white cell distribution.

	Bronchoalveolar lavage fluid			
Treatment	Total cells × 10^6	AM × 10^6	PMNs × 10^6	PLASMA PMNs × 10^6
Untreated	17.3 ± 1.2	17.2 ± 1.2	0.1 ± 0.6	10.3 ± 0.9
PBS 0.25 ml	18.7 ± 1.6	18.6 ± 0.9	0.1 ± 0.9	10.9 ± 1.1
DMSO 0.2 ml	16.9 ± 0.9*	16.8 ± 0.8	0.1 ± 0.8	11.2 ± 1.7
L-NIL 10 mg/kg	18.8 ± 3.9	18.6 ± 0.2	0.2 ± 0.5	10.8 ± 1.4
L-NA 20 mg/kg	14.0 ± 0.4*	14.0 ± 0.7	0.0 ± 0.7	7.6 ± 1.2*
7-NI 25 mg/kg	16.8 ± 1.1	16.7 ± 0.1	0.1 ± 0.1	10.9 ± 1.4
7-NI + L-NA	28.4 ± 2.3	15.9 ± 0.2	12.5 ± 0.4	7.1 ± 1.8
ETOH 4.5 g/kg	13.9 ± 1.2*	13.9 ± 1.2*	0.0 ± 0.0	13.1 ± 1.3*
LPS 0.6 mg/kg	52.8 ± 5.2*	17.3 ± 0.4*	35.5 ± 3.2*	4.9 ± 1.2*
DMSO + LPS	40.7 ± 7.3*	13.3 ± 0.3*	27.4 ± 2.5*	4.2 ± 1.5*
L-NIL + LPS	61.1 ± 5.9**	19.7 ± 1.4*	40.2 ± 1.4*	4.3 ± 1.3*
L-NA + LPS	46.3 ± 3.7**	16.3 ± 0.2*	30.0 ± 0.2*	3.5 ± 0.7**
7-NI + LPS	70.9 ± 4.6**	16.1 ± 4.9**	54.8 ± 3.5**	3.4 ± 0.8**
ETOH + LPS	36.7 ± 6.9**	15.8 ± 1.7*	20.9 ± 1.6**	7.4 ± 1.6*

PBS, sterile phosphate-buffered saline; DMSO, dimethylsulfoxide; LNA, N^G-nitro-L-arginine; L-NIL, L-N^Giminoethyllysine; 7-NI, 7-nitroindazole; ETOH, ethanol; LPS, *Escherichia coli* endotoxin.
Bronchoalveolar lavage fluid and plasma cells were collected 5 hours after administration of the vehicle or drugs individually or concomitantly. All drugs except ETOH were administered 30 minutes prior to intratracheal administration of LPS or PBS. Ethanol was given by intraperstoned injection 30 minutes before intratracheal administration of LPS. *Differs from PBS (0.05); **differs from LPS in PBS-4 or DMSO-pretreated rats. From Greenberg et al. (1999a,b).

or peripheral content of PMNs, coadministration of 7-NI with L-NA enhanced the ability of L-NA to decrease the peripheral plasma content of circulating PMNs when compared with administration of L-NA alone. Moreover, the coadministration of these drugs increased the basal PMN content of the BAL fluid (Table 11.1). Thus, inhibition of nNOS with 7-NI has little effect on the basal adhesion of circulating PMNs to the vascular endothelium of the lung or peripheral circulation. However, when combined with inhibition of constitutive eNOS-derived NO, the PMNs become more adherent to the peripheral circulation and begin to become adherent within the lung.

Effect of Drugs on Lipopolysaccharide-Induced Changes in Polymorphonuclear Neutrophils

Although inhibition of nNOS alone produced little effect on the basal PMN content of plasma or the BAL fluid, inhibition of nNOS significantly modified the PMN response to endotoxin. Intratracheal administration of LPS stimulated PMN recruitment into the alveolar space of the rat lung and decreased the number of recoverable circulating PMNs in the plasma (Table 11.1). The latter effect appears to result from an early effect of LPS-mediated suppression of constitutuive eNOS activity in the region of the lung circulation (Parker et al., 1994; Wylam et al., 1990; Fugatsu et al., 1997; Liu et al., 1996; Greenberg et al., 1995). Five hours after intratracheal administration of LPS, the lung of the rat was congested, the alveolar space within each bronchus was reduced, and more PMNs were present when compared with rats given intratracheal PBS (Figure 11.7A,B). Pretreatment of rats with ethanol, which stimulates nNOS activity and inhibits LPS-inducible NOS transcription (Greenberg et al., 1997, 1998, 1999a,b; Zhao et al., 1999; Xie et al., 1995), attenuated LPS-stimulated PMN recruitment into the alveoli and lung interstitium (Table 11.1; Figure 11.7c) and thus masked the severity of the pneumonia produced by LPS. These effects of ethanol were not secondary to inhibition of NOS II, since pretreatment of rats with a dose of L-NIL sufficient to inhibit NOS II activity slightly enhanced LPS-stimulated lung recruitment of PMNs (Table 11.1; Figure 11.7E). Moreover, inhibition of constitutive eNOS by pretreatment of rats with L-NA prior to LPS administration also attenuated LPS-stimulated PMN recruitment into the lung (Figure 11.7D) and enhanced LPS-mediated decreases of circulating PMNs in peripheral plasma (Table 11.1). Since dimethylsulfoxide (DMSO) was the solvent for 7-NI, we compared the effects of DMSO on the PMN content of BAL fluid and plasma obtained from PBS- and LPS-treated rats. DMSO significantly attenuated LPS-mediated PMN recruitment into the rat lung (Table 11.1). However, systemic administration of DMSO did not affect LPS-mediated decreases in the circulating content of PMNs in peripheral plasma (Table 11.1). In contrast, inhibition of nNOS with 7-NI prior to intratracheal administration of LPS enhanced LPS-mediated lung recruitment of PMNs (Figure 11.7F) and decreases in the circulating PMNs present in peripheral plasma (Table 11.1).

Effect of Lipopolysaccharide and 7-NI in NOS I Knockout Mice

The previous experiments supported the concept that nNOS-derived NO modulated LPS-stimulated lung recruitment of PMNs. However, it was also possible that the changes in PMN recruitment reflected a nonspecific pharmacologic effect of 7-NI. Although transgenic mice may compensate with the up-regulation of one physiologic system for the deletion of another, we tested the concept that nNOS played a role in LPS-stimulated lung recruitment of PMNs in female homozygous NOS I knockout mice (BALB/B-6-129 NOS-1) and their wild-type controls (Jackson Laboratories, Bar Harbor, ME). Pretreatment of wild-type mice with LPS (1 mg/kg, intratracheally) produced significantly less lung recruitment of PMNs than

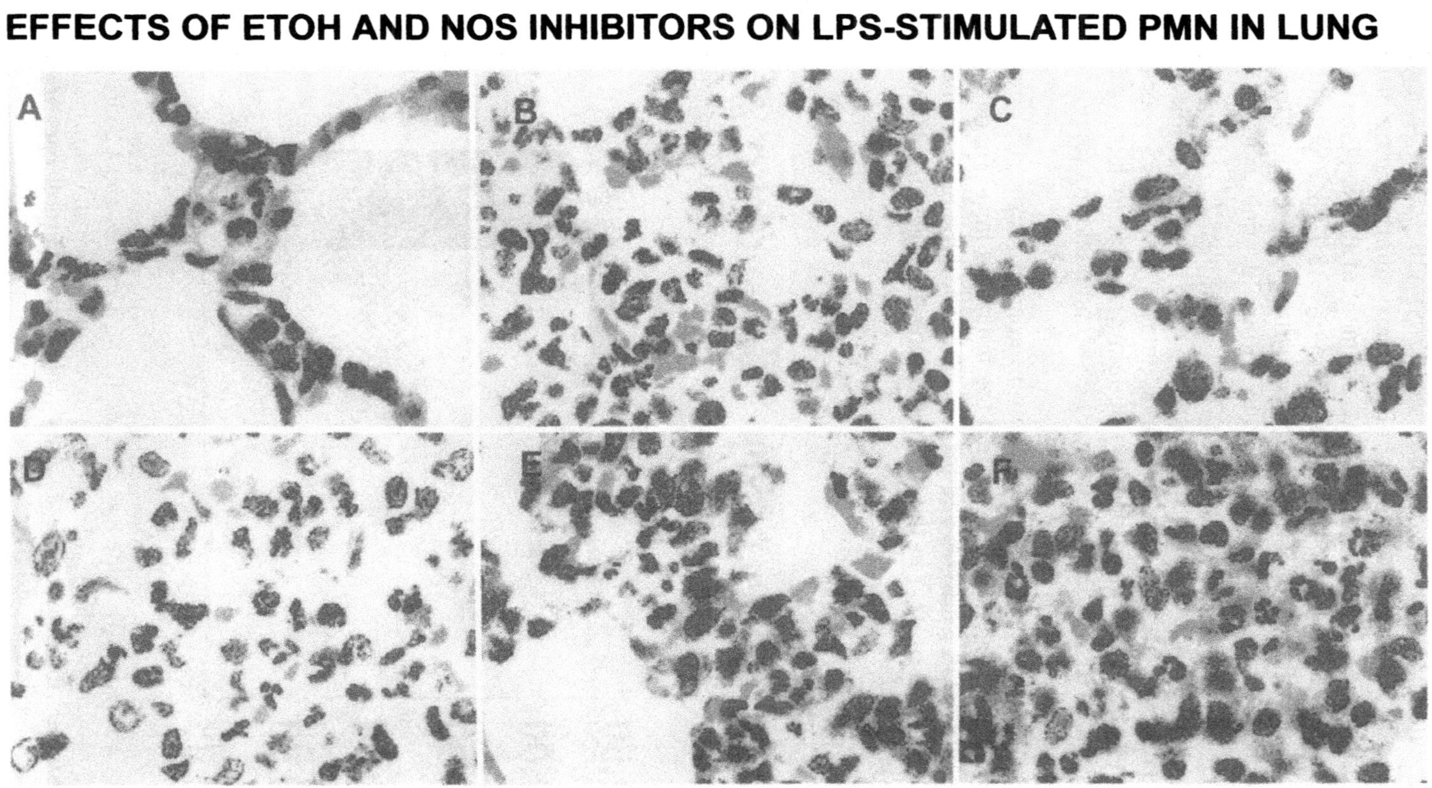

FIGURE 11.7. Hematoxylin and eosin-stained longitudinal sections of rat lungs with cellular inclusions obtained 5 hours after treatment of rats with (A) PBS (0.25 ml, intratracheal) (B) LPS (0.6 mg/kg, intratracheal), (C) LPS pretreated with ethanol (4.5 g/kg, intraperitoneal), (D) L-NA (20 mg/kg, intravenous), (E) L-NIL (10 mg/kg, intravenous), or (F) 7-NI (25 mg/kg, intravenous) given 30 minutes before LPS (0.6 mg/kg, intratracheal). Magnification ×160. From Greenberg et al. (1999a,b).

that observed in the female homozygous nNOS knockout mice (Figure 11.8). Pretreatment of wild-type mice with 7-NI (10, 25, or 40 mg/kg, intraperitoneally) 30 minutes prior to intratracheal administration of LPS (1 mg/kg) enhanced LPS-mediated lung recruitment of PMNs. None of the mice died from the 7-NI injections given alone or with LPS. Pretreatment of female homozygous nNOS knockout mice with 7-NI (25 and 40 mg/kg, intraperitoneally) killed 5 mice per group within 10 minutes after drug administration. The mice died from cyanosis and respiratory failure. In contrast, 7 mice per group tolerated treatment with 7-NI (10 mg/kg, intraperitoneally). However, pretreatment of female homozygous nNOS knockout with 7-NI (10 mg/kg, intraperitoneally) 30 minutes prior to intratracheal administration of LPS, failed to enhance LPS-stimulated lung recruitment of PMNs (Figure 11.9).

To verify the concept that these mice were homozygous relative to the deletion of the NOS gene, we examined the brain and lung homogenates of the mice for NOS I mRNA and protein. The wild-type mice expressed NOS I mRNA and protein in the brain and lung homogenates. However,

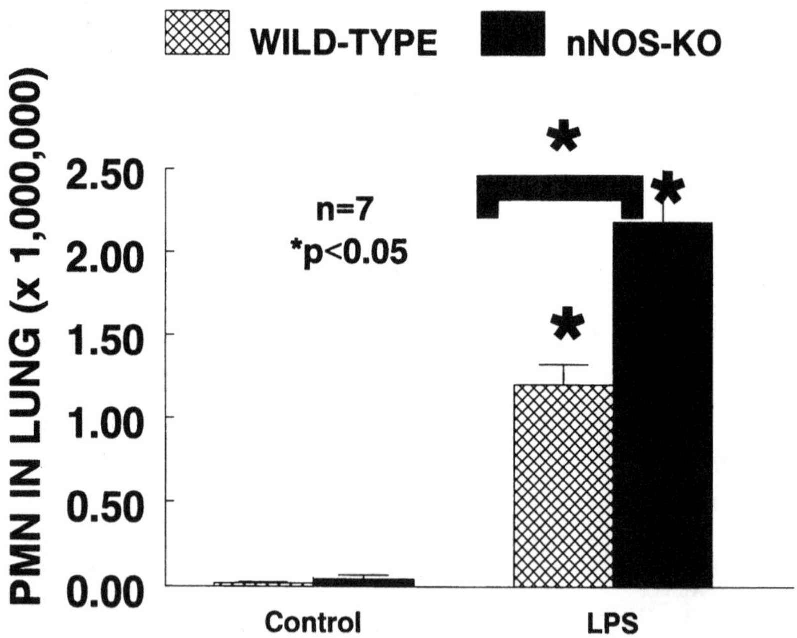

FIGURE 11.8. Mean and SD of the lung content of PMNs prior to and 4 hours after administration of intratracheal LPS (1 mg/kg) to wild-type mice and corresponding mice homozygous for nNOS gene deletion (nNOS-KO). Vertical bars are the numbers of PMNs (in millions) from 7 mice per group per time period. An asterisk denotes that the responses differ from those of the wild-type mice ($P < 0.05$).

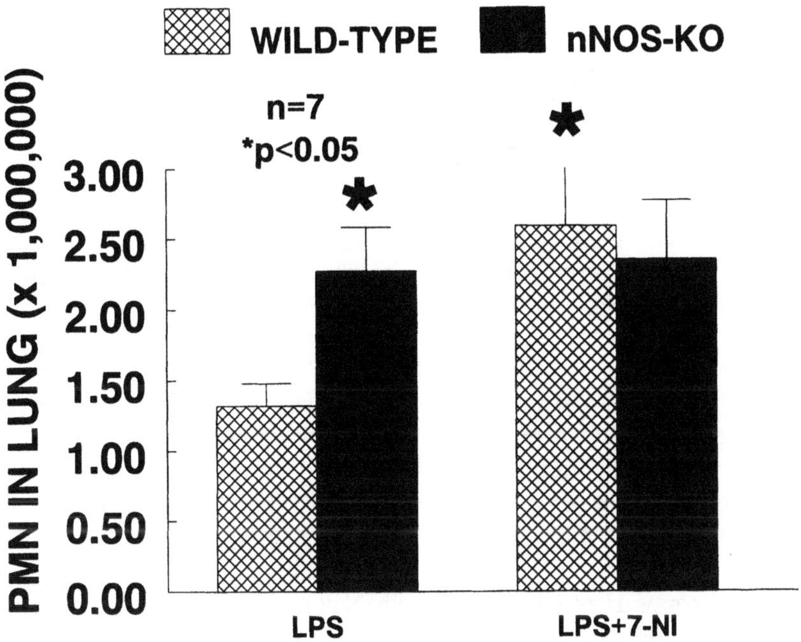

FIGURE 11.9. Mean and SD of the lung content of PMNs prior to and 4 hours after administration of intratracheal LPS (1 mg/kg) to wild-type mice and corresponding mice homozygous for nNOS gene deletion (nNOS-KO) pretreated with 7-NI (10, 25, or 40 mg/kg, intraperitoneal) 30 minutes prior to LPS administration. Vertical bars are the numbers of cells (in millions) from 5–7 mice per group per time period. An asterisk denotes that the responses in the presence of 7-NI differ from those in the absence of 7-NI ($P < 0.05$).

neither NOS I mRNA nor protein was detectable in brain or lung homogenate (Figure 11.10).

Neural Nitric Oxide Synthase and Polymorphonuclear Neutrophil Inactivation of *Klebsiella pneumoniae*

Neural NOS-derived NO appears to modulate lung PMN recruitment in response to LPS. But does this source of NO modulate the phagocytic function of the PMNs? Pretreatment of rats with L-NA inhibited nNOS activity within the lung, but not within PMNs, and failed to enhance LPS-stimulated lung recruitment of PMNs. Pretreatment of rats with 7-NI inhibited NOS l-derived NO production in PMNs and enhanced LPS-stimulated lung recruitment of PMNs. Thus, the data also suggest that nNOS in PMNs, rather than in the neurons of the lung, modulates LPS-stimulated recruitment of PMNs into the lung. Therefore, we would predict that unless

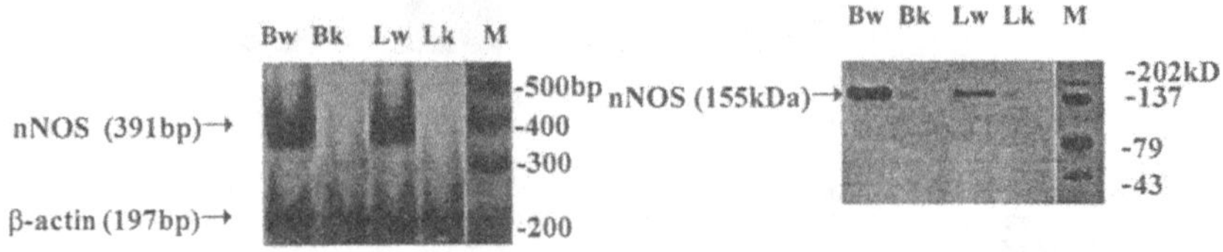

FIGURE 11.10. NOS I mRNA and NOS I protein were present in the brain and lung homogenates of wild-type mice but were undetectable in the brain and lung homogenates of the NOS I knockout mice.

nNOS within the PMN modulated PMN phagocytic activity, an increased lung content of PMNs would be associated with increased bacterial phagocytosis and killing. Therefore, we evaluated the ability of the lung to clear an infection of *Klebsiella pneumoniae.* Inhalation of aerosolized *K. pneumoniae* for 45 minutes did not result in a significant difference in the mean value of the initial colony-forming units of *K. pneumoniae* present in the two groups. However, the lung PMN content was higher in the female homozygous nNOS knockout mice than in the wild-type controls 4 hours after inhalation of the K. *pneumoniae.* Four hours after inhalation of the bacteria, the number of colony-forming units in the lungs of the wild-type mice was decreased compared with time zero. This indicated that the lungs were clearing the bacteria (Figure 11.11). In contrast, the lung content of *K. pneumoniae* increased in the knockout mice over the same time period (Figure 11.11). Thus, mice deficient in NOS I exhibited impaired lung clearance of *K. pneumoniae*, despite a greater content of PMNs within their lungs. This suggests that nNOS within the PMNs modulates not only PMN recruitment but also the phagocytic or bactericidal function of the PMNs.

Role of Neural Nitric Oxide Synthase in the Polymorphonuclear Neutrophils

Neural NOS, initially thought to be present only in nitric oxidergic nerves in the bronchial and airway smooth muscle of the lung, was believed to primarily modulate airflow, intrapulmonary oxygen distribution, and airway reactivity by opposing the bronchial constriction produced by endogenous circulating and neurogenic substances (Rand and Li, 1995). However, a complete and active nNOS system has now been described in rat, mouse, and human PMNs (Chen and Mehta, 1996; Greenberg et al., 1998, 1999a,b; Zhao et al., 1999). Moreover, the data suggest that endogenous nNOS present within the lungs and PMNs plays important roles as a modulator of PMN adhesion and recruitment and lung clearance of bacteria. Finally, as

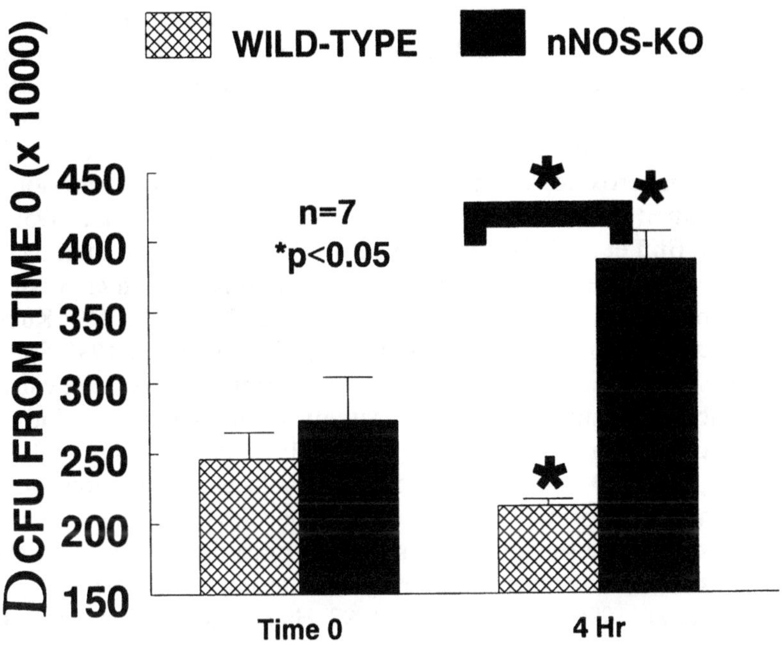

FIGURE 11.11. Deletion of NOS I gene impairs lung clearance of *Klebsiella pneumoniae*. The ordinate shows the colony-forming units of *K. pneumoniae* ($\times 10^3$) in the lung of wild-type mice and corresponding mice homozygous for nNOS gene deletion (nNOS-KO) 45 minutes and 4 hours 45 minutes hr after an aerosol challenge with *K. pneumoniae*. An asterisk indicates the mean values differ from the wild-type ($P < 0.05$).

described above, since L-NA-mediated inhibition of nNOS and NOS III within the lung and its endothelium attenuated LPS-induced lung recruitment of PMNs, whereas 7-NI-mediated inhibition of nNOS within PMNs augmented this effect of LPS, the data suggest that nNOS activity within PMNs, rather than nNOS-derived NO within the peripheral nerves and airways of the lung, modulates LPS-induced recruitment of PMNs within the capillary bed of the arteriolar microcirculation of the lung (Geiger et al., 1997). Speculatively, constitutive NO produced by nNOS within the PMNs may serve the function of the NO derived from constitutive eNOS in the postcapillary venules of the systemic circulation. Neural NOS-derived NO may prevent LPS-induced adherence of PMNs and monocytes to the pulmonary microvascular endothelium, in part by inducing the synthesis and nuclear translocation of the inhibitory transcription factor kappa B alpha (IκBα) within the PMNs. This would initially limit transcription of the β2-integrin CD18 (Zhang et al., 1998) and the selectins ICAM-1 and ICAM-2 on the PMNs, and thus prevent expression of these adhesion molecules within the endothelium and PMNs (Hogg and Duerscuk, 1995; De Caterina et al., 1995; Peng et al., 1995; Biffl et al., 1996; Spiecker et al., 1997). However,

LPS stimulates cytokine and chemokine formation, especially macrophage inflammatory proteins (MIP-1α, MIP-2), tumor necrosis factor alpha (TNFα), and interleukins. These substances can then release additional cytokines which up-regulate transcription of adhesion molecules. Alternatively, the cytokines may up-regulate the production of oxygen free radicals and hydrogen peroxide, which, in turn, can induce NFκB and adhesion molecule formation (Hogg and Duersuck, 1995; De Caterina et al., 1995; Peng et al., 1995; Biffl et al., 1996; Spiecker et al., 1997). This could then override the suppressant effect of endogenous nNOS-derived NO on expression of adhesion molecules within the PMNs (Hogg and Duerscuk, 1995; Kubes et al., 1991; De Caterina et al., 1995; Peng et al., 1995; Biffl et al., 1996; Spiecker et al., 1997; Hickey and Kubes, 1997). In vivo administration of LPS can completely inhibit constitutive NOS III activity in the vascular endothelium (Wang and Marsden, 1995; Xie and Nathan, 1994; Forstermann and Kleinert, 1995), and NO can prevent adhesion molecule expression, but NO cannot inhibit the formation of adhesion molecules once their synthesis is activated or destroy the adhesion molecules once they are formed (Hogg and Duerscuk, 1995; Kubes et al., 1991; De Caterina et al., 1995; Peng et al., 1995; Biffl et al., 1996; Spiecker et al., 1997; Hickey and Kubes, 1997). Thus, nNOS-derived NO within the PMNs may take over, in the lung, the function of the deficient eNOS and modulate adhesion molecule formation within the PMNs and endothelium. Thus, nNOS-derived NO, probably within the PMNs, appears to be involved in the lung recruitment of PMNs. PMNs can generate toxic oxygen radicals, as well as inflammatory oxidants derived from NO, which can damage the lung (Gross and Wolin, 1995; Hogg and Duerscuk, 1995; McCall et al., 1989; Swallow et al., 1991). Thus, we postulate that endogenous nNOS-derived NO in the PMNs may limit bacteria- and endotoxin-mediated lung recruitment of PMNs and thereby limit the tissue damage associated with excessive PMN activity during infection and inflammation. Conversely, impairment of nNOS activity during lung infections or lung inflammation may result in excessive PMN migration into the lung and exacerbation of lung injury.

Interaction of Neural and Endothelial Nitric Oxide Synthase in Lung Recruitment of Polymorphonuclear Neutrophils

Pretreatment of rats with L-NA does not inhibit nNOS within the PMNs, despite its ability to inhibit nNOS activity in neurons within the lung and airways and constitutive eNOS in the vascular endothelium in the absence or presence of LPS (Parker et al., 1994; Wylam et al., 1990; Greenberg et al., 1995). Nevertheless, inhibition of nNOS and eNOS within the pulmonary vascular endothelium would be expected to enhance PMN adhesion to the pulmonary vascular endothelium and increase both basal and LPS-stimulated lung recruitment of PMNs (Wang and Marsden, 1995; Xie and Nathan, 1994; Forstermann and Kleinert, 1995; Gross and Wolin, 1995; Hogg and Duerscuk, 1995; Kubes et al., 1991; Hickey and Kubes, 1997). Although pretreatment of rats with L-NA decreased the circulating PMNs in normal rats, it did not affect the lung content of PMNs. Moreover, pretreatment of rats with L-NA enhanced LPS-mediated decreases in circulating PMNs but paradoxically attenuated LPS-stimulated lung recruitment of PMNs. This can only be explained by suggesting that L-NA-mediated inhibition of constitutive eNOS in the systemic circulation increased adhesion of the PMNs to the peripheral vascular postcapillary endothelium and therefore decreased the content of circulating PMNs subject to lung recruitment in the LPS-treated rats. However, since LPS also inhibits constitutive eNOS activity and increases PMN adhesion to the peripheral vascular endothelium, L-NA-mediated enhancement of LPS-stimulated decreases in circulating and lung PMNs could only be possible if 1) a deficiency of NO derived from eNOS existed in the lung microcirculation (Geiger et al., 1997); 2) an endothelial-derived antiadhesion mediator (e.g., a prostanoid) was unregulated and assumed the role of endogenous NO (Hogg and Duerscuk, 1995; Fukatsu et al., 1997); or 3) nNOS-derived NO within the lung and/or PMNs functioned to down-regulate adhesion molecule formation within both the PMNs and the pulmonary microvascular endothelium. Coadministration of 7-NI and L-NA, to inhibit both nNOS within PMNs and nNOS and eNOS within the lung and systemic circulation, resulted in spontaneous accumulation of PMNs within the normal lung. Moreover, coadministration of these drugs enhanced LPS-stimulated lung recruitment of PMNs more than that observed upon pretreatment of LPS-treated rats with either 7-NI or L-NA alone. Thus, the data clearly suggest that nNOS-derived NO within the PMNs prevents PMN sequestration in the periphery and lung. However, when peripheral eNOS activity is inhibited, lung recruitment of PMNs may be impaired because of the peripheral sequestration of large numbers of PMNs. Thus, endogenous nNOS-derived NO in the PMNs may limit PMN adhesion to the pulmonary endothelium and thereby lung recruitment of PMNs when eNOS activity is compromised by LPS. Further studies are required to test this postulate.

Effect of Ethanol on NOS I-Derived Nitric Oxide

Neural NOS was the only NO-producing enzyme present in circulating PMNs in the absence of LPS treatment (Chen and Mehta, 1996; Greenberg et al., 1998). In vivo and in vitro administration of ethanol increased the spontaneous production of nitrate and nitrite anions by PMNs. However, ethanol did not induce nNOS mRNA or protein. Moreover, this effect of ethanol was inhibited after coincubation of PMNs with 7-NI and ethanol but not after coincubation of PMNs with L-NIL and ethanol. Therefore, we must conclude that ethanol up-regulated the basal enzymatic activity of nNOS in the PMN. This is in agreement with the ability of ethanol to up-regulate nNOS activity in neurons (Chandler et al., 1997; Fitzgerald et al., 1995). Ethanol may increase nNOS activity in the PMNs by increasing protein kinase cδ (PKCδ) in the membrane of PMNs, which then can increase the intracellular concentration of calcium ion by stimulation of calcium influx through L-type calcium channels (Blackwell et al., 1996; Charness, 1996). Alternatively, ethanol may decrease the inactivation of NOS I by suppression of superoxide formation in PMNs (Nagy, 1996). Further studies are required to resolve these and other possible mechanisms of ethanol-mediated up-regulation of nitrate and nitrite anions by circulating PMNs.

Lessons from Neural NOS Knockout Mice

The data obtained from the rat studies with 7-NI and ethanol clearly support a role for nNOS-derived NO from PMNs as a modulator of stimulated lung recruitment of PMNs and PMN function. However, the use of inhibitors and activators always raises the caveat that the observed effects may reflect idiosyncratic or nonspecific effects of the drugs under study. Although transgenic and knockout mice come complete with their own set of caveats and problems, they are usually looked at as a model in which there is no longer need for the study of antagonists. However, we have come to view the knockout mouse as a model in which the specificity and selectivity of a pharmacologic antagonist can also be evaluated. This is because the purported system through which the antagonist functions should no longer exist in the knockout animal. Thus, the use of a knockout animal should aid in defining the effects of a drug unrelated to its primary pharmacologic action. The data obtained from the studies with the homozygous nNOS knockout mice clearly support a role for nNOS as a modulator of LPS-stimulated lung recruitment of PMNs. This conclusion is based on the findings that intratracheal administration of LPS in homozygous nNOS knockout mice increased the BAL recoverable PMNs from the lung significantly more than in wild-type mice. Moreover, 7-NI-mediated enhancement of LPS-stimulated lung recruitment of PMNs was related to inhibition of NOS I, since this effect occurred in the wild-type mice but was not observed in the nNOS knockout mice. Also, the data suggest that since the lung clearance of *K. pneumoniae* was significantly reduced in the nNOS knockout mice relative to the wild-type controls, absence of nNOS in the recruited PMNs impairs the ability of the PMNs to carry out phagocytosis and bacterial killing. This conclusion is based on the observation that the PMN content of the NOS I knockout mice was greater than that of the wild-type mice. However, the mechanism of these effects must be defined by future studies.

Potential Problems with Studies

For a hypothesis to be acceptable, it should explain disparate findings of studies that appear to be in conflict. Previous studies demonstrated that intravenous administration of L-arginine protected the lung against both LPS-stimulated lung recruitment of PMNs and endothelial dysfunction (Sheridan et al., 1996, 1998) and enhanced bacterial phagocytosis in human PMNs (Moffat et al., 1996). This would appear to be in conflict with our findings that administration of L-NA, a mixed inhibitor of constitutive and iNOS, inhibited LPS-induced lung recruitment of PMNs. L-Arginine, the amino acid precursor for NO synthesis by each of the three isozymes of NOS, prevents PMN accumulation in the lung only when a deficiency of L-arginine occurs in the plasma or endothelium as a result of increased utilization or decreased production of L-arginine (Wang and Marsden, 1995; Xie and Nathan, 1994; Forstermann and Kleinert, 1995; Gross and Wolin, 1995; Rand and Li, 1995). The plasma and tissue concentrations of L-arginine are rate-limiting factors in eNOS-derived NO synthesis during vascular injury induced by substances such as TNFα, cytokines, LPS, and other inflammatory stimuli (Wang and Marsden, 1995; Xie and Nathan, 1994; Forstermann and Kleinert, 1995; Gross and Wolin, 1995; Rand and Li, 1995; Kubes et al., 1991; Hickey and Kubes, 1997). Administration of L-arginine usually has little effect on eNOS-derived NO synthesis in blood vessels and smooth muscle with an intact endothelium obtained from normal animals and humans (Wang and Marsden, 1995; Xie and Nathan, 1994; Forstermann and Kleinert, 1995; Gross and Wolin, 1995; Rand and Li, 1995; Kubes et al., 1991; Hickey and Kubes, 1997). Thus, L-arginine-mediated increases in NO and suppression of lung recruitment of PMNs would not be unexpected under these conditions. This finding is not in conflict with L-NA-mediated suppression of LPS-stimulated lung recruitment of PMNs. Pretreatment of rats with L-NA appeared to cause the peripheral sequestration of PMNs. Since L-NA cannot inhibit nNOS in circulating PMNs, this system should remain active and attenuate LPS-mediated adhesion of PMNs to the pulmonary microvascular endothelium. However, nNOS-derived NO appeared insufficient to attenuate adhesion of PMNs to the systemic vascular endothelium, which appears to be controlled primarily by eNOS-derived NO 1 within the vascular endothelium (Wang and Marsden, 1995; Xie and Nathan, 1994; Forstermann and Kleinert, 1995; Kubes et al., 1991; Hickey and Kubes, 1997). This supposition is also demonstrated by the finding that 7-NI-mediated inhibition of basal PMN production of nitrate and nitrite anion failed to stimulate lung recruitment of PMNs or decrease circulating PMNs.

Endothelial NOS is limited in the pulmonary microvascular circulation and more abundant in the systemic arterial and venous circulation. In the lung, constitutive nNOS exists within the PMNs and in the nitric oxidergic nerves innervating the bronchi and pulmonary vasculature. Constitutive eNOS is abundantly present in the pulmonary and bronchial macrovascular endothelium and the systemic macro- and microvascular endothelium, but not in the pulmonary microvascular endothelium (Geiger et al., 1997). Thus, systemic administration of L-NA probably decreases LPS-stimulated lung recruitment of PMNs due to a physical decrease in the number of circulating PMNs as a result of the inhibition of eNOS within, and sequestration of PMNs by, the peripheral vascular endothelium. Although the lung is constantly exposed to bacteria and inhaled microparticulate matter (Hogg and Duerscuk, 1995), these foreign substances are usually engulfed and inactivated by the front-line defense mechanism of the lung, the alveolar macrophage. This spares the constant activation of PMNs (Hogg and Duerscuk, 1995). Thus, in the absence of an inflammatory stimulus, the synthesis of adhesion molecules within the PMNs in the lung is usually not upregulated. Thus, inhibition of nNOS alone should not result in PMN adhesion as long as eNOS is continually produced in the systemic vasculature. This conclusion is supported by the studies in which coadministration of 7-NI and L-NA to PBS-treated rats stimulated lung PMN recruitment and enhanced L-NA-mediated decreases in the amount of recoverable circulating PMNs in plasma.

The conclusion that ethanol directly increases nNOS enzymatic activity in control rats cannot be challenged, since no other measurable NOS isozyme exists. However, the conclusion that this down-regulates adhesion molecule formation on the PMNs and accounts, in part, for ethanol-mediated suppression of LPS-stimulated lung recruitment of PMNs may be challenged, since we have not measured adhesion molecule formation. However, Zhang et al. (1998) demonstrated that acute ethanol administration decreases PMN β2-integrin formation following induction of endotoxemia with LPS. Further studies are now required to demonstrate that this effect of ethanol is mediated through nNOS-derived NO. In addition, our conclusions that ethanol increases nNOS activity in PMNs may be challenged on the basis that pretreatment of rats with ethanol slightly decreased the NOS I mRNA and protein content of LPS-stimulated PMNs in the lung. Ethanol, however, increased the basal production of nitrate and nitrite anion by PMNs. These events are not mutually exclusive, since constitutive nNOS enzymatic activity can be significantly increased by elevations in intracellular calcium ion, tetrahydrobiopterin, and arginine, as well as by decreases in oxygen free radicals (Wang and Marsden, 1995; Xie and Nathan, 1994; Forstermann and Kleinert, 1995; Gross and Wolin, 1995). Ethanol can also increase intracellular calcium ion and alter intracellular amino acid concentrations as well as decrease superoxide formation in

PMNs (Blackwell et al., 1996; Charness, 1996; Nagy, 1996; Dorio and Forman, 1988). Thus, the effects of ethanol on NOS I mRNA and protein in LPS-stimulated PMNs are not inconsistent with ethanol-mediated up-regulation of NOS l-derived enzyme activity.

Conclusions

The data presented here clearly support our postulate that PMN-derived NO modulates the recruitment and sequestration of PMNs into the lung during lung infections and inflammation. Although under normal conditions either neural or endothelial NOS-derived NO appears sufficient to prevent or limit PMN sequestration by the lung, peripheral sequestration of PMNs appears dependent on eNOS-derived NO in the vascular endothelium. However, under conditions of infection and inflammation, nNOS-derived NO, probably produced by the PMNs, appears to limit PMN recruitment into the lung and thereby the tissue toxicity associated with PMN-generated toxic oxygen and NO-derived inflammatory oxidants, which can damage the lung.

References

Barna, M., Komatsu, T., and Reiss, C.S. 1996. Activation of type III nitric oxide synthase in astrocytes following a neurotropic viral infection. *Virology* 223:331–343.

Biffl, W.L., Moore, E.E., Moore, F.A., and Barnett, C. 1996. Nitric oxide reduces endothelial expression of intercellular adhesion molecule (ICAM)-1. *J. Surg. Res.* 63:328–332.

Blackwell, T.S., Blackwell, T.R., Holden, E.P., Christman, B.W., and Christman, J.W. 1996. In vivo antioxidant treatment suppresses nuclear factor-kappa B activation and neutrophilic lung inflammation. *J. Immunol.* 157:1630–1637.

Bredt, D.S., and Schmidt, H.H.H.W. 1996. The citrulline assay. In: *Methods in Nitric Oxide Research* (eds. Feelisch, M. and Stamler, J.S.), pp. 249–258. John Wiley and Sons, New York.

Bryant, J.L., Jr., Mehta, P., Von der Porten, A., and Mehta, J.L. 1992. Co-purification of 130 kD nitric oxide synthase and a 22 kD link protein from human neutrophils. *Biochem. Biophys. Res. Commun.* 189:558–564.

Capanni, C., Squarzoni, S., Petrini, S., Villanova, M., Muscari, C., Maraldi, N.M., Guarnieri, C., and Caldarera, C.M. 1998. Increase of neuronal nitric oxide synthase in rat skeletal muscle during ageing. *Biochem. Biophys. Res. Commun.* 245:216–219.

Chandler, L.J., Sutton, G., Norwood, D., Sumners, C., and Crews, F.T. 1997. Chronic ethanol increases *N*-methyl-D-aspartate-stimulated nitric oxide formation but not receptor density in cultured cortical neurons. *Mol. Pharmacol.* 51:733–740.

Charness, M.E. 1996. Direct effects of ethanol on signaling proteins. *Alcohol. Clin. Exp. Res.* 20:157A–161A.

Chen, L.Y., Lawson, D.L., and Mehta, J.L. 1994. Reduction in human neutrophil superoxide anion generation by n-3 polyunsaturated fatty acids: role of cyclooxygenase products and endothelium-derived relaxing factor. *Thrombosis Res.* 76:317–322.

Chen, L.Y., and Mehta, J.L. 1996. Variable effects of L-arginine analogs on L-arginine-nitric oxide pathway in human neutrophils and platelets may relate to different nitric oxide synthase isoforms. *J. Pharmacol. Exp. Ther.* 276:253–257.

Clancy, R.M., Leszczynska-Piziak, J., and Abramson, S.B. 1992. Nitric oxide, an endothelial cell relaxation factor, inhibits neutrophil superoxide anion production via a direct action on the NADPH oxidase. *J. Clin. Invest.* 90:1116–1121.

De Caterina, R., Libby, P., Peng, H.-B., Thannickal, V.J., Rajavashisth, T.B., Gimbrone, M.A., Jr., Shin, W.S., and Liao, J.K. 1995. Nitric oxide decreases cytokine-induced endothelial activation. Nitric oxide selectively reduces endothelial expression of adhesion molecules and proinflammatory cytokines. *J. Clin. Invest.* 96:60–68.

Dietzschold, B. 1995. The role of nitric oxide in the pathogenesis of virus-induced encephalopathies. *Curr. Top. Microbiol. Immunol.* 196:51–56.

Dorio, R.J., and Forman, H.J. 1988. Ethanol inhibition of signal transduction in superoxide production by rat alveolar macrophages. A proposed mechanism for ethanol related pneumonia. *Ann. Clin. Lab. Sci.* 18:190–194.

El-Dwairi, Q., Comtois, A., Guo, Y., and Hussain, S.N.A. 1998. Endotoxin-induced skeletal muscle contractile dysfunction: contribution of nitric oxide synthases. *Am. J. Physiol. Cell Physiol.* 274:C770–C779.

Fitzgerald, L.W., Charlton, M.E., Duman, R.S., and Nestler, E.J. 1995. Regulation of neuronal nitric oxide synthase by chronic ethanol ingestion. *Synapse* 21:93–95.

Forstermann, U., and Kleinert, H. 1995. Nitric oxide synthase: expression and expressional control of the three isoforms. *Naunyn-Schmiedebergs Arch. Pharmacol.* 352:351–364.

Fukatsu, K., Saito, H., Han, I., Furukawa, S., Hashiguchi, Y., Lin, M.T., Matsuda, T., Inaba, T. Inoue, T., Ikeda, S., Yasuhara, H., and Muto, T. 1997. Nitric oxide inhibition decreases neutrophil adhesion at the inflammatory site, while increasing adhesion in remote organs in peritonitis. *J. Surg. Res.* 68:79–86.

Geiger, M., Stone, A., Mason, S.N., Oldham, K.T., and Guice, K.S. 1997. Differential nitric oxide production by microvascular and macrovascular endothelial cells. *Am. J. Physiol. Lung Cell. Mol. Physiol.* 273:L275–L281.

Greenberg, S.S., Ouyang, J., Zhao, X., and Giles, T.D. 1998. Human and rat neutrophils constitutively express neural nitric oxide synthase (nNOS) mRNA. *Nitric Oxide Biol. Chem.* 2:162–172.

Greenberg, S.S., Ouyang, J., Zhao, X., Parrish, C., Nelson, S., Giles, T. 1999a. Effect of ethanol on neutrophil recruitment and lung host defense in nitric oxide synthase I and nitric oxide synthase II knockout mice. *Alcoholism, Clinical & Experimental Pesearch* 23:1435–1445.

Greenberg, S.S., Ouyang, J., Zhao, X., Wang, J.F., and Giles, T.D. 1999b. Effects of NO synthase inhibitors and NOS-I deficient mice on lung neutrophils and clearance of *K. pneumoniae*. *Am. J. Physiol. Lung Cell. Molec. Physiol.* 275:L••–L••, accepted.

Greenberg, S.S., Ouyang, J., Zhao, X., Xie, J., Wang, J.F., and Giles. T.D. 1997. Interaction of ethanol with inducible NOS mRNA and protein: direct effects on autacoids and endotoxin in vivo. *J. Pharmacol. Exp. Ther.* 282:1044–1054.

Greenberg, S.S., Xie, J., Joseph, K.O., Kolls, J., and Summer, W.R. 1995. In vivo administration of endotoxin and tumor necrosis factor-alpha produce different effects on constitutive and inducible nitric oxide synthase activity in rat neutrophils and aorta ex vivo. *Proc. Soc. Exp. Biol. Med.* 208:199–208.

Greenberg, S.S., Xie, J., Zhao, X., Ouyang, J., and Giles, T.D. 1996. An in vivo cytokine and endotoxin-independent pathway for induction of nitric oxide synthase II mRNA, enzyme, and nitrate/nitrite in alveolar macrophages. *Biochem. Biophys. Res. Commum.* 327:160–167.

Gross, S.S., and Wolin, M.S. 1995. Nitric oxide: pathophysiologic mechanisms. *Annu. Rev. Physiol.* 57:737–770.

Hickey, M.J., and Kubes, P. 1997. Role of nitric oxide in regulation of leucocyte-endothelial cell interactions. *Exp. Physiol.* 82:339–348.

Hogg, J.C., and Duerscuk, C.M. 1995. Leukocyte traffic in the lung. *Annu. Rev. Physiol.* 57:97–112.

Hoshiga, M., Alpers, C.E., Smith, L.L., Giachelli, C.M., and Schwartz, S.M. 1995. $\alpha_v\beta_3$ Integrin expression in normal and atherosclerotic artery. *Circ. Res.* 77:1129–1135.

James, S.L. 1995. Role of nitric oxide in parasitic infections. *Microbiol. Rev.* 59:533–547.

Kadota, K., Yui, Y., Hattori, R., Uchizumi, H., and Kawai, C. 1991. A new relaxing factor in supernatant of incubated rat peritoneal neutrophils *Am. J. Physiol.* 260:H967–H972.

Kaplan, S.S., Lancaster, J.R., Jr., Basford, R.E., and Simmons, R.L. 1996. Effect of nitric oxide on staphylococcal killing and interactive effect with superoxide. *Infect. Immun.* 64:69–76.

Kubes, P., Suzuki, M., and Granger, D.N. 1991. Nitric oxide: an endogenous modulator of leukocyte adhesion. *Proc. Natl. Acad. Sci. USA* 88:4651–4655.

Kuo, H.P., Hwang, K.H., Lin, H.C., Wang, C.H., and Lu, L.C. 1997. Effect of endogenous nitric oxide on tumour necrosis factor-alpha-induced leukosequestration and IL-8 release in guinea-pigs airways in vivo. *Br. J. Pharmacol.* 122:103–111.

Kwon, G., Corbett, J.A., and McDaniel, M.L. 1996. Interleukin 1-induced Fos and Jun do not regulate inducible nitric oxide synthase in rat islets of Langerhans and RINm5F cells. *Endocrinology* 137:825–830.

Lee, D.K., Faunce, D., Henry, D., Sturm, R.J., and Rimele, T. 1990. Neutrophil-derived relaxing factor relaxes vascular smooth muscle through a cGMP-mediated mechanism *Life Sci.* 46:1531–1538.

Liu, J.J., Chen, J.R., Bradley, C.J., Xie, B., Johnston, C.I., and Buxton, B.F. 1994. Autologous neutrophil derived supernatants inhibit endothelium dependent relaxation in human coronary bypass graft. *Cardiovasc. Res.* 28:1353–1359.

Liu, S.F., Adcock, I.M., Old, R.W., Barnes, P.J., and Evans, T.W. 1996. Differential regulation of the constitutive and inducible nitric oxide synthase mRNA by lipopolysaccharide treatment in vivo in the rat. *Crit. Care Med.* 24:1219–1225.

Lu, Y.T., Liu, S.F., Mitchell, J.A., Malik, A.B., Hellewell, P.G., and Evans, T.W. 1998. The role of endogenous nitric oxide in modulating ischemia-reperfusion injury in the isolated, blood-perfused rat lung. *Am. J. Respir. Crit. Care Med.* 157:273–279.

Maines, M.D. 1996. Nitric oxide synthase: characterization and functional analyses. Academic Press, San Diego, pp. 3–345.

McCall, T.B., Boughton-Smith, N.K., Palmer, R.M., Whittle, B.J., and Moncada, S. 1989. Synthesis of nitric oxide from L-arginine by neutrophils. Release and interaction with superoxide anion. *Biochem. J.* 261:293–296.

McCall, T.B., Palmer, R.M., and Moncada, S. 1991. Induction of nitric oxide synthase in rat peritoneal neutrophils and its inhibition by dexamethasone. *Eur. J. Immunol.* 21:2523–2527.

McIlveen, S., White, S., and Parsons, G. 1997. Autonomic control of bronchial circulation in awake sheep during rest and behavior. *Clin. Exp. Pharmacol. Physiol.* 24:940–947.

Miles, P.R., Bowman, L., Rengasamy, A., and Huffman, L. 1998. Constitutive nitric oxide production by rat alveolar macrophages. *Am. J. Physiol. Lung Cell. Mol. Physiol.* 274:L360–L368.

Moffat, F.L., Jr., Han, T., Li, Z.M., Peck, M.D., Jy, W.C., Ahn, Y.S., Chu, A.J., and Bourguignon, L.Y.W. 1996. Supplemental L-arginine HCl augments bacterial phagocytosis in human polymorphonuclear leukocytes. *J. Cell. Physiol.* 168:26–33.

Moore, P.K., and Bland-Ward, P.A. 1996. 7-Nitroindazole: an inhibitor of nitric oxide synthase. *Meth. Enzymol.* 268:393–398.

Moore, W.M., Webber, R.K., Jerome, G.M., Tjoeng, F.S., Misko, T.P., and Currie, M.G. 1994. L-*N*6-(1-iminoethyl)lysine: a selective inhibitor of inducible nitric oxide synthase. *J. Med. Chem.* 37:3886–3888.

Murad, F. 1994. Cyclic GMP: synthesis, metabolism, and function. Introduction and some historical comments. *Adv. Pharmacol.* 26:1–5.

Murad, F., Forstermann, U., Nakane, M., Pollock, J., Tracey, R., Matsumoto, T., and Buechler, W. 1993. The nitric oxide-cyclic GMP signal transduction system for intracellular and intercellular communication. *Adv. Second Messenger Phosphoprotein Res.* 28:101–109.

Nagy, L.E. 1996. Alcohol and tyrosine kinase/phosphatase mediated processes. *Alcohol. Clin. Exp. Res.* 20:106A–108A.

Nichols, W.W., Mehta, J.L., Donnelly, W.H., Lawson, D., Thompson, L., and ter Riet, M. 1988. Reduction in coronary vasodilator reserve following coronary occlusion and reperfusion in anesthetized dog: role of endothelium-derived relaxing factor, myocardial neutrophil infiltration and prostaglandins. *J. Molec. Cell. Cardiol.* 20:943–954.

Parker, J.L., Myers, P.R., Zhong, Q., Kim, K., and Adams, H.R. 1994. Inhibition of endothelium-dependent vasodilation by *Escherichia coli* endotoxemia. *Shock* 2:451–458.

Peng, H.B., Libby, P., and Liao, J.K. 1995. Induction and stabilization of I kappa B alpha by nitric oxide mediates inhibition of NF-kappa B. *J. Biol. Chem.* 270:14214–14219.

Rand, M.J., and Li, C.G. 1995. Nitric oxide as a neurotransmitter in peripheral nerves: nature of transmitter and mechanism of transmission. *Annu. Rev. Physiol.* 57:659–683.

Rubanyi, G.M., Ho, E.H., Cantor, E.H., Lumma, W.C., and Botelho, L.H. 1991. Cytoprotective function of nitric oxide: inactivation of superoxide radicals produced by human leukocytes. *Biochem. Biophys. Res. Commun.* 181:1392–1397.

Schütte, H., Mayer, K., Gessler, T., Rühl, M., Schlaudraff, J., Burger, H., Seeger, W., and Grimminger, F. 1998. Nitric oxide biosynthesis in an exotoxin-induced septic lung model—role of cNOS and impact on pulmonary hemodynamics. *Am. J. Respir. Crit. Care Med.* 157:498–504.

Schwartz, S.M., Reidy, M.A., and De Blois, D. 1996. Factors important in arterial narrowing. *J. Hypertens.* 14:S71–S81.

Sheridan, B.C., McIntyre, R.C., Jr., Agrafojo, J., Meldrum, D.R., Meng, X., and Fullerton, D.A. 1996. Neutrophil depletion attenuates endotoxin-induced dysfunction of cGMP-mediated pulmonary vasorelaxation. *Am. J. Physiol. Lung Cell. Mol. Physiol.* 271:L820–L828.

Sheridan, B.C., McIntyre, R.C., Jr., Meldrum, D.R., and Fullerton, D.A. 1998. L-Arginine prevents lung neutrophil accumulation and preserves pulmonary endothelial function after endotoxin. *Am. J. Physiol. Lung Cell. Mol. Physiol.* 274:L337–L342.

Spiecker, M., Peng, H.B., and Liao, J.K. 1997. Inhibition of endothelial vascular cell adhesion molecule-1 expression by nitric oxide involves the induction and nuclear translocation of IkappaBα. *J. Biol. Chem.* 272:30969–30974.

Swallow, C.J., Grinstein, S., Sudsbury, R.A., and Rotstein, O.D. 1991. Nitric oxide derived from L-arginine impairs cytoplasmic pH regulation by vacuolar-type H^+ ATPases in peritoneal macrophages. *J. Exp. Med.* 174:1009–1021.

Wang, Y., and Marsden, P.A. 1995. Nitric oxide synthases: gene structure and regulation. *Adv. Pharmacol.* 34:71–90.

Wood, K.S., Buga, G.M., Byrns, R.E., and Ignarro, L.J. 1990. Vascular smooth muscle-derived relaxing factor (MDRF) and its close similarity to nitric oxide. *Biochem. Biophys. Res. Commun.* 170:80–88.

Wylam, M.E., Samsel, R.W., Umans, J.G., Mitchell, R.W., Leff, A.R., and Schumacker, P.T. 1990. Endotoxin in vivo impairs endothelium-dependent relaxation of canine arteries in vitro. *Am. Rev. Resp. Dis.* 142:1263–1267.

Xie, J., Kolls, J.K., Bagby, G., and Greenberg, S.S. 1995. Independent suppression of nitric oxide and TNF-α in the lung of conscious rats by ethanol. *FASEB J.* 9:253–261.

Xie, Q., and Nathan, C. 1994. The high-output nitric oxide pathway: role and regulation. *J. Leukocyte Biol.* 56:576–582.

Yui, Y., Hattori, R., Kosuga, K., Eizawa, H., Hiki, K., Ohkawa, S., Ohnishi, K., Terao, S., and Kawai, C. 1991. Calmodulin-independent nitric oxide synthase from rat polymorphonuclear neutrophils. *J. Biol. Chem.* 266:3369–3371.

Zaharchuk, G., Hara, H., Huang, P.L., Fishman, M.C., Moskowitz, M.A., Jenkins, B.G., and Rosen, B.R. 1997. Neuronal nitric oxide synthase mutant mice show smaller infarcts and attenuated apparent diffusion coefficient changes in the peri-infarct zone during focal cerebral ischemia. *Magn. Reson. Med.* 37:170–175.

Zhang, P., Bagby, G.J., Xie, M., Stoltz, D.A., Summer, W.R., and Nelson, S. 1998. Acute ethanol intoxication inhibits neutrophil β_2-integrin expression in rats during endotoxemia. *Alcohol. Clin. Exp. Res.* 22:135–141.

12
Nitric Oxide Toxicity in Neuronal Injury and Degeneration

ALVARO G. ESTÉVEZ, LILIANA VIERA, ANDRÉS KAMAID, AND JOSEPH S. BECKMAN

The aim of this chapter is to analyze growing evidence suggesting that the interaction between oxidative stress, nitric oxide (NO), and peroxynitrite has a role in the induction of motor neuron death during development, after injury, and in pathological conditions such as amyotrophic lateral sclerosis (ALS). We also speculate about the mechanisms of motor neuron death induced by ALS mutant superoxide dismutases (SOD).

Nitric oxide is a widely distributed intercellular messenger with a variety of physiological functions (Ignarro, 1990). Nitric oxide was first identified as the endothelial factor responsible for the relaxation of smooth muscle (Ignarro et al., 1987; Palmer et al., 1987). Shortly after, NO was linked to glutamatergic neurotransmission in the nervous system (Garthwaite et al., 1988) in regulating processes such as long-term potentiation and depression (Schuman and Madison, 1994; Zhang and Snyder, 1995). Nitric oxide also has a role in excitotoxicity (Dawson et al., 1991, 1993). Although a role for NO in neuron death was established early (Dawson et al., 1991), the mechanism for induction of cell death has been largely debated and is still controversial. The most generally accepted mechanism for NO toxicity is the formation of the strong oxidant peroxynitrite (Crow and Beckman, 1995; Dawson et al., 1993). Peroxynitrite is formed from the diffusion-limited reaction of NO with superoxide (Beckman et al., 1990). In fact, NO is by far the most effective biological molecule in competing with SOD for superoxide. The rate of reaction of superoxide with nitric oxide is three to five times faster than the reaction with SOD (Huie and Padmaja, 1993; Kissner et al., 1997). Superoxide dismutase is an abundant enzyme in cells, with concentrations ranging from 1 to 30 μM, depending on the cell type. Under normal physiological conditions, NO is produced in nanomolar concentrations, which do not significantly compete with SOD for superoxide.

In pathological conditions, NO and superoxide production are both increased, allowing the formation of peroxynitrite.

Peroxynitrite is a short-lived molecule with a half-life of less than 1 second at physiological pH (Beckman and Koppenol, 1996), making its direct measurement in biological systems extremely difficult, if not impossible. However, peroxynitrite permanently modifies tyrosine to form nitrotyrosine (Beckman et al., 1992; Ischiropoulos et al., 1992). Nitrotyrosine is formed by peroxynitrite and possibly other NO-derived oxidants but is not produced by NO itself in relevant biological conditions (Beckman, 1996). Although other NO-derived oxidants may also form a fraction of nitrotyrosine, particularly around neutrophils, nitrotyrosine is still a useful footprint for peroxynitrite in the nervous system.

Nitric Oxide Synthase Is Induced During Motor Neuron Programmed Cell Death and After Axotomy

Normal adult motor neurons do not express neuronal nitric oxide synthase (nNOS). However, conditions that compromise motor neuron survival involve the induction of NOS, and NO production seems to play a major role in the regulation of the mechanism of motor neuron death (Abe et al., 1997; Chou et al., 1996a,b; Wu, 1993). Motor neurons express NOS during the period of programmed cell death in the rat (Bruning and Mayer, 1996). More recently, inhibition of NOS has been reported to prevent programmed cell death of motor neurons in chicken embryos (Prevette et al., 1998). Nitric oxide synthase is induced in motor neurons by ventral root avulsion in the rat and mouse followed by motor neuron degeneration over 6 weeks (Wu, 1993; Wu and Li, 1993; Yu, 1994). Inhibition of NOS activity with N^{G}-nitro-L-arginine methylester (L-NAME) or prevention of NOS expression with brain-derived neurotrophic factor (BDNF) largely prevents motor neuron death after ventral root avulsion (Novikov et al., 1995, 1997; Wu and Li, 1993). Similarly, axotomy of facial motor neurons results in the induction of NOS and motor neuron degeneration (deBilbao and Dubois-Dauphin, 1996; Dubois-Dauphin et al., 1994; Ruan et al., 1995). Treatment with L-NAME after axotomy also improves the survival of facial motor neurons (Casanovas et al., 1996; Ruan et al., 1995). These results strongly suggest that impairment of trophic factor availability is sufficient to induce NOS and motor neuron death.

Similar mechanisms seem to be activated in adult and embryonic motor neurons during programmed cell death and injury. Requirement for NO production during motor neuron apoptosis in rat, mouse, and chicken embryonic development suggests the mechanism may be conserved during evolution (Prevette et al., 1998; Wu and Li, 1993; Wu et al., 1994; Yu, 1994).

Nitric Oxide Synthase and Nitrotyrosine in Human Pathology and Animal Models of Motor Neuron Disease

Amyotrophic lateral sclerosis is a progressive neurodegenerative disease characterized by the specific loss of spinal motor neurons and cortical pyramidal motor neurons. Motor neuron death is followed by the loss of muscular mass and death typically between 1 and 5 years after diagnosis. Mutations in SOD are the cause in 20%–25% of familial ALS (FALS) cases (Brown, 1996). Further evidence has accumulated showing that FALS linked to mutations in SOD is a consequence of a gain in function rather than a loss of superoxide-scavenging activity. Transgenic mice carrying the G93A, G85R, and G37R FALS-SOD mutations develop an ALS-like pathology (Bruijn et al., 1997; Dal Canto and Gurney, 1995; Gurney et al., 1994), in spite of expressing normal levels of endogenous mouse SOD. Furthermore, SOD knockout mice do not develop motor neuron pathology, although motor neurons are more vulnerable to noxious stimuli (Reaume et al., 1996).

The gain in function by SOD does not require creation of an entirely new function, but could result from augmention of a preexisting toxic activity. We suggested that the toxicity of mutant FALS-SOD is due to a gain in function related to the capability of the enzyme to catalyze the nitration of tyrosine residues in proteins by peroxynitrite (Beckman et al., 1993). Neuronal NOS immunoreactivity is augmented in both sporadic and familial forms of ALS (Figure 12.1A). Increased NOS expression is associated with increased nitrotyrosine immunoreactivity (Figure 12.1B), particularly in protein aggregates, which are positive for nNOS, neurofilaments, SOD, and nitrotyrosine (Abe et al., 1995, 1997; Beal et al., 1997; Chou et al., 1996a,b; Ferrante et al., 1997a and b). Increased levels of free nitrotyrosine were also reported in the spinal cord from ALS patients (Beal et al., 1997; Ferrante et al., 1997a).

Inhibition of NO production with 7-nitroindazole delays motor dysfunction and motor neuron death in the wobbler mouse, another animal model of motor neuron degeneration (Ikeda et al., 1998). Neuronal NOS immunoreactivity is present in spinal motor neurons from both transgenic mice carrying wild-type human SOD and in mice carrying the G93A FALS-SOD mutation, but not in nontransgenic littermates, suggesting that expression of nNOS is not sufficient to induce degeneration (unpublished observations). Spinal motor neurons from transgenic mice carrying the G93A mutant SOD are immunoreactive for nitrotyrosine (Ferrante et al., 1997b). Nitrotyrosine immunoreactivity could also be detected in the neuropil, but it was not present in the spinal cord from transgenic mice overexpressing wild-type human SOD. Increased levels of free nitrotyrosine were reported, coinciding with the appearance of the symptoms in the trans-

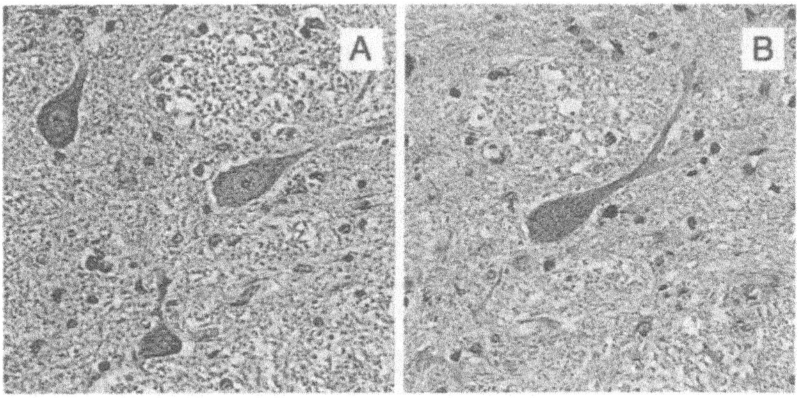

FIGURE 12.1. Neuronal NOS and nitrotyrosine immunoreactivity is increased in the spinal cord from ALS patient. Panel A shows motor neurons immunoreactive for nNOS. Panel B shows the same motor neuron from a consecutive section stained for nitrotyrosine.

genic mice carrying the G93A (Bruijn et al., 1997; Ferrante et al., 1997b) and G37R mutant SOD, but in the latter strain the authors were unable to find increased protein-linked nitrotyrosine by Western blot (Bruijn et al., 1997). Increasing evidence indicates that production of peroxynitrite as evidenced by nitrotyrosine formation is a common event in human ALS and the transgenic mice for the FALS-SOD mutants.

Motor Neuron Cultures

In spite of the growing evidence from human disease and animal models of neurodegeneration linking motor neuron death with oxidative stress and NO production, the cellular and subcellular mechanisms that regulate motor neuron death remain elusive. To study the mechanisms involved in motor neuron death in vivo is difficult. The recent development by Henderson et al. (Camu and Henderson, 1992; Henderson et al., 1993, 1995) of motor neuron cultures with an outstanding level of purity has provided a powerful tool for the study of the cellular and molecular mechanisms involved in the induction of motor neuron death.

Cultured motor neurons were mostly used as a model to test trophic factors. All the trophic factors supporting motor neuron survival in culture also prevent motor neuron death induced after axotomy (deLapeyriere and Henderson, 1997; Henderson, 1996; Henderson et al., 1994; Hughes et al., 1993; Oppenheim, 1996; Pennica et al., 1996). Trophic factors from different groups support motor neuron survival in culture. Motor neurons express the high-affinity receptors for neurotrophins trk B and trk C as well as the p75 low-affinity neurotrophin receptor ($p75^{LNTR}$). BDNF as well as neurotrophins 3 and 4/5 support motor neuron survival in culture (Estévez et al., 1998a,b; Henderson et al., 1993; Oppenheim et al., 1992). Cytokines such as the ciliary neurotrophic factor (CNTF) and cardiotrophin-1 also prevent motor neuron death in culture (Henderson et al., 1993; Hughes et al., 1993; Pennica et al., 1996). Cardiotrophin-1 is the most powerful factor for long-term culture of motor neurons, supporting motor neuron survival for over 14 days in culture (Pennica et al., 1996). The glial-derived neurotrophic factor (GDNF), a member of the tumor growth factor-β family, is the most potent short-term trophic factor for motor neurons in culture (Henderson et al., 1994).

Some trophic factors, which are produced by different cell types, have synergic effects on motor neuron survival (Arce et al., 1998), whereas others do not (Henderson et al., 1993). The effects of trophic factors are also dependent on the density and conditions of culture. A recent report by Hanson et al. (1998) showed that the interactions between trophic factors in motor neurons cultured at high density were stronger than those previously reported in low-density cultures (Arce et al., 1998; Henderson et al., 1993; Pennica et al., 1996). Motor neuron survival in low-density cultures is improved by cGMP but not by cAMP analogues (Estévez et al., 1998), whereas in high-density cultures, elevation of cAMP is sufficient to maintain motor neuron survival (Hanson et al., 1998).

Other trophic factors, such as fibroblast growth factor (FGF), nerve growth factor (NGF), and insulin growth factor-1 (IGF-1), stimulate neurite outgrowth by an indirect mechanism that requires the presence of astrocytes (Ang et al., 1993). However, whether these factors have a direct effect on motor neuron survival is still controversial. NGF, IGF-1, and basic FGF are unable to support the survival of rat motor neurons cultured at low density (Henderson et al., 1993). Conversely, the same factors have a synergic effect on the survival of motor neurons cultured at high density (Hanson et al., 1998). A possible explanation for these apparent contradictions is that the interactions between motor neurons may affect their requirements to survive. As for most cell types in culture, the specific conditions of the culture will affect the response of motor neurons to the same stimulus. We have shown that motor neurons are vulnerable to excitotoxic insults when plated on poly-L-lysine (Estévez et al., 1995), but the excitatory amino acids were less effective when the motor neurons were plated on poly-ornithine-laminin (A.G. Estévez and L. Barbeito, unpublished observations).

Spinal cord astrocytes in culture produce and release to the culture medium a diffusible factor that supports motor neuron survival (Peluffo et al., 1997). Production of trophic factors by astrocytes can be further stimulated by riluzole (Peluffo et al., 1997), which is currently the only FDA-approved drug for ALS (Bensimon et al., 1994; Lacomblez et al., 1996; Riviere et al., 1998).

Induction of Motor Neuron Apoptosis by Trophic Factor Deprivation

Neurons in culture have been effectively used to study the cellular and molecular mechanisms of neurodegeneration (for review see Deckwerth, 1997). Much is known about the trophic requirements for motor neuron survival both in vivo and in culture (deLapeyriere and Henderson, 1997; Henderson, 1996; Hughes et al., 1993; Oppenheim, 1996). A complex image is emerging that motor neuron trophic requirements change temporally during development and differ spatially across the developing spinal cord (Mettling et al., 1995). Some cellular messengers, such as NO, may deliver opposing signals, depending on the metabolic conditions of the cell (Estévez et al., 1998a,b,c). Every cell type interacting with motor neurons in vivo provides different trophic factors, and motor neuron survival is affected by the interaction between the multiple trophic signals (Arce et al., 1998; deLapeyriere and Henderson, 1997; Hanson et al., 1998; Oppenheim, 1996).

Motor neurons cultured in the absence of trophic factors survive and extend neurites in a way that is almost indistinguishable from parallel cultures plated with BDNF or GDNF during the first 15 to 18 hours after plating. However, 50% to 70% of the cells will then degenerate and die with a morphology characteristic of apoptosis during the next 6 to 8 hours (Estévez et al., 1998b). Motor neuron death induced by trophic factor deprivation requires the expression of new genes (Milligan et al., 1994). Inhibition of caspases, a family of cysteine proteases involved in the regulation of apoptosis (Nicholson and Thornberry, 1997), also prevented induction of motor neuron death by trophic factor deprivation (Estévez et al., 1998a,b, 2000; Milligan et al., 1995). Motor neuron death induced by deprivation of trophic factors is characterized by neurite withdrawal, soma shrinkage, and chromatin condensation associated with nuclear fragmentation as well as nicking of DNA strands visualized by the TUNEL method (Estévez et al., 1998b). All these characteristics strongly suggest that motor neurons deprived of trophic support undergo apoptosis.

Motor Neurons Express Endothelial Nitric Oxide Synthase

Growing evidence previously summarized links the induction of motor neuron death with the induction of NOS and NO production. However, whether NO affects motor neuron survival by acting directly on motor neurons or indirectly through other cell types is not clear. We tested whether cultured motor neurons express NOS using immunofluorescence and the reverse transcription polymerase chain reaction (RT-PCR) (Estévez et al., 1998b,c).

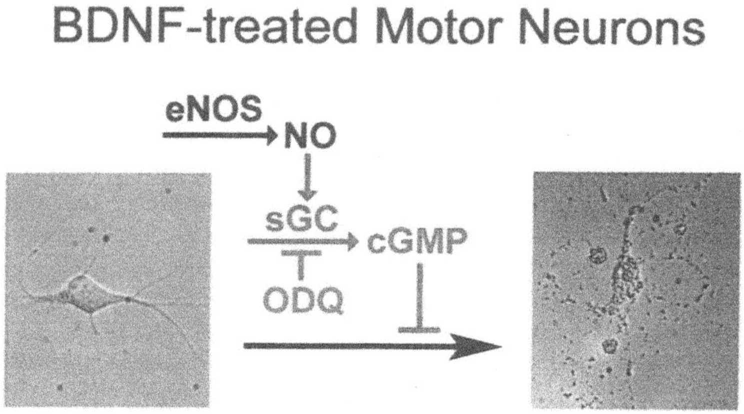

FIGURE 12.2. Proposed role for NO in the prevention of motor neuron apoptosis by BDNF. Our current hypothesis is that NO, produced by the constitutively expressed eNOS, stimulates the synthesis of cGMP by soluble guanylate cyclase (sGC).

Motor neurons in culture showed immunoreactivity for endothelial NOS (eNOS) in every condition tested (Estévez et al., 1998c). The same result was obtained with three different antibodies against eNOS. The expression of eNOS was confirmed by RT-PCR in BDNF-treated motor neurons and in cultures deprived of trophic factors (Estévez et al., 1998c). Surprisingly, inhibition of NO production with L-NAME induced apoptosis in BDNF-treated motor neurons. The toxic effects of L-NAME were prevented by generation of low steady-state concentrations of NO (<100 nM) from DETA-NONOate and cGMP analogues. 1H-[1,2,4]oxadiazolo[4,3-a]quinoxalin-1-one (ODQ), an inhibitor of soluble guanylate cyclase (Garthwaite et al., 1996), also induced motor neuron apoptosis (Estévez et al., 1998c). Apoptosis induced by ODQ was prevented by cGMP analogues but not by equivalent concentrations of cAMP analogues or extracellular production of NO (Estévez et al., 1998c). These results strongly suggest that NO itself is not toxic for motor neurons but can help to maintain motor neuron survival in the presence of trophic factors (Figure 12.2).

Role of Nitric Oxide in Trophic Factor-Induced Motor Neuron Apoptosis

Trophic factor deprivation of motor neurons resulted in the induction of nNOS mRNA and protein (Estévez et al., 1998a). In addition, inhibition of NO production with L-NAME prevented motor neuron apoptosis induced by trophic factor deprivation. Again, the effects of L-NAME were reversed by low steady-state production of extracellular NO from DETA-NONOate. The inactive isomer D-NAME did not prevent motor neuron apoptosis (Estévez et al., 1998a,b,c; 2000). N^G-nitro-L-arginine (L-NA) and N^G-monomethyl-L-arginine (L-NMMA) also prevented induction of apoptosis (Estévez et al., 1998a,b,c). These results suggest that NO produced by the motor neurons acts on the same cell to induce apoptosis. These results seem to be in contradiction with the effect of NO on motor neuron survival in the presence of BDNF. However, cGMP analogues prevented the induction of nNOS and motor neuron apoptosis, indicating that neither NO itself nor the cGMP pathway is involved in the induction of apoptosis (Estévez et al., 1998c).

Increased concentrations of superoxide or NO may lead to peroxynitrite formation. In addition, peroxynitrite induces apoptosis in PC12 cells (Estévez et al., 1995) and primary cultures of cortical neurons (Bonfoco et al., 1995). Manganese (III) 5, 10, 15, 20-tetrakis (4-benzoic acid) porphyrin TBAP, a low-molecular-weight, membrane-permeant superoxide dismutase mimetic (Faulkner et al., 1994), prevents motor neuron apoptosis induced by trophic factor deprivation (Estévez et al., 1998b,c). Similar results were obtained by intracellular SOD delivery using pH-sensitive liposomes (Estévez et al., 2000). These results indicate that both superoxide and NO must be formed at the same time to stimulate motor neuron apoptosis, in agreement with a mechanism involving peroxynitrite formation. Increased peroxynitrite immunoreactivity was found in trophic-factor-deprived motor neurons. Nitric oxide synthase inhibitors, manganese TBAP, and liposome-released human wild-type Cu/Zn SOD prevented the increase in nitrotyrosine immunoreactivity, mimicking the effects on motor neuron survival (Estévez et al., 2000). These results further support peroxynitrite mediating motor neuron apoptosis induced by trophic factor deprivation (Figure 12.3).

An Oxidative Hypothesis for Motor Neuron Apoptosis

Evidence from in vivo and in culture models suggests that during embryonic development and after injury, motor neurons induce de novo synthesis of nNOS. The subsequent NO production is involved in the regulation of apoptosis. However, NO is not toxic by itself, indicating that another factor is necessary to stimulate motor neuron death. We propose that the other factor is the formation of peroxynitrite by the reaction of NO with superoxide (Figure 12.3). To determine what the source of superoxide is in embryonic motor neurons or after injury will require further investigation. Monitoring the formation of nitrotyrosine may be useful to uncover the source of superoxide.

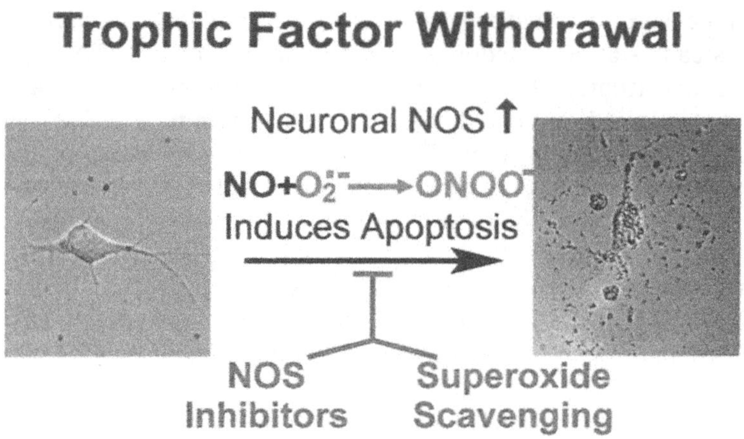

FIGURE 12.3. Induction of motor neuron apoptosis by trophic factor deprivation requires peroxynitrite formation. Trophic factor deprivation induces de novo synthesis of nNOS and increased nitrotyrosine immunoreactivity followed by apoptosis 24 hours after plating. Either inhibition of NO synthesis or increased superoxide scavenging prevents apoptosis.

Familial ALS-linked Superoxide Dismutase Mutations Have Decreased Affinity for Zinc

As previously described, 20% of cases of FALS are linked to more than 67 different point mutations in SOD, most of which have little effect on enzyme activity (Brown, 1996). Because the mutations reduce the affinity of SOD for zinc, we have proposed that the gain in function results from a reduced affinity for zinc, and that altered redox properties of zinc-deficient SOD are the cause of motor neuron death.

The biochemical properties of Cu/Zn wild-type and ALS mutant SODs are almost indistinguishable. However, ALS mutations in SOD weaken the structure of the enzyme, decreasing its affinity for zinc by 5 to 50 times when compared with wild-type human SOD. The affinity of SOD for copper is 7000 times stronger than for zinc and is not affected by the mutations (Crow et al., 1997). Then, the accumulation of zinc-deficient SOD may be the critical factor in developing ALS. Zinc-deficient SOD is about twice as efficient as a catalyst of tyrosine nitration by peroxynitrite than the fully metallated enzyme (Crow et al., 1997).

Zinc-deficient ALS-SOD mutants are not biochemically different from zinc-deficient wild-type SOD. Zinc-deficient wild-type and ALS mutant SODs are efficiently reduced by ascorbate at a rate that is 10^6 faster than the rate of reduction of wild-type Cu/Zn SOD. Other biologically relevant reductants, such as urate or glutathione, can reduce the enzyme, but more slowly than ascorbate. Reduced SOD may react with oxygen to produce superoxide, regenerating the oxidized copper, which can react with other molecules of ascorbate to produce more superoxide. Consequently, zinc-deficient SOD can catalyze the formation of superoxide at the expense of the cellular antioxidants. In addition, the dismutase activity of zinc-deficient SOD is about a third as fast as that of the Cu/Zn SOD (Forman and Fridovich, 1973; Pantoliano et al., 1979).

All these results suggest that the toxic activity of SOD could be due to the loss of zinc from the active site. They also suggest that the gain in function may involve the combination of several different activities. We hypothesize that the source of superoxide in ALS is zinc-deficient SOD, which also catalyzes the nitration of tyrosine residues by peroxynitrite. This hypothesis explains the late onset of the disease, since the ALS-SOD mutant does not become toxic until it becomes zinc-deficient (Figure 12.4). It also may explain the sporadic form of ALS, since wild-type zinc-deficient SOD may be equally as toxic as ALS-SOD mutants.

Recently Bruijn et al. (1998) reported that increasing or decreasing SOD activity does not affect the development of the disease in transgenic mice carrying ALS-SOD mutations. They conclude that oxidants are not involved in the pathogenesis of ALS. We found that zinc-deficient SOD oxidized dichlorodihydrofluoresceine (DCDHF) in the presence of ascorbate and NO. The enzyme alone or with either ascorbate or NO did not oxidize DCDHF. These results suggest that peroxynitrite is formed

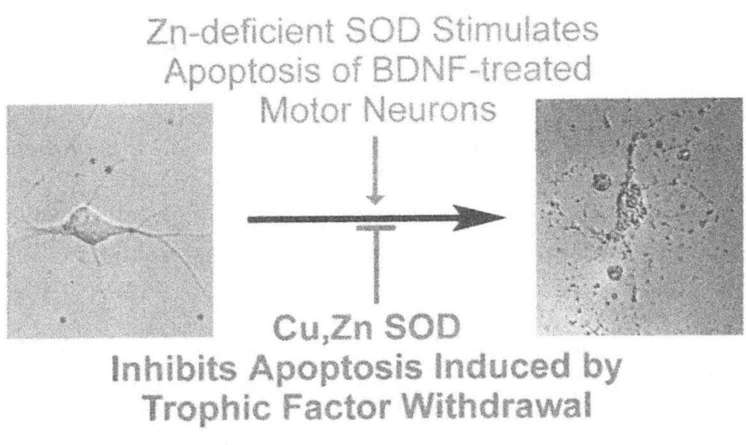

FIGURE 12.4. The scheme shows a hypothesis that explains the mechanism that makes SOD toxic to motor neurons. We propose that both wild-type and ALS-mutant Cu/Zn SOD protect trophic factor-deprived motor neurons by increasing superoxide scavenging, which prevents peroxynitrite formation. However, zinc-deficient wild-type SOD was as toxic as zinc-deficient ALS-SOD mutant for BDNF-treated motor neurons.

(Crow, 1997; Kooy et al., 1997). However, addition of an equimolar concentration of Cu/Zn SOD did not affect the rate of oxidation of DCDHF by zinc-deficient SOD plus ascorbate or NO, suggesting that the oxidizing species may be formed at the active site of zinc-deficient SOD. These results indicate that zinc-deficient SOD does not necessarily have to release free superoxide, and gives a new light to the results of Bruijn et al. (1998), since the formation of an oxidant species may occur even in the presence of Cu/Zn SOD.

Zinc-Deficient Superoxide Dismutase Induces Motor Neuron Apoptosis

To test the hypothesis that zinc-deficient SOD induces motor neuron apoptosis, we delivered human wild-type and ALS-mutant SODs with a defined content of copper and zinc intracellularly to cultured motor neurons using liposomes (Figure 12.4). Motor neuron apoptosis induced by trophic factor deprivation was prevented by either Cu/Zn ALS-mutant or wild-type SOD, whereas the zinc-deficient SODs did not affect motor neuron survival in the same conditions (Estévez et al., 2000)

The survival of BDNF-treated motor neurons was not affected by fully metallated wild-type or ALS mutant SODs. However, either wild-type or ALS mutants deficient in zinc induced apoptosis of 50% of BDNF-treated motor neurons after 24 hours. Motor neuron death induced by zinc-deficient SODs was prevented by induction of nNOS and increased nitrotyrosine immunoreactvity. Controls were performed by liposome delivery of apo-SOD and free and bovine serum albumin (BSA)-bound copper, which did not affect the survival of motor neurons cultured with or without BDNF. Inhibition of NO production by L-NAME or L-NMMA prevented both motor neuron death and the increase in nitrotyrosine immunoreactivity. Similar results were obtained with the superoxide mimetic and peroxynitrite scavenger manganese TBAP. These results strongly suggest that SOD becomes toxic when zinc-deficient by a mechanism involving peroxynitrite formation.

Conclusions and Perspectives

The presence of nNOS, nitrotyrosine, and other markers of oxidative damage is firmly associated with the induction of motor neurons in almost every model tested. However, whether oxidants are a cause or a consequence in the induction of motor neuron apoptosis, especially in ALS, remains controversial. Growing evidence suggests that a common stereotyped mechanism involving the production of free radicals and oxidants is involved in the induction of motor neuron apoptosis. The production of peroxynitrite and nitration of tyrosine residues are also common to motor neuron death in different conditions. However, whether free or protein-linked nitrotyrosine is just a coproduct of peroxynitrite formation, which induces apoptosis by other reactions, or plays a central role in motor neuron death remains to be elucidated. The mechanism of toxicity of ALS-SOD mutations may differ for different mutations, and other hypotheses to explain their toxicity have been proposed, including the formation of toxic aggregates (Bruijn et al., 1998; Durham et al., 1997). We are presently developing new experimental strategies to test these different hypotheses.

References

Abe, K., Pan, L.-H., Watanabe, M., Kato, T., and Itoyama, Y. 1995. Induction of nitrotyrosine-like immunoreactivity in the lower motor neuron of amyotrophic lateral sclerosis. *Neurosci. Lett.* 199:152–154.

Abe, K., Pan, L.-H., Watanabe, M., Konno, H., Kato, T., and Itoyama, Y. 1997. Upregulation of protein-tyrosine nitration in the anterior horn cells of amyotrophic lateral sclerosis. *Neurol. Res.* 19:124–128.

Ang, L.C., Bhaumick, B., and Juurlink, B.H.J. 1993. Neurite promoting activity of insulin, insulin-like growth factor I and nerve growth factor on spinal motoneurons is astrocyte dependent. *Dev. Brain Res.* 74:83–88.

Arce, V., Pollock, R., Philippe, J., Pennica, D., Henderson, C., and deLapeyrière, O. 1998. Synergistic effects of Schwann- and muscle-derived factors on motoneuron survival involved GDNF and cardiotrophin- (CT-1). *J. Neurosci.* 18:1440–1448.

Beal, M.F., Ferrante, R.J., Browne, S.E., Matthews, R.T., Kowall, N.W., and Brown, R.H., Jr. 1997. Increased 3-nitrotyrosine in both sporadic and familial amyotrophic lateral sclerosis. *Ann. Neurol.* 42:646–654.

Beckman, J.S. 1996. Oxidative damage and tyrosine nitration from peroxynitrite. *Chem. Res. Toxicol.* 9:836–844.

Beckman, J.S., Beckman, T.W., Chen, J., Marshall, P.M., and Freeman, B.A. 1990. Apparent hydroxyl radical production by peroxynitrite: implications for endothelial injury by nitric oxide and superoxide. *Proc. Natl. Acad. Sci. (USA)* 87: 1620–1624.

Beckman, J.S., Carson, M., Smith, C.D., and Koppenol, W.H. 1993. ALS, SOD and peroxynitrite. *Nature* 364:584.

Beckman, J.S., Ischiropoulos, H., Zhu, L., van der Woerd, M., Smith, C., Chen, J., Harrison, J., Martin, J.C., and Tsai, M. 1992. Kinetics of superoxide dismutase- and iron-catalyzed nitration of phenolics by peroxynitrite. *Arch. Biochem. Biophys.* 298:438–445.

Beckman, J.S., and Koppenol, W.H. 1996. Nitric oxide, superoxide, and peroxynitrite—the good, the bad, and the ugly. *Am. J. Physiol.* 271 (*Cell Physiol.* 40): C1424–C1437.

Bensimon, G., Lacomblez, L., and Meininger, V. 1994. A controlled trial of riluzole in amyotrophic lateral sclerosis. *N. Engl. J. Med.* 330:585–591.

Bonfoco, E., Krainc, D., Ankarcrona, M., Nicotera, P., and Lipton, S.A. 1995. Apoptosis and necrosis: two distinct events induced, respectively, by mild and intense insults with *N*-methyl-D-aspartate or nitric oxide/superoxide in cortical cell cultures. *Proc. Natl. Acad. Sci. USA* 92:7162–7166.

Brown, R.H., Jr. 1996. Superoxide dismutase and familial amyotrophic lateral sclerosis: new insights into mechanisms and treatments. *Ann. Neurol.* 39:145–146.

Bruijn, L.I., Beal, M.F., Becher, M.W., Schulz, J.B., Wong, P.C., Price, D.L., and Cleveland, D.W. 1997. Elevated free nitrotyrosine levels, but not protein-bound nitrotyrosine or hydroxyl radicals, throughout amyotrophic lateral sclerosis (ALS)-like disease implicate tyrosine nitration as an aberrant in vivo property of one familial ALS-linked superoxide dismutase 1 mutant. *Proc. Natl. Acad. Sci. USA* 94:7606–7611.

Bruijn, L.I., Houseweart, M.K., Kato, S., Anderson, K.L., Anderson, S.D., Ohama, E., Reaume, A.G., Scott, R.W., and Cleveland, D.W. 1998. Aggregation and motor neuron toxicity of an ALS-linked SOD1 mutant independent from wild-type SOD1. *Science* 281:1851–1854.

Bruning, G., and Mayer, B. 1996. Prenatal development of nitric oxide synthase in the mouse spinal cord. *Neurosci. Lett.* 202:189–192.

Camu, W., and Henderson, C.E. 1992. Purification of embryonic rat motoneurons by panning on a monoclonal antibody to the low-affinity NGF receptor. *J. Neurosci. Meth.* 44:59–70.

Casanovas, A., Ribera, J., Hukkanen, M., Riveros-Moreno, V., and Esquerda, J.E. 1996. Prevention by Lamotrigine, MK-801 and *N* omega-nitro-L-arginine methyl ester of motoneuron cell death after neonatal axotomy. *Neuroscience* 71: 313–325.

Chou, S.M., Wang, H.S., and Komai, K. 1996a. Colocalization of NOS and SOD1 in neurofilament accumulation within motor neurons of amyotrophic lateral sclerosis: an immunohistochemical study. *J. Chem. Neuroanat.* 10:249–258.

Chou, S.M., Wang, H.S., and Taniguchi, A. 1996b. Role of SOD-1 and nitric oxide/cyclic GMP cascade on neurofilament aggregation in ALS/MND. *J. Neurol. Sci.* 139 (Suppl.):16–26.

Crow, J.P. 1997. Dichlorodihydrofluorescein and dihydrorhodamine 123 are sensitive indicators of peroxynitrite in vitro: implications for intracellular measurement of reactive nitrogen and oxygen species. *Nitric Oxide*: 1:145–157.

Crow, J.P., and Beckman, J.S. 1995. The role of peroxynitrite in nitric oxide-mediated toxicity. *Curr. Top. Microbiol. Immunol.* 196:57–73.

Crow, J.P., Sampson, J.B., Zhuang, Y., Thompson, J.A., and Beckman, J.S. 1997. Decreased zinc affinity of amyotrophic lateral sclerosis-associated superoxide dismutase mutants leads to enhanced catalysis of tyrosine nitration by peroxynitrite. *J. Neurochem.* 69:1936–1944.

Dal Canto, M.C., and Gurney, M.E. 1995. Neuropathological changes in two lines of mice carrying a transgene for mutant human Cu/Zn SOD, and in mice overexpressing wild type human SOD: a model of familial amyotrophic lateral sclerosis (FALS). *Brain Res.* 676:25–40.

Dawson, V., Dawson, T., Bartley, D., Uhl, G., and Snyder, S. 1993. Mechanisms of nitric oxide-mediated neurotoxicity in primary brain cultures. *J. Neurosci.* 13: 2651–2661.

Dawson, V.L., Dawson, T.M., London, E.D., Bredt, D.S., and Snyder, S.H. 1991. Nitric oxide mediates glutamate neurotoxicity in primary cortical cultures. *Proc. Natl. Acad. Sci. USA* 88:6368–6371.

deBilbao, F., and Dubois-Dauphin, M. 1996. Time course of axotomy-induced apoptotic cell death in facial motoneurons of neonatal wild type and bcl-2 transgenic mice. *Neuroscience* 71:1111–1119.

Deckwerth, T.L. 1997. Molecular mechanisms of neuroprotection from neuronal death by trophic factor deprivation. In: (ed. M.P. Mattson) *Neuroprotective Signal Transduction*, pp. 61–82. Humana Press, Totowa, NJ.

deLapeyriere, O., and Henderson, C. 1997. Motoneuron differentiation, survival and synaptogenesis. *Curr. Opin. Genet. Dev.* 7:642–650.

Dubois-Dauphin, M., Frankowski, H., Tsujimoto, Y., Huarte, J., and Martinou, J. 1994. Neonatal motoneurons overexpressing the bcl-2 protooncogene in transgenic mice are protected from axotomy-induced cell death. *Proc. Natl. Acad. Sci. USA* 91:3309–3313.

Durham, H.D., Roy, J., Dong, L., and Figlewicz, D.A. 1997. Aggregation of mutant Cu/Zn superoxide dismutase proteins in a culture model of ALS. *J. Neuropathol. Exp. Neurol.* 56:523–530.

Estévez, A.G., Stutzmann, J.-M., and Barbeito, L. 1995. Protective effect of riluzole on excitatory amino acid-mediated neurotoxicity in motoneuron-enriched cultures. *Eur. J. Pharmacol.* 280:47–53.

Estévez, A.G., Spear, N., Manuel, S.M., Barbeito, L., Radi, R., and Beckman, J.S. 1998a. Role of endogenous nitric oxide and peroxynitrite formation in the survival and death of motor neurons in culture. In: (eds. T.M. Dawson, V.L. Dawson, and M.J. Friedlander), pp. 269–280. *Prog. Brain Res.* Elsevier, NY.

Estévez, A.G., Spear, N., Manuel, S.M., Radi, R., Henderson, C.E., Barbeito, L., and Beckman, J.S. 1998b. Nitric oxide and superoxide contribute to motor neuron apoptosis induced by trophic factor deprivation. *J. Neurosci.* 18:923–931.

Estévez, A.G., Spear, N., Thompson, J.A., Cornwell, T.L., Radi, R., Barbeito, L., and Beckman, J.S. 1998c. Nitric oxide dependent production of cGMP supports survival of rat embryonic motor neurons cultured with brain derived neurotrophic factor. *J. Neurosci.* 18:3708–3714.

Estévez, A.G., Crow, J.P., Sampson, J.B., Reiter, C., Zhuang, Y.-X., Richardson, G.J., Tarpey, M.M., Barbeito, L., and Beckman, J.S. 1999. Induction of nitric oxide-dependent apoptosis in motor neurons by zinc-deficient superoxide dismutase. *Science* 286:2498–2500.

Estévez, A.G., Sampson, J.B., Zhuang, Y., Spear, N., Richardson, G.J., Crow, J.P., Tarpey, M.M., Barbeito, L., and Beckman, J.S. 2000. Liposome-delivered superoxide dismutase prevents nitric-oxide dependent motor neuron death induced by trophic factor withdrawal. *Free Rad. Biol. Med.* 28(3):437–446.

Faulkner, K.M., Liochev, S.I., and Fridovich, I. 1994. Stable Mn(III) porphyrins mimic superoxide dismutase in vitro and substitute for it in vivo. *J. Biol. Chem.* 269:23471–23476.

Ferrante, R.J., Browne, S.E., Shinobu, L.A., Bowling, A.C., Baik, M.J., MacGarvey, U., Kowall, N.W., Brown, R.H., Jr., and Beal, M.F. 1997a. Evidence of increased oxidative damage in both sporadic and familial amyotrophic lateral sclerosis. *J. Neurochem.* 69:2064–2074.

Ferrante, R.J., Shinobu, L.A., Schulz, J.B., Matthews, R.T., Thomas, C.E., Kowall, N.W., Gurney, M.E., and Beal, M.F. 1997b. Increased 3-nitrotyrosine and oxidative damage in mice with a human copper/zinc superoxide dismutase mutation. *Ann. Neurol.* 42:326–334.

Forman, J.H., and Fridovich, I. 1973. On the stability of bovine superoxide dismutase. *J. Biol. Chem.* 248:2645–2649.

Garthwaite, J., Charles, S.L., and Chess-Williams, R. 1988. Endothelium-derived relaxing factor release on activation of NMDA receptors suggests role as intercellular messenger in the brain. *Nature (Lond.)* 336:385–388.

Garthwaite, J., Sotham, E., Boulton, C.L., Nielsen, E.B., Schmidt, K., and Mayer, B. 1996. Potent and selective inhibition of nitric oxide-sensitive guanylyl cyclase by 1*H*[1,2,4]oxadiazolo[4,3-*a*]quinoxalin-1-one. *Biochem. Pharmacol.* 48: 184–188.

Gurney, M.E., Pu, H., Chiu, A.Y., Dal Corto, M.C., Polchow, C.Y., Alexander, D.D., Caliendo, J., Hentati, A., Kwon, Y.W., Deng, H.-X., Chen, W., Zhai, P., Sufit, R.L., and Siddique, T. 1994. Motor neuron degeneration in mice that express a human Cu/Zn superoxide dismutase mutation. *Science* 264:1772–1775.

Hanson, M.G., Shen, S., Wiemelt, A.P., McMorris, F.A., and Barres, B.A. 1998. Cyclic AMP elevation is sufficient to promote the survival of spinal motor neurons in vitro. *J. Neurosci.* 18:7361–7371.

Henderson, C. 1996. Role of neurotrophic factors in neuronal development. *Curr. Opin. Neurobiol.* 6:64–70.

Henderson, C.E., Bloch-Gallego, E., and Camu, W. 1995. Purification and culture of embryonic motorneurons. In: (eds. J. Cohen and G. Wilkin) *Neural Cell Culture: A Practical Approach*, pp. 69–81. IRL Press, Oxford.

Henderson, C.E., Camu, W., Mettling, C., Gouin, A., Poulsen, K., Karihaloo, M., Rullamas, J., Evans, T., McMahon, S.B., Armanini, M.P., Berkemeier, L., Phillips, H.S., and Rosenthal, A. 1993. Neurotrophins promote motor neuron survival and are present in embryonic limb bud. *Nature* 363:266–270.

Henderson, C.E., Phillips, H.S., Pollock, R.A., Davies, A.M., Lemeulle, C., Armanini, M., Simpson, L.C., Moffet, B., Vandlen, R.A., Koliatsos, V.E., and Rosenthal, A. 1994. GDNF: a potent survival factor for motoneurons present in peripheral nerve and muscle. *Science* 266:1062–1064.

Hughes, R.A., Sendtner, M., and Thoenen, H. 1993. Members of several gene families influence survival of rat motoneurons in vitro and in vivo. *J. Neurosci. Res.* 36:663–671.

Huie, R.E., and Padmaja, S. 1993. The reaction rate of nitric oxide with superoxide. *Free Rad. Res. Commun.* 18:195–199.

Ignarro, L.J. 1990. Biosynthesis and metabolism of endothelium-derived nitric oxide. *Annu. Rev. Pharmacol. Toxicol.* 30:535–560.

Ignarro, L.J., Buga, G.M., Wood, K.S., Byrns, R.E., and Chaudhuri, G. 1987. Endothelium-derived relaxing factor produced and released from artery and vein is nitric oxide. *Proc. Natl. Acad. Sci. USA* 84:9265–9269.

Ikeda, K., Iwasaki, Y., and Kinoshita, M. 1998. Neuronal nitric oxide synthase inhibitor, 7-nitroindazol, delays motor dysfunction and spinal motoneuron degeneration in the wobbler mouse. *J. Neurol. Sci* 160:9–15.

Ischiropoulos, H., Zhu, L., Chen, J., Tsai, H.M., Martin, J.C., Smith, C.D., and Beckman, J.S. 1992. Peroxynitrite-mediated tyrosine nitration catalyzed by superoxide dismutase. *Arch. Biochem. Biophys.* 298:431–437.

Kissner, R., Nauser, T., Bugnon, P., Lye, P.G., and Koppenol, W.H. 1997. Formation and properties of peroxynitrite as studied by laser flash photolysis, high-pressure stopped-flow technique, and pulse radiolysis. *Chem. Res. Toxicol.* 10:1285–1292.

Kooy, N.W., Royall, J.A., and Ischiropoulos, H. 1997. Oxidation of 2′,7′-dichlorofluorescin by peroxynitrite. *Free Rad. Res.* 27:245–254.

Lacomblez, L., Bensimon, G., Leigh, P.N., Guillet, P., and Meininger, V. 1996. Dose-ranging study of riluzole in amyotrophic lateral sclerosis. Amyotrophic lateral sclerosis/riluzole study group II. *Lancet* 347:1425–1431.

Mettling, C., Gouin, A., Robinson, M., M'Hamdi, H.E., Camu, W., Bloch-Gallego, E., Buisson, B., Tanaka, H., Davies, A.M., and Henderson, C.E. 1995. Survival of newly postmitotic motoneurons is transiently independent of exogenous trophic support. *J. Neurosci.* 15:3128–3137.

Milligan, C.E., Oppenheim, R.W., and Schwartz, L.M. 1994. Motoneurons deprived of trophic support in vitro require new gene expression to undergo programmed cell death. *J. Neurobiol.* 25:1005–1016.

Milligan, C.E., Prevette, D., Yaginuma, H., Homma, S., Cardwell, C., Fritz, L.C., Tomaselli, K.J., Oppenheim, R.W., and Scwartz, L.M. 1995. Peptide inhibitors of the ICE protease family arrest programmed cell death of motoneurons in vivo and in vitro. *Neuron* 15:385–393.

Nicholson, D.W., and Thornberry, N.A. 1997. Caspases: killer proteases. *TIBS* 22:299–306.

Novikov, L., Novikova, L., and Kellerth, J.-O. 1995. Brain-derived neurotrophic factor promotes survival and blocks nitric oxide synthase expression in adult rat spinal motoneurons after ventral root avulsion. *Neurosci. Lett.* 200:45–48.

Novikov, L., Novikova, L., and Kellerth, J.-O. 1997. Brain-derived neurotrophic factor promotes axonal regeneration and long-term survival of adult rat spinal motoneurons in vivo. *Neuroscience* 79:765–774.

Oppenheim, R. 1996. Neurotrophic survival molecules for motor neurons: an embarrassment of riches. *Neuron* 17:195–197.

Oppenheim, R.W., Qin-Wei, Y., Prevett, D., and Yan, Q. 1992. Brain-derived neurotrophic factor rescues developing avian motoneurons from cell death. *Nature* 360:755.

Palmer, R.M.J., Ferrige, A.G., and Moncada, S. 1987. Nitric oxide release accounts for the biological activity of endothelium-derived relaxing factor. *Nature (Lond.)* 327:523–526.

Pantoliano, M.W., McDonnell, P.J., and Valentine, J.S. 1979. Reversible loss of metal ions from the zinc binding site of copper-zinc superoxide dismutase. The low pH transition. *J. Am. Chem. Soc.* 101:6454–6456.

Peluffo, H., Estévez, A., Barbeito, L., and Stutzmann, J.M. 1997. Riluzole promotes motoneuron survival by stimulating trophic activity produced by spinal astrocyte monolayers. *Neurosci. Lett.* 228:207–211.

Pennica, D., Arce, V., Swanson, T.A., Vejsada, R., Pollock, R.A., Armanini, M., Dudley, K., Phillips, H.S., Rosenthal, A., Kato, A.C., and Henderson, C.E. 1996. Cardiotrophin-1, a cytokine present in embryonic muscle, supports long-term survival of spinal motoneurons. *Neuron* 17:63–74.

Prevette, D., Wang, S., and Oppenheim, R.W. 1998. Nitric oxide and the programmed cell death of developing avian motoneurons. *Soc. Neurosci. Abst.* 13:708.

Reaume, A.G., Elliott, J.L., Hoffman, E.K., Kowall, N.W., Ferrante, R.J., Siwek, D.F., Wilcox, H.M., Flood, D.G., Beal, M.F., Brown, R.H., Jr., Scott, R.W., and Snider, W.D. 1996. Motor neurons in Cu/Zn superoxide dismutase-deficient mice develop normally but exhibit enhanced cell death after axonal injury. *Nature Genet.* 13: 43–47.

Riviere, M., Meininger, V., Ziesser, P., and Munsat, T. 1998. An analysis of extended survival in patients with amyotrophic lateral sclerosis treated with riluzole. *Arch. Neurol.* 55:526–528.

Ruan, R.-S., Leong, S.-K., and Yeoh, K.-H. 1995. The role of nitric oxide in facial motoneuronal death. *Brain Res.* 698:163–168.

Schuman, E.M., and Madison, D.V. 1994. Nitric oxide and synaptic function. *Annu. Rev. Neurosci.* 17:153–183.

Wu, W. 1993. Expression of nitric-oxide synthase (NOS) in injured CNS neurons as shown by NADPH diaphorase histochemistry. *Exp. Neurol.* 120:153–159.

Wu, W., and Li, L. 1993. Inhibition of nitric oxide synthase reduces motoneuron death due to spinal root avulsion. *Neurosci. Lett.* 153:121–124.

Wu, W., Liuzzi, F.J., Schinco, F.P., Dpeto, A.S., Li, Y., Mong, J.A., Dawson, T.M., and Snyder, S.H. 1994. Neuronal nitric oxide synthase is induced in spinal neurons by traumatic injury. *Neuroscience* 61:719–726.

Yu, W.-H.A. 1994. Nitric oxide synthase in motor neurons after axotomy. *J. Histochem. Cytochem.* 42:451–457.

Zhang, J., and Snyder, S.H. 1995. Nitric oxide in the nervous system. *Annu. Rev. Pharmacol. Toxicol.* 35:213–233.

13
Peroxynitrite and Nitrergic Neural Transmission: Pathophysiological Implications

GERT FOLKERTS, AXEL FISCHER, RICHARD B.R. MUIJSERS, AND FRANS P. NIJKAMP

Introduction

Up to 5% of the life-sustaining, inspired oxygen (O_2) is converted to oxygen free radicals. The addition of a single electron to O_2 produces the superoxide anion radical ($O_2^{\bullet-}$); superoxide can be converted spontaneously or by the catalytic superoxide dismutase (SOD) to the nonradical molecule hydrogen peroxide (H_2O_2), which can be toxic at high concentrations and, more importantly, can be reduced to the hydroxyl radical ($^{\bullet}OH$) (Reiter et al., 1995). Superoxide is released during the respiratory burst of granulocytes and macrophages by NADPH-oxidase activity, in response to several stimuli (Babior, 1978; Baggiolini and Wymann, 1990). Superoxide is also produced during the auto-oxidation of hemoglobin, myoglobin, and cytochrome *c*. Furthermore, enzymes like xanthine oxidase, aldehyde oxidase, and a variety of flavin dehydrogenases are sources of superoxide. Superoxide is therefore produced by virtually all aerobic cells (McCord and Fridovich, 1968).

Nitric oxide (NO) is another radical molecule that has important roles in many physiological and pathophysiological processes. It can be formed in a wide range of cells by the conversion of L-arginine to L-citrulline and NO by inducible and constitutive forms of nitric oxide synthase (NOS) (Moncada et al., 1989). Nitric oxide is a messenger molecule involved in paracrine mechanisms, such as vasorelaxation regulated by the endothelium, and has a function in neurotransmission (Kuriyama and Ohkuma, 1995; Snyder and Bredt, 1991). Furthermore, it is important in inflammatory processes as a free radical, but possibly also as a messenger molecule. Deleterious and beneficial effects may be strictly dependent on the concentration and the biological microenvironment in which NO is released. Nitric oxide decomposes into other nitrogen oxides, such as nitrite (NO_2^-) and nitrate (NO_3^-). Furthermore, NO rapidly reacts with superoxide, yielding peroxynitrite ($ONOO^-$) (reaction 1) (Huie and Padmaja, 1993), and this reaction occurs in vivo (Beckman et al., 1990).

$$^{\bullet}NO + O_2^{\bullet-} \rightarrow ONOO^- \quad [1]$$

The rate constant of this reaction is near the diffusion controlled limit (4–$7 \times 10^9 M^{-1} s^{-1}$) (Goldstein and Czapski, 1995; Huie and Padmaja, 1993). The half-life of peroxynitrite at 37°C and pH 7.4 is approximately 1 second (Beckman et al., 1990). Peroxynitrite is in equilibrium with peroxynitrous acid (reaction 2).

$$ONOO^- + H^+ \Leftrightarrow ONOOH \quad [2]$$

Nitrogen oxides react readily with oxygen, superoxide, water, nucleotides, metalloproteins, thiols, amines, and lipids to form products with biochemical actions such as bronchodilatation and bacteriostasis to cytotoxicity and pulmonary capillary leak, as well as those with frank mutagenic potential. It appears that overproduction or uncontrolled formation of peroxynitrite is an important factor in tissue-damaging mechanisms during pathological situations such as chronic inflammation. Recent discoveries demonstrating the relevance of these species in the body have provided new insights into the pathophysiology of many different diseases, and have opened a new horizon of therapeutic possibilities (Gaston et al., 1994).

Nitric Oxide Production

Nitrergic Neural Transmission (nNOS = type I)

Neuronal NOS (nNOS) has been localized in nerves innervating tissues from humans (Diaz et al., 1993; Fischer and Hoffmann, 1996; Kobzik et al., 1993) as well as from other species (Dey et al., 1993; Diaz et al., 1993; Grozdanovic et al., 1992; Kobzik et al., 1993; Shimosegawa and Toyota, 1994). Substantial species differences became apparent with regard to the extent of the innervation and the origin of nerve fibers. In humans, nerve fibers containing nNOS were shown by both immunohistochemistry and NADPH-diaphorase histochemistry (Fischer and Hoffmann, 1996; Kobzik et al., 1993; Ward et al., 1995). These nerve fibers are present in the airway smooth muscle, where NO has been shown to be the major mediator of neural smooth muscle relaxation (Belvisi et al., 1992; Li and Rand, 1991). The density of these nerve fibers decreases from trachea to small bronchi (Fischer and Hoffmann, 1996), which is associated with a reduced neural bronchodilation (Ellis and Undem, 1992; Ward et al., 1995) mediated by the inhibitory nonadrenergic, noncholinergic (iNANC) system (for review see Stretton, 1991). Colocalization with vasoactive intestinal peptide (VIP) is frequently observed (Figure 13.1) (Kummer et al., 1992c). In humans, NOS-containing nerve fibers are present around submucosal glands (Fischer and Hoffmann, 1996), although their functional

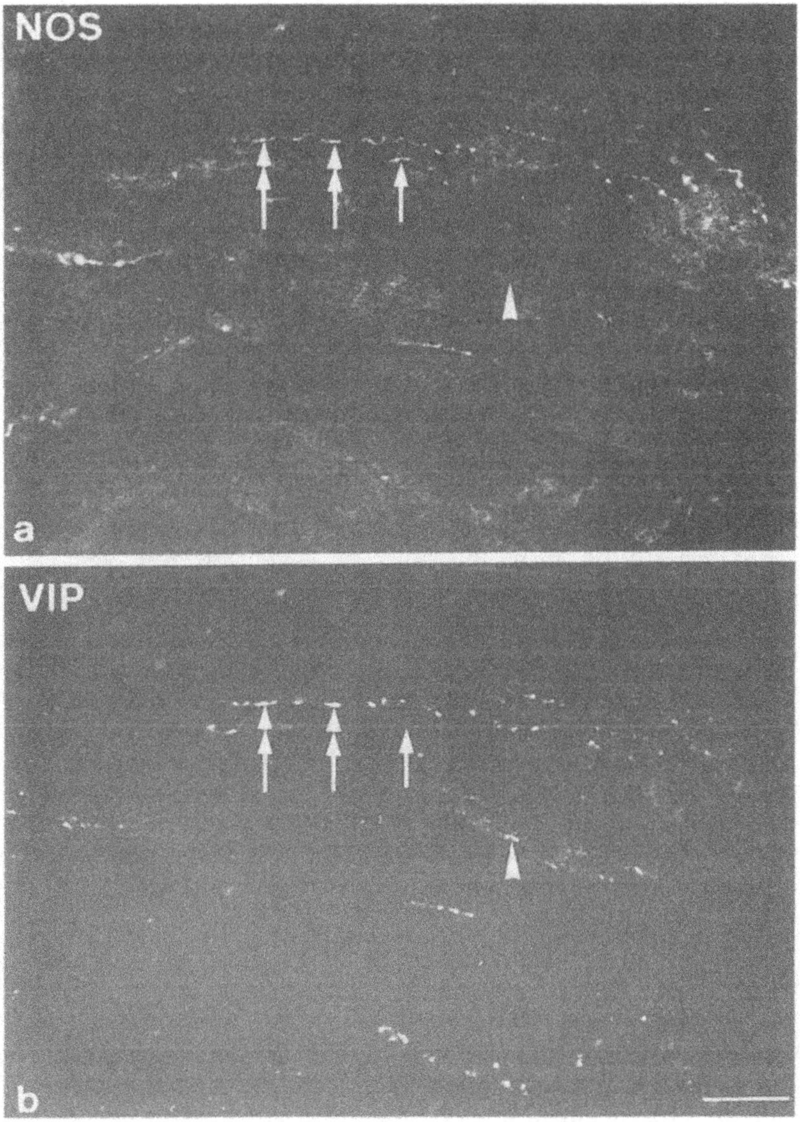

FIGURE 13.1. Most of the cell bodies of the (a) nNOS (NOS type I)-immunoreactive nerve fibers of human airways are localized in the local parasympathetic ganglia. Many of the cells are also immunoreactive for (b) vasoactive intestinal peptide (VIP) (double arrowheads). Some of the cells display either NOS (arrowhead) or VIP (arrow) immunoreactivity. Note the VIP-immunoreactive nerve fibers (small arrows) innervating the cell bodies. Scale bar represents 20 μm.

role in the regulation of glandular secretion is not yet clear. In the guinea pig, this type of nerve fiber has not been found; however a substantial number of nerve fibers immunoreactive for NOS have been found in the lamina propria (Figure 13.2) and occasionally also in the respiratory epithelium (Fischer et al., 1993). In the lamina propria, NO was shown to have potent effects on blood vessels, regulating plasma extravasation (Erjefalt et al., 1994).

The cell bodies of these neurons innervating tissues in humans (Diaz et al., 1993; Fischer and Hoffmann, 1996), ferrets (Dey et al., 1993, 1996), and piglets (Diaz et al., 1993) have been localized predominantly to the local parasympathetic ganglia (Figure 13.3). Contradictory results have been reported for guinea pig airways. Shimosegawa and Toyota (1994) have reported some NADPH-diaphorase-stained neurons innervating the airways, whereas in studies from our laboratory (Canning and Fischer, 1997; Fischer et al., 1993, 1996a), NOS immunoreactivity was seen only in neurons related to the pulmonary artery and vein, and the airway intrinsic ganglia were devoid of NOS-staining. In this species, a projection of the relaxant innervation from the adjacent esophagus was demonstrated (Canning and Undem, 1993a,b). Additional sources (Kummer et al., 1992a) of NOS-immunoreactive nerve fibers were found in vagal sensory and sympathetic ganglia (Fischer et al., 1996b; Olry et al., 1994). Nitric oxide synthase-immunoreactive neurons have been demonstrated in vagal sensory ganglia in humans (Fischer and Hoffmann, 1996; Kummer et al., 1992b, 1994) and rats (Aimi et al., 1991). In sensory neurons, NO could act as a neuromediator both at the peripheral and at the central endings (Ruggiero et al., 1996). A substantial NO-immunreactive innervation is present in trunks of the guinea pig pulmonary artery and vein (Klimaschewski et al., 1992) as well as in vessels with smaller diameters (Haberberger et al., 1996).

The subcellular localization of the neuronal isoform of NOS in nerves has not yet been clarified. In the central nervous system, early reports have identified NOS activity in the cytosolic fraction after preparative centrifugation (for review see Förstermann et al., 1991b). However, in the N-terminus of nNOS a domain has been identified that is responsible for membrane attachment of nNOS by interaction with postsynaptic density (PSD) proteins 95 and 93 (Brenman et al., 1996). Ultrastructural studies to localize NOS using colloidal gold markers on postembedding or ultrathin cryostat sections have been reported for the gastrointestinal tract (Berezin et al., 1994).

The presence of nNOS has also been shown in non-neuronal tissues (Erjefalt et al., 1994; Fischer et al., 1993; Kobzik et al., 1993). Neuronal NOS was also demonstrated as a constitutive isoform in normal endothelial cells (Loesch and Burnstock, 1995), and in pulmonary arteries and veins of rats of all ages, over 50% of endothelial cells displayed a cytoplasmatic immunoreactivity (Loesch and Burnstock, 1996).

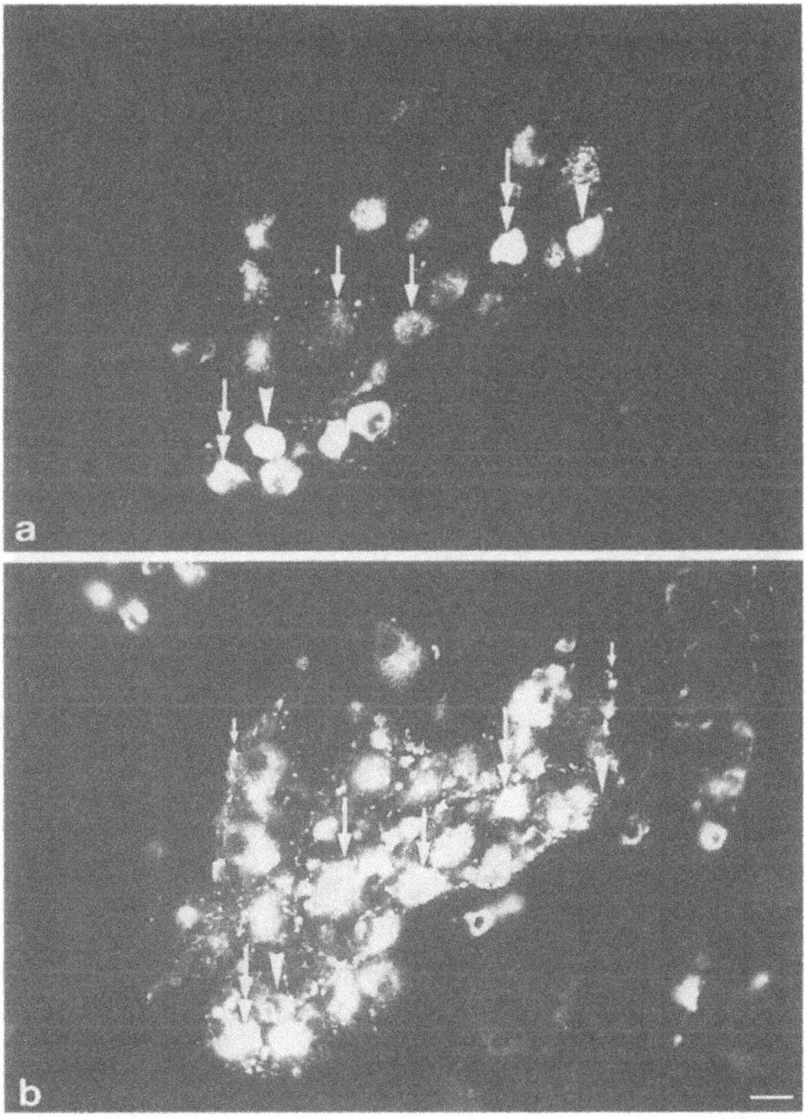

FIGURE 13.2. Localization of (a) nNOS (NOS type I) immunoreactivity in nerve fibers in the airway smooth muscle in a human bronchus and correlation with (b) vasoactive intestinal peptide (VIP). With the use of a confocal laser scanning microscope, in these thin optical sections NOS appears to be frequently colocalized with VIP (double arrows), but NOS can also be seen without VIP (arrow) and VIP without NOS (arrowhead). Scale bar represents 20 μm.

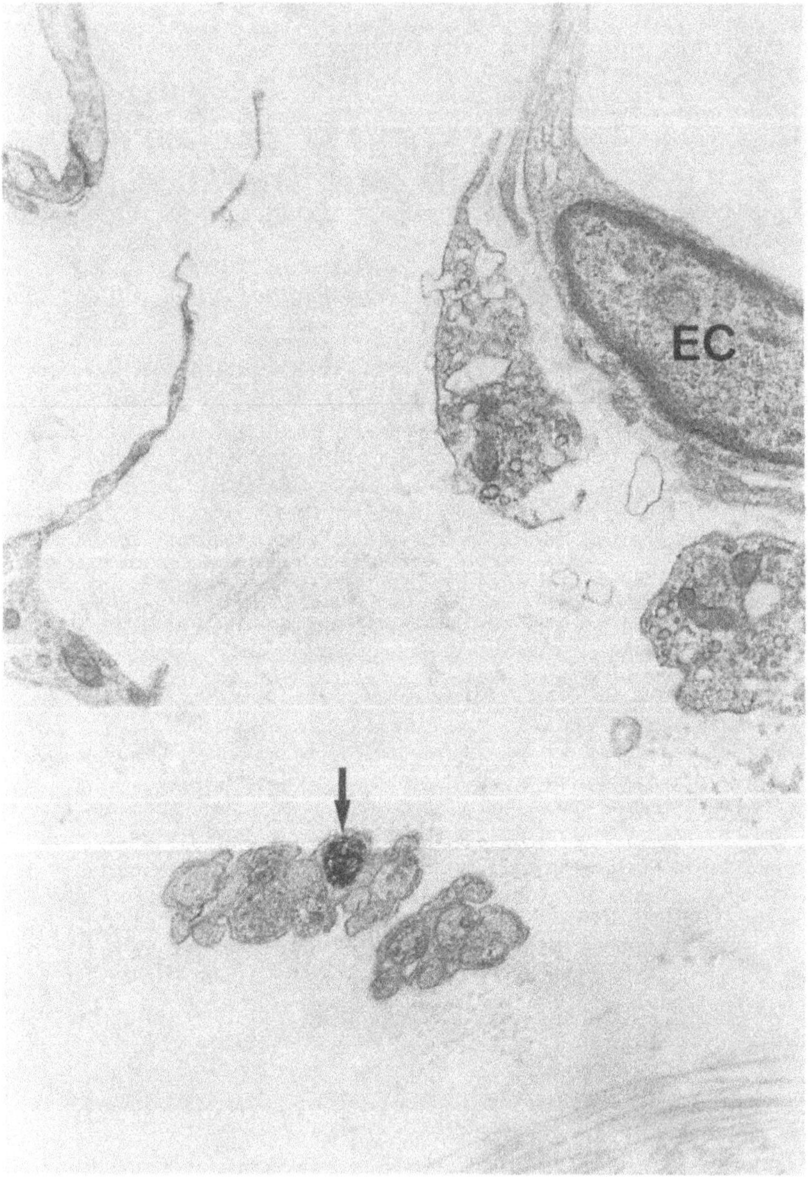

FIGURE 13.3. Electron microscopic micrograph of a guinea pig bronchus. A nerve fiber bundle in the lamina propria innervates a mucosal blood vessel endothelial cell (EC). The nerve fiber contains several axons; only one axon is immunoreactive for neuronal NOS (arrow). Magnification ×14,500.

Inducible Nitric Oxide Synthase (type II)

The inducible isoform of NOS (iNOS) has been identified as a separate, calcium-independent isoform, which could be detected after endotoxin or cytokine treatment (Knowles et al., 1990). Although a constitutive nitric oxide synthase /NADPH-diaphorase-activity in macrophages has been reported, cloning and sequencing from macrophages (Lowenstein et al., 1992; Lyons et al., 1992; Xie et al., 1992) has revealed that the iNOS isoform is expressed de novo at the transcriptional level. Soon it became clear that this isoform is not only localized to macrophages (Figure 13.4) but can be induced in many cells (Nathan, 1992), such as alveolar type II epithelial cells (Warner et al., 1995), fibroblasts (Jorens et al., 1992), airway and vascular smooth muscle cells (Griffith et al., 1995; Thomae et al., 1993; Xue et al., 1994), epithelial cells (Adcock et al., 1994; Hoffmann et al., 1995; Robbins et al., 1994a,b), endothelial cells (Sato et al., 1995), neutrophils (Blackford et al., 1994), and chondrocytes (Kobzik et al., 1993). The stimuli that cause transcriptional activation in these cells vary widely and include endogenous mediators (such as chemokines and cytokines) as well as exogenous factors such as bacterial toxins, viral infection, allergens, environmental pollutants (ozone, oxidative stress, silica), hypoxia, and tumors (Goldman et al., 1996; Yan et al., 1995; Yeadon and Price, 1995). Inducible NOS has been shown to be up-regulated in many disease states (Turnage et al., 1995). A "constitutive" expression of the inducible isoform has been observed at the level of mRNA (Guo et al., 1995) and protein (Kobzik et al., 1993). However, under normal conditions most investigators could not find an expression of the inducible isoform (Buttery et al., 1994).

Biochemical studies have suggested a cytoplasmic localization of iNOS (for review see Haberberger et al., 1996). Morphological observations on the ultrastructural location of the enzyme have not been reported to date.

Endothelial Nitric Oxide Synthase (type III)

Soon after the identification of NO as a messenger molecule generated by endothelial cells (Moncada et al., 1988; Palmer et al., 1987), a calcium- and L-arginine-dependent enzyme (Palmer et al., 1988) was proposed, and more than 95% of its activity was localized to the particulate fraction of the endothelial cells (Förstermann et al., 1991a). Indeed, when the enzyme had been cloned and sequenced (Janssens et al., 1992; Lamas et al., 1992; Nishida et al., 1992), and specific antisera for the endothelial isoform of NOS (eNOS) had become available, abundant eNOS immunoreactivity was demonstrated in endothelial cells of pulmonary vessels (Figure 13.5). In endothelial cells of submucosal blood vessels in the gastrointestinal tract, eNOS has been localized to the Golgi apparatus and cytoplasmatic vesicles by immunohistochemistry at the electron microscopic level (O'Brien et al., 1995). As demonstrated for endothelial cells from rat and bovine

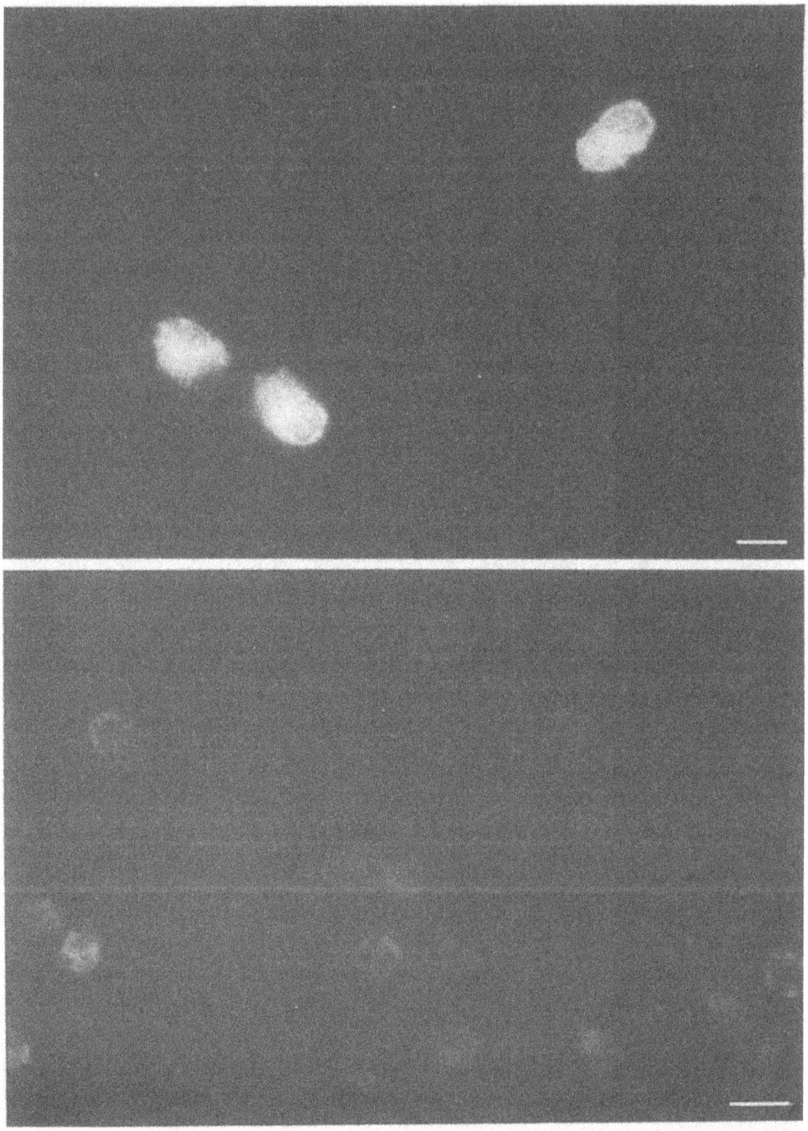

FIGURE 13.4. Alveolar macrophages from the guinea pig express iNOS (type II) after stimulation with lipopolysaccharide and interferon (upper panel). Unstimulated macrophages (lower panel) display no immunoreactivity. Scale bar represents 10 μm in the upper panel and 20 μm in the lower panel.

pulmonary artery, eNOS is targeted to endothelial caveolae by palmitoylation (Garcia-Cardena et al., 1996).

Quantitative developmental studies of mRNA and protein expression as well as immunohistochemical examination have shown that the endothelial

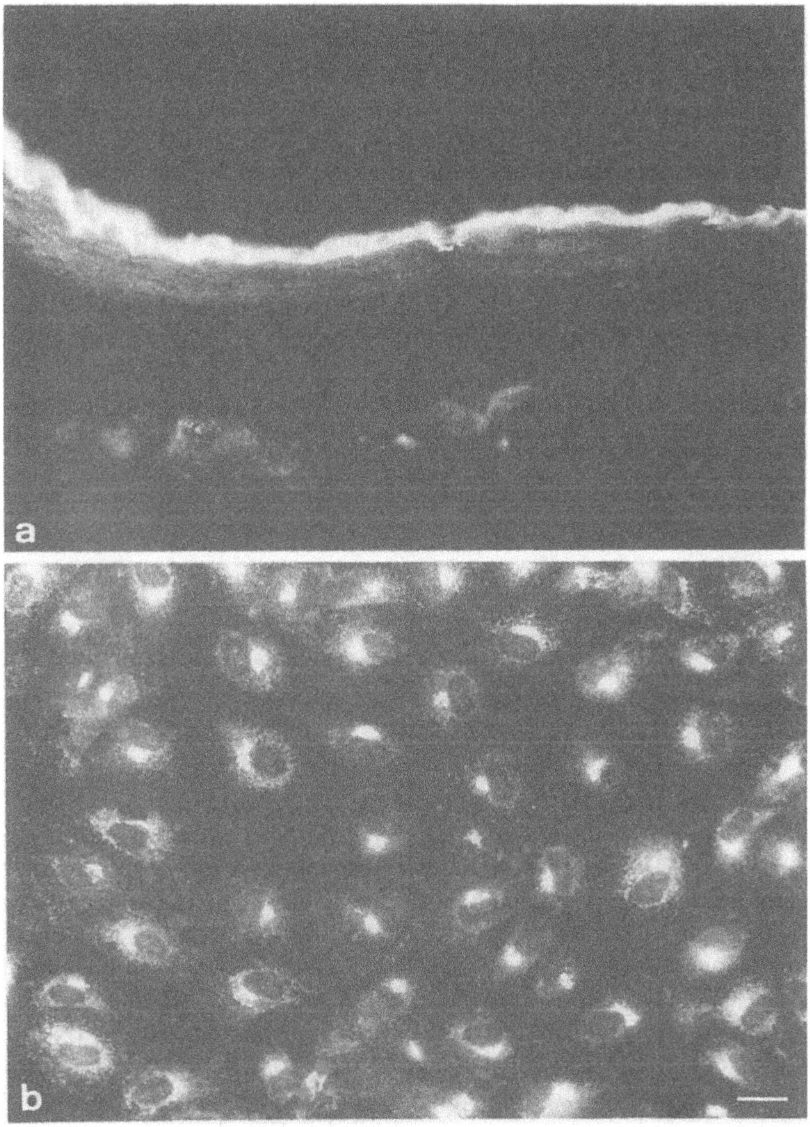

FIGURE 13.5. Immunohistochemical localization of eNOS (type III) in the guinea pig pulmonary artery (a) and in cultured endothelial cells from porcine pulmonary artery (b). (a) Most of the endothelial cells display immunoreactivity for the endothelial isoform. (b) The cytoplasmatic granular staining indicates that the endothelial isoform is located in the membrane of the Golgi apparatus. Scale bar represents 20 μm.

isoform increases during fetal development and reaches a maximum at the time of birth, followed by a postnatal decrease (Halbower et al., 1994; Hislop et al., 1995; Kawai et al., 1995; Norht et al., 1994). These changes have been largely attributed to the enzyme localized in endothelial cells and have been functionally related to the changes in the pulmonary vascular resistance occurring at the transition from fetal to neonatal life.

In addition to endothelial localization, this isoform is constitutively expressed in respiratory epithelial cells (Shaul et al., 1994). Ultrastructural studies have revealed that eNOS is localized at the basal membrane of ciliary microtubules (Xue et al., 1996), where it is thought to contribute to the regulation of ciliary beat frequency (Jain et al., 1993). Thus, all three isoforms are localized in the respiratory epithelium (Asano et al., 1994), where they might be involved in the regulation of airway smooth muscle tone (Morrison et al., 1990; Nijkamp et al., 1993; Rengasamy et al., 1994; Yan et al., 1994).

Peroxynitrite

Formation

Inflammatory cells, such as neutrophils (Carreras et al., 1994; McCall et al., 1989; Salman-Tabcheh et al., 1995; Sutherland et al., 1993) and macrophages (Assreuy et al., 1994; Avron and Gallily, 1995; Blackford et al., 1994; Ischiropoulos et al., 1992; Rodenas et al., 1995; Szabo and Salzman, 1995; Szabo et al., 1996; Wizemann et al., 1994), but also endothelial cells (Kooy and Royall, 1994; Szabo et al., 1995), can release superoxide and/or NO, potentially leading to peroxynitrite formation. Their release is mediated and regulated by several cytokines, such as interferon-γ (IFN-γ) and tumor necrosis factor α (TNFα), by bacteria-derived compounds like f-Met-Leu-Phe (fMLP) and lipopolysaccharide (LPS), and by mycoplasma membranes (Avron and Gallily, 1995; Szabo et al., 1995; Wizemann et al., 1994). The nitration of tyrosine residues in proteins by peroxynitrite to 3-nitrotyrosine is an indication of the presence of peroxynitrite (Ischiropoulos and al-Mehdi, 1995; Ischiropoulos et al., 1992; Vliet et al., 1994a, 1995a,b). An increased immunofluorescence to nitrotyrosine residues was detected at the sites of inflammation in acute lung injury in rats and humans (Haddad et al., 1994; Royall et al., 1995). Recent work, however, demonstrated that nitrite, the autooxidation product of NO, reacts with hypochlorous acid. This reaction yields reactive intermediates that, peroxynitrite independently, can nitrate tyrosine residues (Eiserich et al., 1996). Hypochlorous acid can be formed by the action of myeloperoxidase. Hence, nitrotyrosine formation does not necessarily imply peroxynitrite formation (Halliwell, 1997). A recent review discusses the measurement of peroxynitrite (Packer et al., 1998).

Inhibition

An increasing number of compounds are available that directly inhibit NOS. L-Arginine-like compounds such as N^G-nitro-L-arginine methylester (L-NAME) and N^G-monomethyl-L-arginine (L-NMMA) (Hibbs et al., 1987; Palmer et al., 1988), isothioureas (Arkovitz et al., 1996; Garvey et al., 1994), selenoalkylguanides (Southan et al., 1996), and aminoguanidines (Arkovitz et al., 1996; Corbett et al., 1992) have been demonstrated to inhibit NOS. Some of these compounds are relatively selective toward the inducible isoform of NOS, which is of particular interest in inflammatory situations. Southan and Szabo (1996) have recently reviewed the selective pharmacological inhibition of NOS. 21-Aminosteroids have been shown to be of value for treatment of silica-induced peroxynitrite-mediated lung disease via the inhibition of iNOS (Antonini et al., 1995). Formation of the other peroxynitrite precursor, superoxide, can be inhibited by apocynin (4-hydroxy-3-methoxyacetophenone) (Simons et al., 1990), which is a selective inhibitor of NADPH oxidase in the neutrophil membrane. Inhibition of peroxynitrite formation by apocynin has been demonstrated in a murine macrophage cell line (Muijsers et al., 1998).

The second level of potential interaction is to scavenge peroxynitrite itself. Cysteine analogues (and other thiol groups) (Althaus et al., 1994) are effective scavengers of peroxynitrite. Moreover, deferoxamine (Ely et al., 1992) and the immunosuppressive drug cyclosporin-A (Packer and Murphy, 1994) can prevent some of the detrimental actions of peroxynitrite. The latter has been demonstrated to inhibit the mitochondrial calcium efflux induced by peroxynitrite in vitro, with possible consequences for cell survival. An endogenous scavenger of peroxynitrite in the central nervous system is melatonin. Since endogenous melatonin levels fall markedly in advanced age, the implication of these findings is that the loss of this antioxidant may contribute to the incidence or severity of some age-associated neurogenerative diseases (Reiter, 1998).

Superoxide has an effect on NO levels because it reacts with the compound, forming peroxynitrite. Therefore, scavenging of superoxide may enhance NO activity and prevent peroxynitrite formation (Beckman et al., 1990). It is hypothesized that this may be one of the major biological actions of SOD (Pryor and Squadrito, 1995). Peroxynitrite, unlike its precursor NO, is probably not an intercellular messenger molecule because of its limited stability and diffusion range (Saran and Bors, 1994), but from the presently available evidence one cannot exclude the possibility that peroxynitrite serves as an intra- or pericellular messenger. Cytotoxic activity toward invading pathogens could therefore be one of the major physiological actions of peroxynitrite, and the compound may account for some of the cytotoxic actions formerly atributed to NO.

Pathophysiology

Cell Function

The mechanisms by which peroxynitrite compromises cell function are diverse (Muijsers et al., 1997). DNA fragmentation by peroxynitrite results in the activation of poly-ADP ribosyl transferase, leading to a futile repair cycle in macrophages (Szabo et al., 1996; Zingarelli et al., 1996), smooth muscle cells (Szabo et al., 1996), and neurons (Dawson, 1995). This subsequently leads to cellular energy depletion and cell death. Moreover, peroxynitrite inhibits mitochondrial electron transport and respiration, with a negative effect on cellular energy production (Radi et al., 1994; Szabo and Salzman, 1995). Further, peroxynitrite causes mitochondrial calcium efflux (Packer and Murphy, 1995), resulting in calcium overload and cell death. Moreover, peroxynitrite inactivates enzymes involved in energy metabolism. For example, aconitase, the first enzyme in the citric acid cycle, is inactivated by peroxynitrite. The prosthetic group of the enzyme is the likely oxidation target (Castro et al., 1994; Hausladen and Fridovich, 1994). Finally, peroxynitrite causes apoptosis, as in the rat thymocyte (Salgo et al., 1995), and oxidizes membrane lipids (Radi et al., 1991a; Rubbo et al., 1994) as well as tissue sulfhydryls (Radi et al., 1991b), and is believed to damage membrane sodium channels in the colon (Bauer et al., 1992) and lung (Hu et al., 1994) and calcium channels in the myocardium (Ishida et al., 1996).

Peroxynitrite might also be involved in *N*-methyl-D-aspartate (NMDA)-receptor mediated neuronal cell death via activation of constitutive NOS (cNOS) (Beckman, 1991; Bonfoco et al., 1995; Gunasekar et al., 1995), whereas high concentrations of peroxynitrite and NMDA led to necrosis, and relatively low concentrations induced apoptosis. In another study, however, NMDA receptor activation resulted in superoxide production, leading to NO-independent cell toxicity, and peroxynitrite was shown to be no more toxic to cultured cerebellar granule cells than superoxide (Lafon-Cazal et al., 1993). Further, peroxynitrite generation might explain elevated glutamate and asparate levels in the cerebrospinal fluid of patients with multiple sclerosis (Giovannoni et al., 1998; Gurwitz and Kloog, 1998). Recently, Sawada et al. (1996) demonstrated that glutamate-mediated neurotoxicity in dopaminergic neurons is NO-independent, whereas NO-mediated neurotoxicity is found in nondopaminergic neurons.

Inflammation

Peroxynitrite is a potent microbicidal compound in vitro. It is far more potent than NO in causing toxicity to *Escherichia coli* (Aimi et al., 1991; Zhu et al., 1992), and it is highly toxic to the protozoan *Trypanosoma cruzi* (Denicola et al., 1993; Dubbo et al., 1994). A possible mechanism by which peroxynitrite kills invading microorganisms is the oxidation of enzymes involved in energy metabolism. In *T. cruzi* the enzymes succinate dehydrogenase and fumarate dehydrogenase are inactivated by peroxynitrite via oxidation of sulfhydryl groups (Dubbo et al., 1994). Moreover, the citric acid cycle enzyme aconitase is a target for peroxynitrite in *E. coli*. Peroxynitrite, but not nitric oxide, has been demonstrated to effectively inactivate this enzyme (MacMicking et al., 1995).

These findings suggest a role for peroxynitrite in defense against microorganisms, in synergism with superoxide, NO, and possibly other radicals. Studies done in iNOS knockout mice demonstrate an essential role of NO in immune responses to particular microorganisms (MacMicking et al., 1995; Wei et al., 1995). It remains unclear, however, whether peroxynitrite is essential in this defense. In the mouse, for example, inhibition of NOS during virus-induced pneumonia increased the survival rate of the animals by limiting peroxynitrite-mediated tissue damage (Akaike et al., 1996). The defense against the virus, however, was not altered appreciably after NO inhibition. Therefore, NO is not required for superoxide to kill virus-infected cells, leaving peroxynitrite as an unwanted by-product. Moreover, the killing of *Leishmania major* by murine macrophages is dependent only on NO (Assreuy et al., 1993; Liew et al., 1990). Nevertheless, it appears that overproduction or uncontrolled formation of peroxynitrite is an important factor in the tissue-damaging mechanisms during pathological situations such as chronic inflammation. Since peroxynitrite is a highly reactive anion, it reacts with and oxidizes many cellular components such as lipids and proteins, thereby disturbing their function and thus disturbing cellular homeostasis. Immunoreactivity to nitrotyrosine residues and iNOS is colocalized in guinea pig ileitis, suggesting that peroxynitrite is formed via iNOS (Miller et al., 1995). Moreover, increased nitrotyrosine residues were detected at the sites of inflammation in acute lung injury in rats and humans (Haddad et al., 1994; Royall et al., 1995), acute endotoxemia in rats (Wizemann et al., 1994), influenza-induced pneumonia in mice (Akaike et al., 1996), rheumatoid, asthma (Saleh et al., 1998), and Alzheimer's patients (Hensley et al., 1998; Kaur and Halliwell, 1994). Furthermore, endotoxin was demonstrated to induce iNOS in rat aorta endothelium and to enhance nitrotyrosine formation (Szabo et al., 1995). It is therefore likely that peroxynitrite is formed during septic shock; an interaction of endothelium-derived NO with superoxide released by, for example, granulocytes is a likely explanation for this phenomenon (Bratt and Gyllenhammer, 1995). Many studies indirectly suggest the involvement of peroxynitrite in inflammation. For example induction of iNOS and concomitant suppression of SOD, favoring peroxynitrite formation, were observed in experimental colitis in rats (Seo et al., 1995). Indeed, NOS inhibition by analogues of L-arginine had beneficial effects in rat colitis by preventing peroxynitrite formation (Pfeiffer and Qiu, 1995). Moreover, part of the damage to myelin in multiple sclerosis is postulated to be due to peroxynitrite (Sherman et al., 1992).

Prolonged exposure to peroxynitrite may impair defense mechanisms in body fluids against peroxynitrite and other radicals, resulting in an increased tissue sensitivity to oxidative damage (Vliet et al., 1994b). Two-electron oxidation of α-tocopherol by peroxynitrite to α-tocopherylquinone depletes the former compound and its protective effects, since α-tocopherylquinone is not easily reduced back to α-tocopherol (Groot et al., 1993; Hogg et al., 1993, 1994). Furthermore, peroxynitrite reacts with and thereby depletes glutathione (GSH), another radical scavenger (Augusto et al., 1994; Phelps et al., 1995). It must be noted however, that prolonged exposure of rat lung fibroblasts to NO in vitro increased GSH levels, which possibly protects against damaging effects of peroxynitrite and free radicals (White et al., 1995). Whether peroxynitrite has a similar effect is as yet undetermined. Finally, carcinogenesis associated with chronic inflammation is postulated to be due to chronic peroxynitrite exposure because of its DNA-damaging, mutagenic properties (Liu and Hotchkiss, 1995). Peroxynitrite oxidizes guanine residues in DNA strands to nitroguanine, which in turn is rapidly depurinated from DNA. Apurinic sites in DNA are potentially mutagenic (Villiotou and Deliconstantinos, 1995; Yermilov et al., 1995). Peroxynitrite also caused DNA damage via formation of 8-hydroxydeoxyguanosine (Inoue and Kawanishi, 1995).

Atherosclerosis and Ischemia–Reperfusion Injury

There are several reports on the involvement of peroxynitrite in ischemia–reperfusion injury (Szabo, 1996), for example, in rat lung (Ischiropoulos et al., 1995), rat brain (Beckman, 1991; Kumura et al., 1996), and dog heart (Pearson et al., 1991). Depletion of oxygen during ischemia results in accumulation of free electrons, which leads to superoxide formation after reoxygenation. A characteristic of stressed cells is an exaggerated calcium influx, which suggests that the other precursor of peroxynitrite, NO, is provided by calcium-dependent cNOS. In cerebellar ischemia, activation of the NMDA receptor by glutamate may generate the calcium excess (Beckman, 1991). Again, inhibition of NOS by L-NAME was beneficial because at prevented peroxynitrite formation. L-NAME decreased lipid peroxidation in gerbils following cerebral ischemia (Caldwell et al., 1995) and protected isolated rabbit heart from ischemia–reperfusion injury (Schulz and Wambolt, 1995). 3-Aminobenzamide, an inhibitor of poly (ADP-ribose) synthetase, protected against myocardial ischemia and reperfusion in a rat model (Zingarelli et al., 1997). In the case of cerebellar ischemia, the administration of glutamate antagonists was also effective in limiting the consequences of infarction (Beckman, 1991).

Interestingly, peroxynitrite has also been suggested to be involved in atherosclerosis, one of the causes of ischemia. Extensive nitration of tyrosine residues in proteins has been detected in human atherosclerosis (White et al., 1994). Furthermore, peroxynitrite has been demonstrated to oxidize lipoproteins (Beckman and Crow, 1993; Darley-Usmar et al., 1992; White et al., 1994). Peroxynitrite may therefore be important in the onset of atherosclerosis.

Airway Diseases

The possible interference of peroxynitrite in airway diseases has recently been reviewed (Sadeghi-Hashjin et al., 1998). The production of significant amounts of NO and/or superoxide has been demonstrated by many different cell types isolated from the airways (Gutierrez et al., 1995, 1996; Punjabi et al., 1994; Szabo et al., 1996). Therefore, it is likely that high concentrations of peroxynitrite can be reached in the respiratory tract. In the isolated perfused guinea pig trachea, peroxynitrite induced epithelial damage (Sadeghi-Hashjin et al., 1996). Epithelial damage is suggested to be one of the possible causes of airway hyperresponsiveness (Folkerts and Nijkamp, 1998). Interestingly, the concentration of NO derivatives in sputum in patients with asthma is significantly correlated with the percentage of shed epithelial cells (Kanazawa et al., 1997). Treatment of human alveolar (A549) and human bronchial (BEAS-2B) epithelial cells with peroxynitrite resulted in ADP-ribosylation, NAD^+ depletion, inhibition of mitochondrial respiration, and increased epithelial paracellular permeability.

On the other hand, NO release has been reported to protect lung epithelial (L2) cells from granulocyte-mediated cytotoxicity (Su et al., 1996). In addition, in fetal rat alveolar epithelial cell monolayers stimulated to release NO and superoxide simultaneously, cell injury was increased after inhibition of NO synthesis. Following exposure to zymosan and/or exogenous superoxide, endogenous NO production and administration of NO donors inhibited the increased permeability induced by exposure to reactive oxygen species. Furthermore, inhalation of NO reduced oxygen toxicity in adult rats (Gutierrez et al., 1996). Apparently, when cultured pulmonary epithelial cells and lung tissue in vivo are subjected to inflammatory mediators or acute oxidative stress, NO can play a protective role by inhibiting superoxide-dependent toxicity.

Exogenously administered peroxynitrite mimics some inflammatory conditions in the respiratory tract. Incubation of isolated perfused guinea pig trachea with peroxynitrite causes a significant hyperresponsiveness to histamine and methacholine; epithelial damage as well as eosinophil destruction and the release of major basic protein were associated with airway hyperresponsiveness (Sadeghi-Hashjin et al., 1996). Moreover, 3 to 10 days after intratracheal instillation of peroxynitrite, a significant enhanced increase in pulmonary resistance to histamine of anaesthetized animals was observed (Sadeghi-Hashjin et al., 1996). These findings may have clinical implications, because airway inflammation, epithelial damage, and hyperresponsiveness are characteristic features in patients with asthma. Increased production of NO and, accordingly, peroxynitrite may be implicated in the pathophysiology of asthma. Indeed, asthmatic patients have higher amounts of NO in the expired air, possibly due to the inflammation, which can be prevented by inhaled corticosteroids (Nijkamp and Folkerts, 1995). Moreover, an increased formation of peroxynitrite in the airways of asthmatic patients was associated with the induction of NOS, and peroxynitrite formation was reduced after corticosteroid treatment (Saleh et al., 1998).

Viral respiratory infections may contribute to the development of airway hyperresponsiveness (Folkerts et al., 1998), and peroxynitrite formation might be involved in this event. Experimental pneumonia can be induced with influenza virus in the mouse. Both enzyme activity and mRNA expression of iNOS were increased, and a significant amount of NO was generated in the infected lungs. Immunohistochemistry with a specific antinitrotyrosine antibody showed intense staining of alveolar phagocytic cells and of intraalveolar exudate in the infected lungs. These results strongly suggest formation of peroxynitrite in the lung. Interestingly, administration of L-NMMA improved the survival rate of infected mice without appreciable suppression of their antiviral defenses (Akaike et al., 1996).

Conclusions

It is more than likely that peroxynitrite can be formed under physiological and pathophysiological conditions. Nitric oxide formation is a prerequisite for the synthesis of peroxynitrite. Abundant evidence has been published that NO can be released by nerves (nitrergic neural transmission). Moreover, NO can be formed by cNOS and by iNOS, which can be expressed in almost every cell. Further, superoxide is necessary, and this oxygen radical can also be formed by many cells throughout the body. Peroxynitrite formation has been demonstrated in a great number of organs. On the one hand, this anion has a protective role and therefore a beneficial function. On the other hand, a great number of studies now point to the deleterious effects of high concentrations of this oxygen species, for instance in neurological disorders and inflammatory process, such as atherosclerosis and airway diseases. In these cases, therapeutical intervention with agents that prevent the formation of peroxynitrite might be helpful.

Acknowledgments. We would like to thank Dr. B. Mayer, Graz, Austria, for donating the anti-nNOS- and anti-eNOS-antisera. Figures was 13.1, 13.3, and 13.5b were kindly provided by Mr. P. König, Prof. W. Kummer, and Dr. B. Höhler, Giessen, Germany. Support by the DFG and the EU (BIOMED II) is gratefully acknowledged.

References

Adcock, I.M., Brown, C.R., Kwon, O., and Barnes, P.J. 1994. Oxidative stress NF kappa B DNA binding and inducible NOS mRNA in human epithelial cells. *Biochem. Biophys. Res. Commun.* 199:1518–1524.

Aimi, Y., Fujimura, M., Vincent, S.R., and Kimura, H. 1991. Localization of NADPH-diaphorase-containing neurons in sensory ganglia of the rat. *J. Comp. Neurol.* 306:382–392.

Akaike, T., Noguchi, Y., Ijiri, S., Setoguchi, K., Suga, M., Zheng, Y.M., Dietzschold, B., and Maeda, H. 1996. Pathogenesis of influenza virus-induced pneumonia: involvement of both nitric oxide and oxygen radicals. *Proc. Natl. Acad. Sci. USA* 93:2448–2453.

Althaus, J.S., Oien, T.T., Fici, G.J., Scherch, H.M., Sethy, V.H., and VonVoightlander, P.F. 1994. Structure activity relationships of peroxynitrite scavengers: an approach to nitric oxide neurotoxicity. *Res. Comm. Chem. Pathol. Pharmacol.* 83:243–254.

Antonini, J.M., DiMatteo, M., Reasor, M.J., and Dyke, K. van 1995. Attenuation of acute inflammatory effects of silica in rat lung by 21-aminosteroid, U74389G. *Inflammation* 19:9–21.

Arkovitz, M.S., Wispe, J.R., Garcia, V.F., and Szabo, C. 1996. Selective inhibition of the inducible isoform of nitric oxide synthase prevents pulmonary transvascular flux during acute endotoxemia. *J. Pediat. Surg.* 31:1009–1015.

Asano, K., Chee, C.B.E., Gaston, B., Lilly, C.M., Gerard, C., Drazen, J.M., and Stamler, J.S. 1994. Constitutive and inducible nitric oxide synthase gene expression, regulation and activity in human lung epithelial cells. *Proc. Natl. Acad. Sci. USA* 91:10089–10093.

Assreuy, J., Cunha, F.Q., Epperlein, M., Noronha-Dutra, A., O'Donnell, C.A., Liew, F.Y., and Moncada, S. 1994. Production of nitric oxide and superoxide by activated macrophages and killing of *Leishmania major. Eur. J. Immunol.* 24:672–676.

Assreuy, J., Cunha, F.Q., Liew, F.Y., and Moncada, S. 1993. Feedback inhibition of nitric oxide synthase activity by nitric oxide. *Br. J. Pharmacol.* 108:833–837.

Augusto, O., Gatti, R.M., and Radi, R. 1994. Spin-trapping studies of peroxynitrite decomposition and of 3-morpholinosydnonimine *N*-ethylcarbamide autooxidation: direct evidence for metal-independent formation of free radical intermediates. *Arch. Biochem. Biophys.* 310:118–125.

Avron, A., and Gallily, R. 1995. Mycoplasma stimulates the production of oxidative radicals by murine peritoneal macrophages. *J. Leukocyte Biol.* 57:264–268.

Babior, B.M. 1978. Oxygen dependent microbial killing by phagocytes. *N. Engl. J. Med.* 298:659–688.

Baggiolini, M., and Wymann, M.P. 1990. Turning on the respiratory burst. *Trends Biochem. Sci.* 15:69–72.

Bauer, M.L., Beckman, J.S., Bridges, R.J., Fuller, C.M., and Matalon, S. 1992. Peroxynitrite inhibits sodium uptake in rat colonic membrane vesicles. *Biochim. Biophys. Acta* 1104:87–94.

Beckman, J.S. 1991. The double-edged role of nitric oxide in brain function and superoxide-mediated injury. *J. Dev. Physiol.* 15:53–59.

Beckman, J.S., Beckman, T.W., Chen, J., Marshall, P.A., and Freeman, B.A. 1990. Apparent hydroxyl radical production by peroxynitrite: implication for endothelial injury from nitric oxide and superoxide. *Proc. Natl. Acad. Sci.* 87:1620–1624.

Beckman, J.S., and Crow, J.P. 1993. Pathological implications of nitric oxide, superoxide and peroxynitrite formation. *Biochem. Soc. Trans.* 21:330–334.

Belvisi, M.G., Stretton, C.D., Yacoub, M., and Barnes, P.J. 1992. Nitric oxide is the endogenous neurotransmitter of bronchodilator nerves in humans. *Eur. J. Pharmacol.* 210:221–222.

Berezin, I., Snyder, S.H., Bredt, D.S., and Daniel, E.E. 1994. Ultrastructural localization of nitric oxide synthase in canine small intestine. *Am. J. Physiol.* 266:C981–C989.

Blackford, J.A., Antonini, J.M.J., Castranova, V., and Dey, R.D. 1994. Intratracheal instillation of silica up-regulates inducible nitric oxide synthase gene expression and increases nitric oxide production in alveolar macrophages and neutrophils. *Am. J. Respir. Crit. Care Med.* 11:426–431.

Bonfoco, E., Krainc, D., Ankarcrona, M., Nicotera, P., and Lipton, S.A. 1995. Apoptosis and necrosis: two distinct events induced, respectively, by mild and intense insults with *N*-methyl-D-aspartate or nitric oxide /superoxide in cortical cell cultures. *Proc. Natl. Acad. Sci. USA* 92:7162–7166.

Bratt, J., and Gyllenhammer, H. 1995. The role of nitric oxide in lipoxin A4-induced polymorphonuclear neutrophil-dependent cytotoxicity to human vascular endothelium in vitro. *Arthritis Rheum.* 38:768–776.

Brenman, J.E., Chao, D.S., Gee, S.H., McGee, A.W., Craven, S.E., Santillano, D.R., Wu, Z., Huang, F., Xia, H., Peters, M.F., Froehner, S.C., and Bredt, D.S. 1996. Interaction of nitric oxide synthase with the postsynaptic density protein PSD-95 and alpha11-syntrophin mediated by PDZ domains. *Cell* 84:757–767.

Buttery, L.D., Evans, T.J., Springall, D.R., Carpenter, A., Cohen, J., and Polak, J.M. 1994. Immunochemical localization of inducible nitric oxide synthase in endotoxin-treated rats. *Lab. Invest.* 71:755–764.

Caldwell, M., O'Neill, M., Earley, B., Kelly, J.P., and Leonard, B.E. 1995. *N*-G nitro-L-arginine methyl ester protects against lipid preoxidation in the gerbil following cerebral ischaemia. *Eur. J. Pharmacol.* 285:203–206.

Canning, B.J., and Fischer, A. 1997. Localization of cholinergic nerves in lower airways of guinea pigs using antisera to choline acetyltransferase. *Am. J. Physiol.* 272:L731–L738.

Canning, B.J., and Undem, B.J. 1993a. Evidence that distinct neural pathways mediate parasympathetic contractions and relaxations of guinea-pig trachealis. *J. Physiol.* 471:25–40.

Canning, B.J., and Undem, B.J. 1993b. Relaxant innervation of the guinea-pig trachealis: demonstration of capsaicin-sensitive vagal pathways. *J. Physiol.* 460:719–739.

Carreras, M.C., Pargament, G.A., Catz, S.D., Poderoso, J.J., and Boveris, A. 1994. Kinetics of nitric oxide and hydrogen peroxide production and formation of peroxynitrite during the respiratory burst of human neutrophils. *FEBS Lett.* 341:65–68.

Castro, L., Rodriguez, M., and Radi, R. 1994. Aconitase is readily inactivated by peroxynitrite, but not by its precursor, nitric oxide. *J. Biol. Chem.* 269:29409–29415.

Corbett, J.A., Tilton, R.G., Chang, K., Hassan, K.S., Ido, Y., Wang, J.L., Sweetland, M.A., Lancaster, J.R., Williamson, J.R., and McDaniel, M.L. 1992. Aminoguanidine, a novel inhibitor of nirtic oxide formation, prevents diabetic vascular dysfunction. *Diabetes* 41:552–556.

Darley-Usmar, V.M., Hogg, N., O'Leary, V.J., Wilson, M.T., and Moncada, S. 1992. The simultaneous generation of superoxide and nitric oxide can initiate lipid peroxidation in human low density lipoprotein. *Free Rad. Biol. Med.* 17:9–20.

Dawson, V.L. 1995. Nitric oxide: role in neurotoxicity. *Clin. Exp. Pharmacol. Physiol.* 22:305–308.

Denicola, A., Rubbo, H., Rodriguez, D., and Radi, R. 1993. Peroxynitrite-mediated cytotoxicity to *Trypanosoma cruzi*. *Arch. Biochem. Biophys.* 304:279–286.

Dey, R.D., Altemus, J.B., Rodd, A., Mayer, B., Said, S.I., and Coburn, R.F. 1996. Neurochemical characterization of intrinsic neurons in ferret tracheal plexus. *Am. J. Respir. Cell Mol. Biol.* 14:207–216.

Dey, R.D., Mayer, B., and Said, S.I. 1993. Colocalization of vasoactive intestinal peptide and nitric oxide synthase in neurons of the ferret trachea. *Neuroscience* 54:839–843.

Diaz, de R.O., Villaro, A.C., Montuenga, L.M., Martinez, A., Springall, D.R., and Polak, J.M. 1993. Nitric oxide synthase-immunoreactive neurons in the human and porcine respiratory tract. *Neurosci. Lett.* 162:121–124.

Dubbo, H., Denicola, A., and Radi, R. 1994. Peroxynitrite inactivates thiol-containing enzymes of *Trypanosoma cruzi* energetic metabolism and inhibits cell respiration. *Arch. Biochem. Biophys.* 308:96–102.

Eiserich, J.P., Cross, C.E., Jones, A.D., Halliwell, B., and van der Vliet, A. 1996. Formation of nitrating and chlorinating species by reaction of nitrite with hypochlorous acid—a novel mechanism for nitric oxide-mediated protein modification. *J. Biol. Chem.* 271:19199–19208.

Ellis, J.L., and Undem, B.J. 1992. Antigen-induced enhancement of noncholinergic contractile responses to vagus nerve and electrical field stimulation in guinea pig isolated trachea. *J. Pharmacol.* 262:646–653.

Ely, D., Dunphy, G., Dollwet, H., Richter, H., Sellke, F., and Azodi, M. 1992. Maintenance of left ventricular function (90%) after twenty-four hour heart preservation with deferoxamine. *Free Rad. Biol. Med.* 12:479–485.

Erjefalt, J.S., Erjefalt, I., Sundler, F., and Persson, C.G. 1994. Mucosal nitric oxide may tonically suppress plasma exudation. *Am. J. Respir. Crit. Care Med.* 150:227–232.

Fischer, A., Canning, B.J., and Kummer, W. 1996a. Correlation of vasoactive intestinal peptide and nitric oxide synthase with choline acetyltransferase in the airway innervation. *Ann. NY Acad. Sci.* 805:717–722.

Fischer, A., and Hoffmann, B. 1996. Nitric oxide synthase in neurons and nerve fibers of lower airways and in vagal sensory ganglia of man. *Am. J. Respir. Crit. Care Med.* 154:209–216.

Fischer, A., Mayer, B., and Kummer, W. 1996b. Nitric oxide synthase in vagal sensory and sympathetic neurons innervating the guinea-pig trachea. *J. Auton. Nerv. Syst.* 56:157–160.

Fischer, A., Mundel, P., Mayer, B., Preissler, U., Phillippin, B., and Kummer, W. 1993. Nitric oxide synthase in guinea pig lower airway innervation. *Neurosci. Lett.* 149:157–160.

Förstermann, U., Pollock, J.S., Schmidt, H.H.H.W., Heller, M., and Murad, F. 1991a. Calmodulin-dependent endothelium-derived relaxing factor/nitric oxide synthase activity is present in the particulate and cytosolic fractions of bovine aortic endothelial cells. *Proc. Natl. Acad. Sci. USA* 88:1778–1792.

Förstermann, U., Schmidt, H.H.H.W., Pollock, J.S., Sheng, H., Mitchell, J.A., Warner, T.D., Nakane, M., and Murad, F. 1991b. Isoforms of nitric oxide synthase. Characterization and purification from different cell types. *Biochem. Pharmacol.* 42:1849–1857.

Folkerts, G., Busse, W.W., Nijkamp, F.P., Sorkness, R., and Gern, J.E. 1998. Virus-induced airway hyperresponsiveness and asthma. *Am. J. Respir. Crit. Care Med.* 157:1708–1720.

Folkerts, G., and Nijkamp, F.P. 1998. Airway epithelium: more than just a barrier! *Trends Pharmacol. Sci.* 19:334–341.

Garcia-Cardena, G., Oh, P., Liu, J., Schnitzer, J.E., and Sessa, W.C. 1996. Targeting of nitric oxide synthase to endothelial cell caveolae via palmitoylation: implications for nitic oxide signaling. *Proc. Natl. Acad. Sci. USA* 93:6448–6453.

Garvey, E.P., Oplinger, J.A., Tanoury, G.J., Sherman, P.A., Fowler, M., Marshall, S., Harmon, M.F., Paith, J.E., and Furfine, E.S. 1994. Potent and selective inhibition of human nitric oxide synthase. Inhibition by non-amino acid isothioureas. *J. Biol. Chem.* 269:26669–26676.

Gaston, B., Drazen, J.M., Loscalzo, J., and Stamler, J.S. 1994. The biology of nitrogen oxides in the airways. *Am. J. Respir. Crit. Care Med.* 149:538–551.

Giovannoni, G., Heales, S.J., Land, J.M., and Thompson, E.J. 1998. The potential role of nitric oxide in multiple selerosis. *Mutat. Res.* 4:212–216.

Goldman, D., Cho, Y., Zhao, M., Casadevall, A., and Lee, S.C. 1996. Expression of inducible nitric oxide synthase in rat pulmonary *Cryptococcus neoformans* granulomas. *Am. J. Pathol.* 148:1275–1282.

Goldstein, S., and Czapski, G. 1995. The reaction of NO with O_2^- and HO_2: a pulse radiolysis study. *Free Rad. Biol. Med.* 19:505–510.

Griffith, M.J., Liu, S., Curzen, N.P., Messent, M., and Evans, T.W. 1995. In vivo treatment with endotoxin induces nitric oxide synthase in rat main pulmonary artery. *Am. J. Physiol.* 268:L509–L518.

Groot, H. de, Hegi, U., and Sies, H. 1993. Loss of alpha-tocopherol upon exposure to nitric oxide or the sydnonimine SIN-1. *FEBS Lett.* 315:139–142.

Grozdanovic, Z., Baumgarten, H.G., and Brüning, G. 1992. Histochemistry of NADPH-diaphase, a marker for neuronal nitric oxide synthase, in the peripheral autonomic nervous system of the mouse. *Neuroscience* 48:225–235.

Gunasekar, P.G., Kanthhasamy, A.G., Borowitz, J.L., and Isom, G.E. 1995. NMDA receptor activation produces concurrent generation of nitric oxide and reactive oxygen species: implication for cell death. *J. Neurochem.* 65:2016–2021.

Guo, F.H., De Raeve, H.R., Rice, T.W., Stuehr, D.J., Thunissen, F.B.J.M., and Erzurum, S.C. 1995. Continuous nitric oxide synthesis by inducible nitric oxide synthase in normal human airway epthelium in vivo. *Proc. Natl. Acad. Sci. USA* 92:7809–7813.

Gurwitz, D., and Kloog, Y. 1998. Peroxynitrite generation might explain elevated glutamate and aspartate levels in multiple sclerosis cerebrospinal fluid. *Eur. J. Clin. Invest.* 28:760–761.

Gutierrez, H.H., Nieves, B., Chumley, P., Rivera, A., and Freeman, B.A. 1996. Nitric oxide regulation of superoxide-dependent lung injury: oxidant-protective actions of endogenously produced and exogenously administered nitric oxide. *Free Rad. Biol. Med.* 21:43–52.

Gutierrez, H.H., Pitt, B.R., Schwarz, M., Watkins, S.C., Lowenstein, C., Caniggia, I., Chumley, P., and Freeman, B.A. 1995. Pulmonary alveolar epithelial inducible NO synthase gene expression: regulation by inflammatory mediators. *Am. J. Physiol.* 268:L501–L508.

Haberberger, R., Schemann, M., Sann, H., and Kummer, W. 1996. The innervation pattem of the pulmonary vascular diameter. *J. Appl. Physiol.*

Haddad, I.Y., Pataki, G., Hu, P., Galliani, C., Beckman, J.S., and Matalon, S. 1994. Quantitation of nitrotyrosine levels in lung sections of patients and animals with acute lung injury. *J. Clin. Invest.* 94:2407–2413.

Halbower, A.C., Tuder, R.M., Franklin, W.A., Pollack, J.S., Förstermann, U., and Abman, S.H. 1994. Maturation-related changes in endothelial nitric oxide synthase immunolocalization in developing ovine lung. *Am. J. Physiol.* 267:L585–L591.

Halliwell, B. 1997. What nitrates tyrosine? Is nitrotyrosine specific as a biomarker of peroxynitrite formation in vivo? *FEBS Lett.* 411:157–160.

Hausladen, A., and Fridovich, I. 1994. Superoxide and peroxynitrite inactivate aconitases, but nitric oxide does not. *J. Biol. Chem.* 269:29405–29408.

Hensley, K., Maidt, M.L., Yu, Z., Sang, H., Markesbery, W.R., and Floyd, R.A. 1998. Electrochemical analysis of protein nitrotyrosine and dityrosine in the Alzheimer brain indicates region-specific accumulation. *Neuroscience* 18:8126–8132.

Hibbs, J.B., Vavrin, Z., and Taintor, R.R. 1987. L-Arginine is required for the expression of the activated macrophage effector mechanism, causing selective metabolic inhibition in target-cells. *J. Immunol.* 138:550–565.

Hislop, A.A., Springall, D.R., Buttery, L.D., Pollock, J.S., and Haworth, S.G. 1995. Abundance of endothelial nitric oxide synthase in newborn intrapulmonary arteries. *Arch. Dis. Child.* 73:F17–F21.

Hoffmann, G., Grote, J., Friedrich, F., Mutz, N., and Schobersberger, W. 1995. The pulmonary epithelial cell line L2 as inducible nitric oxide synthase expressing distal airway epithelial cell. *Biochem. Biophys. Res. Commun.* 217:575–583.

Hogg, N., Darley-Usmar, V.M., Wilson, M.T., and Moncada, S. 1993. The oxidation of alpha-tocopherol in human low-density lipoprotein by the simultaneous generation of superoxide and nitric oxide. *FEBS Lett.* 326:199–203.

Hogg, N., Joseph, J., and Kalyanaraman, B. 1994. The oxidation of alpha-tocopherol and trolox by peroxynitrite. *Arch. Biochem. Biophys.* 314:153–158.

Hu, P., Ischiropoulos, H., Beckman, J.S., and Matalon, S. 1994. Peroxynitrite inhibition of oxygen consumption and sodium transport in alveolar type II cells. *Am. J. Physiol.* 266:L628–L634.

Huie, R.E., and Padmaja, S. 1993. The reaction of NO with superoxide. *Free Rad. Biol. Med.* 18:195–199.

Inoue, S., and Kawanishi, S. 1995. Oxidative DNA damage induced by simultaneous generation of nitric oxide and superoxide. *FEBS Lett.* 371:86–88.

Ischiropoulos, H., and al-Mehdi, A.B. 1995. Peroxynitrite-mediated oxidative protein modifications. *FEBS Lett.* 364:279–282.

Ischiropoulos, H., al-Mehdi, A.B., and Fisher, A.B. 1995. Reactive species in ischemic rat lung injury: contribution of peroxynitrite. *Am. J. Physiol.* 269:L158–L164.

Ischiropoulos, H., Zhu, L., Chen, J., Tsai, M., Martin, J.C., Smith, C.D., and Beckman, J.S. 1992. Peroxynitrite-mediated tyrosine nitration catalyzed by superoxide dismutase. *Arch. Biochem. Biophys.* 298:431–437.

Ishida, H., Ichimori, K., Hirota, Y., Fukahori, M., and Nakazawa, H. 1996. Peroxynitrite-induced cardiac myocyte injury. *Free Rad. Biol. Med.* 20:343–350.

Jain, B., Lubinstein, I., Robbins, R.A., Leise, K.L., and Sisson, J.H. 1993. Modulation of airway epithelial cell ciliary beat frequency by nitric oxide. *Biochem. Biophys. Res. Commun.* 191:83–88.

Janssens, S.P., Shimouchi, A., Quertermous, T., Bloch, D.B., and Bloch, K.D. 1992. Cloning and expression of a cDNA encoding human endothelium-derived relaxing factor/nitric oxide synthase. *J. Biol. Chem.* 267:14519–14522.

Jorens, P.G., Van Overveld, F.J., Vermeire, P.A., Bult, H., and Herman, A.G. 1992. Synergism between interleukin-1 beta and interferon-gamma, an inducer of nitric oxide synthase, in rat lung fibroblasts. *Eur. J. Pharmacol.* 224:7–12.

Kanazawa, H., Shoji, S., Yamada, M., Fujii, T., Kawaguchi, T., Kudoh, S., Hirata, K., and Yoshikawa, J. 1997. Increased levels of nitric oxide derivatives in induced sputum in patients with asthma. *J. Allergy Clin. Immunol.* 99:624–629.

Kaur, H., and Halliwell, B. 1994. Evidence for nitric-oxide-mediated oxidative damage in chronic inflammation, nitrotyrosine in serum and synovial fluid from rheumatoid patients. *FEBS Lett.* 350:9–12.

Kawai, N., Bloch, D.B., Fillipov, G., Rabkina, D., Suen, H.C., Losty, P.D., Janssens, S.P., Zapol, W.M., de la Monte, S., and Bloch, K.D. 1995. Constitutive endothelial nitric oxide synthase gene expression is regulated during lung development. *Am. J. Physiol.* 268:L589–L595.

Klimaschewski, L., Kummer, W., Mayer, B., Couraud, J.Y., Preissler, U., Philippin, B., and Heym, C. 1992. Nitric oxide synthase in cardiac nerve fibres and neurons of rat and guinea pig heart. *Circ. Res.* 71:1533–1537.

Knowles, R.G., Merrett, M., Salter, M., and Moncada, S. 1990. Differential induction of brain, lung and liver nitric oxide synthase by endotoxin in rat. *Biochem. J.* 270:833–836.

Kobzik, L., Bredt, D.S., Lowenstein, C.J., Drazen, J., Gaston, B., Sugarbaker, D., and Stamler, J.S. 1993. Nitric oxide synthase in human and rat lung: immunocytochemical and histochemical localization. *Am. J. Respir. Cell Mol. Biol.* 9: 371–377.

Kooy, N.W., and Royall, J.A. 1994. Agonist-induced peroxynitrite production from endothelial cells. *Arch. Biochem. Biophys.* 310:352–359.

Kummer, W., Fischer, A., Kurkowski, R., and Heym, C. 1992a. The sensory and sympathetic innervation of guinea-pig lung and trachea as studied by retrograde neuronal tracing and double-labelling immunohistochemisrty. *Neuroscience* 49:715–737.

Kummer, W., Fischer, A., and Lang, R.E. 1992b. Substance P and nitric oxide: participation in airway innervation. *Regul. Peptides* S1–S92.

Kummer, W., Fischer, A., Lang, R.E., Koesling, D., Mayer, B., and Olry, R. 1994. Nitric oxide and guanylyl cyclases: correlation with neuropeptides. *Neuropeptides Respir. Med.* 12:641–652.

Kummer, W., Fischer, A., Mundel, P., Mayer, B., Hora, B., and Philippin, B. 1992c. Nitric oxide synthase in VIP-containing vasodilator nerve fibres in the guinea pig. *Neuro. Rep.* 3:653–655.

Kumura, E., Yoshimine, T., Iwatsuki, K., Yamanaka, K., Tanaka, S., Hayakawa, T., Shiga, T., and Kosaka, H. 1996. Generation of nitric oxide and superoxide during reperfusion after focal cerebral ischemia in rats. *Am. J. Physiol.* 39:C748–C752.

Kuriyama, K., and Ohkuma, S. 1995. Role of nitric oxide in central synaptic transmission: effects on neurotransmitter release. *Jpn. J. Pharmacol.* 69:1–8.

Lafon-Cazal, M., Culcasi, M., Gaven, F., Pietri, S., and Bockaert, J. 1993. Nitric oxide, superoxide and peroxynitrite: putative mediators of NMDA-induced cell death in cerebellar granule cells. *Neuropharmacology* 32:1259–1266.

Lamas, S., Marsden, P.A., Li, G.K., Tempst, P., and Michel, T. 1992. Endothelial nitric oxide synthase: molecular cloning and characterization of a distinct constitutive molecular isoform. *Proc. Natl. Acad. Sci. USA* 89:6348–6352.

Li, C.G., and Rand, M.J. 1991. Evidence that part of the NANC relaxant response of guinea pig trachea to electrical field stimulation is mediated by nitric oxide. *Br. J. Pharmacol.* 102:91–94.

Liew, F.Y., Li, Y., and Millott, S. 1990. Tumor necrosis factor-alpha synergizes with IFN-gamma in mediating killing of *Leishmania major* through the induction of nitric oxide. *J. Immunol.* 145:4306–4310.

Liu, R.H., and Hotchkiss, J.H. 1995. Potential genotoxicity of chronically elevated nitric oxide: a review. *Mutat. Res.* 339:73–89.

Loesch, A., and Burnstock, G. 1995. Electroimmunocytochemical localization of endothelial and neuronal isoforms of nitric oxide synthase in rat cerebral basilardely artery. *Acta Neurobiol. Exp.* 55:45.

Loesch, A., and Burnstock, G. 1996. Ultrastructural localization of nitric oxide synthase and endothelin in rat pulmonary artery and vein during postnatal development and ageing. *Cell Tissue Res.* 283:355–365.

Lowenstein, C.J., Glatt, C.S., Bredt, D.S., and Snyder, S.H. 1992. Cloned and expressed macrophage nitric oxide synthase contrasts with the brain enzyme. *Proc. Natl. Acad. Sci. USA* 89:6711–6715.

Lyons, C.R., Orloff, G.J., and Gunningham, J.M. 1992. Molecular cloning and functional expression of an inducible nitirc oxide synthase from a murine macrophage cell line. *J. Biol. Chem.* 267:6370–6374.

MacMicking, J.D., Nathan, C., Hom, G., Chartrain, N., Fletcher, D.S., Trumbauer, M., Stevens, K., Xie, Q., Sokol, K., and Hutchinson, N. 1995. Altered responses to bacterial infection and endotoxic shock in mice lacking inducible nitric oxide synthase. *Cell* 81:641–650.

McCall, T.B., Boughton-Smith, N.K., Palmer, R.M.J., Whittle, B.J.R., and Moncada, S. 1989. Synthesis of nitric oxide from L-arginine by neutrophils. Release and interaction with superoxide anion. *Biochem. J.* 261:293–296.

McCord, J.M., and Fridovich, I. 1968. The reduction of cytochrome *c* by milk xanthine oxidase. *J. Biol. Chem.* 243:5753–5760.

Miller, M.J.S., Thompson, J.H., Zhang, X.J., Sadowska-Krowicka, H., Kakkis, J.L., Munshi, U.K., Sandoval, M., Rossi, J.L., Eloby-Childress, S., Beckman, J.S., Ye, Y.Z., Rodi, C.P., Manning, P.T., Currie, M.G., and Clark, D.A. 1995. Role of inducible nitric oxide synthase expression and peroxynitrite formation in guinea pig ileitis. *Gastroenterology* 109:1475–1483.

Moncada, S., Palmer, R.M.J., and Higgs, E.A. 1989. Biosynthesis of nitric oxide from L-arginine: a pathway for the regulation of cell function and communication. *Biochem. Pharmacol.* 38:1709–1715.

Moncada, S., Radomski, M.W., and Palmer, R.M.J. 1988. Endothelium derived relaxing factor identification as nitric oxide and role in the control of vascular tone and platelet function. *Biochem. Pharmacol.* 37:2495–2501.

Morrison, K.J., Gao, Y., and Vanhoutte, P.M. 1990. Epithelial modulation of airway smooth muscle. *Am. J. Physiol.* 258:L254–L262.

Muijsers, R.B.R., Folkerts, G., Henricks, P.A.J., Sadeghi-Hashjin, G., and Nijkamp, F.P. 1997. Peroxynitrite: a two faced metabolite of nitric oxide. *Life Sci.* 60:1833–1845.

Muijsers, R.B.R., Folkerts, G., Van den Worm, E., Beukelman, C.J., Postma, D.S., and Nijkamp, F.P. 1998. Inhibition of peroxynitrite formation by apocynin in a murine macrophage cell-line. *Br. J. Pharmacol.* 123:41p.

Nathan, C. 1992. Nitric oxide as a secretory product of mammalian cells. *FASEB* 6:3051–3064.

Nijkamp, F.P., and Folkerts, G. 1995. Nitric oxide and bronchial hyperresponsiveness. *Arch. Int. Pharmacodyn. Ther.* 329:81–96.

Nijkamp, F.P., Van der Linde, H.J., and Folkerts, G. 1993. Nitric oxide synthesis inhibitors induce airway hyperresponsiveness in the guinea pig in vivo and in vitro. *Am. Rev. Resp. Dis.* 148:727–734.

Nishida, K., Harrison, D.G., Navas, J.P., Fischer, A.A., Dockery, S.P., Uematsu, M., Nerem, R.M., Alexander, R.W., and Murphy, T.J. 1992. Molecular cloning and characterization of the constitutive bovine aortic endothelial cell nitric oxide synthase. *J. Clin. Invest.* 90:2092–2096.

Norht, A.J., Star, R.A., Brannon, T.S., Ujiie, K., Wells, L.B., Lowenstein, C.J., Snyder, S.H., and Shaul, P.W. 1994. Nitric oxide type I and type III gene expression are developmentally regulated in rat lung. *Am. J. Physiol.* 266:L635–L641.

O'Brien, A.J., Young, H.M., Povey, J.M., and Furness, J.B. 1995. Nitric oxide synthase is localized predominantly in the Golgi apparatus and cytoplasmatic vesicles of vascular endothelial cells. *Histochemistry* 103:221–225.

Olry, R., Mayer, B., and Kummer, W. 1994. Nitric oxide synthase in the pre- and post-ganglionic axis of the sympathetic nervous system in the guinea pig. In: *Biology of Nitric Oxide* (eds. S. Moncada, M. Feelisch, R. Busse, and E.A. Higgs), pp. 330–334. Portland Press, London.

Packer, M.A., and Murphy, M.P. 1994. Peroxynitrite causes calcium efflux from mitochondria which is prevented by cyclosporin A. *FEBS Lett.* 345:237–240.

Packer, M.A., and Murphy, M.P. 1995. Peroxynitrite formed by simultaneous nitric oxide and superoxide generation causes cyclosporin-A-sensitive mitochondrial calcium effux and depolarisation. *Eur. J. Biochem.* 234:231–239.

Packer, M.A., Scarlett, J.L., and Martin, S.W. 1998. Peroxynitrite: a biologically significant oxidant. *Gen. Pharmacol.* 31:179–186.

Palmer, R.M.J., Ashton, D.S., and Moncada, S. 1988. Vascular endothellial cells synthesize nitric oxide from L-arginine. *Nature* 333:664–666.

Palmer, R.M.J., Ferrige, A.G., and Moncada, S. 1987. Nitric oxide release accounts for the biological activity of endothelium-derived relaxing factor. *Nature* 327:524–526.

Palmer, R.M.J., Ress, D.D., Ashton, D.S., and Moncada, S. 1988. L-Arginine is the physiological precursor for the formation of nitric oxide in endothelium-dependent relaxation. *Biochem. Biophys. Res. Commun.* 153:1251–1256.

Pearson, P.J., Lin, P.J., and Schaff, H.V. 1991. Production of endothelium-derived contracting factor is enhanced after coronary reperfusion. *Ann. Thorac. Surg.* 51:788–793.

Pfeiffer, C.J., and Qiu, B.S. 1995. Effects of chronic nitric oxide synthase inhibition on TNB-induced colitis in rats. *J. Pharm. Pharmacol.* 47:827–832.

Phelps, D.T., Ferro, T.J., Higgins, P.J., Shankar, R., Parker, D.M., and Johnsin, A. 1995. TNF-alpha induces peroxynitrite-depletion of lung endothelial glutathione via protein kinase C. *Am. J. Physiol.* 269:L551–L559.

Pryor, W.A., and Squadrito, G.L. 1995. The chemistry of peroxynitrite: a product from the reaction of nitric oxide with superoxide. *Am. J. Physiol.* 268:L699–L722.

Punjabi, C.J., Laskin, J.D., Pendino, K.J., Goller, N.L., Durham, S.K., and Laskin, L. 1994. Production of nitric oxide by rat type II pneumocytes: increased expression of inducible nitric oxide synthase following inhalation of a pulmonary irritant. *Am. J. Respir. Cell Mol. Biol.* 11:165–172.

Radi, R., Beckman, J.S., Bush, K.M., and Freeman, B.A. 1991a. Peroxynitrite-induced membrane lipid peroxidation: the cytotoxic potential of superoxide and nitric oxide. *Arch. Biochem. Biophys.* 288:481–487.

Radi, R., Beckman, J.S., Bush, K.M., and Freeman, B.A. 1991b. Peroxynitrite oxidation of sulfhydryls. The cytotoxic potential of superoxide and nitric oxide. *J. Biol. Chem.* 266:4244–4250.

Radi, R., Rodriguez, M., Castro, L., and Telleri, R. 1994. Inhibition of mitochondrial electron transport by peroxynitrite. *Arch. Biochem. Biophys.* 308:89–95.

Reiter, R.J. 1998. Oxidative damage in the central nervous system: protection by melatonin. *Prog. Neurobiol.* 56:359–384.

Reiter, R.J., Melchiorri, D., Sewerynek, E., Poeggeler, B., Barlow-Walden, L., Chuang, J., Ortiz, G.G., and Acuna-Castroviejo, D. 1995. A review of the evidence supporting melatonin's role as an antioxidant. *J. Pineal Res.* 18:1–11.

Rengasamy, A., Xue, C., and Johns, R.A. 1994. Immunohistochemical demonstration of a paracrine role of nitric oxide in bronchial function. *Am. J. Physiol.* 267:L704–L711.

Robbins, R.A., Barnes, P.J., Springall, D.R., Warren, J.B., Kwon, O.J., Buttery, L.D., Wilson, A.J., Geller, D.A., and Polak, J.M. 1994a. Expression of inducible nitric oxide in human lung epithelial cells. *Biochem. Biophys. Res. Commun.* 203:209–218.

Robbins, R.A., Springall, D.R., Warren, J.B., Kwon, O.J., Buttery, L.D.K., Wilson, A.J., Adcock, I.M., Riveros-Moreno, V., Moncada, S., Polak, J., and Barnes, P.J. 1994b. Inducible nitric oxide synthase is increased in murine lung epithelial cells by cytokine stimulation. *Biochem. Biophys. Res. Commun.* 198:835–843.

Rodenas, J., Mitjavila, M.T., and Carbonell, T. 1995. Stimultaneous generation of nitric oxide and superoxide by inflammatory cells in rats. *Free Rad. Biol. Med.* 18:869–875.

Royall, J.A., Kooy, N.W., and Beckman, J.S. 1995. Nitric oxide-related oxidants in acute lung injury. *New Horiz.* 3:113–122.

Rubbo, H., Radi, R., Trujillo, M., Telleri, R., Kalyanaraman, B., Barnes, S., Kirrk, M., and Freeman, B.A. 1994. Nitric oxide regulation of superoxide and peroxynitrite-dependent lipid peroxidation. Formation of novel nitrogen-containing oxidized lipid derivatives. *J. Biol. Chem.* 269:26066–26075.

Ruggiero, D.A., Mtui, K., Otake, K., and Anwar, M. 1996. Central and primary visceral afferents to nucleus tractus solitarii may generate nitric oxide synthase as a membrane-permeant neuronal messeger. *J. Comp. Neurol.* 364:51–67.

Sadeghi-Hashjin, G., Folkerts, G., Henricks, P.A.J., Muijsers, R.B.R., and Nijkamp, F.P. 1998. Peroxynitrite in airway diseases. *Clin. Exp. Allergy* 28:1464–1473.

Sadeghi-Hashjin, G., Folkerts, G., Henricks, P.A.J., Verheyen, A.K.C.P., van der Linde, H.J., van Ark, I., Coene, A., and Nijkamp, F.P. 1996. Peroxynitrite induces airway hyperresponsiveness in guinea pigs in vitro and in vivo. *Am. J. Respir. Crit. Care Med.* 153:1697–1701.

Saleh, D., Ernst, P., Lim, S., Barnes, P.J., and Giaid, A. 1998. Increased formation of the potent oxidant peroxynitrite in the airways of asthmatic patients is associated with induction of nitric oxide synthase: effect of inhaled glucocorticoid. *FASEB* 12:929–937.

Salgo, M.G., Squadrito, G.L., and Pryor, W.A. 1995. Peroxynitrite causes apoptosis in rat thymocytes. *Biochem. Biophys. Res. Commun.* 215:1111–1118.

Salman-Tabcheh, S., Guerin, M.C., and Torreilles, J. 1995. Nitration of tyrosyl-residues from extra- and intracellular proteins in human whole blood. *Free Rad. Biol. Med.* 19:695–698.

Saran, M., and Bors, W. 1994. Signalling by O_2^- and NO: how far can either radical, or any specific reaction product, transmit a message under in vivo conditions? *Chem. Biol. Int.* 90:35–45.

Sato, K., Miyakawa, K., Takeya, M., Hattori, R., Yui, Y., Sunamoto, M., Ichimori, Y., Ushio, Y., and Takahashi, K. 1995. Immunohistochemical expression of inducible nitric oxide synthase (iNOS) in reversible endotoxic shock studied by a novel monoclonal antibody against rat iNOS. 57(1):36–44.

Sawada, H., Kawamura, T., Shimohama, S., Akaike, A., and Kimura, J. 1996. Different mechanisms of glutamate-induced neuronal death between dopaminergic and non-dopaminergic neurons in rat mesencephalic culture. *J. Neurosci. Res.* 43:503–510.

Schulz, R., and Wambolt, R. 1995. Inhibition of nitric oxide synthesis protects the isolated working rabbit heart from ischaemia-reperfusion injury. *Cardiovasc. Res.* 30:432–439.

Seo, H.G., Takata, I., Nakamura, M., Tatsumi, H., Suzuki, K., Fujii, J., and Taniguchi, N. 1995. Induction of nitric oxide synthase and concomitant suppression of superoxide dismutases in experimental colitis in rats. *Arch. Biochem. Biophys.* 324:41–47.

Shaul, P., North, A.J., Wu, L.C., Wells, L.B., Brannon, T.S., Lau, S., and Michel, T.B. 1994. Endothelial nitric oxide synthase is expressed in cultured human bronchial epithelium. *J. Clin. Invest.* 94:2231–2236.

Sherman, M.P., Griscavage, J.M., and Ignarro, L.J. 1992. Nitric oxide-mediated neuronal injury in multiple sclerosis. *Med. Hypotheses* 39:143–146.

Shimosegawa, T., and Toyota, T. 1994. NADPH-or diaphorase as a marker for nitric oxide synthase in neurons of the guinea pig respiratory tract. *Am. J. Respir. Crit. Care Med.* 150:1402–1410.

Simons, J.M., 't Hart, B.A., van Dijk, H., and Labadie, R.P. 1990. Metabolic activation of natural phenols into selective oxidative burst antagonists by activated human neutrophils. *Free Rad. Biol. Med.* 8:251–258.

Snyder, S.H., and Bredt, D.S. 1991. Nitric oxide as a neuronal messenger. *Trends Pharmacol. Sci.* 12:125–127.

Southan, G.J., Salzman, A.L., and Szabo, C. 1996. Potent inhibition of the isoform of nitric oxide synthase by aminoethylisoselenourea and related compounds. *Life Sci.* 58:1139–1148.

Southan, G.J., and Szabo, C. 1996. Selective pharmacological inhibition of distinct nitric oxide synthase isoforms. *Biochem. Pharmacol.* 51:383–394.

Stretton, D. 1991. Nonadrenergic, noncholinergic neural control of the airways. *Clin. Exp. Pharmacol. Pathol.* 18:675–684.

Su, W.Y., Day, B.J., Kang, B.H., Crapo, J.D., Huang, Y.C.T., and Chang, L.Y. 1996. Lung epithelial cell-released nitric oxide protects against PMN-mediated cell injury. *Am. J. Physiol.* 15:L581–L586.

Sutherland, K., Mahoney, J.R., Coury, A.J., and Eaton, J.W. 1993. Degradation of biomaterials by phagocyte-derived oxidants. *J. Clin. Invest.* 92:2360–2367.

Szabo, C. 1996. The pathophysiological role of peroxynitrite in shock, inflammation, and ischemia-reperfusion injury. *Shock* 6:1–10.

Szabo, C., and Salzman, A.L. 1995. Endogenous peroxynitrite is involved in the inhibition of mitochondrial respiration in immuno-stimulated J774.2 macrophages. *Biochem. Biophys. Res. Commun.* 209:739–743.

Szabo, C., Salzman, A.L., and Ischiropoulos, H. 1995. Endotoxin triggers the expression of an inducible isoform of nitric oxide synthase and the formation of peroxynitrite in the rat aorta in vivo. *FEBS Lett.* 363:235–238.

Szabo, C., Zingarelli, B., O'Connor, M., and Salzman, A.L. 1996. DNA strand breakage, activation of poly-ADP ribosyl synthetase, and cellular energy depletion are involved in the cytotoxicity in macrophages and smooth muscle cells exposed to peroxynitrite. *Proc. Natl. Acad. Sci. USA* 93:1753–1758.

Thomae, K.R., Geller, D.A., Billiar, T.R., Davies, P., Pitt, B.R., Simmons, R.L., and Nakayama, D.K. 1993. Antisense oligodeoxynucleotide to inducible nitric oxide synthase muscle cells in culture. *Surgery* 114:272–277.

Turnage, R.H., Kadesky, K.M., Bartula, L., and Myers, S.I. 1995. Intestinal reperfusion up-regulates inducible nitric oxide synthase activity within the lung. *Surgery* 118:288–293.

Villiotou, V., and Deliconstantinos, G. 1995. Nitric oxide, peroxynitrite and nitroso compounds formation by ultraviolet A (UVA) irradiated human squamous cell carcinoma: potential role of nitric oxide in cancer prognosis. *Anticancer Res.* 15:931–942.

Vliet, A. van der, Eiserich, J.P., Halliwell, B., and Cross, C.E. 1995a. Modification of aromatic amino acids by reactive nitrogen species. *Biochem. Soc. Trans.* 23:237S.

Vliet, A. van der, Eiserich, J.P., O'Neill, C.A., Halliwell, B., and Cross, C.E. 1995b. Tyrosine modification by reactive nitrogen species: a closer look. *Arch. Biochem. Biophys.* 319:341–349.

Vliet, A. van der, O'Neill, C.A., Halliwell, B., Cross, C.E., and Kaur, H. 1994a. Aromatic hydroxylation and nitration of phenylalanine and tyrosine by peroxynitrite. Evidence for hydroxyl radical production from peroxynitrite. *FEBS Lett.* 339:89–92.

Vliet, A. van der, Smith, D., O'Neill, C.A., Kaur, H., Darley-Usmar, V., Cross, C.E., and Halliwell, B. 1994b. Interactions of peroxynitite with human plasma and its constituents: oxidative damage and antioxidant depletion. *Biochem. J.* 303:295–301.

Ward, J.K., Barnes, P.J., Springall, D.R., Abelli, L., Tadjkarimi, S., Yacoub, M.H., Polak, J.M., and Belvisi, M.G. 1995. Distribution of human i-NANC bronchodilator and nitric oxide-immunoreactive nerves. *Am. J. Respir. Cell Mol. Biol.* 13:175–184.

Warner, R.L., Paine, R., Christensen, P.J., Marletta, M.A., Richards, M.K., Wilcosen, S.E., and Ward, P.A. 1995. Lung sources and cytokine requirements for in vivo expression of inducible nitic oxide synthase. *Am. J. Respir. Cell Mol. Biol.* 12:649–661.

Wei, X.Q., Charles, I.G., Smith, A., Ure, J., Feng, G.J., Huang, F.P., Xu, D., and Muller, W. 1995. Altered immune responses in mice lacking inducible nitric-oxide synthase. *Nature* 375:408–411.

White, A.C., Maloney, E.K., Boustani, M.R., Hassoun, P.M., and Fanburg, B.L. 1995. Nitric oxide increases cellular glutathione levels in rat lung fibroblasts. *Am. J. Respir. Cell Mol. Biol.* 13:442–448.

White, C.R., Brock, T.A., Chang, L.Y., Crapo, J., Briscoe, P., Ku, D., Bradley, W.A., Gianturco, S.H., Gore, J., and Freeman, B.A. 1994. Superoxide and peroxynitrite in athosclerosis. *Proc. Natl. Acad. Sci. USA* 91:1044–1048.

Wizemann, T.M., Gardner, C.R., Laskin, J.D., Quinones, S., Durham, S.K., Goller, N.L., Ohnishi, S.T., and Laskin, D.L. 1994. Production of nitric oxide and peroxynitrite in the lung during acute endotoxemia. *J. Leukocyte Biol.* 56:759–768.

Xie, Q., Cho, H.J., Calaycay, J., Mumford, R.A., Swiderek, K.M., Lee, T.D., Ding, A., Troso, T., and Nathan, C. 1992. Cloning and characterization of inducible nitric oxide synthase from mouse macrophages. *Science* 256:225–228.

Xue, C., Botkin, S.J., and Johns, R.A. 1996. Localization of endothelial NOS at the basal microtubule membrane in ciliated epithelium of rat lung. *J. Histochem. Cytochem.* 44:463–471.

Xue, C., Rengasamy, A., Le-Cras, T.D., Koberna, P.A., Dailey, G.C., and Johns, R.A. 1994. Distribution of NOS in normoxic vs. hypoxic rat lung: upregulation of chronic hypoxia. *Am. J. Physiol.* 267:L667–L678.

Yan, Z.Q., Hansson, G.K., Skoogh, B.E., and Lotvall, J.O. 1995. Induction of nitric oxide synthase in a model of allergic occupational asthma. *Allergy* 50:760–764.

Yan, Z.Q., Kramer, K., Bast, A., and Timmerman, H. 1994. The involvement of nitric oxide synthase in the effect of histamine on guinea-pig airway smooth muscle tone in vitro. *Agents Actions* 41:C111–C112.

Yeadon, M., and Price, R. 1995. Induction of calcium-independent nitric oxide synthase by allergen challenge in sensitized rat lung in vivo. *Br. J. Pharmacol.* 116: 2545–2546.

Yermilov, V., Rubio, J., and Ohshima, H. 1995. Formation of 8-nitroguanine in DNA treated with peroxynitrite in vitro and its rapid removal from DNA by depurination. *FEBS Lett.* 376:207–210.

Zhu, L., Gunn, C., and Beckman, J.S. 1992. Bactericidal activity of peroxynitrite. *Arch. Biochem. Biophys.* 298:452–457.

Zingarelli, B., Cuzzocrea, S., Zsengeller, Z., Salzman, A.L., and Szabo, C. 1997. Protection against myocardial ischemia and reperfusion injury by 3-aminobenzamide, an inhibitor of poly (ADP-ribose) synthase. *Cardiovasc. Res.* 36:205–215.

Zingarelli, B., O'Connor, M., Wong, H., Salzman, A.L., and Szabo, C. 1996. Peroxynitrite-mediated DNA strand breakage activates poly-adenosine diphosphate ribosyl synthase and causes cellular energy depletion in macrophages stimulated with bacterial lipopolysaccharide. *J. Immunol.* 156:350–358.

14 Nitric Oxide, Peripheral Neuropathy, and Diabetes

Norman E. Cameron and Mary A. Cotter

Introduction

Diabetes mellitus (DM) is characterized by abnormal carbohydrate and fat metabolism. The major classification is into type 1, or insulin-dependent DM, which results from autoimmune destruction of pancreatic β cells, and type 2, or non-insulin-dependent DM, which is associated with varying degrees of reduced insulin secretion and elevated tissue insulin resistance. Regardless of DM type, there is an elevated risk of atherosclerotic macrovascular disease, as well as microvascular complications affecting many vascular beds that are particularly significant for kidney, peripheral nerve, and retina. The combination of neuropathy (both sensory and autonomic) and impaired circulation also causes complications such as the "diabetic foot," which leads to ulceration and amputation (Boulton, 1997).

Diabetes mellitus affects all nerve fiber types; early functional alterations include reduced nerve conduction velocity (NCV) in large myelinated motor and sensory fibers (reviewed in Cameron and Cotter, 1994). These preclinical changes progress to fiber degeneration as the disease proceeds to clinical neuropathy. Clinical signs derived from large fiber damage include reduced or absent vibration sensitivity in distal limbs (particularly toes and feet) and impaired ankle reflexes. Small fibers and the autonomic nervous system are also greatly affected. Early changes include a reduction in heart rate variability to deep breathing or the Valsalva maneuver, indicating impaired autonomic reflexes. Thermal discrimination thresholds are elevated, but pain thresholds may be decreased in some patients who experience severe pain and hyperesthesia. The development of autonomic neuropathy involves several systems. This may cause postural hypotension; impotence and retrograde ejaculation; bladder distension; gastrointestinal effects such as delayed stomach emptying, diarrhea, or constipation; gustatory sweating; and dry feet (reviewed in Dyck and Dyck, 1999).

Population-based studies on the prevalence of clinical polyneuropathy give values of around 30%, although this varies with the diagnostic criteria employed. Major risk factors and associations with neuropathy include age and DM duration, height, glycemic control (glycated hemoglobin levels), microvascular abnormalities (retinopathy and nephropathy), hypertension, macrovascular disease, lipid abnormalities [elevated plasma triglycerides and low-density lipoprotein (LDL) cholesterol, decreased high-density lipoprotein (HDL) cholesterol], elevated fibrinogen, smoking, and high alcohol consumption (Melton and Dyck, 1999). Some of these factors are linked to endothelial abnormalities in other disease states, and therefore it is not surprising that endothelial dysfunction is a hallmark of DM.

Nitric oxide (NO) is implicated in the etiology of neuropathy and neuropathy-linked complications of DM in at least two ways. First, impaired blood flow makes a major contribution to the etiology of diabetic polyneuropathy. Defects in the production, release, or action of NO by endothelium have been generally implicated in micro- and macroangiopathy, and this applies to the vasa nervorum. Second, NO is a neurotransmitter in some autonomic fibers; impaired function of these nitrergic nerves makes a major contribution to autonomic symptoms, including impotence and delayed gastric emptying.

Nitric Oxide and Neurovascular Dysfunction in Diabetes Mellitus

Nerve Blood Flow and Neuropathy

Tuck et al. (1984) first showed a reduction in nerve perfusion as a primary factor in the etiology of neuropathic changes in the streptozotocin-induced diabetic rat model. This is an early phenomenon, occurring within 1 week of DM induction (Cameron et al., 1991b), which causes a marked reduction in nerve endoneurial oxygen tension (Tuck et al., 1984; Cameron et al., 1994a). Evidence in support of the relationship between nerve dysfunction in DM and impaired perfusion comes from experiments that mimic the ischemic/hypoxic conditions in nondiabetic rats and that correct the perfusion deficit without correcting the hyperglycemic state in diabetic rats. Thus, NCV is compromised when nondiabetic rats are reared under chronic hypoxia (Low et al., 1986), when hind limb ischemia is produced by an arteriovenous shunt (Sladsky et al., 1991), during chronic pharmacological manipulation [combined treatment with low doses of NO synthase (NOS) and cyclooxygenase inhibitors] to reduce endoneurial perfusion (Cameron and Cotter, 1994). Conversely, NCV is improved when diabetic rats are exposed to chronic hyperbaric oxygenation (Low et al., 1988), or when nerve blood flow deficits are corrected by vasodilators (Cameron and Cotter, 1994). Parallels may be drawn with the limited human DM literature, where impaired nerve perfusion and endoneurial hypoxia were found in neuropathic patients, and chronic vasodilator treatment gave some functional improvements (reviewed in Tesfaye et al., 1994).

Nitric Oxide and Vascular Changes in Diabetes

The reduction in nerve blood flow in experimental DM has been linked to an impairment of the vasa nervorum NO system. Evidence for this includes a marked diabetic deficit in the effect of N^G-nitro-L-arginine (NOLA) suffusion of epi- and perineurial vessels on resting endoneurial blood flow (Figure 14.1a), which is derived primarily from spontaneously liberated and flow-induced NO production (Kihara and Low, 1995). Furthermore, norepinephrine preconstricted vasa nervorum from DM rats show reduced NO vasodilatation (Figure 14.1b) in response to acetylcholine suffusion. This was not due to changes in vasa nervorum smooth muscle NO sensitivity, since responses to the donor, glyceryl trinitrate, did not differ in diabetic and nondiabetic rats (Figure 14.1c), which implicates the endothelium as a major site of dysfunction in DM. The decreased vasa nervorum NO enhances smooth muscle reactivity to vasoconstrictors such as norepinephrine, and this is exacerbated by elevated production of the potent vasoconstrictors endothelin 1, acting on vasa nervorum ET_A receptors, and angiotensin II, acting on AT_1 receptors (Cameron and Cotter, 1994). Thus, a dual effect of depressed vasodilatation and enhanced vasoconstriction is primarily responsible for the nerve ischemia in diabetic neuropathy.

Data from drug suffusion experiments show that the vasa nervorum NO deficit is relative (Figure 14.1), and flow- and acetylcholine-induced NO production, albeit of reduced magnitude, probably remains important for the maintenance of blood flow in DM rats. Perfusion and NCV defects in diabetic rats may be corrected by α_1-adrenoceptor antagonists acting on smooth muscle receptors, or vasodilator prostanoid precursors such as γ-linolenic acid (GLA) acting via the endothelium (Cameron and Cotter, 1994). However, cotreatment with a low dose of NOLA which does not appreciably alter neurovascular function in nondiabetic rats, negates the beneficial effects of α_1-adrenoceptor antagonists or GLA (Cameron et al., 1996; Cotter and Cameron, 1998).

The NO-related effects seen for vasa nervorum, namely a reduction in endothelium production, release, or action, coupled with unchanged sensitivity of vascular smooth muscle to nitrodilators, are similar to findings for several other vascular beds and large and small vessels in experimental models and DM patients (reviewed in Cameron and Cotter, 1994; Pieper, 1998). However, other factors can complicate interpretation, because, in some species and vascular beds, DM causes an increase in acetylcholine-stimulated vasoconstrictor prostanoid release, which opposes NO-mediated responses, further exacerbating the deficit in endothelium-dependent relaxation (Cohen, 1993). Moreover, in some vessels, endothelium-derived hyperpolarizing factor (EDHF) is also involved in acetylcholine-induced relaxation, and recent studies have shown that DM markedly impairs EDHF responses in the rat mesenteric vasculature (Fukao et al., 1997; Jack et al., 1998).

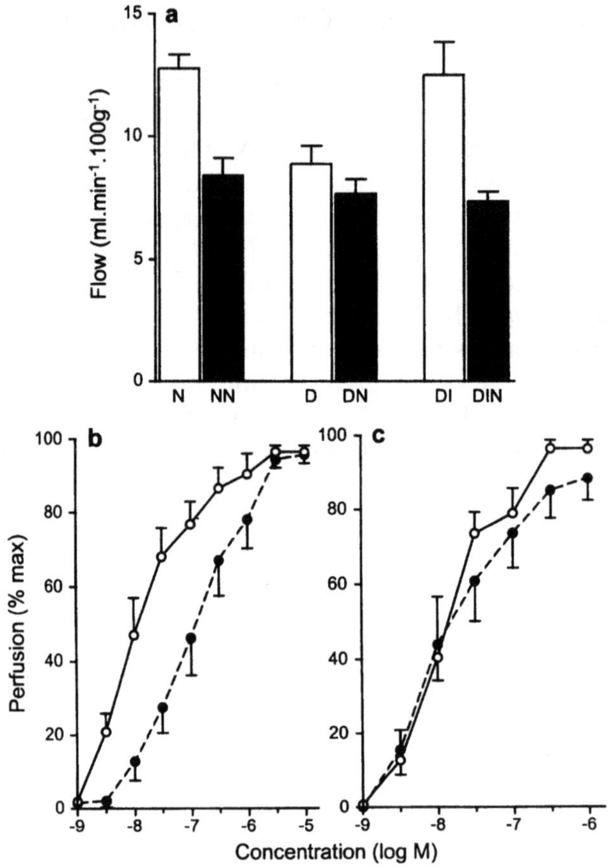

FIGURE 14.1. Effects of diabetes on vasa nervorum in rats. Sciatic endoneurial blood flow (a), measured by hydrogen clearance microelectrode polarography, for nondiabetic rats in the absence (N) or presence (NN) of N^G-nitro-L-arginine (NOLA) suffusion of epi- and perineurial vessels and the effects of diabetes (D and DN) and insulin treatment (DI and DIN). NOLA suffusion reduced blood flow in controls. Flow was diminished by diabetes, and NOLA suffusion did not cause a further decrease, suggesting a relative NO deficit for resting blood flow. Insulin treatment corrected the diabetic flow reduction and restored responses to NOLA (after Kihara and Low 1995). Laser-Doppler techniques were used to monitor increases in sciatic nerve perfusion, after preconstriction with norepinephrine, for suffusion of acetylcholine (b) and glyceryl trinitrate (c). Compared with nondiabetic rats (open circles), the dose-response curve for acetylcholine in diabetic rats (closed circles) was shifted approximately 1 log unit to the right, indicating reduced sensitivity. This contrasts with responses to glyceryl trinitrate, which were similar in diabetic and nondiabetic groups, indicating that the sensitivity of smooth muscle itself to NO was not altered by diabetes (after Maxfield et al., 1997). Data are means ± SEM.

Studies on patients with type 1 and type 2 DM generally reveal NO-related vascular deficits from in vivo measurements and in vitro experiments on biopsy material, although where metabolic control was good and in the absence of other markers of microvascular disease, abnormalities were minimal (Smits et al., 1993). When microalbuminuria was present, however, the endothelial NO system was impaired (Elliott et al., 1993). Endothelial function is also compromised by lipid abnormalities (elevated triglycerides and LDL cholesterol) in patients with poorly controlled diabetes (Watts et al., 1996). The situation regarding sensitivity to nitrodilators is less clear than in animal models. Several human studies showed no changes with DM, but a similar number found attenuated responses (Pieper, 1998).

Thus, DM causes defective endothelium-dependent vasodilatation in experimental models and in patients, particularly, if they have poor glycemic control and evidence of microvascular disease. The NO system is badly affected, giving rise to impaired perfusion of several vascular beds, including nerve and retina (Cameron and Cotter, 1994; Tesfaye et al., 1994; Koya and King, 1998). An understanding of the pathogenesis of NO defects in DM is, therefore, of vital importance to guide the development of therapeutic interventions against the major complications of this disease.

Mechanisms of Nitric Oxide-Related Endothelial and Nerve Dysfunction

Hyperglycemia is thought to be the main cause of eNO defects in DM. Acute exposure to elevated glucose in vitro and in vivo can impair vasodilatation to acetylcholine in some vessels and vascular beds of nondiabetic animals (Cohen, 1993; Mayhan and Patel, 1995). It also results in reduced nerve blood flow (Cameron et al., 1991b); however, other vessels, such as rat aorta or corpus cavernosum, are resistant to acute glucose effects (Archibald et al., 1996; Keegan et al., 1999). In nondiabetic humans, 24 hours of local forearm hyperglycemia had no effect on acetylcholine-stimulated endothelium-dependent vasodilation (Houben et al., 1996). Insulin may be a confounding factor under some conditions, because it stimulates eNO production to increase blood flow (Steinberg et al., 1994). In nondiabetic rats, acute insulin treatment (with glucose infusion to prevent hypoglycemia) increased vasa nervorum arteriovenous shunting, and this was accompanied by a reduction in endoneurial oxygen tension (Kihara et al., 1994). This phenomenon could be important for insulin neuritis in patients, where a painful neuropathy can rapidly develop following instigation of tight metabolic control with insulin (Tesfaye et al., 1996). However, in the longer term, insulin treatment that produces good glycemic control prevents nerve blood flow and NCV deficits in experimental models and slows the progression of neuropathy in patients (Cameron and Cotter, 1994).

One factor of potential importance for endothelial dysfunction in DM is reduced availability of the substrate for NOS, L-arginine. For unknown reasons and despite hyperphagia, plasma L-arginine levels are decreased in diabetic rats. The deficit in endothelium-dependent NO-mediated relaxation of aorta from DM rats, measured in vitro, was corrected by adding L-arginine (but not D-arginine) to the bathing fluid (Pieper, 1998). However, in other studies, no corrective effect of L-arginine was apparent (Archibald et al., 1996; Mayhan et al., 1996). It is unlikely that a vasa nervorum L-arginine deficiency hypothesis can explain neuropathy. Thus, NCV deficits in DM rats are prevented by NO donor treatment (Cameron and Cotter, 1995a), indicating that vasa nervorum smooth muscle retains NO sensitivity, whereas a fivefold increase in dietary L-arginine intake was without effect (K.C. Dines, N.E. Cameron, and M.A. Cotter, unpublished observations).

The major hypotheses of nerve dysfunction in DM involve mechanisms strongly or entirely linked to the metabolic consequences of hyperglycemia. Those that impinge directly on the vasa nervorum NO-system include enhanced polyol pathway activity, oxidant stress, advanced glycation reactions, and elevated protein kinase C (PKC) activity. Non-NO-related vascular mechanisms include impaired ω-6 essential fatty acid and prostanoid metabolism, which interacts synergistically with the NO mechanism (Cameron and Cotter, 1999). There may also be nonvascular effects on nerve fiber growth and maintenance caused by deficiencies of trophic agents (Feldman and Windebank, 1999).

Polyol Pathway

Hyperglycemia increases polyol pathway flux, causing elevated production of the sugar alcohol, sorbitol, catalyzed by aldose reductase (AR). Sorbitol is then further metabolized to fructose by sorbitol dehydrogenase. A considerable number of studies over the past 20 years have shown that aldose reductase inhibitors (ARIs) can prevent or correct motor and sensory NCV deficits in DM models and in the longer term attenuate some of the abnormalities in nerve morphology, such as axon dwindling and dystrophic changes in autonomic nerves (reviewed in Greene et al., 1992; Cameron and Cotter, 1994; Yagihashi, 1995). Clinical trials of ARIs have shown modest improvements in neuropathic symptoms, nerve function, and morphometry (Sima et al., 1993). The precise function of the polyol pathway in most tissues is unknown, although in the renal medulla, production of sorbitol (which does not readily cross cell membranes) may have an osmolyte function to maintain cell volume in the hypertonic environment produced by the loop of Henle (Burg and Kador, 1988).

Neurochemical hypotheses of the mechanism connecting polyol pathway activity with nerve dysfunction have focused on osmotic damage due to nerve sorbitol accumulation, or on a concomitant reduction in nerve content of another putative osmolyte, myoinositol (Stevens et al., 1993). In turn, this deficit was thought to reduce phosphoinositide turnover and decrease the availability of second messengers such as diacylglycerol (DAG), which, via stimulation of PKC, would normally maintain a high level of nerve Na^+-K^+-ATPase pump activity. The compromised axon and Schwann cell ionic homeostasis would lead to reduced NCV and neurodegeneration (Greene et al., 1992). These hypotheses, however, have been unable to explain several findings. Thus, sorbitol accumulation in nerve does not appear to be important; interventions that do not alter sorbitol and myoinositol levels, for example using vasodilators or antioxidants, nonetheless correct NCV deficits in diabetic rats (Cameron et al., 1991a). Furthermore, although ARI treatment does correct nerve Na^+-K^+-ATPase activity, this has been dissociated from NCV changes (Cameron et al., 1991a; Stevens et al., 1996). An alternative hypothesis is that the polyol pathway contributes to impaired nerve blood flow in diabetic rats. Treatment with ARIs corrects nerve blood flow and endoneurial oxygen tension deficits (Cameron et al., 1994b; Hotta et al., 1995). In other vessels, ARI treatment prevents the development of impaired NO-mediated endothelium-dependent relaxation (Cameron and Cotter, 1992; Cohen, 1993; Otter and Chess-Williams, 1994). Compared to ARIs, sorbitol dehydrogenase inhibitors have little effect on NCV and endoneurial blood flow in diabetic rats (Cameron et al., 1997); therefore, the first half of the polyol pathway is most important for the etiology of neurovascular dysfunction. This conclusion is reinforced by data from another model of polyol pathway hyperactivity, the galactose-fed rat. Galactose is a better substrate for AR than glucose, and high levels of galactitol accumulate in tissues when rats are fed a diet containing 20%–50% galactose. Galactitol is a poor substrate for sorbitol dehydrogenase, so only the AR step is stimulated. Furthermore, insulin and glucose levels remain normal. Galactosemic rats develop similar changes in nerve function, blood flow, and morphometry to those with DM (Myers and Powell, 1984; Kamijo et al., 1994) and also have an ARI-preventable deficit in aorta endothelium-dependent NO-mediated relaxation (Cameron and Cotter, 1993).

In addition to Schwann cell cytoplasm, AR is found in endothelium and smooth muscle of arterial vessels in the vasa nervorum (Powell et al., 1991). The polyol pathway could theoretically modulate cell function via effects on NADPH and NADH availability. Thus, NADPH is a cofactor for AR, NOS, and glutathione reductase. Competition between aldose reductase and NOS for NADPH may contribute to reduced endothelial NO production. However, the affinity of NOS for NADPH is an order of magnitude greater than that of AR, so the latter would have to be markedly in excess in endothelial cells to account for the impaired NO-mediated vasodilation. A second candidate for an NADPH competitive effect, glutathione reductase, catalyzes the conversion of glutathione from its oxidized (GSSG) to reduced (GSH) form. GSH is a major endogenous antioxidant, used to neutralize peroxides, and it is decreased by DM (and galactosemia) in several tissues, including lens and nerve, which both contain high levels of AR (Lou et al., 1988; Nagamatsu et al., 1995; Hohman et al., 1997). However, at least for nerve, the effect of DM is not as predicted from simple interference with glutathione reductase action. Thus, the GSH/GSSG ratio would be diminished, but total glutathione (GSSG + GSH) would be unchanged. However, DM causes a reduction in total glutathione, which is restored by ARI treatment (Hohman et al., 1997), implying effects on glutathione synthesis. The mechanism responsible is unknown, but it does not depend on improving nerve blood flow, because vasodilator treatment does not restore nerve GSH content.

A third possibility, with vascular implications, concerns flux through the second half of the polyol pathway, which produces NADH. An increase in the cellular NADH/NAD ratio has been suggested to cause a state of "reductive stress" or "pseudohypoxia" (Williamson et al., 1993). This could stimulate the production of reactive oxygen species (ROS) in endothelial cells by NADH oxidase (Harrison, 1997), which may cause cellular damage as well as destroying NO. For nerve and vasa nervorum, however, the lack of effect of sorbitol dehydrogenase inhibitor treatment on blood flow or NCV in DM rats argues against the relevance of this putative mechanism. Thus, although there are theoretical possibilities whereby polyol pathway activity contributes to the etiology of diabetic neuropathy by modulating NADPH and NADH levels, currently there is little hard evidence to support this view. However, the effects of DM and ARI treatment on tissue GSH levels, coupled with evidence that ARIs reduce markers of oxidant stress such as malondialdehyde (Lowitt et al., 1995), suggest an indirect antioxidant effect likely to be relevant for neurovascular dysfunction.

Oxidant Stress

The GSH deficit, consequent on polyol pathway hyperactivity, reduces tissue antioxidant defenses in DM. Against this backdrop, there is increased ROS formation. Sources of ROS include autoxidation of glucose, the advanced glycation process, altered prostanoid production, inefficient mitochondrial function, and the NAD(P)H oxidase system (Baynes, 1991; Wolff, 1993; Harrison, 1997). Reactive oxygen species also result from the blood-flow problems they cause during ischemia–reperfusion episodes. Another potential source that may be relevant during infection and inflammatory disease is the macrophage respiratory burst. Nitric oxide is an important target for ROS; superoxide neutralizes NO to form peroxynitrite. This reduces endothelium-dependent relaxation; the peroxynitrite also nitrosylates proteins, altering their function, and breaks down to form hydroxyl radicals that damage the endothelium (Beckman et al., 1990). Chronic treatment with the ROS scavengers vitamin E, butylated hydroxytoluene, acetylcysteine, α-lipoic acid, or the hydroxyl radical scavenger, dimethylthiourea, protects the vascular NO system in DM rats (Rösen et al., 1995; Archibald et al., 1996; Jack et al., 1998; Pieper, 1998). Many studies have shown that ROS scavenger treatment prevents and reverses defective nerve blood flow, endoneurial oxygen tension, NCV changes, and morphological changes in DM rats (reviewed in Cameron and Cotter, 1994; Low et al., 1997). The scavengers used include those listed above for endothelial dysfunction, as well as GSH, probucol (Cameron et al., 1994a), vitamin C, and β-carotene (Cotter et al., 1995). Generally, both lipophilic and hydrophilic scavengers were effective, although oral GSH only gave partial protection. The dose-response curve for vitamin C had an inverted "U" shape; decreased efficacy at high doses probably revealed a prooxidant action due to autoxidation. Extremely high doses of scavengers are necessary to preserve neurovascular function in DM rats; for example 1 g/kg/day of vitamin E was needed for 80% protection (Cotter et al., 1995). This would limit clinical usage, although α-lipoic acid trials revealed some benefits for symptomatic and autonomic neuropathy in patients (Ziegler et al., 1997).

LDL is oxidized in DM (Baynes, 1991), the lipid content undergoing peroxidation chain reactions that could be an important target for lipophilic scavengers. This source of hydroxyl radicals causes endothelial damage (Morel and Chisolm, 1989). Endothelium-dependent relaxation in diabetic mice was protected by treatment with the lipid-lowering drugs cholestyramine and taurine (Kamata et al., 1996a,b). Elevated LDL cholesterol is a risk factor for neuropathy and endothelial dysfunction in patients (Watts et al., 1996; Melton and Dyck, 1999). The lipid-lowering approach has not been applied to vasa nervorum; however, taurine prevents NCV deficits in DM rats (Stevens et al., 1993).

An alternative approach to the use of scavengers is to prevent ROS formation by autooxidation and the Fenton reaction, which are catalyzed by free transition metal ions. This can be accomplished using transition metal chelators. Low doses of deferoxamine and trientine (which chelate iron and copper) completely corrected sciatic nerve blood flow and NCV deficits in DM rats (Cameron and Cotter, 1995b). For vasa nervorum, the diabetic NO deficit markedly increases reactivity to norepinephrine (Maxfield et al., 1997), and this was corrected by deferoxamine. Chronic deferoxamine treatment prevented defective aorta endothelium-dependent relaxation (Pieper, 1998).

Thus, studies with scavengers and chelators highlight the importance of ROS for diabetic neuropathy and defects in the vascular NO system. Antioxidant treatment has a general protective effect on the endothelium; other vasodilator systems (prostacyclin and EDHF) deleteriously affected by DM also benefit (Karpen et al., 1982; Jack et al., 1998). The magnitude of chelator effects emphasizes the role of transition metal-catalyzed reactions for ROS production in DM.

Advanced Glycation

Advanced glycation end products (AGEs) are produced in several ways. The nonenzymatic reaction of glucose or other reducing sugars with protein amino groups gives rise to the formation of reactive dicarbonyl intermediates by autooxidation, which then further react with protein amino groups to form a variety of AGEs (Brownlee, 1992; Bucala et al., 1995). It is difficult to separate the effects of oxidant stress from changes due to the advanced glycation process, because the latter produces ROS and is accelerated by ROS. Transition metal-catalyzed autooxidation is also involved in AGE formation (Baynes, 1991). AGEs cause pathological changes by ROS generation; increased protein crosslinking; rapid modification of intracellular protein properties by highly reactive metabolic intermediates of glucose, including those related to polyol pathway activity (Hamada et al., 1996); and binding to specific cell-surface receptors for AGEs. Once activated, AGE receptors mediate signal transduction through ROS generation to activate the transcription factor, nuclear factor κB, which is involved in cell injury responses. AGEs affect the NO system, acting via ROS generation and also directly to quench NO, which impairs vascular endothelium-dependent relaxation to acetylcholine (Bucala et al., 1995). Antioxidant treatments inhibit AGE formation and the effects of AGE receptor activation (Baynes, 1991). Another compound, aminoguanidine, reacts with dicarbonyl intermediates to inhibit AGE formation. In DM rats, aminoguanidine prevents and corrects nerve blood flow and NCV deficits and prevents deficits in neuronal morphology (Yagihashi, 1995; Cameron and Cotter, 1996a; Low et al., 1997). It also prevents impaired endothelium-dependent relaxation to acetylcholine in DM rat aorta (Archibald et al., 1996). Aminoguanidine is also a relatively specific iNOS inhibitor, although this does not explain the beneficial neurovascular effects, which are abolished by cotreatment with a low dose of NOLA (Cameron and Cotter, 1996a).

It is clear that oxidant stress, polyol pathway, and advanced glycation are intimately linked in the pathogenesis of neural and vascular changes in DM, which is schematized in Figure 14.2. Another mechanism, PKC activation, may in part act to mediate oxidant stress-related neurovascular effects.

Protein Kinase C

Diabetes mellitus activates PKC in vascular tissue; directly by de novo DAG synthesis from glucose and indirectly by oxidant stress (Figure 14.2). In the retina, aorta, and heart, vitamin E treatment decreases DAG levels and PKC activation in experimental DM (reviewed in Koya and King, 1998). There is an early reduction in retinal blood flow paralleling that for nerve in DM rats. In both tissues, blood flow was restored by PKC inhibitors, including a specific PKCβ blocker, and NCV deficits in nerve were corrected (Cameron et al., 1998b; Koya and King, 1998; Nakamura et al., 1998; Jack et al., 1999). The beneficial effects of PKC inhibition are unlikely to be due to a direct action on the neurons (as opposed to the vasa nervorum), because nerve DAG levels and PKC activity are either unchanged or reduced by DM (Zhu and Eichberg, 1990; Borghini et al., 1994). Protein

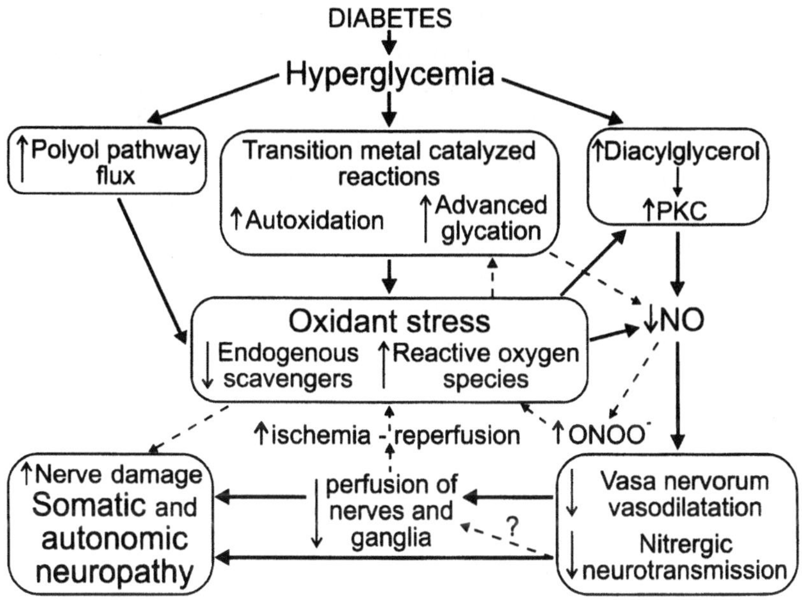

FIGURE 14.2. Scheme depicting the NO-related mechanisms that may be respon sible for impaired nerve blood flow and nitrergic neurotransmission, which leads to neuropathy in diabetes. PKC, protein kinase C; $ONOO^-$, peroxynitrite.

kinase C inhibition also attenuated the development of defective NO- and EDHF-mediated relaxation in mesenteric vessels of DM rats (Jack et al., 1998). Protein kinase C effects are complex; they include increased expression of endothelin and reduced expression of consitative NOS (cNOS), and phosphorylation to decrease cNOS activity (Ohara et al., 1995).

Interactions with Essential Fatty Acid–Prostanoid Mechanisms

Polyunsaturated fatty acids are a major target for ROS damage. This includes the ω-6 essential components derived from dietary linoleic acid, which is metabolized by a series of desaturation and elongation steps to produce the important prostanoid precursor arachidonic acid. Diabetes mellitus impairs the rate-limiting desaturation steps, particularly Δ-6 desaturation, which is responsible for GLA synthesis. This reduces vascular endothelium prostacyclin production and has deleterious effects on vasodilatation in concert with the NO and EDHF deficits. GLA treatment improves NCV and nerve blood flow and restores vasa nervorum prostacyclin production in DM rats (reviewed in Cameron and Cotter, 1999).

The endothelial vasodilator systems do not act in isolation, but rather are mutually interactive. In an experiment where DM rats were given combined treatment with low doses of evening primrose oil (a source of GLA) to stimulate the prostacyclin system and an ARI to correct NO deficits, the effects of combined treatment on blood flow and NCV greatly exceeded those predicted for an additive drug effect. The data showed an approximately 10-fold amplification of drug efficacy (Cameron et al., 1996). There were similar findings using antioxidants (probucol analogue or α-lipoic acid) in combination with GLA (Cameron and Cotter, 1996b; Cameron et al., 1998a). Recent studies showed that the latter combination prevented deficits in corpus cavernosum NO-mediated endothelium-dependent relaxation in DM rats (A. Keegan, M.A. Cotter, and N.E. Cameron, unpublished observations). Thus, it may be possible to harness these marked synergies using polypharmacy to gain a therapeutic advantage in the treatment of diabetic neuropathy.

Effects of Diabetes on Small Nerve Fiber Function and Neurotransmission

The mechanisms responsible for the deleterious effects of DM have been studied in much less detail in autonomic and small sensory fibers than in the large fibers. It is plausible that reduced blood flow contributes to small fiber dysfunction, linked to oxidant stress as outlined in Figure 14.2. The cell bodies of these neurons may be particularly vulnerable. Thus, blood flow is normally three- to fivefold higher in autonomic and dorsal root ganglia than in peripheral nerve, to match the metabolic demands of cell bodies. In DM rats there were 40% deficits in the perfusion of dorsal root and superior cervical ganglia and neuronal cell body mitochondria were damaged (Sasaki et al., 1997), and there was a reduction in the number of nNOS-containing neurons (Schmidt et al., 1993).

Corpus cavernosum innervation has been investigated because of the problem of impotence in DM. Saenz De Tejada et al. (1989) found impaired endothelium-dependent and neurogenic relaxation of corpus cavernosum tissue obtained from impotent diabetic men undergoing penile prosthesis implantation. Similar findings have been made for corpus cavernosum from DM rabbits and rats studied in vitro (Azadzoi and Saenz De Tejada, 1992; Keegan et al., 1999). In rats, this is accompanied by impaired penile reflexes in vivo (Vernet et al., 1995; Rehman et al., 1997).

The effects of short-term (1–2 months) DM in rats were selective on corpus cavernosum innervation. Thus, contractile responses mediated by noradrenergic nerves were normal, whereas relaxation to stimulation of nonadrenergic, noncholinergic (NANC) fibers (Figure 14.3) was markedly impaired. These NANC fibers contain nNOS, often colocalized with vasoactive intestinal peptide (VIP). In rats, NO appears to be the main neuro-

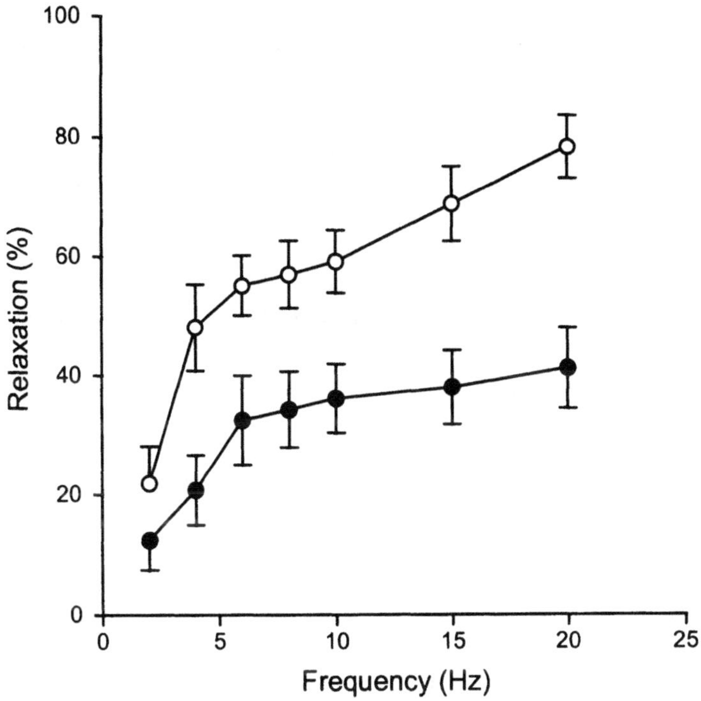

FIGURE 14.3. Effects of 8 weeks of diabetes on corpus cavernosum relaxation to transmural electrical field stimulation. Corpora cavernosa were precontracted with phenylephrine (30 μM), and preincubated with guanethidine (5 μM) and atropine (1 μM) to inhibit noradrenergic and cholinergic fiber responses, respectively. Data are means ± SEM ($n = 10$). Unpaired Student's t tests showed that responses were significantly ($P < 0.05$) greater in nondiabetic (*open circles*) than diabetic (*closed circles*) rats at frequencies above 2 Hz (A. Keegan, M.A. Cotter, and N.E. Cameron, unpublished observations).

transmitter (Ignarro et al., 1990), and responses are completely blocked by NOS inhibitors (Keegan et al., 1999); therefore, these nerves are primarily nitrergic. Penile nNOS levels and total NOS activity were 33%–55% reduced after 4 to 8 months in rat genetic models of type 1 and type 2 DM (Vernet et al., 1995). Similarly, nNOS immunoreactivity was reduced in penile tissue from DM patients with neurogenic impotence (Ehmke et al., 1995). However, the question of how DM changes nNOS or cNOS expression in corpus cavernosum, or indeed other vascular tissues (reviewed in Pieper, 1998), remains unsettled. Thus, in streptozotocin-induced DM rats, total penile NOS activity and cNOS levels were increased over the first 3 months (Elabbady et al., 1995; Sullivan et al., 1996), when nitrergic nerve (Figure 14.3) and endothelial NO responses were degraded (Keegan et al.,

1999). However, even if NOS levels are unchanged or enhanced in experimental models, the increased ROS in DM would reduce endothelium-dependent and nitrergic vasodilatation (Figure 14.2). Antioxidant treatment with α-lipoic acid prevented the effects of 2 months of DM, and there was partial (60%) correction of existing deficits in NO-mediated responses for treatment during the second month (Keegan et al., 1999). Recent experiments have also shown protection of endothelium and nitrergic nerves by an ARI and the metal chelator, trientine (A. Keegan, M.A. Cotter, and N.E. Cameron, unpublished observations). Thus, at least some of the ROS-related mechanisms outlined in Figure 14.2 may be responsible for penile dysfunction in DM rats.

Nitrergic neurotransmission is also impaired in anococcygeus muscle, stomach, and duodenum of DM rats (Way and Reid, 1994; Jenkinson and Reid, 1995; Martinezcuestav et al., 1995), and gastric myenteric plexus nNOS expression is reduced (Takahashi et al., 1997). The gastrointestinal NO system is important for controlling motility and sphincter relaxation, and therefore dysfunction probably contributes to autonomic symptoms such as delayed stomach emptying in DM patients. Attempts at preventing the anococcygeus NANC deficit in DM rats using the ARI sorbinil or the AGE inhibitor aminoguanidine were unsuccessful (Way and Reid, 1994, 1996). This could indicate a difference in pathogenic mechanisms compared with corpus cavernosum or nerve vascular supply, although sorbinil is not a very potent ARI and it is unclear whether the high-level blockade needed to protect nerve function was attained (Cameron et al., 1994b). Moreover, it is not known whether the polar molecule aminoguanidine can penetrate axon terminals to prevent the rapid intracellular AGE and ROS formation that may compromise NO release. Thus, further studies are needed to determine the extent to which the etiology of endothelial and neural NO deficits differs in DM.

Conclusions

Over the last 10 years, work on animal models has identified a defective NO mechanism in the etiology of diabetic neuropathy. Endothelial dysfunction in the vasa nervorum, linked to impaired NO-mediated vasodilatation, leads to ischemic nerve damage. Although a specific defect in vasa nervorum NO production has not been identified in DM patients, it is highly plausible, since NO deficits have been found in other vascular beds. Nitrergic nerve dysfunction appears similar in patients and animal models. Dysfunction is closely linked to hyperglycemia, oxidant stress, and PKC activation, and a greater understanding of these mechanisms and their interrelations provides some hope that it will be soon be possible to find effective therapeutic interventions against diabetic polyneuropathy.

References

Archibald, V., Cotter, M.A., Keegan, A., and Cameron, N.E. 1996. Contraction and relaxation of aortas from diabetic rats: effects of chronic anti-oxidant and aminoguanidine treatments. *Naunyn-Schmeideberg's Arch. Pharmacol.* 353: 584–591.

Azadzoi, K.M., and Saenz De Tejada, I. 1992. Diabetes mellitus impairs neurogenic and endothelium-dependent relaxation of rabbit corpus cavernosum smooth muscle. *J. Urol.* 148:1587–1591.

Baynes, J.W. 1991. Role of oxidative stress in the development of complications in diabetes. *Diabetes* 40:405–412.

Beckman, J.S., Beckman, T.W., Chen, J., Marshall, P.A., and Freeman, B.A. 1990. Apparent hydroxyl radical production by peroxynitrite: implications for endothelial injury from nitric oxide and superoxide. *Proc. Natl. Acad. Sci. USA* 87:1620–1624.

Borghini, I., Ania-Lahuerta, A., Regazzi, R., Ferrari, G., Gjinovci, A., Wollheim, C.B., and Pralong, W.-F. 1994. α, βI, βII, δ, and ε protein kinase C isoforms and compound activity in the sciatic nerve of normal and diabetic rats. *J. Neurochem.* 62:686–696.

Boulton, A.J.M. 1997. Late sequelae of diabetic neuropathy. In: *Diabetic Neuropathy* (ed. A.J.M. Boulton), pp. 63–76. Marius Press, Carnforth.

Brownlee, M. 1992. Glycation products and the pathogenesis of diabetic complications. *Diabetes Care* 15:1835–1843.

Bucala, R., Cerami, A., and Vlassara, H. 1995. Advanced glycosylation end products in diabetic complications. Biochemical basis and prospects for therapeutic intervention. *Diabetes Rev.* 3:258–268.

Burg, M.B., and Kador, P.F. 1988. Sorbitol, osmoregulation and the complications of diabetes. *J. Clin. Invest.* 81:635–640.

Cameron, N.E., and Cotter, M.A. 1992. Impaired contraction and relaxation in aorta from streptozotocin-diabetic rats: role of the polyol pathway. *Diabetologia* 35:1011–1019.

Cameron, N.E., and Cotter, M.A. 1993. Contraction and relaxation of aortas from galactosaemic rats and the effects of polyol pathway inhibition. *Eur. J. Pharmacol.* 243:47–53.

Cameron, N.E., and Cotter, M.A. 1994. The relationship of vascular changes to metabolic factors in diabetes mellitus and their role in the development of peripheral nerve complications. *Diabetes Metab. Rev.* 10:189–224.

Cameron, N.E., and Cotter, M.A. 1995a. Effects of chronic treatment with a nitric oxide donor on nerve conduction abnormalities and endoneurial blood flow in streptozotocin-diabetic rats. *Eur. J. Clin. Invest.* 25:19–24.

Cameron, N.E., and Cotter, M.A. 1995b. Neurovascular deficits in diabetic rats: potential contribution of autoxidation and free radicals examined using transition metal chelating agents. *J. Clin. Invest.* 96:1159–1163.

Cameron, N.E., and Cotter, M.A. 1996a. Rapid reversal by aminoguanidine of the neurovascular effects of diabetes in rats: modulation by nitric oxide synthase inhibition. *Metabolism* 45:1147–1152.

Cameron, N.E., and Cotter, M.A. 1996b. Interaction between oxidative stress and γ-linolenic acid in impaired neurovascular function of diabetic rats. *Am. J. Physiol.* 271:E471–E476.

Cameron, N.E., and Cotter, M.A. 1999. Role of linolenic acid in diabetic polyneuropathy. In: *Diabetic Neuropathy*, 2nd ed. (eds. P.J. Dyck and P.K. Thomas), pp. 359–367. Saunders, Philadelphia.

Cameron, N.E., Cotter, M.A., Archibald, V., Dines, K.C., and Maxfield, E.K. 1994a. Anti-oxidant and pro-oxidant effects on nerve conduction velocity and endoneurial blood flow and oxygen tensions in non-diabetic and streptozotocin-diabetic rats. *Diabetologia* 37:449–459.

Cameron, N.E., Cotter, M.A., Basso, M., and Hohman, T.C. 1997. Comparison of the effects of inhibitors of aldose reductase and sorbitol dehydrogenase on neurovascular function, nerve conduction and tissue polyol pathway metabolites in streptozotocin-diabetic rats. *Diabetologia* 40:271–281.

Cameron, N.E., Cotter, M.A., Dines, K.C., Maxfield, E.K., Carey, F., and Mirrlees, D.J. 1994b. Aldose reductase inhibition, nerve perfusion, oxygenation and function in streptozotocin-diabetic rats: dose-response considerations and independence from a myo-inositol mechanism. *Diabetologia* 37:651–663.

Cameron, N.E., Cotter, M.A., Ferguson, K., Robertson, S., and Radcliffe, M.A. 1991a. Effects of chronic α-adrenergic receptor blockade on peripheral nerve conduction, hypoxic resistance, polyols, Na-K ATPase activity and vascular supply in streptozocin-diabetic rats. *Diabetes* 40:1652–1658.

Cameron, N.E., Cotter, M.A., and Hohman, T.C. 1996. Interactions between essential fatty acid, prostanoid, polyol pathway and nitric oxide mechanisms in the neurovascular deficit of diabetic rats. *Diabetologia* 39:172–182.

Cameron, N.E., Cotter, M.A., Horrobin, D.H., and Tritschler, H.J. 1998a. Effects of α-lipoic acid on neurovascular function in diabetic rats: interaction with essential fatty acids. *Diabetologia* 41:390–399.

Cameron, N.E., Cotter, M.A., and Low, P.A. 1991b. Nerve blood flow in early experimental diabetes in rats: relation to conduction deficits. *Am. J. Physiol.* 261:E1–E8.

Cameron, N.E., Jack, A., Ways, D.K., and Cotter, M.A. 1998b. Effects of the protein kinase Cβ inhibitor, LY333531, on nerve and vascular function in diabetic rats. *Diabetologia* 41 (S1):A54.

Cohen, R.A. 1993. Dysfunction of vascular endothelium in diabetes mellitus. *Circulation* 87 (SV):V67–V76.

Cotter, M.A., and Cameron, N.E. 1998. Correction of neurovascular deficits in diabetic rats by β_2 agonist and α_1 adrenoceptor antagonist treatment: interactions with the nitric oxide system. *Eur. J. Pharmacol.* 343:217–223.

Cotter, M.A., Love, A., Watt, M.J., Cameron, N.E., and Dines, K.C. 1995. Effects of natural free radical scavengers on peripheral nerve and neurovascular function in diabetic rats. *Diabetologia* 38:1285–1294.

Dyck, J.B., and Dyck, P.J. 1999. Diabetic polyneuropathy. In: *Diabetic Neuropathy*, 2nd ed. (eds. P.J. Dyck and P.K. Thomas), pp. 279–295. Saunders, Philadelphia.

Ehmke, H., Junemann, K.P., Mayer, B., and Kummer, W. 1995. Nitric oxide synthase and vasoactive intestinal polypeptide colocalization in neurons innervating the human penile circulation. *Int. J. Impotence Res.* 7:147–156.

Elabbady, A.A., Gagnon, C., Hassouna, M.M., Begin, L.R., and Elhilali, M.M. 1995. Diabetes mellitus increases nitric oxide synthase in penises but not in major pelvic ganglia of rats. *Br. J. Urol.* 76:196–202.

Elliott, T.G., Cockcroft, J.R., Groop, P.H., Viberti, G.C., and Ritter, J.M. 1993. Inhibition of nitric oxide synthesis in forearm vasculature of insulin-dependent diabetic patients: blunted vasoconstriction in patients with microalbuminuria. *Clin. Sci.* 85:687–693.

Feldman, E.L., and Windebank, A.J. 1999. Growth factors and peripheral neuropathy. In: *Diabetic Neuropathy*, 2nd ed. (eds. P.J. Dyck and P.K. Thomas), pp. 377–386. Saunders, Philadelphia.

Fukao, M., Hattori, Y., Kanno, M., Sakuma, I., and Kitabatake, A. 1997. Alterations in endothelium-dependent hyperpolarization and relaxation in mesenteric arteries from streptozotocin-induced diabetic rats. *Br. J. Pharmacol.* 121: 1383–1391.

Greene, D.A., Sima, A.A.F., Stevens, M.J., Feldman, E.L., and Lattimer, S.A. 1992. Complications: neuropathy, pathogenetic considerations. *Diabetes Care* 15: 1902–1925.

Hamada, Y., Araki, N., Horiuchi, S., and Hotta, N. 1996. Role of polyol pathway in nonenzymatic glycation. *Nephrol. Dialysis Transplant.* 11:95–98.

Harrison, D.G. 1997. Endothelial function and oxidant stress. *Clin. Cardiol.* 20 (S2):11–17.

Hohman, T.C., Banas, D., Basso, M., Cotter, M.A., and Cameron, N.E. 1997. Increased oxidative stress in experimental diabetic neuropathy. *Diabetologia* 40 (S1):A549.

Hotta, N., Kaskuta, H., Fukasawa, H., Koh, N., Sakakibara, F., Nakamura, J., Hamada, Y., Wakao, T., Hara, T., Mori, K., Naruse, K., Nakashima, E., Inukai, S., and Sakamoto, N. 1995. Effect of a potent new aldose reductase inhibitor (5-(3-thienyl)tetrazol-1-yl)acetic acid (TAT), on diabetic neuropathy in rats. *Diabetes Res. Clin. Pract.* 27:107–117.

Houben, A.J.H.M., Schaper, N.C., De Haan, C.H.A., Huvers, F.C., Slaaf, D.W., DeLeeuw P.W., and Niewenhuijzen-Kruseman, A.C. 1996. Local 24-hr hyperglycemia does not affect endothelium-dependent or -independent vasoreactivity in humans. *Am. J. Physiol.* 270:H2014–H2020.

Ignarro, L.J., Bush, P.A., Buga, G.M., Wood, K.S., Fukoto, J.M., and Rajfer, J. 1990. Nitric oxide and cyclic GMP formation upon electrical field stimulation cause relaxation of corpus cavernosum smooth muscle. *Biochem. Biophys. Res. Commun.* 170:843–850.

Jack, A., Cameron, N.E., and Cotter, M.A. 1999. Effects of a diacylglycerol complexing agent on nerve conduction velocity and perfusion in diabetic rats. *J. Diabet. Complications* 13:2–9.

Jack, A., Cotter, M.A., and Cameron, N.E. 1998. Antioxidant treatment effects on defective endothelium-derived hyperpolarizing factor and relaxation in the mesenteric vasculature of diabetic rats. *Diabetes* 47 (S1):A54.

Jenkinson, K.M., and Reid, J.J. 1995. Effect of diabetes on relaxations to non-adrenergic, non-cholinergic nerve stimulation in longitudinal muscle of the rat gastric fundus. *Br. J. Pharmacol.* 116:1551–1556.

Kamata, K., Sugiura, M., Kojima, S., and Kasuya, Y. 1996a. Preservation of endothelium-dependent relaxation in cholesterol-fed and streptozotocin-induced diabetic mice by the chronic administration of cholestyramine. *Br. J. Pharmacol.* 118:385–391.

Kamata, K., Sugiura, M., Kojima, S., and Kasuya, Y. 1996b. Restoration of endothelium-dependent relaxation in both hypercholesterolemia and diabetes by chronic taurine. *Eur. J. Pharmacol.* 303:47–53.

Kamijo, M., Basso, M., Cherian, P.V., Hohman, T.C., and Sima, A.A.F. 1994. Galactosemia produces ARI-preventable nodal changes similar to those of diabetic neuropathy. *Diabetes Res. Clin. Pract.* 25:117–129.

Karpen, C.W., Pritchard, K.A., Arnold, J.H., Cornwell, D.G., and Pangonamala, R.V. 1982. Restoration of the prostacyclin/thromboxane A_2 balance in the diabetic rat: influence of vitamin E. *Diabetes* 31:947–951.

Keegan, A., Cotter, M.A., and Cameron, N.E. 1999. Effects of diabetes and treatment with the antioxidant, α-lipoic acid, on endothelial and neurogenic responses of corpus cavernosum in rats. *Diabetologia* 42:343–350.

Kihara, M., and Low, P.A. 1995. Impaired vasoreactivity to nitric oxide in experimental diabetic neuropathy. *Exp. Neurol.* 132:180–185.

Kihara, M., Zollman, P.J., Smithson, I.L., Lagerlund, T.D., and Low, P.A. 1994. Hypoxic effect of exogenous insulin on normal and diabetic peripheral nerve. *Am. J. Physiol.* 266:E980–E985.

Koya, D., and King, G.L. 1998. Protein kinase C activation and the development of diabetic complications. *Diabetes* 47:859–866.

Lou, M.F., Dickerson, J.E., Garadi, R., and York, B.M. 1988. Glutathione depletion in the lens of galactosemic and diabetic rats. *Exp. Eye Res.* 46: 517–530.

Low, P.A., Nickander, K.K., and Tritschler, H.J. 1997. The roles of oxidative stress and antioxidant treatment in experimental diabetic neuropathy. *Diabetes* 46 (S2):S38–S42.

Low, P.A., Schmelzer, J.D., Ward, K.K., Curran, G.L., and Poduslo, J.F. 1988. Effect of hyperbaric oxygenation on normal and chronic streptozotocin diabetic peripheral nerves. *Exp. Neurol.* 99:201–212.

Low, P.A., Schmelzer, J.D., Ward, K.K., and Yao, J.K. 1986. Experimental chronic hypoxic neuropathy: relevance to diabetic neuropathy. *Am. J. Physiol.* 250: E94–E99.

Lowitt, S., Malone, J.I., Salem, A.F., Korthals, J., and Benford, S. 1995. Acetyl-L-carnitine corrects the altered peripheral nerve function of experimental diabetes. *Metabolism* 44:677–680.

Martinezcuestav, M.A., Massuda, H., Whittle, B.J.R., and Moncada S. 1995. Impairment of nitrergic-mediated relaxation of rat isolated duodenum by experimental diabetes. *Br. J. Pharmacol.* 114:919–924.

Maxfield, E.K., Cameron, N.E., and Cotter, M.A. 1997. Effect of diabetes on reactivity of sciatic vasa nervorum in rats. *J. Diabet. Complications* 11:47–55.

Mayhan, W.G., Didion, S.P., and Patel, K.P. 1996. L-Arginine does not restore dilatation of the basilar artery during diabetes mellitus. *J. Cereb. Blood Flow Metab.* 16:500–506.

Mayhan, W.G., and Patel, K.P. 1995. Acute effects of glucose on reactivity of cerebral microcirculation: role of activation of protein kinase C. *Am. J. Physiol.* 269:H1297–H1302.

Melton, L.J. III, and Dyck P.J. 1999. Epidemiology. In: *Diabetic Neuropathy*, 2nd ed. (eds. P.J. Dyck and P.K. Thomas), pp. 239–254. Saunders, Philadelphia.

Morel, D.W., and Chisolm, G. 1989. Antioxidant treatment of diabetic rats inhibits lipoprotein oxidation and cytotoxicity. *J. Lipid Res.* 30:1827–1834.

Myers, R.R., and Powell, H.C. 1984. Galactose neuropathy—impact of chronic endoneurial edema on nerve blood flow. *Ann. Neurol.* 16:587–594.

Nagamatsu, M., Nickander, K.K., Schmelzer, J.D., Raya, A., Wittrock, A., Tritschler, H., and Low, P.A. 1995. Lipoic acid improves nerve blood flow, reduces oxidative stress and improves distal nerve conduction in experimental diabetic neuropathy. *Diabetes Care* 18:1160–1167.

Nakamura, J., Koh, N., Hamada, Y., Naruse, K., Chaya, S., Kato, K., Ienaga, K., and Hotta, N. 1998. Effect of a protein kinase C-β specific inhibitor on diabetic neuropathy in rats. *Diabetes* 47 (S1):A70.

Ohara, Y., Sayegh, H.S., Yamin, J.J., and Harrison, D.G. 1995. Regulation of endothelial constitutive nitric oxide synthase by protein kinase C. *Hypertension* 25:415–420.

Otter, D.J., and Chess-Williams, R. 1994. The effects of aldose reductase inhibition with ponalrestat on changes in vascular function in streptozotocin diabetic rats *Br. J. Pharmacol.* 113:576–580.

Pieper, G.M. 1998. Review of alterations in endothelial nitric oxide production in diabetes: protective role of arginine on endothelial dysfunction. *Hypertension* 31:1047–1060.

Powell, H.C., Garrett, R.S., Kador, P.F., and Mizisin, A.P. 1991. Fine structural localization of aldose reductase and ouabain-sensitive, K^+-dependent *p*-nitrophenylphosphatase in rat peripheral nerve. *Acta Neuropathol.* 81:529–539.

Rehman, J., Chenven, E., Brink, P., Peterson, B., Walcott, B., Wen, Y.P., Melman, A., and Christ, G. 1997. Diminished neurogenic but not pharmacological erections in the 2- to 3-month experimentally diabetic F-344 rat. *Am. J. Physiol.* 272:H1960–H1971.

Rösen, P., Ballhausen, T., Bloch, W., and Addicks, K. 1995. Endothelial relaxation is disturbed by oxidative stress in the diabetic rat heart: influence of tocopherol as antioxidant. *Diabetologia* 38:1157–1168.

Saenz De Tejada, I., Goldstein, I., Azadzoi, K., Krane, R.J., and Cohen, R.A. 1989. Impaired neurogenic and endothelium-mediated relaxation of penile smooth muscle from diabetic men with impotence. *N. Engl. J. Med.* 320:1025–1030.

Sasaki, H., Schmelzer, J.D., Zollman, P.J., and Low, P.A. 1997. Neuropathology and blood flow of nerve, spinal roots and dorsal root ganglia in longstanding diabetic rats. *Acta Neuropathol.* 93:118–128.

Schmidt, R.E., Dorsey, D.A., McDaniel, M.L., and Corbett, J.A. 1993. Characterization of NADPH diaphorase activity in rat sympathetic autonomic ganglia: effect of diabetes and aging. *Brain Res.* 617:343–348.

Sima, A.A.F., Greene, D.A., Brown, M.B., Hohman, T.C., Hicks, D., Graepel, G.J., Bochenek, W.J., Beg, M., Gonen, B., and the Tolrestat Study Group. 1993. Effect of hyperglycemia and the aldose reductase inhibitor tolrestat on sural nerve biochemistry and morphometry in advanced diabetic peripheral polyneuropathy. *J. Diabet. Complications* 7:157–169.

Sladsky, J.T., Tschoepe, R.L., Greenberg, J.H., and Brown, M.J. 1991. Peripheral neuropathy after chronic endoneurial ischemia. *Ann. Neurol.* 29:272–278.

Smits, P., Kapma, J.A., Jacobs, M.C., Lutterman, J., and Thien, T. 1993. Endothelium-dependent vascular relaxation in patients with type I diabetes. *Diabetes* 42:148–153.

Steinberg, H.O., Brechtel, G., Johnson, A., Fineberg, N., and Baron, A.D. 1994. Insulin mediated skeletal muscle vasodilation is nitric oxide dependent: a novel action of insulin to increase nitric oxide release. *J. Clin. Invest.* 94:1172–1179.

Stevens, M.J., Lattimer, S.A., Feldman, E.L., Helton, E., Millington, D.S., Sima, A.A.F., and Greene, D.A. 1996. Acetyl-L-carnitine deficiency as a cause of altered nerve myo-inositol content, Na,K-ATPase activity and motor conduction velocity in the streptozotocin-diabetic rat. *Metabolism* 45:865–872.

Stevens, M.J., Lattimer, S.A., Kamijo, M., Van Huysen, C., Sima, A.A.F., and Greene D.A. 1993. Osmotically induced nerve taurine depletion and the compatible osmolyte hypothesis in experimental diabetic neuropathy. *Diabetologia* 36:608–614.

Sullivan, M.E., Bell, C.R.W., Dashwood, M.R., Miller, M.A.W., Thompson, C.S., Mikhailidis, D.P., and Morgan, R.J. 1996. Autoradiographic localization of nitric oxide synthase binding-sites in normal and diabetic rat corpus cavernosum. *Eur. Urol.* 30:506–511.

Takahashi, T., Nakamura, K., Itoh, H., Sima, A.A.F., and Owyang, C. 1997. Impaired expression of nitric oxide synthase in the gastric myenteric plexus of spontaneously diabetic rats. *Gastroenterology* 113:1535–1544.

Tesfaye, S., Malik, R., Harris, N., Jakubowski, J.J., Mody, C., Rennie, I.G., and Ward, J.D. 1996. Arterio-venous shunting and proliferating new vessels in acute painful neuropathy of rapid glycaemic control (insulin neuritis). *Diabetologia* 39:329–335.

Tesfaye, S., Malik, R., and Ward, J.D. 1994. Vascular factors in diabetic neuropathy. *Diabetologia* 37:847–854.

Tuck, R.R., Schmelzer, J.D., and Low, P.A. 1984. Endoneurial blood flow and oxygen tension in the sciatic nerves of rats with experimental diabetic neuropathy. *Brain* 107:935–950.

Vernet, D., Cai, L., Garban, H., Babbitt, M.L., Murray, F.T., Rajfer, J., and Gonzalez-Cadavid, N.F. 1995. Reduction of penile nitric oxide synthase in diabetic BB/WORdp (type I) and BBZ/WORdp (type II) rats with erectile dysfunction. *Endocrinology* 136:5709–5717.

Watts, G.F., O'Brien, S.F., Silvester, W., and Millar, J.A. 1996. Impaired endothelium-dependent and -independent dilatation of forearm resistance arteries in men with diet-treated non-insulin-dependent diabetes: role of dyslipidaemia. *Clin. Sci.* 91:567–573.

Way, K.J., and Reid, J.J. 1994. Effect of aminoguanidine on the impaired nitric oxide-mediated neurotransmission in anococcygeus muscle from diabetic rats. *Neuropharmacology* 33:1315–1322.

Way, K.J., and Reid, J.J. 1996. The aldose reductase inhibitor sorbinil does not prevent the impairment in nitric oxide-mediated neurotransmission in anococcygeus muscle from diabetic rats. *Eur. J. Pharmacol.* 318:101–108.

Williamson, J.R., Chang, K., Frangos, M., Hasan, K.S., Ido, Y., Kawamura, T., Nyengaard, J.R., Van Den Enden, M., Kilo, C., and Tilton, R.G. 1993. Hyperglycemic pseudohypoxia and diabetic complications. *Diabetes* 42:801–813.

Wolff, S.P. 1993. Diabetes mellitus and free radicals. *Br. Med. Bull.* 49:642–652.

Yagihashi, S. 1995. Pathology and pathogenetic mechanisms of diabetic neuropathy. *Diabetes Metab. Rev.* 11:193–225.

Zhu, X., and Eichberg, J. 1990. 1,2-Diacylglycerol content and its arachidonyl-containing molecular species are reduced in sciatic nerve from streptozotocin-induced diabetic rats. *J. Neurochem.* 55:1087–1090.

Ziegler, D., Schatz, H., Gries, F.A., Ulrich, H., and Reichel, G. 1997. Effects of treatment with the antioxidant α-lipoic acid on cardiac autonomic neuropathy in NIDDM patients. A 4-month randomized controlled multicenter trial (DEKAN Study). *Diabetes Care* 20:369–373.

15 Potential Clinical Impact of Nitric Oxide and Nitric Oxide Inhibitors in the Periphery

Ilari Paakkari, Asko Riutta, and Ilkka Pörsti

Overview of Nitric Oxide-Based Pharmacotherapy

The goal of nitric oxide (NO)-based pharmacotherapy is to attain proper homeostasis of NO metabolism in the target tissue or tissues where the endogenous production of NO is either too weak or excessively increased.

Nitric oxide levels in target tissues can be increased directly by administration of gaseous NO in the respiratory system or by giving systemic drugs that release NO (NO donors). Nitric oxide levels may be increased by administration of the NO synthase (NOS) substrate, L-arginine. Nitric oxide production increases when constitutive nitric oxide synthase (cNOS) is stimulated via cell membrane receptors or the inducible enzyme (iNOS) protein increasingly expressed by exposure to cytokines or bacterial lipopolysaccharides (LPS). The effects of NO can be amplified by blocking its second transmitter, i.e., cGMP breakdown. This has been already proven to be a very useful approach in increasing the effect of neuronally released NO. Affecting the second transmitter system of neuronally released NO makes it possible to modify the pharmacological effect only in the presence of a physiological signal and offers a possibility for selective treatment due to diversity of the second transmitters in various target tissues.

Nitric oxide levels can be decreased by blocking the catalytic activity of NOS with nonselective inhibitors, whereby endothelial NOS (eNOS), neuronal NOS (nNOS), and, in various tissues, iNOS are equally blocked. An improved approach is to selectively inhibit iNOS, which offers the benefit of treating overt and usually harmful production of NO while the constitutive physiological NO production is preserved.

Problems in NO-based pharmacotherapy include the toxicity of the compound itself, which at sufficient concentrations is capable of blocking various enzymes involved in DNA and mitochondrial metabolism. Moreover, as a radical with an unpaired electron, NO or its metabolites may also cause direct structural damage. Although the toxic nature of NO is physiologically utilized by the body's defense cells, it has no pharmacotherapeutic use at the moment. The short lifetime and the gaseous form of pure

NO limit its use but are of definite advantage when it is given locally by inhalation. An ideal NO donor should exert target tissue selectivity and then slowly release the active agent. Future prospects may feature local transfer of the cNOS gene for treatment of cardiovascular diseases such as essential hypertension.

Nitric Oxide: Physical Properties and Metabolism

The small and uncharged NO molecule diffuses freely across cell membranes but as a very reactive molecule is rapidly destroyed. The biological half-life of NO varies from 2 to 30 seconds, and after transmitting a signal it decays into nitrite (Lowenstein et al., 1994; Marletta, 1994a). With respect to many physiological functions of NO, it can be speculated that the half-life of NO is relatively short for a paracrine mediator but long for a neurotransmitter. Therefore, it is no surprise that peripheral nitrinergic neurotransmission involves the modulation of relatively slowly initiated and persistent smooth muscle events.

Nitric oxide synthase produces NO from L-arginine, whereby a by-product L-citrulline is again converted into L-arginine in the urea cycle (Stryer, 1995). The constitutive production of NO by the endothelium (eNOS) and nerves (nNOS) is stimulated by the presence of Ca^{2+} ions, and it can to some extent be increased by abundant availability of the substrate L-arginine. The external influences that activate eNOS include the shear force (distension) caused by blood flow and stimulation of many cell surface receptors, such as those for acetylcholine, bradykinin, and serotonin. In addition to depolarization, the external influences to activate nNOS include stimulation of cell surface receptors, for example, for glutamate. The third NO-producing inducible isoform (iNOS) is not dependent on the Ca^{2+} signal but is expressed in large amounts in various tissues in the presence of proinflammatory cytokines and bacterial lipopolysaccharides. The genetic expression of iNOS, but not of eNOS or nNOS, is blocked by glucocorticoids.

Means of Increasing Nitric Oxide Levels in the Target Tissue

Nitric Oxide Gas

Nitric oxide is a powerful vasodilator in the pulmonary circulation, where it also inhibits hypoxic vasoconstriction (Liu et al., 1991), whereas inhibition of NO synthesis enhances it (Archer et al., 1989; Robertson et al., 1990). The short half-life of inhaled NO is an advantage in the treatment of acute respiratory distress syndrome (ARDS) and pulmonary hypertension, since the use of common long-acting vasodilators—even if they could be administered locally—would cause severe concurrent systemic hypotension (Frostell et al., 1991). The therapeutic window of inhaled NO is narrow and caution is required, but promising results in newborns, children, and adults have been reported (Foubert et al., 1992; Pepke-Zaba et al., 1991; Roberts et al., 1992; Pearl, 1993; Dellinger et al., 1998). Interestingly, inhaled NO could also be useful in the treatment of septic shock with generalized severe hypotension, where it can counteract the pulmonary hypertension caused by NOS inhibitors (Weitzberg et al., 1993; Offner et al., 1996).

Arginine Supplementation

L-Arginine, the substrate for NOS, is a nonessential amino acid that is abundantly synthesized by the urea cycle (Stryer, 1995). Therefore, it is rather surprising to learn that supplementation of L-arginine increases NO production and thus improves, for example, endothelially dependent vasodilatation. One explanation might be that in addition to the substrate induction of NOS, L-arginine may compete with the endogenous arginine analogue inhibitors that have been described in humans (Vallance et al., 1992a). Regardless of the somewhat uncertain mechanism, administration of L-arginine restores impaired NO availability in coronary arteries of hypercholesterolemic patients (Drexler et al., 1991) as well as preserving endothelium-dependent vasodilator function and reducing the progression of atherosclerosis in hypercholesterolemic rabbits (Girerd et al., 1990; Boger et al., 1997). Dietary supplementation of L-arginine attenuated myocardial ischemia–reperfusion injury in cats and pigs (Weyrich et al., 1992; Pernow et al., 1994). In hypertensive patients, L-arginine decreased blood pressure and improved endothelium-dependent vasodilation to acetylcholine in the forearm vasculature (Imaizumi et al., 1992). Also, in human heart failure, supplemental oral L-arginine increased peripheral blood flow and improved functional status (Rector et al., 1996).

Receptor-Mediated Increase in Nitric Oxide Synthesis

Furchgott and Zawadzki (1980) were the first to show that cholinergic vasodilatation was mediated via muscarinic receptors and required the presence of intact endothelium, which in turn released smooth muscle relaxing factor, later identified as NO. Subsequently, a variety of agents, including bradykinin, adenine nucleotides, substance P, histamine, serotonin, and thrombin, have been shown to mediate endothelium-dependent relaxation (Furchgott, 1984). Hypoxia and increased flow also cause endothelium-dependent relaxation of vascular tissue (Moncada et al., 1991;

Hecker et al., 1993a). Thus, the stimuli of NO release are diverse, ranging from small molecules to large peptides, which should provide new pharmacological possibilities to enhance endogenous NO release by the endothelium.

Antihypertensive angiotensin-converting enzyme (ACE) inhibitors exert a twofold action: they inhibit the synthesis of the vasoconstricting angiotensin II, but concomitantly prevent the breakdown of vasodilating plasma kinins, because ACE also exhibits a kininase II property. Accordingly, ACE inhibitors increase tissue concentrations and/or potentiate the actions of endogenous bradykinin (Bönner et al., 1990; Hecker et al., 1994). This peptide is a potent stimulus to cNOS and releases both NO and prostacyclin from the endothelium (Busse and Fleming, 1993). Thus, in addition to the hypotensive effect, bradykinin may also contribute to the antiproliferative and cytoprotective effects of ACE inhibitors via the NO pathway (Linz and Schölkens, 1992; Kiowski et al., 1992; Hecker et al., 1993b; Ishigai et al., 1997; Berkenboom et al., 1997; Kichuk et al., 1997). Interestingly, ACE inhibitors have been shown to prevent and even reverse endothelial dysfunction in experimental and essential hypertension (Arvola et al., 1993; Hirooka et al., 1992) and reduce myocardial infarct size in experimental animals, an effect that is abolished by a bradykinin antagonist (Martorana et al., 1990).

Drugs Releasing Nitric Oxide (NO Donors)

Nitric oxide-releasing nitrovasodilators are important pharmacological tools in the treatment of angina pectoris and congestive heart failure (Warren et al., 1994; Elkayam, 1996; Abrams, 1996). Organic nitrocompounds such as nitroglycerin release NO via a metabolic process that requires thiol groups, whose shortage in the target tissue may contribute at least in part to the development of nitrate tolerance (Newman et al., 1990). In this respect, inorganic nitrovasodilators of the sydnonimide type (e.g., SIN-1) that do not require thiol groups for NO release may thus induce less or no tolerance (Bates et al., 1991; Feelisch et al., 1989; Rehse et al., 1994). In the future, the tissue site of metabolism of nitrovasodilators could be altered, thus targeting the site of action of these compounds. Incorporating sulfhydryl groups in the molecular structure of the drug would favor the formation of nitrosothiol, which is active and stable (Myers et al., 1990; Langford et al., 1994).

Increasing Nitric Oxide Synthase Expression

Increased expression of iNOS by many inductors such as cytokines and LPS is well known. However, with the exception of the efficacy of the defense cells such as macrophages against infectious agents or tumor cells, increased expression of iNOS seems only to contribute to untoward effects such as septic shock, asthma, and other inflammatory conditions. Expression of constitutive endothelial cNOS is regulated by intraarterial pressure and blood flow-induced shear stress (Black et al., 1998; Ziegler et al., 1998). Estrogens increase expression of eNOS (MacRitchie et al., 1997; Caulin-Glaser et al., 1997; Vagnoni et al., 1998), which may contribute to their protective effect on the cardiovascular system during postmenopausal hormone replacement therapy. Other drugs that increase the expression of eNOS include plasma lipid-lowering 3-hydroxy-3-methylglutaryl coenzyme A (HMG-CoA) reductase inhibitors (Laufs and Liao, 1998; Endres et al., 1998; Laufs et al., 1998; Treasure et al., 1995), calcium channel blockers (Ding and Vaziri, 1998), and the macrolide antibiotic erythromycin (Mitsuyama et al., 1998).

Means of Decreasing Nitric Oxide Levels in the Target Tissue

From the theoretical point of view, inhibition of the constitutive synthesis of NO by cNOS or nNOS may only cause adverse effects, such as general vasoconstriction, hypertension, tissue remodeling, and impairment of nervous function. Therefore, most efforts have been directed toward selective inhibition of iNOS that could be used in the treatment of endotoxic shock, inflammation, bronchial asthma, and other clinical problems related to inflammation and abnormally high levels of NO.

Nonselective Inhibition of Nitric Oxide Synthase

Substitution of one or both of the two terminal guanidino nitrogen atoms of L-arginine results in analogues that inhibit NOS. The compounds N^G-nitro-L-arginine (L-NNA), N^G-nitro-L-arginine methylester (L-NAME), N^G-monomethyl-L-arginine (L-NMMA), and *N*-iminoethylornithine (L-NIO) inhibit both the constitutive and inducible forms of NOS, i.e., cNOS and iNOS (Moncada et al., 1991; Marletta, 1994b), and their likely mechanism of action is competitive inhibition by serving as a false substrate for NOS.

Selective Inhibition of iNOS

The need for the design of specific arginine-based analogues that show selective inhibition of the iNOS isoform clearly exists. The rationale would be selective inhibition of the often deleterious iNOS induction while the function of constitutional eNOS and nNOS would remain intact. Selective iNOS inhibition could mean a new approach to treat septic shock and inflammatory conditions such as rheumatoid arthritis and bronchial asthma, while the probable adverse effect would be the impairment of immunological defense. L-*N*-6(1-iminoethyl)lysine (L-NIL) and aminoguadinine (Wolff and Lubeskie, 1995) are NOS inhibitors with selectivity toward iNOS. The primary mechanism of iNOS inactivation by L-NIL is the destruction and loss of heme at the active site of the enzyme, whereas aminoguadinine inactivation proceeds through covalent modification of the iNOS protein and the heme residue (Bryk and Wolff, 1998). Other selective iNOS inhibitors are L-canavanine (Liaudet et al., 1998), guanidinoethylsulfide (Szabo et al., 1996), and 2,4-diamino-6-hydroxy-pyrimidine (Schwartz et al., 1997).

Inhibition of iNOS Expression

Glucocorticoids selectively inhibit the expression of iNOS without affecting the constitutive form of the enzyme. Inhibition of iNOS suppresses the increased microvascular blood flow in several models of inflammation (Pons et al., 1993; Warren et al., 1993), especially if glucocorticoids are given as pretreatment before NOS is induced et al., Clinically in asthmatic patients, the glucocorticoid-induced inhibition of iNOS is seen as decreased NO content of the expired air (Springall et al., 1995; Kharitonov et al., 1996).

Interactions of Nitric Oxide with Eicosanoids (Prostanoids and Leukotrienes)

Increased production of NO and prostanoids seems to coexist in inflamed tissues. The inducible form of the enzyme cyclooxygenase (COX-2) produces proinflammatory prostanoids in the same fashion as iNOS produces NO after stimulation by several inflammatory mediators (Masferrer et al., 1992). Interestingly, selective inhibitors of iNOS have been found also to inhibit prostaglandin production (Salvemini et al., 1995). In bovine coronary arteries, prostaglandin D releases NO (Braun and Schrör, 1992). Inhibition of prostacyclin production with indomethacin enhanced NO release in cultured human endothelial cells (Bolz and Pohl, 1997), and acetylsalicylic acid treatment resulted in stimulation of NO production in rabbits (Lopez-Farre et al., 1996). On the other hand, NO donors activate cyclooxygenase (Riutta et al., 1994; Alanko et al., 1995) and inactivate leukotriene-producing lipoxygenase enzymes (Riutta et al., 1994; Wiesner et al., 1996; Nakatsuka and Osava, 1994). Enhanced production of antiaggregatory prostacyclin caused by the NO-mediated cyclooxygenase activation (Salvemini et al., 1996), and phosphorylation of the thromboxane receptor (Wang et al., 1998) may be the key events in the antiplatelet effects of nitrovasodilators.

Possibilities for Nitric Oxide-Based Therapy in the Cardiovascular System

In the cardiovascular system, constitutively produced NO is the major opponent of numerous endogenous vasoconstrictor mediators, such as noradrenaline, angiotensin II, and endothelins (Moncada et al., 1991; Nathan, 1992). It also acts synergistically with prostacyclin to inhibit platelet aggregation, thus contributing to the antiadhesive and antiaggregatory property of the endothelium (Radomski et al., 1987a, b). Sources of NO within the cardiovascular system include blood vessels, perivascular nerves, myocardium, and endocardium (Moncada et al., 1991; Nathan, 1992; Smith et al., 1991).

Myocardial Ischemia

Brief periods of myocardial ischemia cause loss of endothelium-dependent vasodilatation and enhance coronary arterial constrictor responses, without detectable histological damage to the vessels (Kim et al., 1992). In rat hearts, loss of endothelial function following ischemia is parallelled by a loss of eNOS activity (Giraldez et al., 1997). The reperfusion period following transient coronary occlusion is associated with accumulation of neutrophils, which probably contribute to myocardial damage (Brady et al., 1992; Tsao and Lefer, 1990). Decreased NO production after ischemia also promotes neutrophil adherence to coronary endothelium (Ma et al., 1993). Nitric oxide donors suppress neutrophil-mediated damage by decreasing adherence to coronary endothelium by reducing endothelial expression of adhesion molecules and proinflammatory cytokines (Delyani et al., 1996; De Caterina et al., 1995; Shin et al., 1996; Takahashi et al., 1996). This effect may explain some of the beneficial effects of nitrates in acute myocardial infarction (Du Toit et al., 1998; Kupatt et al., 1996; Pabla et al., 1996).

Heart Failure

Elevated plasma concentrations of nitrate, the stable metabolite of NO, have been found in patients with congestive heart failure (Winlaw et al., 1994). Recent reports have shown that in patients with heart failure the expression of cNOS is decreased (Drexler et al., 1998; Smith et al., 1996) but that of iNOS is increased (Riede et al., 1998), suggesting that the elevated plasma nitrate concentrations are due to iNOS-catalyzed NO production. In failing human hearts, high iNOS activity is associated with attenuated positive inotropic response to β-adrenergic stimulation (Riede et al., 1998). Moreover, inhibition of iNOS activity was observed to enhance the positive response to β-adrenergic stimulation in failing hearts but not in control subjects (Hare et al., 1998; Riede et al., 1998).

Platelet Aggregation

Nitric oxide inhibits aggregation by acting directly on platelets by diffusing from endothelial cells into the vessel lumen, where it enters platelets. Constitutive NOS is present in platelets, and its stimulation reduces aggregation by inhibiting both platelet adhesion (Sneddon and Vane, 1988; Radomski et al., 1987a) and aggregation (Radomski et al., 1990; Vallance et al., 1992a). The underlying mechanisms probably include activation of platelet guanylate and adenylate cyclase (Nolte et al., 1991; Yamakado et al., 1982), cyclooxygenase activation (Salvemini et al., 1996), phosphorylation of thromboxane receptor (Wang et al., 1998), and inhibition of phospholipase C (Durante et al., 1992). Nitric oxide donors also regulate blood clotting by inhibiting clotting factor XIII activity (Catani et al., 1998) and decreasing the expression of platelet membrane surface glycoprotein IIb-IIIa (Keh et al., 1996). Nitric oxide donors also inhibit the release of plasminogen activator inhibitor (Korbut et al., 1995), which may contribute to prolonged bleeding times observed after nitroglycerin administration (Ring et al., 1983).

Hypertension

In experimental animal models, inhibition of NOS by arginine analogues increases mean arterial blood pressure by approximately 40%, suggesting that the resistance vessels are in a state of continuous nitrergic vasodilation (Rees et al., 1989). The blood pressure of healthy individuals increases when NOS inhibitors are infused, whereas the blood pressure of those with hypertension changes much less, suggesting reduced production of NO in hypertension (Panza et al., 1993). It appears that in hypertension, there is a deficiency in NO production or an impairment in the function of its second transmitter system (Lüscher, 1992; Calver et al., 1992; Lüscher et al., 1992; Panza et al., 1990). Somatic delivery of the human eNOS gene induces a prolonged reduction of high blood pressure in spontaneously hypertensive rats (Lin et al., 1997), a finding that suggests the possibility of eNOS gene therapy for hypertension.

Atherosclerosis

Several studies have reported attenuation of endothelium-mediated vasodilatation in atherosclerotic coronary arteries (Lüscher et al., 1993; Chester et al., 1990) and also in the forearm vasculature of hypercholesterolemic humans (Creager et al., 1990; Chowienczynk et al., 1990). Oxidized low-density lipoprotein has been found to inhibit the relaxation response to exogenous and endothelium-derived NO in experimental animals (Jacobs et al., 1990; Plane et al., 1992). Thus, endothelial dysfunction may contribute to inappropriate constrictor responses (vasospasms) of atherosclerotic coronary arteries (Ross, 1993; Vanhoutte, 1988). In addition to lipid-lowering effects, HMG-CoA reductase inhibitors augment expression of eNOS (Laufs and Liao, 1998; Endres et al., 1998; Laufs et al., 1998), which may contribute to their therapeutic effects.

Endotoxic Shock

Excess generation of NO by iNOS, induced by bacterial LPS, plays a major role in the hypotension and systemic vasodilatation characteristic of sepsis (Petros et al., 1994; Gunnett et al., 1998). This is also indicated by the failure of LPS to impair vasomotor function in carotid arteries from iNOS-deficient mice (Gunnett et al., 1998). Inhibition of NOS with analogues of L-arginine can reduce the hyporesponsiveness of vascular tissue to pressor agents in experimental animals and elevate blood pressure in humans with septic shock (Thiemermann and Vane, 1990; Petros et al., 1991; Petros et al., 1994). However, in experimental animals and patients with septic shock, nonselective NOS inhibitors produced a fall in cardiac output, which resulted in decreased tissue perfusion (Klabunde et al., 1991; Petros et al., 1994). Selective iNOS inhibition prevents reductions in glomerular filtration rate (GFR), whereas nonselective inhibition of NOS decreases GFR. These findings suggest that the GFR after LPS decreases due to a local inhibition of eNOS by iNOS, possibly via NO autoinhibition (Schwartz et al., 1997). Moreover, inhibition of cNOS potentiates, whereas inhibition of iNOS markedly attenuates, losses in plasma volume and albumin escape elicited by LPS, suggesting that selective inhibitors of iNOS may be more effective than nonselective inhibitors of NOS in the therapy of septic shock (Filep et al., 1997). It seems that the treatment of septic shock by means of nonspecific inhibition of both eNOS and iNOS with compounds such as L-NAME offers only a limited benefit because of increase in systemic and pulmonary vascular resistance, reduced cardiac output, and increased platelet aggregation (Avontuur et al., 1998). In this respect promising approaches include selective inhibition of iNOS, for example with L-canavanine, which leaves the constitutive eNOS viable (Liaudet et al., 1998; Parrat, 1998), perhaps combined with gaseous NO inhalation to decrease pulmonary arterial resistance (Weitzberg et al., 1993; Offner et al., 1996).

Erectile Dysfunction

Nitric oxide is involved in penile erection (Burnett et al., 1992). Immunohistochemical studies have localized NOS to the nerves in the pelvic plexus and the nerve endings of the arteries of the corpus cavernosum. In addition, electrical stimulation of the pelvic nerves of the rat causes erection and also relaxes isolated smooth muscle preparations from the corpus cavernosum, both of which effects are blocked by NOS inhibition (Bush et al., 1992; Trigo-

Rocha et al., 1993). Thus, in penile erection NO functions as both a neurotransmitter and a vasodilator. Nitric oxide appears to be the main proerectile neurotransmitter, and noradrenaline is considered to be the major antierectile agent. Increasing knowledge of the significance of NO in the urogenital region initiated several pilot studies reporting efficacy of topical application of the NO donor nitroglycerin in the treatment of erectile dysfunction (Sonksen and Biering-Sorensen, 1992; Meyhoff et al., 1992; Nunez and Anderson, 1993). However, topically applied nitroglycerin proved to have less efficacy than the topical vasodilator minoxidil (Cavallini, 1994) or intracavernously injected papaverine (Renganathan et al., 1997). A new therapeutic principle was discovered in sildenafil, which was the first compound to increase nitrinergic vasodilatation in the cavernous erectile tissue by inhibiting the breakdown of the NO-coupled second transmitter cGMP (Boolell et al., 1996). Sildenafil preferentially inhibits the enzyme phosphodiesterase-5 (PDE-5), which inactivates cGMP. Moreover, sildenafil offers some degree of selectivity, since PDE-5 is the prevailing phosphodiesterase in the penile tissue (Park et al., 1998). Such a therapy has the benefit of being administered orally, whereas the effect is largely selective to penile tissue and occurs only upon sexual arousal (Goldstein et al., 1998).

Inflammatory Bowel Disease

In patients with ulcerative colitis, increased concentrations of citrulline indicate enhanced NO production. This has also been demonstrated in biopsy specimens, suggesting that increased NO synthesis in colonic mucosa may have a pathogenic role (Middleton et al., 1993). Recently, mucosal microelectrode studies, biopsies, and direct measurements from the gut lumen from patients with ulcerative colitis have shown greatly increased NO production and induced iNOS expression in the colonic epithelium (Iwashita, 1998; Kolios et al., 1998; Herulf et al., 1998). Although the modes of therapeutic effect of glucocorticoids in inflammatory bowel disease are manifold, some part of it may be due to suppression of iNOS. In experimental models of inflammatory bowel disease, selective inhibitors of iNOS have been tested, with variable results (Ribbons et al., 1997; Laszlo and Whittle, 1997).

Nitric Oxide in Respiratory Distress Syndrome and Asthma

Therapeutic Significance of Nitric Oxide

In patients with ARDS, inhaled gaseous NO was well tolerated and induced a significant improvement in oxygenation and cardiac output (Johannigman et al., 1997; Dellinger et al., 1998; Benzing et al., 1997). For this indication the kinetic properties of the NO molecule are ideal, i.e., good membrane penetration but a very short half-life, particularly in the presence of hemoglobin, whose iron-containing heme group readily binds and inactivates NO. Thus, inhaled NO decreases pulmonary resistance without causing severe systemic hypotension. Moreover, short-acting gaseous NO can be dosed more accurately than any long-acting drug. In addition to the adrenergic β_2-bronchodilatory effect, NO released from nonadrenergic, noncholinergic (NANC) nerves may contribute to physiological bronchodilatation (Barnes, 1992; Belvisi et al., 1992; Ellis and Undem, 1992). Decreased constitutive NO release in the bronchi may have a causal relationship to the increased bronchial excitability (Gaston et al., 1998). However, according to in vitro experiments, the bronchorelaxing effect of various NO donors was clearly inferior to the β_2-agonist salbutamol (Vaali et al., 1996). In healthy volunteers inhaled NO had no effect on airway tone but modulated the response to methacholine toward bronchodilatation. In patients with asthma, but not in those with chronic obstructive pulmonary disease (COPD), NO exerted only a weak bronchodilatory effect (Hogman et al., 1993). According to another report, inhaled NO worsened hypoxemia induced by methacholine in healthy and asthmatic individuals (Takahashi et al., 1998). In pediatric subjects with asthma and mild airway disease, NO inhaled at 40 ppm had no apparent bronchodilatory effect (Pfeffer et al., 1996). Thus, it can be concluded that exogenous NO or NO donors seem to add very little to the bronchodilatory therapy of asthma.

Role of Induced Nitric Oxide Production in the Lungs

In the rat model, airway epithelial cell iNOS immunoreactivity and mRNA were increased 8 hours after antigen challenge (Renzi et al., 1997). Bronchial biopsies from asthma patients show expression of iNOS in the epithelium (Hamid et al., 1993). In asthma patients, but not in healthy subjects, exhaled NO is reduced dose dependently by topical glucocorticoids, along with decreased iNOS activity in the airways (Springall et al., 1995; Kharitonov et al., 1996; ten Hacken et al., 1998; Persson et al., 1994). Moreover, inhalation of a selective iNOS inhibitor resulted in decreased exhaled NO in asthmatic patients but not in healthy individuals, whereas a nonselective NOS inhibitor reduced exhaled NO in both groups (Yates et al., 1996). Accordingly, it is concluded that increased expression of iNOS is the source of high exhaled NO concentrations in asthma patients, and exhaled NO could serve as a marker for asthmatic inflammation (Sovijärvi et al., 1998). Exhaled NO is not increased in COPD, and its levels are not affected by methacholine or β-adrenergic sympathomimetics (Garnier et al., 1996; Persson et al., 1994; Yates et al., 1997). Increased NO production by iNOS may contribute to the epithelial damage and cause microvascular plasma leakage in the respiratory mucosa (Bernareggi et al., 1997; Kageyama et al., 1997). Oral administration of the NOS substrate L-arginine potentiated allergen-induced airway inflammation in mice (Takano et al., 1998). Likewise, in asthmatic subjects inhaled L-arginine increased exhaled NO concomitantly with an increase in bronchial resistance (fall in FEV_1) (Sapienza et al., 1998). According to recent clinical opinion, asthmatic inflammation should be detected and treated as early as possible (Laitinen et al., 1993; Haahtela, 1995) to prevent bronchial damage. Thus, selective topical iNOS inhibition gives promise of a new nonsteroidal way to treat asthmatic inflammation.

References

Abrams, J. 1996. Beneficial actions of nitrates in cardiovascular disease. *Am. J. Cardiol.* 77:31C–37C.

Alanko, J., Sievi, E., Lähteenmäki, T., Mucha, I., Riutta, A., and Vapaatalo, H. 1995. Effects of NO-donors, SIN-1 and GEA 3175 on prostacyclin and cGMP synthesis in cultured rat endothelial cells. *Agents Actions* 45:S195–S199.

Archer, S.L., Tolins, J.P., Raij, L., and Weir, K.E. 1989. Hypoxic pulmonary vasoconstriction is enhanced by inhibition of the synthesis of an endothelium derived relaxing factor. *Biochem. Biophys. Res. Commun.* 164:1198–1205.

Arvola, P., Ruskoaho, H., Wuorela, H., Pekki, A., Vapaatalo, H., and Pörsti, I. 1993. Quinapril treatment and arterial smooth muscle responses in spontaneously hypertensive rats. *Br. J. Pharmacol.* 108:980–990.

Avontuur, J.A., Tutein-Nolthenius, R.P., van Bodegom, J.W., and Bruning, H.A. 1998. Prolonged inibition of nitric oxide synthesis in severe septic shock: a clinical study. *Crit. Care Med.* 26:660–667.

Barnes, P.J. 1992. Neural mechanisms in asthma. *Br. Med. Bul.* 48:149–168.

Bates, J.N., Baker, M.T., Guerra, R., and Harrison, D.G. 1991. Nitric oxide generation from nitroprusside by vascular tissue. Evidence that reduction of the nitroprusside anion and cyanide loss are required. *Biochem. Pharmacol.* 42: S157–S165.

Belvisi, M.G., Stretton, C.D., Yacoub, M., and Barnes, P.J. 1992. Nitric oxide is the endogeneous neurotransmitter of bronchodilator nerves in humans. *Eur. J. Pharmacol.* 210:221–222.

Benzing, A., Mols, G., Beyer, U., and Geiger, K. 1997. Large increase in cardiac output with ARDS and acute right heart failure during inhalation of nitric oxide. *Acta Anaesthesiol. Scand.* 41:643–646.

Berkenboom, G., Langer, I., Carpentier, Y., Grosfils, K., and Fontaine, J. 1997. Ramipril prevents endothelial dysfunction induced by low-density lipoproteins: a bradykinin-dependent mechanism. *Hypertension* 30:371–376.

Bernareggi, M., Mitchell, J.A., Barnes, P.J., and Belvisi, M.G. 1997. Dual action of NO on airway plasma leakage. *Am. J. Respir. Crit. Care Med.* 155:869–874.

Black, S.M., Fineman, J.R., Steinhorn, R.H., Bristow, J., and Soifer, S.J. 1998. Increased endothelial NOS in lambs with increased pulmonary blood flow and pulmonary hypertension. *Am. J. Physiol.* 275:H1643–H1651.

Boger, R.H., Bode-Boger, S.M., Brandes, R.P., Phivthong-ngam, L., Bohme, M., Nafe, R., Mugge, A., and Frölich, J.C. 1997. Dietary L-arginine reduces the progression of atherosclerosis in cholesterol-fed rabbits: comparison with lovastatin. *Circulation* 96:282–1290.

Bolz, S.S., and Pohl, U. 1997. Indomethacin enhances endothelial NO release—evidence for a role of PGI_2 in the autocrine control of calcium-dependent autacoid production. *Cardiovasc. Res.* 36:437–444.

Bönner, G., Preis, S., Schunk, U., Toussaint, C., and Kaufmann, W. 1990. Hemodynamic effects of bradykinin on systemic and pulmonary circulation in healthy and hypertensive humans. *J. Cardiovasc. Pharmacol.* 15:S175–S180.

Boolell, M., Gepi-Attee, S., Gingell, J.C., and Allen, M.J. 1996. Sildenafil, a novel oral therapy for male erectile dysfunction. *Br. J. Urol.* 78:257–261.

Brady, A.J., Williams, F.M., and Williams, T.J. 1992. Inflammatory injury in myocardial ischaemia. *Clin. Sci.* 83:511–518.

Braun, M., and Schrör, K. 1992. Prostaglandin D_2 relaxes bovine coronary arteries by endothelium-dependent nitric oxide-mediated cGMP formation. *Circ. Res.* 71:1305–1313.

Bredt, D.S., Hwang, P.M., Glatt, C.E., Lowenstein, C., Reed, R.R., and Snyder, S.H. 1991. Cloned and expressed nitric oxide synthase structurally resembles cytochrome P-450 reductase. *Nature* 351:714–718.

Bryk, R., and Wolff, D.J. 1998. Mechanism of inducible nitric oxide synthase inactivation by aminoguanidine and L-N6-(1-iminoethyl)lysine. *Biochemistry* 37:4844–4852.

Burnett, A.L., Lowenstein, C.J., Bredt, D.S., Chang, T.S., and Snyder, S.H. 1992. Nitric oxide: a physiologic mediator of penile erection. *Science* 257:401–403.

Bush, P.A., Aronson, W.J., Buga, G.M., Rajfer, J., and Ignarro, L.J. 1992. Nitric oxide is a potent relaxant of human and rabbit corpus cavernosum. *J. Urol.* 147:1650–1655.

Busse, R., and Fleming, I. 1993. The endothelial organ. *Curr. Opin. Cardiol.* 8:719–727.

Calver, A., Collier, J., Moncada, S., and Vallance, P. 1992. Effect of local intraarterial N^G-monomethyl-L-arginine in patients with hypertension: the nitric oxide dilator mechanism appears abnormal. *J. Hypertens.* 10:1025–1031.

Catani, M.V., Bernassola, F., Rossi, A., and Melino, G. 1998. Inhibition of clotting factor XIII activity by nitric oxide. *Biochem. Biophys. Res. Commun.* 249:275–278.

Caulin-Glaser, T., Garcia-Cardena, G., Sarrel, P., Sessa, W.C., and Bender, J.R. 1997. 17 beta-estradiol regulation of human endothelial cell basal nitric oxide release, independent of cytosolic Ca^{2+} mobilization. *Circ. Res.* 81:885–892.

Cavallini, G. 1994. Minoxidil versus nitroglycerine: a prospective, double-blind, controlled trial in transcutaneous therapy for organic impotence. *Int. J. Impotence Res.* 6:205–212.

Chester, A.H., O'Neil, G.S., Moncada, S., Tadjkarimi, S., and Yacoub, M.H. 1990. Low basal and stimulated release of nitric oxide in atherosclerotic epicardial coronary arteries. *Lancet* 336:897–900.

Chowienczyk, P.J., Watts, G.F., Cockroft, J.R., and Ritter, J.M. 1992. Impaired endothelium-dependent vasodilation of forearm resistance vessels in hypercholesterolaemia. *Lancet* 340:1430–1432.

Creager, M.A., Cooke, J.P., Mendelsohn, M.E., Callagoer, S.J., Coleman, S.M., Loscalzo, J., and Dzau, V.J. 1990. Impaired vasodilation of forearm resistance vessels in hypercholesterolemic humans. *J. Clin. Invest.* 86:228–234.

De Caterina, R., Libby, P., Peng, H.B., Thannickal, V.J., Rajavashisth, T.B., Gimbrone, M., Jr., Shin, W.S., and Liao, J.K. 1995. Nitric oxide decreases cytokine-induced endothelial activation. Nitric oxide selectively reduces endothelial expression of adhesion molecules and proinflammatory cytokines. *J. Clin. Invest.* 96:60–68.

Dellinger, R.P., Zimmerman, J.L., Taylor, R.W., Straube, R.C., Hauser, D.L., Criner, G.J., Davis, K., Jr., Hyers, T.M., and Papadakos, P. 1998. Effects of inhaled nitric oxide in patients with acute respiratory distress syndrome: results of a randomized phase II trial. *Crit. Care Med.* 26:15–23.

Delyani, J.A., Nossuli, T.O., Scalia, R., Thomas, G., Garvey, D.S., and Lefer, A.M. 1996. *S*-Nitrosylated tissue-type plasminogen activator protects against myocardial ischemia/reperfusion injury in cats: role of the endothelium. *J. Pharmacol. Exp. Ther.* 279:1174–1180.

Ding, Y., and Vaziri, N.D. 1998. Calcium channel blockade enhances nitric oxide synthase expression by cultured endothelial cells. *Hypertension* 32:718–723.

Drexler, H., Kastner, S., Strobel, A., Studer, R., Brodded, O.E., and Hasenfuss, G. 1998. Expression, activity and functional significance of inducible nitric oxide synthase in the failing human heart. *J. Am. Coll. Cardiol.* 32:955–963.

Drexler, H., Zeiher, A.M., Meinzer, K., and Just, H. 1991. Correction of endothelial dysfunction in coronary microcirculation of hypercholesterolaemic patients by L-arginine. *Lancet* 338:1546–1550.

Durante, W., Kroll, M.H., Vanhoutte, P.M., and Schafer, A.I. 1992. Endothelium derived relaxing factor inhibits thrombin-induced platelet aggregation by inhibiting platelet phospholipase C. *Blood* 79:110–116.

Du Toit, E.F., McCarthy, J., Miyashiro, J., Opie, L.H., and Brunner, F. 1998. Effect of nitrovasodilators and inhibitors of nitric oxide synthase on ischaemic and reperfusion function of rat isolated hearts. *Br. J. Pharmacol.* 123:1159–1167.

Elkayam, U. 1996. Nitrates in the treatment of congestive heart failure. *Am. J. Cardiol.* 77:41C–51C.

Ellis, J.L., and Undem, B.J. 1992. Inhibition by L-N^G-nitro-L-arginine of nonadrenergic-noncholinergic-mediated relaxations of human isolated central and peripheral airway. *Am. Rev. Respir. Dis.* 146:1543–1547.

Endres, M., Laufs, U., Huang, Z., Nakamura, T., Huang, P., Moskowitz, M.A., and Liao, J.K. 1998. Stroke protection by 3-hydroxy-3-methylglutaryl (HMG)-CoA reductase inhibitors mediated by endothelial nitric oxide synthase. *Proc. Natl. Acad. Sci. USA* 95:8880–8885.

Feelisch, M., Ostrowski, J., and Noack, E. 1989. On the mechanism of NO release from sydnonimines. *J. Cardiovasc. Pharmacol.* 14:S13–S22.

Filep, J.G., Delalandre, A., and Beauchamp, M. 1997. Dual role for nitric oxide in the regulation of plasma volume and albumin escape during endotoxin shock in conscious rats. *Circ. Res.* 81:840–847.

Foubert, L., Fleming, B., Latimer, R., Jonas, M., Oduro, A., Borland, C., and Higenbottam, T. 1992. Safety guidelines for use of nitric oxide. *Lancet* 339:1615–1616.

Frostell, C., Fratacci, M.-D., Wain, J.C., Jones, R., and Zapol, W.M. 1991. Inhaled nitric oxide: a selective pulmonary vasodilator reversing hypoxic pulmonary vasoconstriction. *Circulation* 83:2038–2047.

Furchgott, R.F. 1984. Role of endothelium in the responses of vascular smooth muscle to drugs. *Annu. Rev. Pharmacol. Toxicol.* 24:175–197.

Furchgott, R.F., and Zawadzki, J. 1980. The obligatory role of endothelial cells in the relaxation of arterial smooth muscle by acetylcholine. *Nature* 288:373–376.

Garnier, P., Fajac, I., Dessanges, J.F., Dall'Ava-Santucci, J., Lockhart, A., and Dinh-Xuan, A.T. 1996. Exhaled nitric oxide during acute changes of airways calibre in asthma. *Eur. Respir. J.* 9:1134–1138.

Gaston, B., Sears, S., Woods, J., Hunt, J., Ponaman, M., McMahon, T., and Stamler, J.S. 1998. Bronchodilator *S*-nitrosothiol deficiency in asthmatic respiratory failure. *Lancet* 351:1317–1319.

Giraldez, R.R., Panda, A., Xia, Y., Sanders, S.P., and Zweier, J.L. 1997. Decreased nitric oxide synthase activity causes impaired endothelium-dependent relaxation in the postischemic heart. *J. Biol. Chem.* 272:21420–21426.

Girerd, X.J., Hirsch, A.T., Cooke, J.P., Dzau, V.J., and Creager, M.A. 1990. L-Arginine augments endothelium-dependent vasodilation in cholesterol-fed rabbits. *Circ. Res.* 67:1301–1308.

Goldstein, I., Lue, T.F., Padma-Nathan, H., Rosen, R.C., Steers, W.D., and Wicker, P.A. 1998. Oral sildenafil in the treatment of erectile dysfunction. *N. Engl. J. Med.* 338:1397–1404.

Gunnett, C.A., Chu, Y., Heistad, D.D., Loihl, A., and Faraci, F.M. 1998. Vascular effects of LPS in mice deficient in expression of the gene for inducible nitric oxide synthase. *Am. J. Physiol.* 275:H416–H421.

Haahtela, T. 1995. The importance of inflammation in early asthma. *Respir. Med.* 89:461–462.

Hamid, Q., Springall, D.R., Riveros-Moreno, V., Chanez, P., Howarth, P., Redington, A., Bousquet, J., Godard, P., Holgate, S., and Polak, J.M. 1993. Induction of nitric oxide synthase in asthma. *Lancet* 342:1510–1513.

Hare, J.M., Givertz, M.M., Creager, M.A., and Colucci, W.S. 1998. Increased sensitivity to nitric oxide synthase inhibition in patients with heart failure: potentiation of beta-adrenergic inotropic responsiveness. *Circulation* 97:161–166.

Hecker, M., Mülsch, A., Bassenge, E., and Busse, R. 1993a. Vasoconstriction and increased flow: two principal mechanisms of shear stress-dependent endothelial autacoid release. *Am. J. Physiol.* 265:H828–H833.

Hecker, M., Fleming, I., Pörsti, I., and Busse, R. 1993b. ACE inhibitors, kinins and endothelial autacoid formation. *Pharm. Pharmacol. Lett.* 3:S64–S67.

Hecker, M., Pörsti, I., Bara, A.T., and Busse, R. 1994. Potentiation by ACE inhibitors of the dilator response to bradykinin in the coronary microcirculation: interaction at the receptor level. *Br. J. Pharmacol.* 111:238–244.

Herulf, M., Ljung, T., Hellstrom, P.M., Weitzberg, E., and Lundberg, J.O. 1998. Increased luminal nitric oxide in inflammatory bowel disease as shown with novel minimally invasive method. *Scand. J. Gastroenterol.* 33:164–169.

Hirooka, Y., Imaizumi, T., Masaki, H., Ando, S., Harada, S., Momohara, M., and Takeshita, A. 1992. Captopril improves impaired endothelium-dependent vasodilation in hypertensive patients. *Hypertension* 20:175–180.

Hogman, M., Frostell, C.G., Hedenstrom, H., and Hedenstierna, G. 1993. Inhalation of nitric oxide modulates adult human bronchial tone. *Am. Rev. Respir. Dis.* 148:1474–1478.

Imaizumi, T., Hirooka, Y., Masaki, H., Harada, S., Momohara, M., Takawa, T., and Takeshita, A. 1992. Effects of L-arginine on forearm vessels and responses to acetylcholine. *Hypertension* 20:511–517.

Ishigai, Y., Mori, T., Ikeda, T., Fukuzawa, A., and Shibano, T. 1997. Role of bradykinin-NO pathway in prevention of cardiac hypertrophy by ACE inhibitor in rat cardiomyocytes. *Am. J. Physiol.* 273:H2659–H2663.

Iwashita, E. 1998. Greatly increased mucosal nitric oxide in ulcerative colitis determined in situ by a novel nitric oxide-selective microelectrode. *J. Gastroenterol. Hepatol.* 13:391–395.

Jacobs, M., Plane, F., and Bruckdorfer, K.R. 1990. Native and oxidized low density lipoproteins have different inhibitory effects on endothelium-derived relaxing factor in the rabbit aorta. *Br. J. Pharmacol.* 100:21–26.

Johannigman, J.A., Davis, K., Jr., Campbell, R.S., Luchette, F., Hurst, J.M., and Branson, R.D. 1997. Inhaled nitric oxide in acute respiratory distress syndrome. *J. Trauma*. 43:904–909.

Kageyama, N., Miura, M., Ichinose, M., Tomaki, M., Ishikawa, J., Ohuchi, Y., Endoh, N., and Shirato, K. 1997. Role of endogeneous nitric oxide in airway microvascular leakage induced by inflammatory mediators. *Eur. Respir. J.* 10:13–19.

Keh, D., Gerlach, M., Kurer, I., Seiler, S., Kerner, T., Falke, K.J., and Gerlach, H. 1996. The effects of nitric oxide (NO) on platelet membrane receptor expression during activation with human alpha-thrombin. *Blood Coagul. Fibrinolysis* 7:615–624.

Kharitonov, S.A., Yates, D.H., and Barnes, P.J. 1996. Inhaled glucocorticoids decrease nitric oxide in exhaled air of asthmatic patients. *Am. J. Respir. Crit. Care Med.* 153:454–457.

Kichuk, M.R., Zhang, X., Oz, M., Michler, R., Kaley, G., Nasjletti, A., and Hintze, T.H. 1997. Angiotensin-converting enzyme inhibitors promote nitric oxide production in coronary microvessels from failing explanted human hearts. *Am. J. Cardiol.* 80:137A–142A.

Kim, Y.D., Fomsgaard, J.S., Heim, K.F., Ramwell, P.W., Thomas, G., Kagan, E., Moore, S.P., Coughlin, S.S., Kuwahara, M., and Amalou, A. 1992. Brief ischemia-reperfusion induces stunning of endothelium in canine coronary artery. *Circulation* 85:1473–1482.

Kiowski, W., Linder, L., Kleinbloesem, C., van Brummelen, P., and Bühler, F.R. 1992. Blood pressure control by the renin-angiotensin system in normotensive subjects. Assessment by angiotensin converting enzyme and renin inhibition. *Circulation* 85:1–8.

Klabunde, R.E., and Ritger, R.C. 1991. N^G-monomethyl-L-arginine (NMA) restores arterial blood pressure but reduces cardiac output in a canine model of endotoxic shock. *Biochem. Biophys. Res. Commun.* 178:1135–1140.

Kolios, G., Rooney, N., Murphy, C.T., Robertson, D.A.F., and Westwick, J. 1998. Expression of inducible nitric oxide synthase activity in human colon epithelial cells: modulation by T lymphocyte derived cytokines. *Gut* 43:56–63.

Korbut, R., Marcinkiewicz, E., Cieslik, K., and Gryglewski, R.J. 1995. The effect of nitric oxide donors on the release of plasminogen activator inhibitor (PAI) from rabbit platelets in vitro. *J. Physiol. Pharmacol.* 46:37–44.

Kupatt, C., Zahler, S., Seligmann, C., Massody, P., Becker, B.F., and Gerlach, E. 1996. Nitric oxide mitigates leukocyte adhesion and vascular leak after myocardial ischemia. *J. Mol. Cell. Cardiol.* 28:643–654.

Laitinen, L., Laitinen, A., and Haahtela, T. 1993. Airway mucosal inflammation even in patients with newly diagnosed asthma. *Am. Rev. Respir. Dis.* 147:697–704.

Langford, E.J., Brown, A.S., Wainwright, R.J., de Belder, A.J., Thomas, M.R., Smith, R.E., Radomsky, M.W., Martim, J.F., and Moncada, S. 1994. Inhibition of platelet activity by S-nitrosoglutathione during coronary angioplasty. *Lancet* 344:1458–1460.

Laszlo, F., and Whittle, B.J. 1997. Actions of isoform-selective and non-selective nitric oxide synthase inhibitors on endotoxin-induced vascular leakage in rat colon. *Eur. J. Pharmacol.* 334:99–102.

Laufs, U., La Fata, V., Plutzky, J., and Liao, K. 1998. Upregulation of endothelial nitric oxide synthase by HMG CoA reductase inhibitors. *Circulation* 97:1129–1135.

Laufs, U., and Liao, J.K. 1998. Post-transcriptional regulation of endothelial nitric oxide synthase mRNA stability by Rho GTPase. *J. Biol. Chem.* 273: 24266–24271.

Liaudet, L., Rosselet, A., Schaller, M.D., Markert, M., Perret, C., and Feihl, F. 1998. Nonselective versus selective inhibition of inducible nitric oxide synthase in experimental endotoxin shock. *J. Infect. Dis.* 177:127–132.

Lin, K.F., Chao, L., and Chao, J. 1997. Prolonged reduction of high blood pressure with human nitric oxide synthase gene delivery. *Hypertension* 30:307–313.

Linz, W., and Schölkens, B.A. 1992. A specific B_2-bradykinin antagonist Hoe-140 abolishes antihypertrophic effect of ramipril. *Br. J. Pharmacol.* 105:771–772.

Liu, S.F., Crawley, D.E., Barnes, P.J., and Evans, T.W. 1991. Endothelium-derived relaxing factor inhibits hypoxic pulmonary vasoconstriction in rats. *Am. Rev. Respir. Dis.* 143:32–37.

Lopez-Farre, A., Riesco, A., Digiuni, E., Mosquera, J.R., Caramelo, C., de Miguel, L.S., Millas, I., de Frutos, T., Cernadas, M.R., Monton, M., Alonso, J., and Casado, S. 1996. Aspirin-stimulated nitric oxide production after acute myocardial ischemia in rabbits. *Circulation* 94:83–87.

Lowenstein, C.J., Dinerman, J.L., and Snyder, S.H. 1994. Nitric oxide: a physiologic messenger. *Ann. Intern. Med.* 120:227–237.

Lüscher, T.F. 1992. Heterogeneity of endothelial dysfunction in hypertension. *Eur. Heart J.* 13:S50–S55.

Lüscher, T.F., Tanner, F.C., and Dohi, Y. 1992. Age hypertension and hypercholesterolaemia alter endothelium-dependent vascular regulation. *Pharmacol. Toxicol.* 70:S32–S39.

Lüscher, T.F., Tanner, F.C., Tschudi, M.R., and Noll, G. 1993. Endothelial dysfunction in coronary artery disease. *Annu. Rev. Med.* 44:395–418.

Ma, X.-I., Weyrich, A.S., Lefer, D.J., and Lefer, A.M. 1993. Diminished basal nitric oxide release after myocardial ischemia and reperfusion promotes neutrophil adherence to coronary endothelium. *Circ. Res.* 72:403–412.

MacRitchie, A.N.,. Jun, S.S., Chen, Z., German, Z., Yuhanna, I.S., Sherman, T.S., and Shaul, P.W. 1997. Estrogen upregulates endothelial nitric oxide synthase gene expression in fetal pulmonary artery endothelium. *Circ. Res.* 81:355–362.

Marletta, M.A. 1994a. Nitric oxide synthase: aspects concerning structure and catalysis. *Cell* 78:927–930.

Marletta, M.A. 1994b. Approaches toward selective inhibition of nitric oxide synthase. *J. Med. Chem.* 37:1899–1907.

Martorana, P.A., Kettenbach, B., Breipohl, G., Linz, W., and Schölkens, B.A. 1990. Reduction of infarct size by local angiotensin-converting enzyme inhibition is abolished by a bradykinin antagonist. *Eur. J. Pharmacol.* 182:395–396.

Masferrer, J.L., Seibert, K., Zweifel, B., and Needleman, P. 1992. Endogenous glucocorticoids regulate an inducible cyclooxygenase enzyme. *Proc. Natl. Acad. Sci. USA* 89:3917–3921.

Meyhoff, H.H., Rosenkilde, P., and Bodker, A. 1992. Non-invasive management of impotence with transcutaneous nitroglycerin. *Br. J. Urol.* 69:88–90.

Middleton, S.J., Shorthouse, M., and Hunter, J.D. 1993. Increased nitric oxide synthesis in ulcerative colitis. *Lancet* 341:465–466.

Mitsuyama, T., Hidaka, K., Furuno, T., and Hara, N. 1998. Release of nitric oxide and expression of constitutive nitric oxide synthase of human endothelial cells: enhancement by a 14-membered ring macrolide. *Mol. Cell Biochem.* 181: 157–161.

Moncada, S., Palmer, R.M.J., and Higgs, E.A. 1991. Nitric oxide: physiology, pathophysiology and pharmacology. *Pharmacol. Rev.* 43:109–141.

Myers, P.R., Minor, R.L.J., Guerra, R.J., Bates, J.N., and Harrison, D.G. 1990. Vasorelaxant properties of the endothelium-derived relaxing factor more closely resemble S-nitrosocysteine than nitric oxide. *Nature* 345:161–163.

Nakatsuka, M., and Osawa, Y. 1994. Selective inhibition of the 12-lipoxygenase pathway of arachidonic acid metabolism by L-arginine or sodium nitroprusside in intact human platelets. *Biochem. Biophys. Res. Commun.* 200:1630–1634.

Nathan, C. 1992. Nitric oxide as a secretory product of mammalian cells. *FASEB J.* 6:3051–3064.

Newman, C.M., Warren, J.B., Taylor, G.W., Boobis, A.R., and Davies, D.S. 1990. Rapid tolerance to the hypotensive effects of glyceryl trinitrate in the rat: prevention by N-acetyl-L- but not N-acetyl-D-cysteine. *Br. J. Pharmacol.* 99:825–829.

Nolte, C., Eigenthaler, M., Schanzenbacher, P., and Walter, U. 1991. Endothelial cell-dependent phosphorylation of a platelet protein mediated by cAMP- and cGMP-elevating factors. *J. Biol. Chem.* 266:14808–14812.

Nunez, B.D., and Anderson, D.C., Jr. 1993. Nitroglycerin ointment in the treatment of impotence. *J. Urol.* 150:1241–1243.

Offner, P.J., Ogura, H., Jordan, B.S., Pruitt, B.A., Jr., and Cioffi, W.G. 1996. Cardiopulmonary effects of combined nitric oxide inhibition and inhaled nitric oxide in porcine endotoxin shock. *J. Trauma* 41:641–646.

Pabla, R., Buda, A.J., Flynn, D.M., Blesse, S.A., Shin, A.M., Curtis, M.J., and Lefer, D.J. 1996. Nitric oxide attenuates neutrophil-mediated myocardial contractile dysfunction after ischemia and reperfusion. *Circ. Res.* 78:65–72.

Panza, J.A., Casino, P.R., Kilcoyne, C.M., and Quyyumi, A.A. 1993. Role of endothelium derived nitric oxide in the abnormal endothelium dependent vascular relaxation of patients with essential hypertension. *Circulation* 87:1468–1474.

Panza, J.A., Quyyumi, A.A., Brush, J.E., and Epstein, S.E. 1990. Abnormal endothelium-dependent vascular relaxation in patients with essential hypertension. *N. Engl. J. Med.* 323:22–27.

Park, K., Moreland, R.B., Goldstein, I., Atala, A., and Traish, A. 1998. Sildenafil inhibits phosphodiesterase type 5 in human clitoral corpus cavernosum smooth muscle. *Biochem. Biophys. Res. Commun.* 249:612–617.

Parrat, J.R. 1998. Nitric oxide in sepsis and endotoxemia. *J. Antimicrobial Chemother.* 41:S31–S39.

Pearl, R.G. 1993. Inhaled nitric oxide. The past, the present, and the future. *Anesthesiology* 78:413–416.

Pepke-Zaba, J., Higenbottam, T.W., Dinh-Xuan, A.T., Stone, D., and Wallwork, J. 1991. Inhaled nitric oxide as a cause of selective pulmonary vasodilatation in pulmonary hypertension. *Lancet* 338:1173–1174.

Pernow, J., Uriuda, Y., Wang, Q.-D., Li, X.-S., Nordlander, R., and Rydén, L. 1994. The protective effect of L-arginine on myocardial injury and endothelial function following ischaemia and reperfusion in the pig. *Eur. Heart J.* 15:1712–1719.

Persson, M.G., Zetterström, O., Argenius, V., Ihre, E., and Gustafsson, L.E. 1994. Single breath nitric oxide measurements in asthmatic patients and smokers. *Lancet* 343:146–147.

Petros, A., Bennet, D., and Vallance, P. 1991. Effect of nitric oxide synthase inhibitors on hypotension in patients with septic shock. *Lancet* 338:1557–1558.

Petros, A., Lamb, G., Leone, A., Moncada, S., Bennett, D., and Vallance, P. 1994. Effects of a nitric oxide synthase inhibitor in humans with septic shock. *Cardiovasc. Res.* 28:34–39.

Pfeffer, K.D., Ellison, G., Robertson, D., and Day, R.W. 1996. The effect of inhaled nitric oxide in pediatric asthma. *Am. J. Respir. Crit. Care Med.* 153:747–751.

Plane, F., Bruckdorfer, K.R., Kerr, P., Steuer, A., and Jacobs, M. 1992. Oxidative modification of low-density lipoproteins and the inhibition of relaxations mediated by endothelium-derived nitric oxide in rabbit aorta. *Br. J. Pharmacol.* 105:216–222.

Pons, F., Williams, T.J., and Warren, J.B. 1993. Nitric oxide, but not interleukin-1, mediates the local blood flow response to lipopolysaccharide in rabbit skin. *Eur. J. Pharmacol.* 239:23–30.

Radomski, M.W., Palmer, R.M., and Moncada, S. 1987a. Endogenous nitric oxide inhibits human platelet adhesion to vascular endothelium. *Lancet* 2:1057–1058.

Radomski, M.W., Palmer, R.M.J., and Moncada, S. 1987b. The anti-aggregating properties of vascular endothelium: interactions between prostacyclin and nitric oxide. *Br. J. Pharmacol.* 92:639–646.

Radomski, M.W., Palmer, R.M., and Moncada, S. 1990. An L-arginine/nitric oxide pathway present in human platelets regulates aggregation. *Proc. Natl. Acad. Sci. USA* 87:5193–5197.

Rector, T.S., Bank, A.J., Mullen, K.A., Tschumperlin, L.K., Sih, R., Pillai, K., and Kubo, S.H. 1996. Randomized, double-blind, placebo-controlled study of supplemental oral L-arginine in patients with heart failure. *Circulation* 93:2135–2141.

Rees, D.D., Palmer, R.M.J., and Moncada, S. 1989. Role of endothelium-derived nitric oxide in the regulation of blood pressure. *Proc. Natl. Acad. Sci. USA* 86: 3375–3378.

Rehse, K., Schleifer, K.J., Martens, A., and Kampfe, M. 1994. New NO-donors with antithrombotic and vasodilating activities. 5. Oligonitroso sydnone imines. *Arch. Pharmazie* 327:393–397.

Renganathan, R., Suranjan, B., and Kurien, T. 1997. Comparison of transdermal nitroglycerin and intracavernous injection of papaverine in the treatment of erectile dysfunction in patients with spinal cord lesions. *Spinal Cord* 35:99–103.

Renzi, P.M., Sebastio, N., al Assaad, A.S., Giaid, A., and Hamid, Q. 1997. Inducible nitric oxide synthase mRNA and immunoreactivity in the lungs of rats eight hours after antigen challenge. *Am. J. Respir. Cell Mol. Biol.* 17:36–40.

Ribbons, K.A., Currie, M.G., Connor, J.R., Manning, P.T., Allen, P.C., Didier, P., Ratterree, M.S., Clark, D.A., and Miller, M.J. 1997. The effect of inhibitors of inducible nitric oxide synthase on chronic colitis in the rhesus monkey. *J. Pharmacol. Exp. Ther.* 280:1008–1015.

Riede, U.N., Forstermann, U., and Drexler, H. 1998. Inducible nitric oxide synthase in skeletal muscle of patients with chronic heart failure. *J. Am. Cell. Cardiol.* 32:964–969.

Ring, T., Knudsen, F., Kristensen, S.D., and Larsen, C.E. 1983. Nitroglycerin prolongs the bleeding time in healthy males. *Thromb. Res.* 29:553–559.

Riutta, A., Alanko, J., Mucha, I., and Vapaatalo, H. 1994. Effects of Trolox C and SIN-1 on arachidonic metabolism and on cyclic GMP formation in leukocytes. *Eur. J. Pharmacol.* 262:217–222.

Roberts, J.D., Polaner, D.M., Lang, P., and Zapol, W.M. 1992. Inhaled nitric oxide in persistent pulmonary hypertension of the newborn. *Lancet* 340:818–819.

Robertson, B.E., Warren, J.B., and Nye, P.C. 1990. Inhibition of nitric oxide synthesis potentiates hypoxic vasoconstriction in isolated rat lungs. *Exp. Physiol.* 75:255–257.

Ross, R. 1993. The pathogenesis of atherosclerosis: a perspective for the 1990's. *Nature* 362:801–809.

Salvemini, D., Currie, M.G., and Mollace, V. 1996. Nitric oxide-mediated cyclooxygenase activation: a key event in the antiplatelet effects of nitrovasodilators. *J. Clin. Invest.* 97:2562–2568.

Salvemini, D., Manning, P.T., Zweifel, B.S., Seibert, K., Connor, J., Currie, M.G., Needleman, P., and Masferrer, J.L. 1995. Dual inhibition of nitric oxide and prostaglandin production contributes to the antiinflammatory properties of nitric oxide synthase inhibitors. *J. Clin. Invest.* 96:301–308.

Sapienza, M.A., Kharitonov, S.A., Horvath, I., Chung, K.F., and Barnes, P.J. 1998. Effects of inhaled L-arginine on exhaled nitric oxide in normal and asthmatic subjects. *Thorax* 53:172–175.

Schwartz, D., Mendonca, M., Schwartz, I., Xia, Y., Satriano, J., Wilson, C.B., and Blantz, R.C. 1997. Inhibition of constitutive nitric oxide synthase (NOS) by nitric oxide generated by inducible NOS after lipopolysaccharide administration provokes renal dysfunction in rats. *J. Clin. Invest.* 100:439–448.

Shin, W.S., Hong, Y.H., Peng, H.B., de Caterina, R., Libby, P., and Liao, J.K. 1996. Nitric oxide attenuates vascular smooth muscle cell activation by interferon-gamma. The role of constitutive NF-kappa B activity. *J. Biol. Chem.* 271:11317–11324.

Smith, C.J., Sun, D., Hoegler, C., Roth, B.S., Zhang, X., Zhao, G., Xu, X.B., Kobari, Y., Pritchard, K., Jr., Sessa, W.C., and Hintze, T.H. 1996. Reduced gene expression of vascular endothelial NO synthase and cyclooxygenase-1 in heart failure. *Circ. Res.* 78:58–64.

Smith, J.A., Shah, A.M., and Lewis, M.J. 1991. Factors released from endocardium of the ferret and pig modulate myocardial contraction. *J. Physiol.* 439:1–14.

Sneddon, J.M., and Vane, J.R. 1988. Endothelium-derived relaxing factor reduces platelet adhesion to bovine endothelial cells. *Proc. Natl. Acad. Sci. USA* 85:2800–2804.

Sonksen, J., and Biering-Soresen, F. 1992. Transcutaneous nitroglycerin in the treatment of erectile dysfunction in spinal cord injured. *Paraplegia* 30:554–557.

Sovijärvi, A.R.A., Saarinen, A., and Helin, T. 1998. Exhaled nitric oxide is increased in patients with asthmatic symptoms not fulfilling the functional criteria of asthma. *Clin. Physiol.* 18:264.

Springall, D.R., Meng, Q., Redington, A., Howarth, P.H., Evens, T.J., and Polak, J.M. 1995. Inducible nitric oxide synthase in asthmatic airway epithelium is reduced by corticosteroid therapy. *Am. J. Respir. Crit. Care Med.* 151:A833.

Stryer, L. 1995. Biosynthesis of amino acids and heme. In: *Biochemistry*. W.H. Freeman, New York, pp. 713–738.

Szabo, C., Bryk, R., Zingarelli, B., Southan, G.J., Gahman, T.C., Bhat, V., Salzman, A.L., and Wolff, D.J. 1996. Pharmacological characterization of guanidinoethyldisulphide (GED), a novel inhibitor of nitric oxide synthase with selectivity towards the inducible isoform. *Br. J. Pharmacol.* 118:1659–1668.

Takahashi, M., Ikeda, U., Masuyama, J., Funayama, H., Kano, S., and Shimada, K. 1996. Nitric oxide attenuates adhesion molecule expression in human endothelial cells. *Cytokine* 8:817–821.

Takahashi, Y., Kobayashi, H., Tanaka, N., Honda, K., Kawakami, T., and Tomita, T. 1998. Worsening of hypoxemia with nitric oxide inhalation during bronchospasm in humans. *Respir. Physiol.* 112:113–119.

Takano, H., Lim, H.B., Miyabara, Y., Ichinose, T., Yoshikawa, T., and Sagai, M. 1998. Oral administration of L-arginine potentiates allergen-induced airway inflammation and expression of interleukin-5 in mice. *J. Pharmacol. Exp. Ther.* 286:767–771.

ten Hacken, N.H., van der Vaart, H., van der Mark, T.W., Koeter, G.H., and Postma, D.S. 1998. Exhaled nitric oxide is higher both at day and night in subjects with nocturnal asthma. *Am. J. Respir. Crit. Care Med.* 158:902–907.

Thiemermann, C., and Vane, J. 1990. Inhibition of nitric oxide synthesis reduces the hypotension induced by bacterial lipopolysaccharides in the rat in vivo. *Eur. J. Pharmacol.* 182:591–595.

Treasure, C.B., Klein, J.L., Weintraub, W.S., Talley, J.D., Stillabower, M.E., Kosiuski, A.S., Zhang, J., Boccuzzi, S.J., Cedarholm, J.C. 1995. Beneficial effects of cholesterol-lowering therapy on the coronary endothelium in patients with coronary arterial disease. *N. Engl. J. Med.* 332:481–487.

Trigo-Rocha, F., Aronson, W.J., Hohenfellner, M., Ignarro, L.J., Rajfer, J., and Lue, T.F. 1993. Nitric oxide and cGMP: mediators of pelvic nerve-stimulated erection in dogs. *Am. J. Physiol.* 264:H419–H422.

Tsao, P.S., and Lefer, A.M. 1990. Time course and mechanism of endothelial dysfunction in isolated ischemic- and hypoxic-perfused rat hearts. *Am. J. Physiol.* 259:H1660–H1666.

Vaali, K., Li, L., Redemann, B., Paakkari, I., and Vapaatalo, H. 1996. In-vitro bronchorelaxing effects of novel nitric oxide donors GEA 3268 and GEA 5145 in guinea-pigs and rats. *J. Pharm. Pharmacol.* 48:1309–1314.

Vagnoni, K.E., Shaw, C.E., Phernetton, T.M., Meglin, B.M., Bird, I.M., and Magness, R.R. 1998. Endothelial vasodilator production by uterine and systemic arteries. III. Ovarian and estrogen effects on NO synthase. *Am. J. Physiol.* 275:H1845–H1856.

Vallance, P., Leone, A., Calver, A., Collier, J., and Moncada, S. 1992a. Endogenous dimethylarginine as an inhibitor of nitric oxide synthesis. *J. Cardiovasc. Pharmacol.* 20:S60–S62.

Vallance, P., Benjamin, N., and Collier, J. 1992b. The effect of endothelium derived nitric oxide on ex vivo whole blood platelet aggregation in man. *Eur. J. Clin. Pharmacol.* 42:37–41.

Vanhoutte, P.M. 1988. The endothelium—modulator of vascular smooth muscle tone. *N. Engl. J. Med.* 319:512–513.

Wang, G.R., Zhu, Y., Halushka, P.V., Lincoln, T.M., and Mendelsohn, M.E. 1998. Mechanism of platelet inhibition by nitric oxide: in vivo phosphorylation of thromboxane receptor by cyclic GMP-dependent protein kinase. *Proc. Natl. Acad. Sci. USA* 95:4888–4893.

Warren, J.B., Loi, R.K., and Coughlan, M.L. 1993. Involvement of nitric oxide synthase in the delayed vasodilator response to ultraviolet light irradiation of rat skin in vivo. *Br. J. Pharmacol.* 109:802–806.

Warren, J.B., Pons, F., and Brady, A.J.B. 1994. Nitric oxide biology: implications for cardiovascular therapeutics. *Cardiovasc. Res.* 28:25–30.

Weitzberg, E., Rudehill, A., and Lundberg, J.M. 1993. Nitric oxide inhalation attenuates pulmonary hypertension and improves gas exchange in endotoxin shock. *Eur. J. Pharmacol.* 233:85–94.

Weyrich, A.S., Ma, X., and Lefer, A.M. 1992. The role of L-arginine in ameliorating reperfusion injury after myocardial ischemia in the cat. *Circulation* 86:279–288.

Wiesner, R., Rathmann, J., Holzhutter, H.G., Stosser, R., Mader, K., Nolting, H., and Kuhn, H. 1996. Nitric oxide oxidises a ferrous mammalian lipoxygenase to a pre-activated ferric state. *FEBS Lett.* 389:229–232.

Winlaw, D.S., Smythe, G.A., Keogh, A.M., Schyvens, C.G., Spratt, P.M., and MacDonald, P.S. 1994. Increased nitric oxide production in heart failure. *Lancet* 344:373–374.

Wolff, D.J., and Lubeskie, A. 1995. Aminoguanidine is an isoform-selective, mechanism-based inactivator of nitric oxide synthase. *Arch. Biochem. Biophys.* 316:290–301.

Yamakado, T., Nishikawa, M., and Hidaka, H. 1982. Stimulation of human platelet guanylate cyclase by nitroso compounds. *Thromb. Res.* 26:135–140.

Yates, D.H., Kharitonov, S.A., and Barnes, P.J. 1997. Effects of short- and longacting beta2-agonists on exhaled nitric oxide in asthmatic patients. *Eur. Respir. J.* 10: 1483–1488.

Yates, D.H., Kharitonov, S.A., Thomas, P.D., and Barnes, P.J. 1996. Endogenous nitric oxide is decreased in asthmatic patients by an inhibitor of inducible nitric oxide synthase. *Am. J. Respir. Crit. Care Med.* 154:247–250.

Ziegler, T., Silacci, P., Harrison, V., and Hayoz, D. 1998. Nitric oxide synthase expression in endothelial cells exposed to mechanical forces. *Hypertension* 32:351–355.

Index

CPSIA information can be obtained
at www.ICGtesting.com
Printed in the USA
BVHW012153070921
616288BV00002B/4